NOISES

in Optical Communications and Photonic Systems

Optics and Photonics

Series Editor

Le Nguyen Binh

Huawei Technologies, European Research Center, Munich, Germany

NOISES

in Optical Communications and Photonic Systems

Le Nguyen Binh

Huawei Technologies, Munich, Germany

CRC Press
Taylor & Francis Group
Boca Raton London New York

CRC Press is an imprint of the
Taylor & Francis Group, an **informa** business

CRC Press
Taylor & Francis Group
6000 Broken Sound Parkway NW, Suite 300
Boca Raton, FL 33487-2742

First issued in paperback 2019

© 2017 by Taylor & Francis Group, LLC
CRC Press is an imprint of Taylor & Francis Group, an Informa business

No claim to original U.S. Government works

ISBN-13: 978-1-4822-4694-0 (hbk)
ISBN-13: 978-0-367-87029-4 (pbk)

Library of Congress Cataloging-in-Publication Data

Names: Binh, Le Nguyen, author.
Title: Noises in optical communications and photonic systems / Le Nguyen Binh.
Description: Boca Raton : Taylor & Francis, a CRC title, part of the Taylor &
Francis imprint, a member of the Taylor & Francis Group, the academic
division of T&F Informa, plc, [2017] | Series: Optics and phototonics ; 14
| Includes bibliographical references and index.
Identifiers: LCCN 2016015995 | ISBN 9781482246940 (acid-free paper)
Subjects: LCSH: Optical communications. | Electronic noise. | Photonics.
Classification: LCC TK5103.59 .B524 2017 | DDC 621.382/7--dc23
LC record available at https://lccn.loc.gov/2016015995

Visit the Taylor & Francis Web site at
http://www.taylorandfrancis.com

and the CRC Press Web site at
http://www.crcpress.com

Dedication

To the memory of my father
To my mother, Nguyen Thi Huong
To Phuong and Lam

Contents

Preface

Optical communications and photonic systems have been dominating the back and front hauling of information in global networks, that is, in the core, metro, and access areas. Noises are always the important topics for communication system engineering to maximize the integrity of transmitting and receiving signals and to minimize the error rate. Traditional telecom networks are proving inadequate to meet the challenges emerging from data center-centric networks to serve the ultrafast and ultrahigh-density demands from global mass users. Thus, the ultimate capacity to be offered from a single-mode fiber or multicore fibers must be examined in which the noise contributions are the principal factors to achieve an optimum signal-to-noise ratio.

The first book on this topic, *Noises in Digital Optical Transmission Systems,* was written by Professor Gunnar Jacobsen of Denmark Technical University.[*] It provided detailed technical presentations in treating noises in digital transmission systems employing light waves as the carriers. However, the digital optical transmission systems have progressed tremendously over the past two decades since the publication of that book, from a bit rate of a few Gbps to 100 Gbps and then Tbps and multi-Tbps in 2015 with a total aggregate bit rates of higher than 25 Tbps over the C-band of silica-based optical fibers.

Thus, the author thinks that new and updated presentations of such treatment of noises in optical transmission systems are necessary; hence, the objectives of this book are (1) to provide the principles of digital communications within the context of optical and photonic transmission technology and the principles of noise generated in the electronic, optical, and photonic processes, including the sensitivity and quantum limits of such modulated light wave systems; (2) to present the noise effect on optical transmission of various modulation formats over optically amplified fiber links; and (3) to identify the roles of the generated noises in photonic amplification processes.

Furthermore, the optical communications systems in the twenty-first century and its photonics have been extensively developed employing both the coherent and noncoherent receptions for long- and short-haul transmission, respectively. The information rate per channel can reach several Tbps by employing multisubcarriers. This high rate can reach even further by pulse shaping, ultra-high sampling such as Nyquist theorem, and processing the channel information by sampling at a rate of a number of samples per symbol less than the required Nyqusit rate of 2, now commonly termed as the faster than Nyquist processing technique. These channels face the interchannel cross talks and intermodulation effects, thus reducing the signal integrity.

The chapters in this book have been organized as follows: Chapter 1 gives generic models of noises of electronic, optical, and photonic types, in particular when reception subsystems employ both direct and coherent detection. The sensitivity limits of such detections are also described. Chapter 2 provides an introduction to the capacity limit of noisy optical transmission systems suffering impairments due to both linear and nonlinear distortions and electronic noises. Chapters 3 and 4 describe the noise processes in both noncoherent and coherent reception systems. Chapter 5 discusses some techniques in the optical domain to suppress the resultant electronic noises in coherent reception. Chapter 6 gives a brief introduction to the digital signal processing techniques for coherent detections and the digital processes to overcome several practical difficulties in non-DSP coherent communications. Chapter 7 provides necessary concepts and techniques in optical modulation so that readers will have some understandings of the modulation of light waves. Readers who are familiar with the optical modulation may by-pass this chapter. Phasor representations of the light waves under modulation assist the understanding of the behavior and any noises superpositioning

---

[*] Jacobsen, G., *Noises in Digital Optical Transmission Systems*, Artech House, London, UK, 1994.

on the modulated waves. Chapters 8 through 11 describe noises in optical transmission systems under the modulation of the amplitude and phases (both continuous and discrete) of the light waves, including the duobinary modulation technique in optical domain to reduce the signal bandwidth for less noisy and efficient transmission. The last chapter of the book, Chapter 12, describes a multi-Tbps optical transmission platform for the ultimate ultrahigh-capacity information networks.

Le Nguyen Binh
Huawei Technologies, European Researh Institute, Munchen, Duetschland

MATLAB® is a registered trademark of The MathWorks, Inc. For product information, please contact:

The MathWorks, Inc.
3 Apple Hill Drive
Natick, MA 01760-2098 USA
Tel: 508-647-7000
Fax: 508-647-7001
E-mail: info@mathworks.com
Web: www.mathworks.com

Acknowledgments

I am grateful to Huawei Technologies for making laboratories and facilities available; the commercial confidentiality of the technical materials related to the chapters has been strictly observed, and only the general principles are outlined. I am grateful to Dr. Thomas Lee of SHF AG of Berlin, Germany for frequent discussions on several topics related to minimum noise/distortion generations of digital optical signals. I am also indebted to the discussions with my colleagues at Huawei European Research Center and Shenzhen headquarters.

Last, but not least, I thank my wife Phuong and my son Lam, who have supported me happily over the years and by putting up with my work hours while writing this book. Over the years, my parents provided me with the education of *learning for life* and made sacrifices for educating all their children.

Author

Le Nguyen Binh earned BE(Hons) and PhD degrees in electronic engineering and integrated photonics in 1975 and 1978, respectively, both from the University of Western Australia, Western Australia. In 1980, he joined the Department of Electrical Engineering of Monash University after a 3-year stint with CSIRO Australia as a research scientist.

He has worked on several major advanced world-class projects, especially the Tb/s and 100 Gb/s DWDM optical transmission systems and networks, employing both direct and coherent detection techniques for the Department of Optical Communications of Siemens AG Central Research Laboratories in Munich, Germany, and the Advanced Technology Centre of Nortel Networks in Harlow, England. He was the Tan-Chin Tuan Professorial Fellow of Nanyang Technological University of Singapore in 2004. In 2008, he was also a visiting professor at the Faculty of Engineering of Christian Albrecht University of Kiel, Germany.

Dr. Binh has authored and coauthored more than 250 papers in leading journals and refereed conferences and 7 books in the field of photonics and digital optical communications, all published by CRC Press (Taylors & Francis Group), Boca Raton, Florida. His current research interests are in advanced modulation formats for super-channel long-haul optical transmission, electronic equalization techniques for optical transmission systems, ultrashort pulse lasers and photonic signal processing, optical transmission systems, and network engineering.

Dr. Binh studied and developed several courses in engineering such as fundamentals of electrical engineering, physical electronics, small signal electronics, large signal electronics, signals and systems, signal processing, digital systems, microcomputer systems, electromagnetism, wireless and guided wave electromagnetics, communications theory, coding and communications, optical communications, advanced optical fiber communications, advanced photonics, integrated photonics, and fiber optics. He has also led several course and curriculum developments in electrical and electronic engineering (EEE) and joint courses in physics and EEE.

Since January 2011, he has been working for the European Research Center of Huawei Co. Ltd. in Munich, Germany as the technical director of the company, working on several engineering aspects of 100G, multi-Tbps long-haul and metro coherent and noncoherent optical transmission networks, and integrated photonic technologies. He is a recipient of three prestigious Huawei gold medals for his work in these fields. Dr. Le Binh has also been a series editor of *Optics and Photonics* (CRC Press) and chaired the Commission D (Electronics and Photonics) of the national Committee of Radio Sciences of the Australian Academy of Sciences.

Dr. Le Binh is an alumnus of the high schools Phan Chu Trinh and Phan Boi Chau of Phanthiet, Vietnam, where, in 1969, he obtained his Bacalaureat Parts I and II.

List of Abbreviations

ADC	analog-to-digital converter
AGC	automatic gain control
AIS	alarm indication signal
APC	automatic power control
ASE	amplified spontaneous emission
B2B	back-to-back
BalDet	balanced detectors or PDP (photodetector pair) connected back to back
BalORx	balanced optical receiver
BCJR	stading for names of inventors L. Bahl, J. Cocke, F. Jelinek and J. Raviv
BER	bit-error rate
BOL	begin of life
CD	chromatic dispersion
CG	comb generator or combgen
CMA	constant modulus algorithm
CMU	clock management unit
Combgen	comb generator
CombgenNL	comb generator using nonlinear driving conditions
Co-OFDM	coherent optical OFDM
Co-ORx	coherent optical receiver
CPE	carrier phase estimation
CR	carrier recovery
CW	continuous wave
DAC	digital-to-analog converter
DGD	differential group delay
DQPSK	differential quadrature phase-shift keying
DSP	digital signal processing
DWDM	dense wavelength division multiplexing
ECL	external cavity laser
EDFA	erbium-doped fiber amplifier
ENOB	effective number of bits
EOF	end of life
FDEQ	frequency domain equalization
FEC	forward error correction
FFT	fast fourier transform
FIR	finite impulse response
FWM	four-wave mixing
Ga/s	giga-samples per second
GB or GBaud	giga-baud or giga-symbols/s
GSys/s	giga-symbols per second
HD	hard decision
ISI	intersymbol interference
iTLA	integrated tunable laser assembler
IXPM	intrachannel cross-phase modulation

LMS	least mean square
LO	local oscillator
LPF	low-pass filter
LSB	least significant bit
LUT	look-up table
MAP	maximum-a-posteriori probability
MIMO	multiple input multiple output
MLD	multilane distribution
MLSE	maximum likelihood sequence estimation
MSB	most significant bit
MSE	mean square error
MZIM	mach–zehnder intensity modulator
MZM	mach–zehnder modulator
NLE	nonlinear performance enhancement
NLSE	nonlinear Schrödinger equation
NLPN	nonlinear phase noise
NRZ	nonreturn-to-zero
Nyquist QPSK	nyquist pulse shaping modulation quadrature phase-shift keying
NZDSF	nonzero dispersion-shifted fiber
OA	optical amplifier
OFDM	orthogonal frequency division multiplexing
OFE	optical front-end
ORx	optical receiver
OSA	optical spectrum analyzer
OSNR	optical signal-to-noise ratio
PD	photodetector
PDL	polarization-dependent loss
PDM	polarization division multiplexing
PDP	photodetector pair—see also balanced detection
PLL	phase-locked loop
PMD	polarization mode dispersion
RF	radio frequency
RFS	recirculating frequency shifting
RFSCG	RFS comb generator
RIN	relative intensity noise
RZ	return-to-zero
SoftD	soft decision
SOP	state of polarization
SPM	self-phase modulation
SSB	single sideband
SSMF	standard single-mode fiber (G.652)
TDEQ	time-domain equalization
TIA	trans-impedance amplifier
VOA	variable optical attenuator
WSS	wavelength selective switch
XPM	cross-phase modulation

1 Introduction

This introductory chapter briefly outlines the historical development, emergence, and merging of fundamental digital communication techniques and optical communications to fully exploit and respond to the challenges of the availability of the ultrahigh frequency and ultra-wideband in the optical spectra of optical fiber communication technology. The organization of the chapters of the book is outlined.

1.1 DIGITAL OPTICAL COMMUNICATIONS AND TRANSMISSION SYSTEMS: A HISTORY OVERVIEW AND CHALLENGING ISSUES

Starting from the proposed dielectric waveguides by Kao and Hockham [1] in 1966, the first research phase attracted intensive interests around the early 1970s in demonstration of fiber optics, and optical communications have tremendously progressed over the last three decades. At the time, the optical waveguide is proposed as a dielectric waveguide for optical frequency operation. The first generation lightwave systems were commercially deployed in 1983 and operated in the first wavelength window of 800 nm over multimode optical fibers (MMFs) at transmission bit rates of up to 45 Mbps [2,3]. After the introduction of ITU-G652 standard single-mode fibers (SSMFs) in the late 1970s [4], the second generation of lightwave transmission systems became available in the early 1980s [5,6]. The operating wavelengths were shifted to the second window of 1300 nm, which offers much lower attenuation for silica-based optical fibers compared to the previous 800 nm region and especially the chromatic dispersion (CD) factor is almost zero. These second generation systems could operate at bit rates of up to 1.7 Gbps and have a repeater-less transmission distance of about 50 km [7]. Further research and engineering efforts were also pushed for the improvement of the receiver sensitivity by coherent detection (CoD) techniques, and the repeater-less distance has reached 60 km in installed systems with a bit rate of 2.5 Gbps. Optical fiber communications then evolved to third generation transmission systems that utilized the lowest attenuation 1550 nm wavelength window and operated up to 2.5 Gbps bit rate [8]. These systems were commercially available in 1990 with a repeater spacing of around 60–70 km [9,10]. At this stage, the generation of optical signals was mainly based on direct modulation of the semiconductor laser source and either direct detection. Since the invention of erbium-doped fiber amplifiers (EDFAs) in the early 1990s [9,11,12], lightwave systems have rapidly evolved to wavelength-division multiplexing (WDM) and shortly after that, dense WDM (DWDM) optically amplified transmission systems that are capable of transmitting multiple 10 Gbps channels. This is due to the fact that the loss is no longer a major issue for external optical modulators that normally suffer an insertion loss of at least 3 dB. These modulators allow the preservation of the narrow linewidth of distributed feedback lasers (DFB). These high-speed and high-capacity systems extensively exploited the external modulation in their optical transmitters. The present optical transmission systems are considered at the fifth generation, having a transmission capacity of a few Tbps [13]. Figure 2.1 shows the evolution of the capacity of information rate transmitted over a single strand of SMF over the last 35 years since the 80's of the last century.

Coherent detection, homodyne or heterodyne, was the focus of extensive research and developments during the 1980s and early 1990s [11,14–18], and was the main detection technique in the first three generations of lightwave transmission systems. At that time, the main motivation for the development of coherent optical systems was to improve the receiver sensitivity, commonly by 3–6 dB [14,17] and the boosting of the received optical channel by the mixing with a high-power local oscillation laser to overcome the limitation of transmission distance by fiber attenuation. This was thought possible due to the fabrication and manufacturing of the single-mode fiber (SMF). The

SMF has a very small index difference between the core and the cladding (about 0.3% difference), thus very weak tightness of the guided mode in the core of the fiber; only 70% of optical power of the guided mode is confined within the core region. This weak tightness of the mode allows the long-distance guidance and low scattering loss at the boundary between the core and the cladding. The guided mode has been termed as the weakly guided mode or the linearly polarized mode. Thence, the two-polarization mode has two very similar modes, the vertical and horizontal polarized modes, that propagate at a very close phase velocity. At that time, the operating speed of the system was still in the MHz region, and no attention was paid to the difference in these polarized guided modes till the last decade of the twentieth century.

The SMF has allowed several proposals on the research and developments of coherent fiber optic communications. There were a number of photonic components required for such coherent transmitters and receivers that are single-frequency laser sources, external modulators so as to preserve the line of the lasers. DFB were successfully developed. $LiNbO_3$ phase and interferometric modulators were available in integrated form with a total insertion loss of 44–6 dB and 3 dB bandwidth of around 8 GHz allowing system operating speed of several hundred MHz and eventually 2.5 Gbps. Modulation formats by the modulation of amplitude, phase, or frequency were investigated and demonstrated. However the insertion loss and the linewidth of the DFB limit the repeater-less transmission to around 60 GHz in a laboratory environment and 40 km of fiber in practice. This is why we can see that there several buildings for housing repeaters along the routes in several terrestrial transmission networks throughout the world. These housing premises were very costly to construct, and network operators, nowadays, still take advantages of these infrastructures.

The repeater-less transmission distance was thus able to be extended to more than 60 km of SSMF (with 0.2 dB/km attenuation factor). However, coherent optical systems suffer severe performance degradation due to fiber dispersion impairments. In addition, the phase coherence for lightwave carriers of the laser source and the local laser oscillator was very difficult to be maintained.

However, in the late 1980s, the invention of optical amplifiers by the successful doping of erbium ions in silica fiber offering an optical gain of 20–30 dB has nullified the prospects of coherent transmission as this gain is sufficient to overcome the loss problems of SMF. Thence, there were tremendous developments in long-haul transmission at 2.5 Gbps with optical amplifiers positioned at a periodic distance. This period has witnessed for the first time that the SMF transmission was no longer limited by the fiber loss but the CD. The CD was due to the waveguide guiding of the mode is a function of wavelength as well as that due to the materials. To resolve this dispersion, dispersion-compensating fiber (DCF) was used to compensate in the photonic domain. At the same time due to the narrowness of the laser linewidth by externally feedback to the DFB cavity and the bandwidth improvement of $LiNbO_3$ modulator, the 10 Gbps DWDM and SMD+EDFA+DCF+EDFA spans have been proven to offer significant increase in the transmission capacity. The evolution of the capacity is shown in Figure 1.1. It is increased 10 times for every 4 years. The 10 Gbps DWDM has beaten the 2.5 Gbps over the long-haul core networks with 44–88 wavelength channels in the C-band. Raman optical amplifiers have also been employed to extend the transmission over the L-band and optically equalized the gain ripple of EDFAs over the entire C-band with multiple wavelength sources.

On the contrary, the incoherent detection technique minimizes the linewidth obstacles of the laser source as well as the local laser oscillator, and thus, relaxes the requirement of the phase coherence. Moreover, incoherent detection mitigates the problem of polarization control in the mixing of transmitted lightwaves and the local laser oscillator at a multi-THz optical frequency range. The invention of EDFAs that are capable of producing optical gains of 20 dB and above has also greatly contributed to the progress of incoherent digital photonic transmission systems up till now. These direction detection receivers are much less expensive so they are currently attracting significant interests for access systems and networks.

In the early 1990s, a huge increase in the demand for broadband communications driven mainly by the rapid growth of multimedia services, peer-to-peer networks, and IP streaming services, in

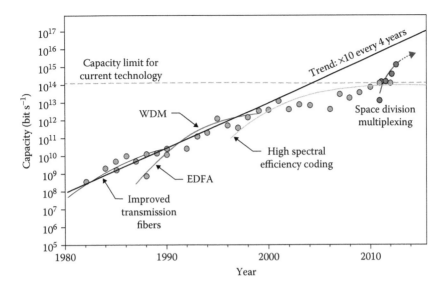

FIGURE 1.1 Evolution of transmission capacity over a single SMF with respect to time and year.

particular IP TV, was experienced. It is most likely that such tremendous growth will continue in the coming years. This is the main driving force for local and global telecommunications service carriers to develop high-performance and high-capacity next generation optical networks. The overall capacity of WDM or DWDM optical systems can be boosted either by increasing the base transmission bit rate of each optical channel, multiplexing more channels in a DWDM system, or preferably, by combining both of these schemes. However, while implementing these schemes, optical transmission systems encounter a number of challenging issues that are outlined in the rest of this section.

The 10 Gbps DWDM transmission systems employ intensity modulation, also known as on-off keying (OOK) and utilize nonreturn-to-zero pulse shapes. The term OOK can also be used interchangeably with amplitude shift keying (ASK).[*] The reception technique is a self-homodyne detection using the same optical signal that split and delay in one arm of an optical Mach–Zehnder delay interferometer with a one bit delay as shown in Figure 1.2(b). The photodetector pair can be used to offer another 33 dB gain in the push mode operation. The processing of the received electronic signals is still in the hardware digital circuitry. Moving toward high bit rate transmission, such as 40 Gbps, the performance of OOK photonic transmission systems is severely degraded due to fiber impairments, including fiber dispersion and fiber nonlinearities. The fiber dispersion is classified into CD and polarization mode dispersion (PMD), causing the intersymbol interference problem. On the other hand, severe deteriorations on the system performance due to fiber nonlinearities are resulted from high-power spectral components at the carrier and signal frequencies of OOK-modulated optical signals. It is also of concern that existing transmission networks comprise millions of kilometers of SSMF that have been installed for approximately two decades. These fibers do not have as advanced properties as state-of-the-art fibers used in recent laboratory "hero" experiments, and they have degraded after many years of use.

The compensation of the fiber dispersion over the entire C-band was done by employing the DCF at the end of the transmission fiber of each span. The rest of the residual dispersion was done in the optical domain such as the array waveguide gratings and phase reflectors or few-mode fiber (FMF) or integrated optical multistage filter [19]. Because of the insertion of DCF, the optical loss becomes high (about 11–12 dB for 22 km DCF for 100 km transmission fiber SSMF); so an

---

[*] The OOK format simply implies the on-off states of the lightwaves where only the optical intensity is considered. On the other hand, the ASK format is a digital modulation technique representing the signals in the constellation diagram by both the amplitude and phase components.

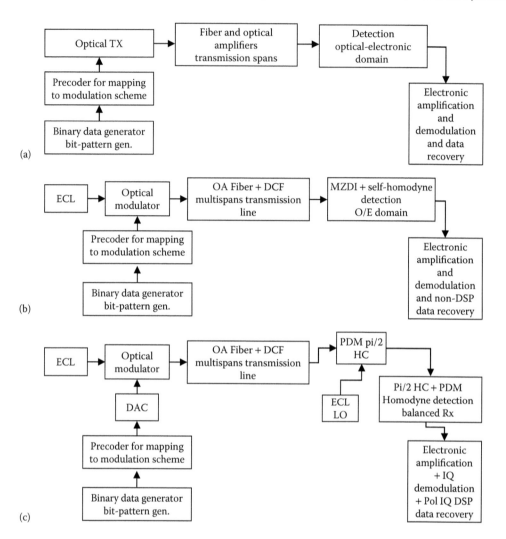

FIGURE 1.2 Schematic diagram of optical fibrer communication systems (a) Direct modulation and direction, (b) External modulation and self-coherent detection and (c) coherent modulation and reception associating with digital signal processing. Legend: Tx = transmitter ; ECL = external cavity; OA = optical amplifier; MZDI = Mach Zehnder delay interferometer; DCF = dispersion compensating fiber; O/E = optical to electronic conversion; DSP = digital signal processing; PDM = polarization division multiplexing; HC = hybrid coupler; DAC = digital to analog converter; IQ = inphase and quadrature phase.

additional optical amplifier (the EDFA) must be used to compensate for this loss. Furthermore, the effective area of the DCF is smaller (~15 μm²), thus the nonlinear effects are enhanced and thus the overall line distance must be reduced.

The accumulation of the amplified stimulated emission optical noises of the dispersion compensated fiber transmission spans has reduced the optical signal-to-noise ratio (OSNR) as the maximum optical power level of individual channels is limited by the nonlinear threshold. This has motivated engineers to innovate novel techniques to extend the transmission distance and capacity of the optical communication systems.

The questions were whether coherent techniques can be reinvestigated to offer the recovery of received optical signals including mitigating the difficulties faced by the former coherent research in the 1980s. Digital signal processing (DSP) with the ultrahigh sampling rates and high-speed application specific integrated circuit processors and novel algorithms. The answers are that it is very much possible to employ DSP to solve the issues of difference in frequency of the local oscillator (LO) and the carriers, equalization of extremely long and highly dispersive signals, and so on.

The total transmission capacity can be enhanced by increasing the number of multiplexed DWDM optical channels. This can be carried out by reducing the frequency spacing between these optical channels, for example, from 100 GHz down to 50 GHz, or even 25 and 12.5 GHz [16,17]. The reduction of the channel spacing also results in narrower bandwidths for the optical multiplexers and demultiplexers. Passing through these narrowband optical filters, signal waveforms are distorted and optical channels suffer the problem of interchannel crosstalks. The narrowband filtering problems are getting more severe at high data bit rate, for example, 56–64 Gbaud, thus degrading the system performance significantly.

Together with the demand of boosting the total system capacity, another challenge for the service carriers is to find cost-effective solutions for the upgrading process. These cost-effective solutions should require minimum renovation to the existing photonic and electronic subsystems, that is, the upgrading should only take place at the transmitter and receiver ends of an optical transmission link. Another possible cost-effective solution is to extend significantly the uncompensated reach of optical transmission links, that is, without using dispersion-compensation fibers (DCFs), thus reducing a considerable number of required inline EDFAs. This network configuration has recently received much interest of both photonic research community as well as service carriers. These additional amplifiers of all the spans superimpose their noises on the optical signals and hence reduce the OSNR.

Taylor [20,21] has been the first to propose and prove the DSP for the coherent reception technique to overcome several severe difficulties of coherent transmission over SMF started in the early 1980s [22,23]. This has enabled several novel techniques and developments of algorithms for compensating dispersion and hence eliminating the use of DCF in each span, thus minimizing the ASE noises reduction of the OSNR [24,25]. Figure 1.2(c) shows the schematic diagram of this coherent DSP-based reception system. The availability of narrow linewidth laser sources and wideband integrated optical modulators as well as high-speed analog-to-digital converters [26–28] has allowed the pushing of the symbol rates and higher order modulation M-ary quadrature amplitude modulation to the highest level. The capacity can now reach close to the Shannon limit.

Huawei Co. Ltd (Shenzhen, China) has also demonstrated its first tera-bps transmission over more than 3,500 km multispan optically amplified fiber transmission in field trial employing multisubcarrier coherent transmitter and coherent DSP-based reception in field trial in Europe [29,30]. Currently, the explosion in demands on data center-centric (DCC) networking and cloud-based computing as well as cloud telecom networks to transform the traditional telecom networking, that is, to flatten the telecom networks to meet the challenges posed by the DCC networking. The transmission capacity is expected to reach the order of several peta-bps and even exa-bps. The capacity for interconnections of data centers (DCs) and access nodes of core telecom networks will reach tremendously high capacity. Furthermore, the infrastructures of the optical access networks are needed to be upgraded to handle this extremely high capacity to the end users. 5G wireless cloud-based networks are expected to be deployed in the year 2020. Several data clouds and distributed DC are to be deployed throughout the networks to the edges of the networks.

Recently, several reports have addressed the peta-bps [31–33] transmission capacity as shown in Figure 1.2. In another dimension, the spatial domain has been added via the use of multicore fibers or FMF and multiple input multiple output (MIMO) processing techniques. The MIMO DSP technique is possible due to the integration of several DSPs in the same chips and the fabrication of the multicore fibers. However, for practical deployment, there would be several issues to be resolved.

The MIMO processing techniques are well established in wireless communications technology and are currently attracting intensive interests in optical communications community for further increasing the transmission capacity. Thus, we have witnessed over the second decade of the twenty-first century the emergence of multicore fibers, the FMF, and even MMF in which separate information optical channels are launched into individual modes. The received modes are separated at the end of the FMF or MMF and fed into multiple optical receivers, sampled, and DSP in the MIMO processors.

We thus now see that the MMFs are again emerged as one of the possible mediums for transmission of ultrahigh capacity, at least in the DC or local access environments prior to the uses in core networks. The impairments due to modal delays between the guided modes and intermodal coupling in MMF can be easily resolved by MIMO processors.

1.2 OBJECTIVES OF THIS BOOK

Therefore, the principal motivations of this book are to discuss the photonic and electronic noises and their effects on transmission capacity in optically amplified multispan coherent transmission and DSP-based reception for ultralong transmission systems without using DCF and additional EDFAs. Furthermore, the basics of noises in terms of definitions, analyses, and impacts on the ultimate capacity of the optical transmission systems are given.

The employment of digital communications in modern optical communications (see Peta-bps information capacity transmission system of Figure 1.3) is expected to be dominating the information transmission networks in cloud flattened telecom networks and DCC networks. The fundamental principles of digital communications, both coherent and incoherent transmission and detection techniques, are described with focus to the technological development and limitations in the optical domain. The enabling technologies and research results and demonstration in laboratory experimental platforms for the development of high-performance and high-capacity next generation optical transmission systems impose significant challenges to the engineering of optical transmission systems in the near future and techniques for network monitoring.

A number of aspects of noises in electronic and photonic domains are collected from the subjects given in chapters of the published books by the author [34–36]. Furthermore, extensive definitions and analyses as well as applications are given to give a complete overview of the noises in optical systems in modern optical long-haul and access networks in the next coming decades.

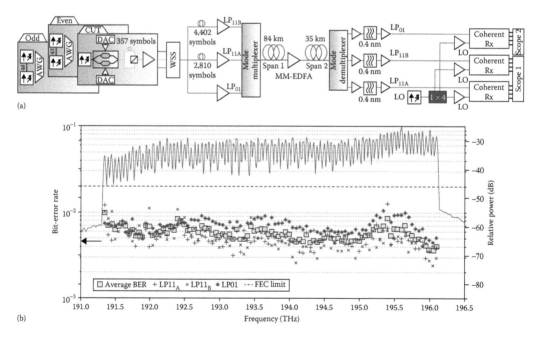

FIGURE 1.3 Experiment schematic arrangement for demonstration of Pbps capacity by space-division multiplexing in optical fibers (a) system coherent reception of multicarrier transmission set-up and (b) BER (vertical axis) versus wavelength of channels in THz. (Extracted from Okoshi, T., *IEEE Trans. Microw. Theory Tech.*, 82, 1138–1149, 1982. With permission.)

1.3 ORGANIZATION OF CHAPTERS

The book subjects are organized with (a) an introduction of capacity and noises and their impacts on capacity; (b) noises in reception systems, both direct and coherent receptions; and (c) impacts of noises on modulation formats in optical transmission systems.

The chapters are then organized as follows:

Chapter 2 gives an introduction of capacity since the extremely innovative works by Shannon since 1948, but special emphases are given to memory-less and memory optical systems described in [37].

Chapter 3 describes the coherent reception and noises in such systems at the physical levels. The DSP-based coherent reception is described in Chapter 6. The coherent reception is presented first before for the noncoherent reception techniques and associate noises in Chapter 4.

Chapter 5 describes techniques for suppression of noises in CoD and an example on the design of electric transimpedance (TI) amplification receivers at an electronic level and results on the noise suppression.

Chapters 7–11 describe the optical modulations and issues of distortions impacts on transmission quality, in particular the linear and nonlinear amplitude and phase distortions as well as the DSP-based techniques to maximize the transmission capacity. Discrete and continuous phase-shift keying formats as well as the quadrature amplitude modulation and duobinary partial response schemes are presented under noisy conditions for optimum transmission capacity.

Chapter 12 describes experimental demonstration of Tbps optical transmission employing multisubcarriers, hence subchannels via the modulation and filtering of comb lasers. This enables novel methods for ultrahigh capacity optical networks in the hexa = use Exa- = 10 to an exponential of 18 range.

REFERENCES

1. K. C. Kao and G. A. Hockham, Dielectric-fibre surface waveguides for optical frequencies, *Proc. IEE* 113(7), 1151–1158, 1966.
2. I. P. Kaminow and T. Li, *Optical Fiber Communications*, Volume IVB, USA: Elsevier Science, 2002.
3. R. S. Sanferrare, Terrestrial lightwave systems, *AT&T Technology Journal*, 66, 95–107, 1987.
4. C. Lin, H. Kogelnik, and L. G. Cohen, Optical pulse equalization and low dispersion transmission in single-mode fibers in the 1.3–1.7 μm spectral region, *Optics Letters*, 5, 476–478, 1980.
5. A. H. Gnauck, S. K. Korotky, B. L. Kasper, J. C. Campbell, J. R. Talman, J. J. Veselka, and A. R. McCormick, Information bandwidth limited transmission at 8 Gb/s over 68.3 km of single mode optical fiber. In *Proceedings of OFC'86*, Paper PDP6, Atlanta, GA, 1986.
6. H. Kogelnik, High-speed lightwave transmission in optical fibers, *Science*, 228, 1043–1048, 1985.
7. G. P. Agrawal, *Fiber-Optic Communication Systems*, 3rd ed., New York: Wiley, 2002.
8. A. R. Chraplyvy, A. H. Gnauck, R. W. Tkach, and R. M. Derosier, 8 × 10 Gb/s transmission through 280 km of dispersion-managed fiber, *IEEE Photonics Technology Letters*, 5, 1233–1235, 1993.
9. P. C. Becker, N. A. Olsson, and J. R. Simpson, *Erbium-Doped Fiber Amplifiers, Fundamentals and Technology*, San Diego: Academic Press, 1999.
10. C. R. Giles and E. Desurvire, Propagation of signal and noise in concatenated erbium-doped fiber amplifiers, *IEEE Journal of Lightwave Technology*, 9(2), 147–154, 1991.
11. T. Okoshi, Heterodyne and coherent optical fiber communications: Recent progress, *IEEE Transactions on Microwave Theory and Techniques*, 82(8), 1138–1149, 1982.
12. M. C. Farries, P. R. Morkel, R. I. Laming, T. A. Birks, D. N. Payne, and E. J. Tarbox, Operation of erbium-doped fiber amplifiers and lasers pumped with frequency-doubled Nd:YAG lasers, *IEEE Journal of Lightwave Technology*, 7(10), 1473–1477, 1989.
13. H. Kogelnik, High-capacity optical communications: Personal recollections, *IEEE Journal on Selected Topics in Quantum Electronics*, 6(6), 1279–1286, November/December 2000.

14. T. Okoshi, Recent advances in coherent optical fiber communication systems, *IEEE Journal of Lightwave Technology*, 5(1), 44–52, 1987.
15. J. Salz, Modulation and detection for coherent lightwave communications, *IEEE Communications Magazine*, 24(6), 34–49, 1986.
16. T. Okoshi, Ultimate performance of heterodyne/coherent optical fiber communications, *IEEE Journal of Lightwave Technology*, 4(10), 1556–1562, 1986.
17. P. S. Henry, *Coherent Lightwave Communications*, New York: IEEE Press, 1990.
18. A. F. Elrefaie, R. E. Wagner, D. A. Atlas, and A. D. Daut, Chromatic dispersion limitation in coherent lightwave systems, *IEEE Journal of Lightwave Technology*, 6(5), 704–710, 1988.
19. L. N. Binh, *Photonic Signal Processing*, Boca Raton, FL: CRC Press, 2007.
20. M. G. Taylor, Coherent detection method using DSP for demodulation of signal and subsequent equalization of propagation impairments, *IEEE Photonics Technology Letters*, 16(2), 674–676, 2004.
21. M. G. Taylor, Coherent detection for optical communications using digital signal processing. In *Proceedings of Conference Optical Fiber Communications National Fiber Optic Engineers, OFC/NFOEC 2007*, IEEE, Anaheim CA, March 2007.
22. K. Iwashita and N. Takachio, Chromatic dispersion compensation in coherent optical communications, *IEEE Journal of Lightwave Technology*, 8(3), 367–375, 1990.
23. D.-S. Ly-Gagnon, S. Tsukamoto, K. Katoh, and K. Kikuchi, Coherent detection of optical quadrature phase-shift keying signals with carrier phase estimation, *IEEE Journal of Lightwave Technology*, 24(1), 12–21, 2006.
24. S. J. Savory, A. D. Stewart, S. Wood, G. Gavioli, M. G. Taylor, R. I. Killey and P. Bayvel, Digital equalisation of 40 Gbit/s per wavelength transmission over 2480 km of standard fibre without optical dispersion compensation. In *Proceedings of ECOC 2006*, IEEE, Cannes, France, 2006.
25. S. J. Savory, G. Gavioli, R. I. Killey and P. Bayvel, Transmission of 42.8 Gbit/s polarization multiplexed NRZ-QPSK over 6400 km of standard fiber with no optical dispersion compensation. In *Proceedings of OFC 2007*, IEEE, Anaheim, CA, 2007.
26. Fujitsu Co. Ltd., UK, *56GSa/s 8-bit Analog-to-Digital Converter*, Factsheet, http://www.fujitsu.com/downloads/MICRO/fma/pdf/56G_ADC_FactSheet.pdf (accessed November 15, 2005).
27. K. Kikuchi, M. Fukase and S.-Y. Kim, Electronic post-compensation for nonlinear phase noise in a 1000-km 20-Gbit/s optical QPSK transmission system using the homodyne receiver with digital signal processing. In *Proceedings of OFC 2007*, IEEE, Anaheim, CA, 2007.
28. Fujitsu Co. Ltd., UK, *The Fujitsu 56GSa/s Analog-to-Digital Converter Enables 100GbE Transport Ultrafast CMOS ADC Provides Technology Breakthrough for Upcoming Telecommunication Applications*, http://www.fujitsu.com/downloads/MICRO/fma/pdf/56G_techback.pdf (accessed July 15, 2016).
29. Y. Zhao, N. Stojanovic, D. Chang, C. S. Xie, B. Mao, L. N. Binh, Z. Xiao and F. Yu, Adaptive joint carrier recovery and turbo decoding for Nyquist terabit optical transmission in the presence of phase noise. In *Proceedings of OFC 2014*, IEEE, San Francisco, CA, March 2014.
30. L. N. Binh, B. Mao, N. Stojanović, C. S. Xie, N. Yang, Synchronous modulator incorporated re-circulating comb laser sources for Tbps superchannel transmission. In *OSA Congress, Proceedings of Conference Advanced Solid-State Lasers*, Paris, France, November 2013, Paper JTh2A.13–1.
31. D. J. Richardson, J. M. Fini, and L. E. Nelson, Space-division multiplexing in optical fibres, *Nature Photonics*, 7, 354–362, 2013. doi:10.1038/nphoton.2013.94, Pub. online: 29 April 2013.
32. B. J. Puttman, R. S. Luis, W. Klaus, T. Hayashi, M. Hirano, and J. Marciante, 2.15 Pb/s transmission using multicore fiber and wideband optical comb. In *Proceedings of ECOC*, Paper PDP 3.1, Valencia, Spain, October 2015.
33. D. Soma, K. Igarashi, Y. Wakayama, K. Takeshima, and K. Kawayuchi, 2.05 Peta-bit/s super-Nyquist-WDM SDM transmission using 9.8-km 6-mode 19-core fiber in full C band. In *Proceedings of ECOC*, Paper PDP3.2, Valencia, Spain, 2015.
34. L. N. Binh, *Advanced Digital Optical Communications: Optics and Photonics*, 2nd edn., Boca Raton, FL: CRC Press, 2015.
35. L. N. Binh, *Digital Processing: Optical Transmission and Coherent Receiving Techniques: Optics and Photonics*, Boca Raton, FL: CRC Press, 2014.
36. L. N. Binh, *Optical Fiber Communication Systems with MATLAB® and Simulink® Models: Optics and Photonics*, 2nd edn., Boca Raton, FL: CRC Press, 2014.
37. E. Agrell, A. Alvarado, G. Durisi and M. Karlsson, Capacity of a nonlinear channel with finite memory, *IEEE Journal of Light Technology*, 32(6), 2862–2876, August 16, 2014.

2 Capacity and Quantum Limits in Optical Systems

This chapter briefly introduces the historical development, emergence, and merging of the fundamental digital communication techniques and optical communications to fully exploit and respond to the challenges of the availability of the ultrahigh frequency and ultra-wideband in the optical spectra of optical fiber communication technology. Furthermore, the quantum limits are given for coherent reception and extended for the case of quantum transmission.

2.1 CAPACITY LIMIT OF THE FIBER OPTICS INFORMATION CHANNEL

The system transmission rates (in Giga-bits/sec) have has been demonstrated over the years, as shown in Figure 2.1. The rate of increase is about quadruple every decade. The first period is the rate of increase via the transmission over single mode type, to avoid the intermodal interferences and improve the transmission media. Dispersion-shifted fibers were employed to minimize the pulse broadening and improvement of laser sources such as DFB types. In this period, coherent transmission and reception were also investigated employing both heterodyne and homodyne techniques. However, there were several practical problems due to the frequency offset between the local oscillator and the signal carriers, making so difficult as well as the phase detection and phase noises and linewidth of the laser sources. Then, till the end of the 1980s, optical amplifiers made by erbium doped in silica core were offering an optical gain reaching 30 dB, making the intensity direct detection and dense wavelength multiplexed transmission and the use of integrated external modulators become possible as their insertion loss can be compensated; hence the problems due to broadening of the direct-modulated lasers were eliminated. The baud rate and advanced modulation formats were exploited to offer the bit rate reaching 80 Gb/s (Gbps) with time division multiplexing. Thus, together with the dense wavelength-division multiplexing (DWDM) the total capacity over a single single-mode fiber (SMF) reached few terabits/second. Transmission systems of 10 Gbps DWDM over ultralong transmission lines were deployed successfully throughout the world by dispersion compensating fibers. During this period, the optically amplified multispan systems faced severe difficulties in both the nonlinear impairments and the accumulated amplification stimulated emission noises of the EDFAs that are placed after each fiber span of the transmission and/or compensating fibers. Since the beginning of the twenty-first century, coherent reception in association with the digital signal processing (DSP) techniques [1–4] has been making possible the reviving of the coherent transmission and reception mainly placed at the receivers. The DSP-based coherent reception is employed together with the multiplexing of the polarized modes of the linearly polarized mode of the SMF, pushing the capacity to a new peak. The capacity is now increased by the use of a new dimension, the multiplexing in spatial domain, the spatial division multiplexed fibers (SDMFs) such as few-mode fibers (FMFs) or multicore fibers (see Figure 2.2) or the orbital angular momentum (OAM) modes that theoretically can offer infinite capacity over a single fiber. Fiber bundles composed of physically independent SMFs with reduced cladding thickness could provide increased core packing densities relative to current fiber cables. However, "in-fiber" SDM is required to achieve the higher core densities and integration levels ultimately desired. Multicore containing multiple independent cores with sufficiently large spacing can be employed to limit the crosstalk. Fibers with up to 19 cores have been demonstrated for long-haul transmission—multicore

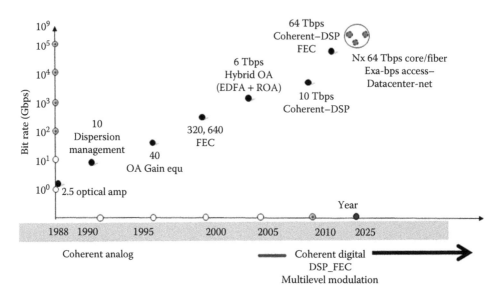

FIGURE 2.1 Evolution of Transmission Capacity with respect to time (year) at which optical coherent techniques influencing.

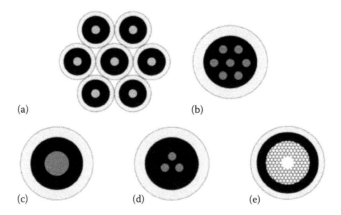

FIGURE 2.2 SDM fibers: (a) Fiber bundles composed of physically independent SMFs with reduced cladding thickness, (b) multicore fiber, (c) few-mode fiber, (d) coupled-core fibers, and (e) photonic bandgap fibers.

fibers are possible for short-haul applications for which higher levels of crosstalk per unit length can be tolerated. FMF can also be employed to guide a restricted number of modes (typically 6–12 distinct modes, including all degeneracy and polarizations) to further increase the capacity over a single fiber. The received optical signals via these modes can be digitally processed by the well-established multiple input multiple output (MIMO) techniques, commonly used in wireless systems. Coupled-core fibers support supermodes that allow higher spatial mode densities than isolated-core fibers. MIMO processing is essential to address the inherent mode coupling. Photonic bandgap fibers guide light in an air core and thus offer ultralow optical nonlinearity and potentially lower losses than solid-core fibers. Work is currently being conducted to determine whether such fibers can support multi-core multiplexing [5].

The principal question related to the capacity is as follows: Under the existing ASE noises, the nonlinear phase noises due to nonlinear effects, the electronic noises of the reception subsystems, and the maximum limited optical powers, what would be the maximum upper and lower bound of the SFS without using dispersion compensating fiber modules?

2.1.1 CHANNEL CAPACITY

The performance of any communication system is ultimately limited by the SNR of the received signal and available bandwidth. This limitation can be stated more formally by using the concept of channel capacity introduced within the framework of information theory [6,7]. The channel capacity is defined as the maximum possible bit rate for error-free (or bit-error rate BER $< 10^{-9}$) transmission in the presence of noise. For a linear communication channel operating under an additive white Gaussian noise (GN), and constraint by a total signal power at the input, the *Gaussian channel capacity* C_N is given by the well-known Shannon theorem and hence the Shannon formula

$$C_N = B \log_{10} \frac{1+P_S}{P_N} \text{ or } C_N = B \log_e\left(1+2\frac{E_S}{N_0}\right) \tag{2.1}$$

where:
  B is the overall channel 3 dB bandwidth
  P_S is the average signal power*
  P_N is the average noise power
  E_S is the signal average energy
  N_0 is the noise energy

For the noises, the noises in quantity of noise spectral density S_N as measured by an electrical or optical spectrum analyzer of units of A^2/Hz (noise current power) or $W/Hz^\dagger$ are commonly specified. Thus, the channel capacity can be expressed as

$$C_N = B \log_e\left(1+2\frac{P_S}{BS_N}\right) \tag{2.2}$$

Roughly, the theorem states that

> There is a maximum rate, the channel capacity, to the rate at which any communication system can operate satisfactorily when constrained in power. Thus operation at a rate greater than the channel capacity condemns the systems to higher probability of error.

This can lead to the two distinct possibilities:

  $C_N > R_S$: With R_S being the rate of information input to the system from the transmitter, commonly known as the symbol rate. There exist sets of M transmitter signals such that the probability of error that can be achieved with optimum receivers is arbitrarily small.
  $C_N < R_S$: The number of equally likely messages, $M = 2^{NR_S}$, is large, and the probability of error is close to unity for every possible set of M transmitted signals.

2.1.2 FIBER OPTIC TRANSMISSION SYSTEMS

Current optical fiber systems operate substantially below the fundamental limitation, imposed by Equation 2.1. However, a considerable improvement in the coding schemes for lightwave communications, expected in the near future, may result in the development of systems, whose efficiency may approach this fundamental limit. However, the representation of the channel capacity in the

---

* Notes on notations: subscripts S are used to indicate signals; it is also used to indicate the signal variables. Later on in the chapter, notations for input signal and outputs are $x(t)$; $y(t)$ so as to follow the common assignment of input and output variables of communication or signal processing systems. The subscripts N are used for channel capacity and variable N indicates noise in voltage. Note that for noise, one must use the square term and the average power—as noises are random variables so they are never subtracted but always superimposed randomly.

† In a spectrum analyzer, the input signal is resolved in the frequency domain with a slide of the spectral window, and the energy contained in such a slide is measure and displayed. So in measuring the total signal power, one has to integrate the entire spectrum. In the optical domain, the resolution is done by a grating and then detected by a PD array and digitally processed to display the optical spectrum.

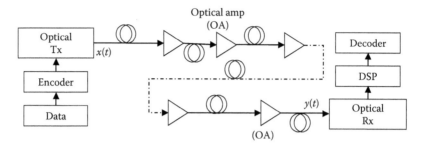

FIGURE 2.3   Schematic of an optically amplified optical transmission system. $x(t)$ = signal input sequence launched into the first fiber span and $y(t)$ = signal output sequence after transmission.

standard form (Equation 2.1) may not be completely suitable for applications to the current actual fiber optics systems. It was obtained based on the assumption of linearity of the communication channel, while the modern fiber optics systems operate in a substantially nonlinear regime, especially when the number of channels of different lightwave carriers are employed, the DWDM with a total transmission capacity reaching more than 100 Tbps over the whole C-band. Because the optical transmission lines must satisfy very strict requirements for a BER of 10^{-12}–10^{-15} or under hard FEC of 1.5×10^{-2} and soft FEC of 2.5×10^{-3}, the pulse amplitude should be large enough so that is can be effectively detectable. The encoder and decoder are required to map the binary sequence to the symbols of a modulation format to increase the capacity (bits/symbol), subject to the lower and upper bound of the channel capacity, especially when the data sequence is independent identically distributed (IID) type, that are presented in Section 2.4.2.

The increase of the number of WDM channels [8,9] in the modern fiber optics communication systems also leads to a substantial increase of the electric field intensity in the fiber. As a consequence, the Kerr nonlinearity of the fiber refractive index $n(I) = n_0 + \gamma |E^2|$ (where $|E^2|$ is the pulse intensity) becomes substantial and should be taken into account.

Consider a typical fiber optics communication system, which consists of a sequence of N fiber spans that consist of a transmission fiber of L kilometers long of the standard SMF followed by an amplifier (see Figure 2.3) to compensate for the power loss in the fiber. An inevitable consequence is the generation of the noise in the system, coming from the spontaneous emission in the optical amplifiers. It is assumed that all the fibers and the amplifiers of the link are identical. The information is encoded in the optical electric field at the *input* of the system, typically using the light pulses sent at different frequencies. The available bandwidth of the amplifiers as well as the increase of the fiber absorption away from the "transparency window" near the wavelength $\lambda = 1.55$ μm, limits the bandwidth of the fiber optic communication system.

The maximum amount of the information that can be transmitted through the communication system per unit time is called the channel capacity C_N. According to Shannon's basic result, this quantity is given by the maximum value of the mutual information per second over all possible input distributions,

$$C_N = \max{}_{p_x} \left\{ H[y] - \left\langle H[y|x] \right\rangle_{p_x} \right\} \tag{2.3}$$

with the mutual information rate $\Re$ given by

$$\Re = H[y(\omega)] - \left\langle H[y(\omega)|x(\omega)] \right\rangle_{p_x} \tag{2.4}$$

which is a functional of the *input distribution* $p_x[x(\omega)]$, representing the encoding of the information via the use of the optical electric field components at different frequencies given as

$$E_i(t) = \int_B x(\omega)\, e^{j\omega t} dx \tag{2.5}$$

13

The entropy $H[y(\omega)]$ is the measure of the information received at the output of the optical communication channel. However, if the channel is noisy, then for any output signal there is some uncertainty in the recovery of what was originally launched into the optical channel (the first span). The *conditional entropy* $H[y(\omega)|x(\omega)]$ at the output for a given input $x(\omega)$ represents this uncertainty. The frequency-dependent entropies $H[y(\omega)]$ and $H[y(\omega)|x(\omega)]$ are defined in terms of the corresponding distributions $p(y)$ and $p(y|x)$ via the standard relation

$$H \triangleq -\int D_y(\omega)p\left[p(y[\omega])\ln\left(p[y(\omega)]\right)\right]$$ (2.6)

where $p \equiv p_y[y]$ for the entropy $H[y(\omega)]$, and $p \equiv p(y|x)$ for the entropy, $H[y(\omega)|x(\omega)]$ and the functional integral is defined in the standard way

$$\int D_\xi(\omega) \equiv \lim_{M\to\infty} c_M\left[\prod_{m=1}^{M}\int d\xi(\omega_m)\right]$$ (2.7)

with c_M being a scaling constant of normalization.

The first step to determine the channel capacity of an optically amplified system is to find the relationship between the input and output optical fields of the lightwaves under some modulation schemes. This is normally governed by the nonlinear Schrödinger equation (NLSE) given by

$$\frac{\partial A}{\partial z} + \beta_1\frac{\partial A}{\partial z} + \frac{j}{2}\beta_2\frac{\partial^2 A}{\partial t^2} + \alpha\frac{A}{2} = j|A|^2 A$$ (2.8)

where the amplitude of the guided lightwave electric field is given by $E(z,t) = A(z,t)e^{j(\beta z - \omega_0 t)}$ + c.c. at a distance z and optical signal envelop $A(z,t)$ and an instant t; β is the effective propagation wave number of the guided lightwave in the fiber; β_1 and β_2 are the first-order and second-order derivatives (or the phase velocity and group velocity, respectively) of the propagation wave number with respect to frequency or wavelength with some conversion constant, that is, the phase velocity and the linear dispersion parameter of the lightwaves. It is noted here that there are a group of lightwaves of different frequencies embedded in the band of the modulated lightwave signal. These lightwaves travel at different phase velocities that are determined by the propagation constant $\beta(\omega)$, the effective wave number of the guided mode of the fiber. This propagation vector can be expanded around the central wavelength by using the Taylor series expansion as a function of the frequency difference from this central wavelength

$$\beta(\omega) = \beta(\omega_0) + \beta_1(\omega) + \frac{\beta_2}{2}\omega^2 + \Theta(\omega^3)$$ (2.9)

with $\Theta(\cdot)$ denoting the residual error. Equation 2.10 neglects the effects such as the stimulated Raman scattering (SRS) and the stimulated Brillouin scattering (SBS) [2], compared to the Kerr nonlinearity γ of the refraction index of the fiber, represented by the term $\gamma|A^2|A$.

The optical amplifiers incorporated into the communication system, at the end of each transmission fiber length (Figure 2.3), compensate for the power losses in the fiber. Because of the spontaneous emission in the optical amplifiers, the amplification spontaneous emission (ASE) noises are inevitably introduced as $n(t) = e^{\int e^{j\omega t} n_\omega d\omega}$ into the channel. Generally, even in a single optical amplifier, the noise distribution at any given frequency $\omega_0 + \omega$ within the channel bandwidth $n(\omega)$ is close to a Gaussian distribution:

$$p_n[n(\omega)] \simeq e^{\left[-\frac{|n(\omega)|^2}{P_N^\omega}\right]}$$ (2.10)

The noise power P_N^ω is assumed to be flat and wideband, that is, uniformly distributed over the whole spectrum of interests. In modern optical fiber transmission systems, the spectral width may be over more than 100 nm.

2.2 OPTICAL COHERENT RECEPTION: CAPACITY AND NONLINEARITY ISSUES

Noises are critical for the transmission capacity and commonly contributed by both linear and nonlinear distortion in amplitude and phases, in optical and electronic domains. Electronic noises are mainly due to the optoelectronic conversion in the photodetector (PD) or photodetector pair (PDP) and the amplifiers following them, the shot noises of the optical-to-electronic (O/E) conversion especially in coherent detection (CoD) in which residual energy of the LO is very significant. Optical noises are due to the accumulated noises of the optical amplification stages, the ASE noises. Linear distortion creates intersymbol interference (ISI) noises. Nonlinear noises are normally known as phase noises due to the random noise features generated by the Kerr or self-phase modulation (SPM) effects or cross phase modulation (XPM), Raman (SRS) and/or Brillouin (SBS) scattering, and so on.

This section briefly describes the contributions of these noises to the reduction of the signal-to-noise ratio (SNR), hence the channel capacity. It especially evaluates the probabilistic distribution and their contribution to the degradation of the capacity when the finite memory is determined as one of the characteristics of the fiber medium.

A coherent optical communication link converts a discrete, complex-valued electric data signal X_k to a modulated, continuous optical signal that is guided and transmitted through an optical fiber, and received coherently, and then converted back to a discrete output sequence Y_k, especially when the samples are produced from the analog-to-digital converter (ADC) and then stored digitally for DSP. For amplitude- and phase-modulated optical signals, there is a $\pi/2$ hybrid coupler that splits the imaginary and real parts which are then detected by a balanced PDP and then electronically amplified through a differential transimpedance amplifier (TIA) (see also Chapter 5). Under the property of the linearly polarized mode of the weakly guided SMF, both polarizations of the guided optical field can be employed to carry optical signals. These polarized channels can be multiplexed and transmitted and split in the optical front unit and then coherently demodulated channel by channel.

It is noted that the received optical signal is mixed with a local laser that beats with it to give a boosted signal as the product between the LO amplitude and the signal amplitude and phases. Of course, there is an offset in phases and hence frequency offset between the LO and the signal carrier (local oscillator frequency offset [LOFO]) [10]. This LOFO can be easily compensated digitally in the DSP. The coherent link is particularly simple theoretically, in that the transmitter and receiver directly map the electric data to the optical field, which is a linear operation, and can ideally be performed without distortions. The channel can then be well described by the propagation of the continuous optical fields in the fiber transmission link, that is, the signals envelop covering the optical carrier. It should be emphasized that this assumes the coherent receiver to be ideal, with perfect synchronization (i.e., compensated LOFO) and negligible optical phase noise, except the linear electronic noises. Two main linear impairment propagation effects in the fiber are the dispersion and the attenuation. The attenuation effects can be overcome by periodic inline optical amplification, at the expense of additive GN from the inline amplifiers, the commonly known amplified stimulated emission noises and its accumulated noises over the whole optically amplified transmission line. The dispersion effects are usually equalized DSP electronically by a linear filter in the coherent receiver. Such a linear optical link can be well described by an arbitrary white Gaussian noise (AWGN) channel, the capacity of which is unbounded with the signal power. However, the fiber Kerr nonlinearity introduces signal distortions, and greatly complicates the transmission modeling. The nonlinear signal propagation in the fiber is described by a nonlinear partial differential equation, the NLSE, which includes dispersion, attenuation, and nonlinearity. At high power levels, the three effects can no longer be conveniently separated.

However, in contemporary coherent links of ultralong distance (long-haul or core networks) at least 500 km and a symbol rate of 28 GBd and above, the nonlinearity can be significantly weaker than the other two effects, and a perturbation approach can be successfully applied to the NLSE [11]. This leads to the GN model, which will be described in Section 2.4.

2.3 GAUSSIAN-NOISY FINITE MEMORY SYSTEM

2.3.1 FINITE MEMORY MODEL

Current optical fiber transmission line used in DSP-based coherent reception and under nondispersion compensating fiber, is highly dispersive, and thus has a finite memory. For example, a signal with a dispersive length $L_D = 1/2(|\beta_2|\Delta\omega)$, where β_2 is the group velocity dispersion (GVD) and $\Delta\omega$ the optical bandwidth (used as the signal channel 3 dB bandwidth), broadens (temporally) a factor L/L_D over a total fiber of length L. With typical propagation lengths of 5–50 km of standard single-mode fiber (SSMF), this broadening factor can correspond to hundreds to thousands of adjacent symbols, a large but finite number. The same will hold for the interaction among WDM channels; if one interprets $\Delta\omega$ as the channel separation, L/L_D will give an approximation on the number of symbols that two WDM channels separate due to walk-off (and hence interact with nonlinearly during transmission). The channel memory is thus even larger in the WDM case, and increase with channel separation, but the nonlinear interaction will decrease due to the shorter L_D. Thus, the principle of a finite channel memory holds also for DWDM signals transmitting over dispersive fiber.

We consider a single, scalar, wavelength channel in this chapter. Extensions to dual polarizations and WDM are straightforward, but involve obscuring complications such as four-dimensional constellation space in the former case and behavioral models in the latter. We can thus say that in an optical link a certain signal may sense the interference from $N \approx L/L_D$ neighboring symbols, which is the physical reason for introducing a finite memory model.

If we let the number N of interfering symbols go to infinity, an even simpler type of model is obtained. The interference is now averaged over infinitely many transmitted symbols.

2.3.2 GAUSSIAN NOISE MODEL

Assuming that an IID or time-invariant sequence is transmitted, the time average converges to a statistical average, which would simplify the analysis. Thus, we can distinguish the contribution to the total accumulated noises in a finite memory digital optical transmission system as follows. First, the linear noises accumulated over many optical amplified fiber spans are due to the noises of the ASE noises and the electronic noises of the reception circuits. Second, contributions are due to the nonlinear phase noises, the XPM, and the quantum shot noises due to the high-power local oscillator via the square-law O/E conversion process. These noises are wideband and superimposed on the spectral components in the signal bands. For the signal, due to the limited memory aspects of the sampled DSP reception subsystem, the signal noises are considered as discrete, and thus we have the additive GN by superimposing the linear ASE noises and the nonlinear noises resulting from [12]

$$N_k = \bar{N}_k \sqrt{P_{\text{ASE}} + \eta\left(\frac{1}{2N+1}\sum_{k-N}^{k+N}|S_i|^2\right)^3} \qquad \forall k \in \text{baud period} \qquad (2.11)$$

where N is the number of samples (one-sided property of the symmetric double-sideband channel memory).

For coherent long-haul fiber-optical links without dispersion compensation, one can derive the models where the nonlinear impairments as noises can be approximated by a Gaussian distribution, whose statistics depend on the transmitted signal power via a cubic relationship. The models assume that the transmitted symbols S_k in time slot $k \in S$ are IID events. In this model, the additive noise in Equation 2.1 is given by

$$N_k = \tilde{N}_k \sqrt{P_{N_{ASE}} + \eta P^3} \tag{2.12}$$

where $\{N_k\}$ is a set of complex Gaussian random variables that are IID zero-mean unit-variance circularly symmetric; P_{ASE} and η are real and nonnegative constants, and $P = E|\tilde{S}|^2$ is the average transmitted power. For distributed amplification, for example, optical amplifiers placed at the end of the transmission fiber span, the factor η can be proposed [12,13]. For a dual polarized modal channel transmitting over M lumped amplified spans can be written as

$$\eta = \frac{3\gamma^2 L}{\alpha} M^{1+\varepsilon} \tan h\left(\frac{\alpha}{4|\beta|^2 R_s^2}\right) \tag{2.13}$$

and for a single polarized channel

$$\eta = \frac{2\gamma^2 M}{\alpha} \tan h\left(\frac{\alpha}{4|\beta|^2 R_s^2}\right) \tag{2.14}$$

where γ is the nonlinear coefficient; α is the fiber attenuation parameter (linear scale); and R_s is the symbol rate. We can see that from Equations 2.13 and 2.14 the coefficient η becomes zero when there is no nonlinear effect or when the total power of all optical channels is below the nonlinear threshold. Furthermore, the noise power given in Equation 2.12 is then contributed only by the linear ASE noises. These noises reach a plateau level when the dispersion becomes reasonably large. Table 2.1 shows typical parameters of optical transmission over multispan EDFA optical amplified systems employing standard SMF and coherent reception in association with DSP.

TABLE 2.1
Typical Parameters of Multispan Fiber Optic Amplified Transmission

Parameter	Value (Nominal)/Units	Meaning
α	0.19 dB/km	Fiber attenuation coefficient in log scale
β_2	−22 ps²/km at 1550 nm	Group velocity coefficient of standard SMF
γ	1.27 (W km)⁻¹ silica–Ge-doped core fiber	Fiber nonlinear coefficient
M	10–40	Number of fiber spans varying from 10 to 40 depending on transmission symbol rate or 800 km to 3500–4000 km
L_S	80 or 100 km SSMF	Length of a fiber span pending on amplification gain of EDFA
L	800–400 km	Total length of the optical transmission line, about 10–50 spans for symbol rate of 28G modulation QPSK
R_S	28–64 GSy/s	Symbol rate
P_{ASE}	4.1 × 10⁻⁶ W	ASE Noise power of EDFA
η	~7244 W⁻²	Nonlinear index coefficient

2.4 CORRECTION OF CHANNEL CAPACITY BY PERTURBATION

2.4.1 CORRECTION UNDER LINEAR DISPERSION EFFECT

The entropy $H\left(y(\omega)\right)$ can be obtained using the principle of superposition in the linear limit when the nonlinear effect is not strong or in the weakly nonlinear effect with respect to its original linear entropy $H_0\left(y(\omega)\right)$,

$$H\left(y(\omega)\right) = H_0\left(y(\omega)\right) - \Delta C_{N1} - \Delta H_1 + \Theta(\gamma^4) \tag{2.15}$$

and the corresponding conditional entropy $H\left(y(\omega)|x(\omega)\right)$ related to the original linear entropy and its reduction of the conditional capacity ΔC_{C2},

$$H\left(y(\omega)|x(\omega)\right) = H_0\left(y(\omega)|x(\omega)\right) + \Delta C_{N2} + \Theta(\gamma^4) \tag{2.16}$$

In this section, we consider corrections to the channel capacity of the optical fiber communication system, originating from the nonlinearity of the fiber. The technique that we use involves a perturbation computational technique of the relevant mutual information and subsequent optimization. Under the perturbation analysis, that is, the weakly nonlinear regime, the channel capacity can be derived and obtained [14] as follows.

The input in the frequency domain of the optical signal at the output of the optical amplifier, or at the launching input to the next span of the optical transmission link can be written as

$$A_\omega^L = e^{\alpha L/2} A_\omega^0 + n_\omega \tag{2.17}$$

in which the superscript L indicates the length of the span of the transmission link and the subscript ω indicates the frequency domain of the amplitude and noise components. As shown in [14], by using the perturbation principles one can derive the transfer function in the frequency domain as the relationship between the output and input of the guided lightwaves over the entire multispan optically amplified link, with an assumption of uniform span length and amplification factor, as

$$\Phi_\omega^n\left(A_\omega^{n-1}\right) = \left[A_\omega^{n-1} + \sum_{l=1}^{\infty} \gamma^l \Im_\omega^l\left(A_\omega^{n-1}\right)\right] e^{-j\kappa_\omega L} + n_\omega \tag{2.18}$$

in which the superscript of the amplitude indicates the order of the span and that of the nonlinear coefficient γ and the transform indicates the lth order of approximation. The parameter κ_ω is the effective propagation with the contribution of the nonlinear effects, that is, the nonlinear phase contributed propagation delay as a function of frequency, given as

$$\kappa_\omega = \beta_1 \omega - \frac{1}{2} \beta_2 \omega^2 \tag{2.19}$$

The transform $\Im_\omega^l$ can be obtained in the frequency domain using nonlinear approximation or the Volterra series of second and third, and the XPM by other wavelength channels copropagating along the optical transmission line.

Further calculations to obtain the mutual information rate $\Re$ involve the following steps: first, iterate Equation 2.18 N times, to obtain the *input–output* relation for the whole communication system $\Phi_\omega^n[x(\omega); n_\omega^{(1)}, n_\omega^{(1)}, \ldots, n_\omega^{(N)}]$ and its influence by nonlinear noises contributed in N dimensions. Second, obtain the conditional probability distribution $p\left(y|x\right)$ and the output distribution $p_y(y)$ in terms of the input distribution $p_x(x)$ as expansions in powers of γ. Then, calculate the entropies $H\left[y(\omega)\right]$ and $H\left[x(\omega)|y(\omega)\right]$ to deduce the mutual information rate $\Re$.

To estimate the aggregate entropy, one can consider that the perturbed probability of error on the output of the span is given in terms of the unperturbed term, $p_y^0(y)$, as

$$p_y(y) = p_y^0(y) \left[1 + \sum_{n=1}^{\infty} p_y^n(y) \right] \tag{2.20}$$

where the unperturbed probability of error is given as

$$p_y^0(y) = \frac{1}{\pi(P_0 + P_N)} e^{-\frac{|y|^2}{P_0 + P_N}} \tag{2.21}$$

The degradation of the channel capacity determined as the correction factor for this channel capacity for a channel of bandwidth B can then be estimated [14] as

$$\Delta C_{N1} = N^2 B \left(\frac{\gamma P_0}{\alpha} \right)^2 Q_1 \left(\alpha L, \frac{\beta_2^2 B^4}{\alpha^2} \right)$$

$$\Delta C_{N2} = \frac{4}{3} \left(N^2 - 1 \right) B \left(\frac{\gamma P_0}{\alpha} \right)^2 Q_2 \left(\alpha L, \frac{\beta_2^2 B^4}{\alpha^2} \right) \tag{2.22}$$

with

$$Q_1(u, z) = f(u, z; x_1, x_2, x) \int_{-1/2}^{1/2} dx_1 \int_{-1/2}^{1/2} dx_2 \int_{1/2}^{\bar{x}} dx$$

$$Q_2(u, z) = f(u, z; x_1, x_2, x) \int_{-1/2}^{1/2} dx_1 \int_{-1/2}^{1/2} dx_2 \int_{1/2}^{1/2} dx$$

$$\bar{x} = x_{max} \left[\frac{1}{2}; \frac{1}{2} + x_1 + x_2 \right] \tag{2.23}$$

$$f(u, z; x_1, x_2, x) = \frac{\left| 1 - e^{-u - jv^2} \right|^2}{1 + v^2}; \quad v \equiv z(x - x_1)(x - x_2)$$

$B = \Delta\omega = 3$ dB signal bandwidth

$u, z; x_1, x_2, x =$ dummy variables

Figure 2.4 plots the variation of the reduction and/or correction factor deviating from the purely linear channel capacity. The deviation can be observed when the linear dispersion over attenuation ratio is about double.

The overall channel capacity taking into account these deviations can be modified as

$$C_N = B \log_e \left(1 + \frac{P_0}{P_N} \right) - \Delta C_{N1} - \Delta C_{N2} + \Theta(\gamma^4) \tag{2.24}$$

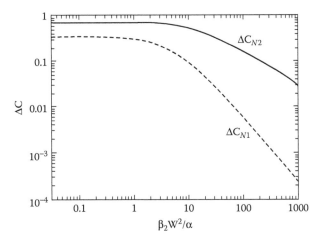

FIGURE 2.4 The correction factors to the channel capacity, ΔC_{N1} and ΔC_{N2}, in units of $BN^2\gamma^2 P_0^2\alpha^{-2}$ ($B = 3$ dB bandwidth—bits/symbol), shown as functions of $|\beta^2| B/\alpha$ (ratio of linear dispersion over attenuation—dimensionless) limit with $\alpha L_s \gg 1$ (L_s = span length); $N \gg 1$. The correction ΔC_{N2} is represented by the solid line, while $\Delta C1$ corresponds to the dashed line. Note that ΔC_{N1}, representing the effect of the power leakage from the bandwidth, is more strongly affected by the dispersion. (Extracted from S. Verdi and T. Sun Han, *IEEE Trans. Inform. Theory*, 40, 1147–1158, July 1994. With permission.)

2.4.2 CORRECTION AND CAPACITY UNDER NONLINEAR PHASE MODULATION EFFECTS AND MEMORY LESS

Agrell et al. [12] have pointed out that under Shannon's coding theorem [6], the channel capacity in bits/symbol of a discrete time memory-less channels with code words subject to constraints of power is given as

$$C_N = \sup\big(I(\underline{X};\underline{Y})\big) \tag{2.25}$$

where sup($\cdot$) indicates the supreme, and the mutual information $I(X;Y)$ can be found by [12,15]

$$I(\underline{X};\underline{Y}) = \iint f_{X,Y}(x,y)\log_e \frac{f_{X,Y}(x,y)}{f_X(x)f_Y(y)}\,dxdy \tag{2.26}$$

$$f_X(x); f_Y(y) \equiv \text{probability density function}$$

These distributions must satisfy the physical definition of the expected power of the channel. That is that there exist code words of $M = 2^{nR}$ of length n that can used to design and obtain the channel capacity, the rate limit C_N for a transmission system. The elements of these code words are IID random samples satisfying the distribution $f_X(x)$ that maximizes the mutual information given in Equation 2.26. At the transmitter, the encoder maps each message j into a unique code word x, which is decoded after the transmission in a decoder at the receiver to recover the code words to its most similar to output Y components. The code word length n can be chosen long enough so as to achieve the error sufficiently small. The interpretation of the paradigm of Shannon can be found in [7].

The noises in both linear and nonlinear effects are statistically independent and additive so the channel capacity can follow the GN model obtained as [12]

$$C_N = \log_2\left(1 + \frac{P_S}{P_{ASE} + \eta P_S^3}\right) \tag{2.27}$$

The capacity of the case of the channel with memory can be given under the assumptions on the stability of information, as

$$C_N = \lim_{n \to \infty} \sup \frac{1}{n} I\left(X_1^n; Y_1^n\right)$$ (2.28)

where $X_1^n; Y_1^n$ are defined as the sampled n-dimensional vector and $I\left(X_1^n; Y_1^n\right)$ as the n-dimensional integral analogous to Equation 2.26. The supremum is evaluated over all joint distributions of $X_1, \ldots, X_n$ satisfying the physical meaning of $E[\| X_1^n \|^2] = nP_S$. The sequence $X_1, \ldots, X_n$ is expected to consist of non-IID elements, and thus, the channel model must satisfy the inputs that are non-IID.

2.5 BER, SER, AND LOWER BOUND

2.5.1 BIT-ERROR RATE (BER) AND SYMBOL-ERROR RATE (SER)

The sequence is noncoded and given by $\{X_1, \ldots, X_n\}$, whose components are independently derived from the discrete constellation points $\{s_1, \ldots, s_n\}$ known as symbols. An example of the binary code words and corresponding symbols is shown in Figure 2.5 for a 16-QAM modulation scheme. The indices of the symbols can be arbitrary, but the code words corresponding to symbols must follow some strict rules such as those by the Gray code in which only one bit can be changed between adjacent symbols.

Under the uncoded case, the symbols are selected so that the Euclidean distances from a considered point to its adjacent points are equal, indicating that the probability of error is the same. Thus, the average transmitted power can be written as

$$P = \left\langle |X|^2 \right\rangle = \frac{1}{2m} \sum_{s \in \{S\}}^{m} |s_i|$$ (2.29)

where the $\langle \cdot \rangle$ indicates the expected average value. For each symbol X_k transmitted, the contributions of $2N$ symbols surrounding it can be $\underline{X}_k^{2N} = \{x_{k-N}, \ldots, x_{k-1}, x_{k+1}, \ldots, x_{k+N}\}$, in which the lower case indicates the contribution parts of the adjacent component vectors at the kth instant. $2N$ is the total memory length of the channel that is expected to be contained in the transmitting fiber and also in the memory storage of the digital processor that follows the DSP-based coherent optical receiver described in Chapters 3 and 4. The symbol vector $\tilde{X}$ is a random vector whose components are

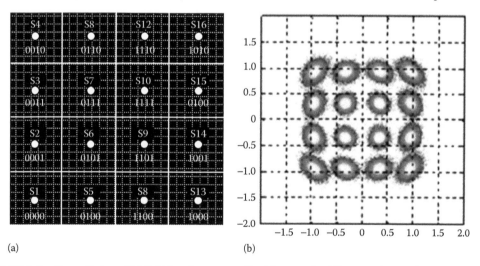

(a)   (b)

FIGURE 2.5 Constellation of 16-QAM and corresponding binary codes. Binary code must have only one bit different between adjacent symbols, numbering of symbols assigned arbitrary. (a) Ideal and (b) experimental generated and transmitted.

made up by the central considered instant and those in the memory that contributes to the interference or unwanted signals to the received signal outputs. Thus, the noise vector $\underset{\sim}{Z}_k$ contributing to the kth symbol can be represented as

$$N_k = \tilde{N}_k \sqrt{P_{N_{ASE}} + \eta\left(\frac{a}{2N+1}\right)^3 + \left\|x_k^{2N}\right\|^2} \tag{2.30}$$

where $\left\|x_k^{2N}\right\|$ represents the nominal Euclidean distance of noises of the other symbol components. The GN probability density function (PDF) can then be written as

$$f_{Y_k|X_k,\underset{\sim}{X}_k^{2N}}(y|\underset{\sim}{x}_k^{2N}) = \frac{1}{\pi\left(\rho|x_k|^2 + \left\|\underset{\sim}{x}_k^{2N}\right\|^2\right)}\exp\left[\frac{|y-x_k|^2}{\left(\rho|x_k|^2 + \left\|\underset{\sim}{x}_k^{2N}\right\|^2\right)}\right] \tag{2.31}$$

with

$$\rho(A) \triangleq P_{ASE} + \eta\left(\frac{A}{2N+1}\right)^3$$

Agrell et al. [12] have derived the SER and BER for 16-QAM with a channel of memory capacity N, under Gray coding, as

$$\text{BER} = \frac{1}{2^{4N+3}}\sum_{l=0}^{4N}\binom{4N}{l}\binom{2\lambda_{l,1,1}+3\lambda_{l,1,5}+\lambda_{l,1,9}+}{\lambda_{l,3,1}+2\lambda_{l,3,5}+\lambda_{l,3,9}-\lambda_{l,5,5}-\lambda_{l,5,9}} \tag{2.32}$$

where

$$\lambda_{l,r,t} = Q\left(\sqrt{\frac{\left(r^2 P/5\right)}{\rho\left((2N+4l+t)/5\right)}}\right);$$

$$Q(x) = \frac{1}{\sqrt{2\pi}}\int_x^\infty e^{-\frac{t^2}{2}}dt; \quad \text{memory} = N$$

$$\text{SER} = \frac{1}{2^{4N+2}}\sum_{l=0}^{4N}\binom{4N}{l}\left(4\lambda_{l,1,1}-4\lambda_{l,1,1}^2+6\lambda_{l,1,5}-4\lambda_{l,1,5}^2+2\lambda_{l,1,9}-\lambda_{l,1,9}^2\right)$$

We can see that when $N \to \infty$, the SER and BER approach the Gaussian model of memory less. Reference [12] provides the plots of the BER and SER variation with respect to the signal average power (in dBm) into the fiber for cases of memory-less receivers and receivers and decoder with limited memory as shown in Figure 2.6.

We can observe the following:

1. The AWGN curve represents the ideal infinite memory or memory-less and GN with an extremely wide and uniform noise property.
2. The GN with memory gives the worst error rate under nonlinear noises contribution.
3. With limited memory length and the most likelihood detection, the BER and SER can become high; thus, it may require forward error coding (FEC) in addition to the coding/mapping.

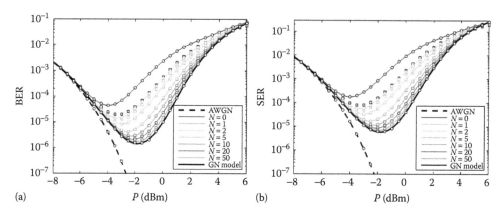

FIGURE 2.6   (a) BER and (b) SER of 16-QAM modulation scheme versus signal average power under the channel with no memory GN, AWGN arbitrary white Gaussian noise and with memory limited storage capacity N. (Extracted from Agrell, E. et al., *IEEE J. Light Tech.*, 32, 2862–2876, 2014, Figure 4. With permission.)

Thus, under nonlinear noises contribution, the error rate can only follow the ideal Shannon limit when the power is below the threshold level. This has been observed in practical transmission systems employing fiber multispan transmission optically amplified systems with DSP-based optical reception.

2.5.2   Lower Bound Capacity for Channel with Memory

The expression for exact capacity of transmission system with memory with GN model has been derived in [12] with a lower bound for any random process $\{X_k\}$ given as

$$C_N \geq \lim_{n \to \infty} \frac{1}{n} I\left(\underline{X}_1^n ; \underline{Y}_1^n\right) \tag{2.33}$$

Under the condition that every joint distribution of the element of $\{X_k\}$ satisfies the energy conservation relation

$$\left\langle \left\| \underline{X}_1^n \right\|^2 \right\rangle = nP,$$

Theorem 3 of [12] states that for every real and positive value $r_1 \geq 0$ and the probability distribution f_R over $\mathbb{R}$ the real space such that

$$\frac{2Nr_1^2 + \left\langle \left[R^2 \right] \right\rangle}{2N + 1} = P,$$

the lower bound of the channel capacity is given as

$$C_N \geq \frac{\left\langle \log_2 f_U(\underline{U}) \right\rangle}{2N + 1} - \int_0^{\infty} f_R(r) \log_2\left(e\rho\left(2Nr_1^2 + r^2\right)\right) dr \tag{2.34}$$

where

$$\underline{U} \triangleq \left[U_{-N}, U_{-N+1}, \ldots, U_{N-1}, U_N\right] \text{ is a random vector;}$$

$$e = \text{Euler number}$$

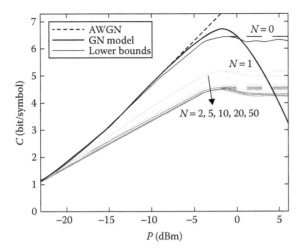

FIGURE 2.7 Lower bounds of the capacity (in bits/symbol) of the finite-memory model by coherent reception for different values of memory N. The exact capacities of the AWGN channel and the GN model are included for comparison. (Extracted from Agrell, E. et al., *IEEE J. Light Tech.*, 32, 2862–2876, 2014 Figure 4. With permission.)

The PDF of the random vector $\underset{\sim}{U}$ elements is given by

$$f_U(\underset{\sim}{u}) = \int_0^\infty f_{\mathbb{R}}(r) \frac{\exp\left(-\dfrac{\displaystyle\sum_{-k=N}^{N} u_k + 2Nr_1^2 + r^2}{\rho(2Nr_1^2 + r^2)}\right)}{\rho(2Nr_1^2 + r^2)^{2N+1}} \times I_0\left(\frac{2r\sqrt{u_0}}{\rho(2Nr_1^2 + r^2)}\right) \times \prod_{\substack{k=-N \\ k\neq 0}}^{N} I_0\left(\frac{2r_1\sqrt{u_k}}{\rho(2Nr_1^2 + r^2)}\right) dr \qquad (2.35)$$

with $I_0(u)$ being the Bessel function of first kind and $\rho(\cdot)$ is defined in Equation 2.31. The capacity (bits/symbol) of the 16-QAM modulation scheme versus the total average power using Equation 2.34 can be numerically calculated and plotted [12] with the memory length as a parameter, as shown in Figure 2.7, for AWGN, GN model with memory-less and with memory cases. Above the nonlinear threshold level, the capacity of the finite memory system is flattened to a saturation level.

2.6 COHERENT STATES AND QUANTUM LIMITS

Since the invention of the optical transmission systems following the proposed guiding of light-waves through a dielectric waveguide [16] in 1966, the detection has been limited to direct detection that normally operates well above the quantum limit due to the high level of thermal and shot noises of the PD. Optical amplification or heterodyne detection of the two quadrature fields has overcome this limitation in long-haul transmission via the guiding through the single linearly polarized mode of the SMF to offer extremely high bit rate or symbol rate in current optical transmissions systems in association with the advances of digital signal processors (DSPs). DSP-based coherent homodyne or intradyne reception has further overcome the 3 dB loss in heterodyne detection and the employment of balanced PDP allowing the reaching of the quantum limit.

In Section 2.1 of this chapter, the capacity limits have been examined for transmission systems with and without memory together with the GN model and modulation formats under linear and nonlinear scenarios.

This section aims to give an overview of different quantum receivers and to compare the minimum signal energy required to achieve a given BER. The examinations here can be applicable in ultrahigh bit rate optical transmission systems as well as quantum communication systems that now find some significant applications in secure point-to-point link. The reception can be in both guided lightwave and free-space media such as transmission of secret quantum keys in those free-space links between satellites to earth stations of approximate distance from 34 to 45 km.

2.6.1 Signal Representations and Receiver Structures

The optical field $E(t)$ of an optical signal at a photon quantum level can be written as

$$E(t) = \sqrt{\frac{h\nu}{T_b}} s(t) e^{j(\omega t + \phi)} \qquad (2.36)$$

where h is the Plank constant, ν is the optical frequency (ω is the angular frequency), thus this is the energy of a photon, and T_b is the observable bit period in which the photon exists. $s(t)$ is the complex slow time-varying envelope whose phase made up by the real and imaginary parts can be coded to represent logical levels. Thus, the slow time average power can be written as

$$\langle P \rangle = h\nu \left\langle |a(t)|^2 \right\rangle = \left\langle |E(t)|^2 \right\rangle \qquad (2.37)$$

The $a(t)$ and $|a(t)|^2$ represent the signal envelop and energy. The phase of the optical waves can be coded logically to represent the information states.

For the optoelectronic direct detection without any noises, the electronic current is proportional to $N_e = aa^*$. The proportional constant is the quantum efficiency of the PD. In terms of power, the current output of the PD can be proportional to the responsivity of the PD, $\mathfrak{R}$. Alternatively if $a(t)$ is represented by the in-phase and quadrature components as $a(t) = a_I(t) + ja_Q(t)$, then we have $N_e = aa^* = a_I^2 + a_Q^2$.

In the case of coherent reception, the signal is mixed with the local oscillator field via an optical coupler as shown in Figure 2.8. The converted and detected photons can be fed into photon counters (PCs) or differential TIAs for both cases of single or dual pair of PDs. The field amplitudes of the fields at the output of the optical 2×2 coupler can be written as

$$\begin{bmatrix} E_3 \\ E_4 \end{bmatrix} = \frac{1}{\sqrt{2}} \begin{bmatrix} a_1 - jE_{LO} \\ E_{LO} - ja_1 \end{bmatrix} \qquad (2.38)$$

This is clear as the $\pi/2$ phase shift in the 2×2 optical coupler rotates the in-phase to the quadrature of the other input port and vice versa. The signal LO fields of the in-phase and quadrature are added vectorially and then detected by the square-detection law. The beating of these components can only be detected by the electronic response of the band-limited electronic systems. Thus, only the products of the amplitudes of the added fields can be resolved. For the DC component of the LO, the quantum shot noise is generated. The oscillating frequency of the CW LO and the optical waves are very high and not detected by the electronic system. The phase states of the "1" and "0" logical states can be the difference in phases or amplitude or both for multilevel modulation schemes.

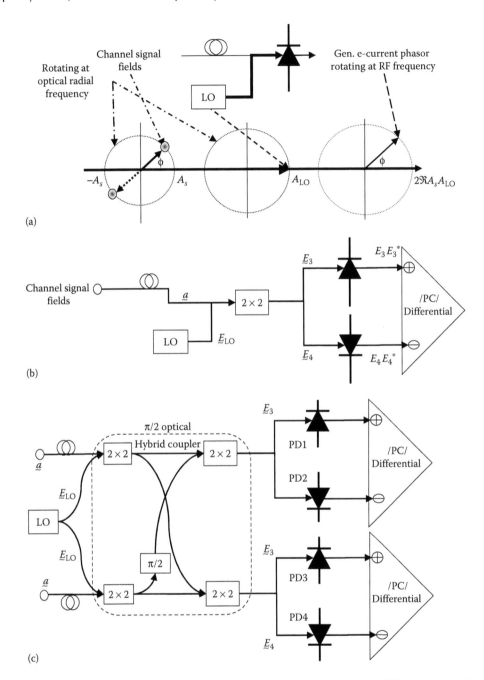

(a)

(b)

(c)

FIGURE 2.8 (a) Single PD for homodyne reception and phasor representation of BPSK optical signals, local oscillator (LO) and detected RF current of the homodyne beating product of the optical fields between LO and signals in the electrical baseband; (b) coherent dual PD balanced differential detection; and (c) coherent complex in-phase-quadrature (IQ) quad balanced PD detection.

For simplicity, let the quantum efficiency be unity; the number of electrons (use subscript e to indicate electrons) at the inputs of the differential amplifier can be written as

$$\begin{bmatrix} N_{e,2} \\ N_{e,1} \end{bmatrix} = \begin{bmatrix} 2a_{1I}E_{LO,Q} \\ 2E_{LO,I}a_{1Q} \end{bmatrix} \tag{2.39}$$

Equation 2.39 represents the number of electrons at the output of the PDs, and if amplification gain is unity, then this can be considered as the current at the output. Similarly for the case of complex modulation with in-phase and quadrature (I and Q) components and quad PDs shown in Figure 2.8c, the currents at the output of the upper and lower electronic amplifiers can be written as

$$N_{e,2,I} - N_{e,1,I} = \text{Re}\left\{aE_{LO}^*\right\}$$

$$N_{e,2,Q} - N_{e,1,Q} = \text{Im}\left\{aE_{LO}^*\right\}$$

(2.40)

Now for the case of polarization multiplexing, there would be optical polarization splitter into a pair of complex optical quad PD detectors shown in Figure 2.8c. These reception systems will be described in detail in Chapter 5. It is observable that the amplitude of the LO acts as the gain coefficient boosting the amplitude of the signal coherent states. So if the signal level is at or below the single photon states, this LO can boost the signal to the level states where the detection can be at an acceptable state for *error free*. This would ensure that the number of photons to be transmitted can be below the *listening* level, hence securing the absolute security states of transmission. The above is the case for strong local oscillator. It is noted that in some cases the local oscillator can be its own signal. This case can be termed as self-coherent detection; then both the signals and oscillator levels are at a quantum level and no gain by LO boosting is available. In the next section, the BER of the quantum limit modulation schemes are investigated.

2.6.2 QUANTUM LIMITS: ASK AND BPSK MODULATED COHERENT STATES

Coherent states are more classical-like signals especially when used in optical transmission systems, and even in cryptography and the contemporary quantum communications [17–19]. For practical and economical reasons, these coherent states can be generated by semiconductor devices and colocated active modulators recovered by high-speed PDs and optically passive and active components. The excess noises generated by the sources are also attenuated in the same amount as signals. However, error-free discrimination of two coherent states is impossible due to their mutual nonorthogonality. To overcome this hurdle, we can represent the nonorthogonal states $|\alpha\rangle$ by a sum of $|n\rangle$ orthogonal states in the familiarized Poisson distribution of independent arrival of photons, as [20]

$$|\alpha\rangle = e^{-\frac{1}{2}|\alpha|^2} \sum_{n=0}^{\infty} \frac{\alpha^n}{\sqrt{n!}} |n\rangle$$

(2.41)

Thus, the reception of the two antipodal coherent states by a square-law PD can be written as

$$|\langle\alpha_1|\alpha_2\rangle|^2 = \left| e^{-\frac{1}{2}|\alpha_1|^2} e^{-\frac{1}{2}|\alpha_2|^2} \sum_{n=0}^{\infty} \sum_{m=0}^{\infty} \frac{\alpha^n}{\sqrt{n!}} \frac{\alpha^m}{\sqrt{m!}} \langle n||m\rangle \right|^2 \cong e^{-\left(|\alpha_1 - \alpha_2|^2\right)}$$

(2.42)

The states $\langle\alpha_1|$ and $|\alpha_2\rangle$ represent the states of the two antipodal phases in a binary phase-shift keying (BPSK) modulation scheme. This can be reinterpreted as $\langle\alpha|$ and $|-\alpha\rangle$ and for a minimum number of photons N_S for the "1" states. The same number of photons would be required for the "−1" states for the antipodal case.

However for the binary ASK scheme, the "0" states could be a vacuum state. Thus, we have

$$\text{ASK} :\rightarrow \langle\alpha|\alpha=0\rangle = e^{-N_P}$$

$$\text{BPSK} :\rightarrow \langle\alpha|-\alpha\rangle = e^{-4N_P}$$

(2.43)

The BER of the received coherent states can be evaluated by the overlapping states of these two states by assuming the equal probability of transmission of the two states $\langle\alpha_1|$ and $|\alpha_2\rangle$ [20–23], as

$$\text{BER} = \frac{1}{2}\left(1 - \sqrt{1 - \langle\alpha_1|\alpha_2\rangle}\right) \tag{2.44}$$

Thus for BPSK, we have

$$\text{BER} = \frac{1}{2}\left(1 - \sqrt{1 - e^{-4N_P}}\right) \approx \frac{1}{4}e^{-4N_P} \tag{2.45}$$

2.6.2.1 Binary ASK "1" and "0"

When binary states arrive at the detector, then the BER, assuming the equal probability of detection of either states, is given by (see more details in Chapter 6)

$$\text{BER} = \frac{1}{2}\left[P_e(1|0) + P_e(0|1)\right] \rightarrow \frac{1}{2}e^{-2\hat{n}_p} \tag{2.46}$$

Thus, for a BER = 10^{-9} using Equation 2.46, the number of photons would be 10. Later on, we will see that this depends also on the time interval in which the photons are placed. This number is reduced to 3–4 for BPSK when the BER is determined by $1/2\,e^{-4\hat{n}_p}$.

2.6.2.2 BPSK under Strong LO Mixing Coherent Detection and PDP

Under strong LO mixing as described above in Section 2.6.2, the BER of a dual PD (PDP) reception is given by $\text{BER} = 1/2\,e^{-2N_S}$. The SNR equals the signal power or the number of signal photons due to the noises imposed on the detector by the strong residual LO power. Thus, we can use the complementary error function model to estimate the BER. This requires nine photons for a 10^{-9} BER.

If the TIA is a nondifferential type, then the two PDs are connected back to back so that the push–pull operation can be achieved with 3 dB gain compared with a single PD. This type is called balanced reception. In the strong LO mixing, the dominant noises are expected to be coming from the quantum shot noise generating from the residual strong LO power.

2.7 QUANTUM LIMIT OF OPTICAL RECEIVERS UNDER DIFFERENT MODULATION FORMATS

Instead of using equivalent noise current density or NEP, optical receiver front ends are sometimes also characterized in terms of their receiver sensitivity. While the receiver sensitivity is undoubtedly of high interest in the optical receiver design, it not only comprises the degrading effects of noise, but also encompasses the essential properties of the received signal, such as extinction ratio, signal distortions, and ISI, generated either within the transmitter or within the receiver itself. Thus, knowledge of the receiver sensitivity alone does not allow trustworthy predictions on how the receiver will perform for other formats.

Although electronics noise usually dominates shot noise, it can be squeezed to zero. The signal-dependent shot noise, however, is fundamentally present. The limit, when only fundamental noise sources determine receiver sensitivity, is called the quantum limit in optical communications. The existence of quantum limits makes the optical receiver design an exciting task, because there is always a fundamental measure against which practically implemented receivers can be compared. Note, however, that each class of receivers in combination with each class of modulation formats has its own quantum limit.

From Equation 2.36, we can observe that when the amplifier is noiseless, the receiver would require $\delta^2 G$ number of photon energy for detection. This is the quantum limit of the receiver. For example if a BER of $10{-}9$ is required for a pin optical receiver under the ASK modulation signal format then we would need at least a total optical energy over one bit period equivalent to that of 36 photons, when both the "1" and "0" are 50:50 randomly received.

Noises play the major parts in the receiving end of any communications systems. In optical communications using coherent or noncoherent detection techniques, noises are contributed by (i) the electronic noises of the electronic amplifications following the optoelectronic processes in the PD, (ii) the quantum shot noises due to the electronic current generated by the optical signals, (iii) the quantum shot noises due to the high power of the local oscillator (additional and dominant source of CoD), and (iv) the beating of the local oscillator and the optical signals.

These noise sources vary from optical receivers to the others depending on their structures whether they consist of a PD and electronic amplifiers with and without optical preamplifications under coherent or noncoherent detection. This section thus gives the fundamental issues of the noise processes and their impact on the sensitivity of the receiving systems. In particular, we examine the quantum limits of the optical receivers, that is, when the electronic noises are considered to be nullified.

Schematic diagrams of coherent and incoherent direct detection receivers are shown in Figure 2.9a and b without using an optical amplifier, while Figure 2.9c and d show their counterparts with optical amplifiers. Figure 2.9f–g show the balanced and fiber version of the detection and receiving systems. The difference between these configurations is the noises generated after the photodetection and at the input of the electronic amplifier. Note that coherent systems are identified with the mixing of the optical signals and the local oscillator whose polarization directions are aligned with each other.

2.7.1 DIRECT DETECTION

Optical detection for optical fiber communications is in the form of direct modulation and direct detection. Direct detection is the simplest form of detection that requires only a PD followed by an electronic amplifier, the decision circuitry, clock recovery, and data recovery. For an ASK system, the signals for the "1" and "0" can be expressed in terms of the number of photon energy over the entire bit period, contained within a bit period T as

$$s_{1,0}(t) = \begin{cases} \dfrac{\hat{n}_p}{T} \text{ for "1"} \\ 0 \text{ for "0"} \end{cases} \qquad 0 \le t \le T \tag{2.47}$$

with $\hat{n}_p$ being the number of photons and the energy of the lightwave is normalized with only one photon energy at the operating wavelength. Thus, the total optical energy is

$$E_s = \hat{n}_p h\upsilon T \tag{2.48}$$

The electronic current generated after the PD, with R the responsivity of the photodetection, is

$$\langle i_s \rangle = \hat{n}_p h\upsilon T \, \Re = \hat{n}_p h\upsilon T \frac{\eta q}{h\upsilon} \tag{2.49}$$

Thus, we could say that the number of photon energy per bit is $\hat{n}_p$ required for the detection of the "1" if there is no noise contributed by the electronic amplifier or detection.

When the probabilities of the "1" and "0" are equal, 50%, then the probability of error of the detection is

$$P_e = \frac{1}{2} e^{-\hat{n}_p} \tag{2.50}$$

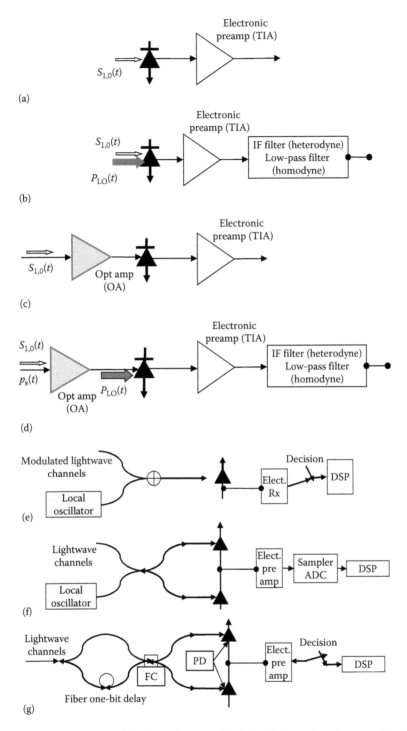

FIGURE 2.9 Schematic diagram of (a) direct detection, (b) CoD, (c) direct detection with OA, (d) CoD with OA, (e) fiber version of coherent receiver. (a) Single-ended, (b) balanced, and (c) self-heterodyne balanced receiver structures. Following the photodetection, various kinds of electronic signal processing can be performed, including equalization. Detection of the (a, f) balanced receiver and (g) self-homodyne reception balanced receiver with a one-bit-delay photonic phase comparator. FC, fiber coupler; PD, photodetector; DSP, digital signal processor; and ADC, analogue to digital converter.

Thus for a BER of 10^{-9}, the argument $\hat{n}_p = \left(3\sqrt{2}\right)^2 = 18$, with the allowance of the 1/2 factor of the single-sided estimation so that $\hat{n}_p = 20$ for the full detection error. This is the super quantum limit. We also assume a unity responsibility of the photodetection.

2.7.2 COHERENT DETECTION

In the case of coherent homodyne detection with a local oscillator whose optical power P_{LO} is very much larger than the signal average power, we have the quantum shot noise current that dominates the noise process. The detected electronic current is the beating current between the local oscillator lightwave and the signal; thus we have

$$\left\langle i_{N(LO)}^2 \right\rangle = 2qP_{LO}\frac{\eta q}{h\upsilon}\frac{1}{T} = 2q^2 P_{LO}\frac{\eta}{h\upsilon}\frac{1}{T} \tag{2.51}$$

and the SNR is given by

$$\text{SNR} = \frac{\left\langle i_s^2 \right\rangle}{\left\langle i_{N(LO)}^2 \right\rangle} = \frac{4\Re^2 p_s(t)P_{LO}/T}{2q\Re P_{LO}/T} = 2\hat{n}_p \tag{2.52}$$

$$P_e = \text{erfc}(2\hat{n}_p) \tag{2.53}$$

Thus, for 10^{-9} BER, we require $\hat{n}_p = \dfrac{3\sqrt{2}}{2}$.

2.7.3 COHERENT DETECTION WITH MATCHED FILTER

Now it is assumed that a matched filter is inserted after the coherent receiver shown in Figure 2.9b and different modulation formats are used for transmission over long-haul optically amplified fiber systems. In general and under the assumption that the noise process in the optical detection is Gaussian, the BER is given by

$$\text{BER} = \text{erfc}\left(\frac{d}{N_0}\right) \tag{2.54}$$

where d is the signal power separation between the average level of the "1" and "0" for a binary system and the equal-distance between the constellation points of the modulation scheme as shown in Figure 2.10. Let E_1 and E_0 be the field amplitudes of the signals "1" and "0"; then the Euclidean distance d is given by

$$d^2 = E_1^2 + E_0^2 - 2\rho_C E_1 E_0 \tag{2.55}$$

with

$$\rho_C = \frac{2}{T}\int_0^T s_1(t)s_0(t)dt$$

ρ is the correlation coefficient between the two logic levels or alternatively the Euclidean angle between the two vector signals as represented on the scattering plane of constellation.

2.7.3.1 Coherent ASK Systems

In the heterodyne detection with an intermediate frequency (IF) region, the two bit signals are given by

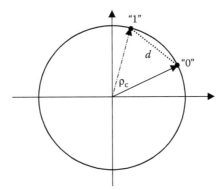

FIGURE 2.10 Signal constellation and energy level and the geometrical distance between "1" and "0."

$$s_{1,0}(t) = \begin{cases} \sqrt{\dfrac{2n_p}{T}} \cos \omega_{IF} t & \text{for "1"} \\ 0 & \text{for "0"} \end{cases} \qquad 0 \le t \le T \qquad (2.56)$$

where the amplitude of the lightwave-modulated signal is expressed in terms of the number of photon energy over the time interval; thus, the square root of this quantity is the amplitude of the field of the lightwave. Naturally, the characteristic impedance of medium is set at unity. The distance is then $d = \hat{n}_p$.

Thus, the BER is given by

$$\text{BER} = \text{erfc}\left(\frac{\sqrt{\hat{n}_p}}{2N_0} \right) \qquad (2.57)$$

By setting $N_0 = 1$, we require, for a BER of 10^{-9}, $\hat{n}_p = 4 \times 9 \times 2 = 72$ photon energy for the heterodyne receiver while a 3 dB improvement for the homodyne receiver, thus the 36 photon energy under the assumption that no electronic noise is contributed by the electronic amplifier. These are the quantum limits of ASK heterodyne and homodyne detection with the power of the local oscillator being much larger than that of the signal (Figure 2.11).

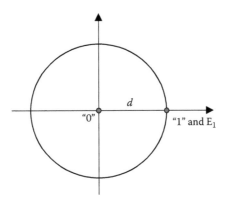

FIGURE 2.11 Signal constellation and energy level and the geometrical distance between "1" and "0" of an ASK system.

2.7.3.2 Coherent Phase and Frequency Shift Keying Systems

$$s_{1,0}(t) = \begin{cases} \sqrt{\dfrac{2\hat{n}_p}{T}} \cos \omega_1 t & \text{for "1"} \\[3mm] \sqrt{\dfrac{2\hat{n}_p}{T}} \cos \omega_0 t & \text{for "0"} \end{cases} \qquad 0 \le t \le T \tag{2.58}$$

The frequency shift keying (FSK) modulation scheme with two distinct frequencies f_1 and f_2 can be represented with a constant envelop and variation of the carrier frequency, or a continuous phase between the two states as shown in Figure 2.12. The modulation index can be defined as

$$m = \frac{|\omega_1 - \omega_0|}{2\pi} T \tag{2.59}$$

and the signal correlation coefficient ρ_C is given by Equation 2.55, under the condition that the two frequencies are large enough and the second and higher harmonics are outside the detection region,

$$\rho = \frac{2}{T} \int_0^T s_1(t) s_0(t) \mathrm{d}t = \frac{2}{T} \int_0^T \cos(\omega_1 t) \cos(\omega_0 t) \mathrm{d}t \tag{2.60}$$

$$\approx \frac{\sin(2\pi m)}{2\pi m}$$

Thus, the BER is given as

$$\mathrm{BER} = \frac{1}{2} \mathrm{erfc}\left(\sqrt{\frac{n_p}{2}\left(1 - \frac{\sin(2\pi m)}{2\pi m}\right)} \right) \tag{2.61}$$

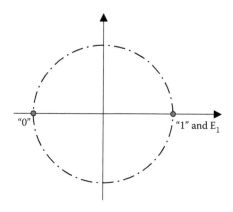

FIGURE 2.12 Signal constellation and energy level evolution of the signal envelop and the continuous phase between the "1" and "0" of an FSK system.

Thus for a BER of 10^{-9} we have

$$\text{BER} = 10^{-9} \longrightarrow \sqrt{\frac{\hat{n}_p}{2}\left(1 - \frac{\sin(2\pi m)}{2\pi m}\right)} = 3\sqrt{2} \tag{2.62}$$

then

$$\hat{n}_p = \frac{36}{1 - \frac{\sin(2\pi m)}{2\pi m}}$$

For MSK, the modulation index is 0.25, leading to the required number of photon energy per bit of 60–70, much higher than that of 0.8 at which only 30 photon energy is required per bit. It is, however, shown that the MSK can be optimum for the transmission over a dispersive medium due to the optimum bandwidth of the modulation scheme and hence minimum dispersive effects on the phase of the carrier.

The FSK can be implemented with continuous-phase frequency-shift keying (CPFSK) that is the phase of the carrier is continuously chirped. Assume that the phase is linearly chirped such that the phase variation for the "1" and "0" is given by

$$s_{1,0}(t) = \begin{cases} \sqrt{\dfrac{2\hat{n}_p}{T}}\cos\theta_1(t) & \text{for "1"} \\[3mm] \sqrt{\dfrac{2\hat{n}_p}{T}}\cos\theta_0(t) & \text{for "0"} \end{cases}$$

where

$$\theta_{1,0}(t) = \begin{cases} \omega_{\text{IF}}t + \dfrac{m\pi}{T}t & \text{for "1"} \\[3mm] \omega_{\text{IF}}t - \dfrac{m\pi}{T}t & \text{for "0"} \end{cases} \qquad 0 \le t \le \frac{T}{2m} \tag{2.63}$$

$$\theta_{1,0}(t) = \begin{cases} \omega_{\text{IF}}t + \dfrac{\pi}{2} & \text{for "1"} \\[3mm] \omega_{\text{IF}}t - \dfrac{\pi}{2} & \text{for "0"} \end{cases} \qquad \frac{T}{2m} \le t \le T$$

The correlation coefficient is given by

$$\text{BER} = 10^{-9} = \frac{1}{2}\text{erfc}\left(\sqrt{\hat{n}_p\left(1 - \frac{1}{4m}\right)}\right) \longrightarrow \sqrt{\hat{n}_p\left(1 - \frac{1}{4m}\right)} = 3\sqrt{2} \tag{2.64}$$

$$\text{then} \quad \hat{n}_p = \frac{18}{1 - \dfrac{1}{4m}}$$

The variation of the photon energy versus the modulation index for this linear CPFSK is shown in Figure 2.13.

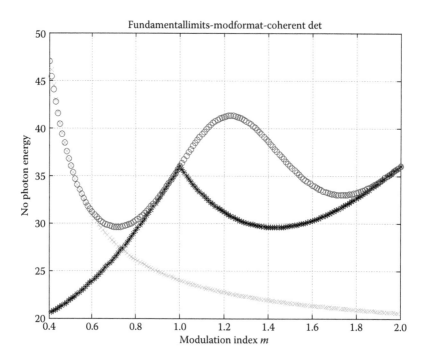

FIGURE 2.13 Fundamental limits of the coherent phase and frequency shift keying detection for (a) medium gray (O) CPFSK, (b) light gray (×) FSK (ω_1 and ω_2) with optimized correlation coefficient with phase control, and (c) black (*) linear DMPSK. For a BER = 10^{-9}.

The FSK can be modified with the control of the relative phase of the two carriers between the two bits; thus the data bits can be written as

$$s_{1,0}(t) = \begin{cases} \sqrt{\dfrac{2\hat{n}_p}{T}}\cos(\omega_1 t + \theta) & \text{for "1"} \\[3mm] \sqrt{\dfrac{2\hat{n}_p}{T}}\cos\omega_0 t & \text{for "0"} \end{cases} \qquad 0 \le t \le T \qquad (2.65)$$

where the phase angle $\theta(t)$ can be chosen as

$$\theta_{1,0}(t) = \begin{cases} \pi - m\pi & \text{for "}p \le m \le 2p+1\text{"} \quad p = 0,1,2,\dots \\ -m\pi & \text{for "}2p+1 \le m \le 2(p+1)\text{"} \end{cases} \qquad (2.66)$$

with the correlation coefficient

$$\rho_C = \frac{2}{T}\int_0^T s_1(t)s_0(t)\,\mathrm{d}t = \frac{2}{T}\int_0^T \cos(\omega_1 t + \theta)\cos(\omega_0 t)\,\mathrm{d}t \approx -\frac{|\sin(\pi m)|}{\pi m} \qquad (2.67)$$

Thus, this gives a BER for a phase-controlled FSK,

$$\text{BER} = \frac{1}{2}\text{erfc}\left(\sqrt{\frac{n_p}{2}\left(1 + \frac{|\sin(\pi m)|}{\pi m}\right)}\right) \rightarrow n_p = \frac{36}{1 + \dfrac{|\sin(\pi m)|}{\pi m}} \quad \text{for BER} = 10^{-9} \qquad (2.68)$$

The required number of photons plotted against the modulation index is shown in Figure 2.13b ("black *").

2.8 BER OF HIGHER ORDER M-ARY QAM

In general, one can deduce the BER of M-ary QAM modulation format corresponding to their constellations under the GN model and are given in Table 2.2 [24]. Further, these BER versus SNR can be obtained for AWGN of different modulation formats that can be generated using bertool.m of the communication tool box of MATLAB®.

Reference [24] has reported a comprehensive spectral efficiency as a function of the SNR for different modulation formats employing M-ary-ASK, M-ary QAM, and M-ary-PSK, as shown in

TABLE 2.2
BER of Popular Modulation Scheme

Modulation Format	BER	Constellation
ASK	$Q\sqrt{\text{SNR}}$	
Bipolar M-ary ASK	$2\dfrac{m-1}{m\log_2(m)}Q\left(\sqrt{\dfrac{6\log_2(m)}{m^2-1}\text{SNR}}\right)$	
BPSK	$Q\sqrt{2\text{SNR}}$	
M-ary PSK	$\dfrac{2}{\log_2(m)}Q\left(\sqrt{2\log_2(m)\text{SNR}}\sin\dfrac{\pi}{m}\right)$	
Rectangular M-ary QAM	$\dfrac{4}{\log_2(m)}\left(1-\dfrac{1}{\sqrt{m}}\right)Q\left(\sqrt{\dfrac{3\log_2(m)}{m-1}\text{SNR}}\right)$	

$Q(x) = \dfrac{1}{2}\text{erfc}\left[\dfrac{x}{\sqrt{2}}\right]$ = complementary error function.

SNR, signal-to-noise ratio.

Note: The SNR is defined here as the ratio between the signal power and the noise power, which is measured over the entire signal 3 dB bandwidth. The signal power can be estimated by the average energy over one period or over the 3 dB bandwidth of the signal spectrum. This is different with optical SNR, which is defined with the noise power measured over 0.1 nm (or 12.5 GHz at 1550 nm wavelength) spectral band. For 28 GSy, the signal bandwidth is wider, about 0.18 nm. Presently, the bit rate can reach 56 or 64 GSy/s (GSy/s = Giga-symbols/second) and for QPSK the signal bandwidth can be around 34 GHz or 0.34 × 0.8 = 0.28 nm.

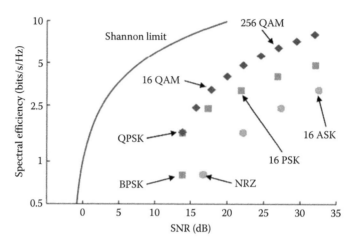

FIGURE 2.14   Spectral efficiency in bits/s/Hz versus SNR of different ASK, M-QAM, and PSK modulation scheme for BER $= 10^{-12}$. Note the linear and nonlinear region. (Adapted from Ellis, A.D. et al., *IEEE J. Lightwave Technol.*, 28, 423–434, 2010, Figure 6; Desurvire, E.B., *IEEE J. Lightwave Technol.*, 24, 4697–4710, 2006; Kahn, J.M. and Ho, K.-P., *IEEE J. Sel. Topics Quantum Electron.*, 10, 259–272, 2004; Shannon, C.E., *Bell Syst. Tech. J.*, 27, 379–423, 623–656, 1948. With permission.)

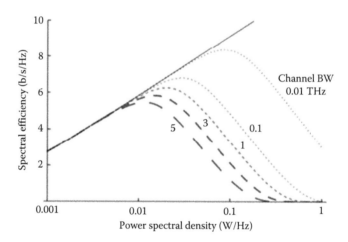

FIGURE 2.15   Spectral efficiency versus power spectral density for channels of bandwidth (BW) variable from 0.01 THz to 5 THz over the transmission overall band of 10 THz. (Adapted from Ellis, A.D. et al., *IEEE J. Lightwave Technol.*, 28, 423–434, 2010. With permission.)

Figure 2.14. The demand of a significantly higher SNR of 25–35 dB places severe restriction on transmission equipment, especially the output power of the lightwave sources in DWDM systems. The spectral efficiency variation with respect to the power density or power required per unit frequency window as plotted in Figure 2.15 indicates the nonlinear limits at which the spectral efficiency deviates or the higher the power embedded in the channel the worst the error rate will be. Furthermore, in long-haul multispan optically amplified link, the contribution of accumulated optical amplification of the cascaded spans is significantly affected by the channel capacity, as shown in Figure 2.16, in which the spectral efficiency is plotted against the power density with the noise figure (NF) of the amplifier operating in the C-band as a parameter.

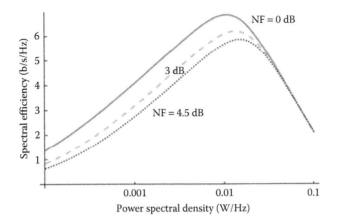

FIGURE 2.16   Spectral efficiency versus power spectral density with noise figure (NF) of optical amplifiers in multispan optical transmission.

2.9   REMARKS

This chapter has addressed the challenging issues of the channel capacity, especially for the two outstanding cases of memory-less and with memory. The latter has been clarified in [7, Chapter 6] for the design of code words and the classic structure of memory. Current ultrahigh sampling rate ADCs and DSP reaching 64 GSy/s ADCs [25,26] are available together with high-speed ASIC DSP, which will allow the transmission reaching the Shannon limit in the linear region. However, the nonlinear region of optical transmission fiber lines is of much important when several wavelength channels are employed, for example, the 96 instead 88 channels in the C-band. In optically amplified multispan transmission systems, the nonlinear effects will limit the capacity. Recent interests and development in the uses of high-order spectral analysis [27] and nonlinear Fourier transform [28–30] have been approached so that coding in the transmitter can be used to mitigate this nonlinearity-limited capacity.

Under coherent states and CoD techniques and mitigation of the nonlinearity in fiber optic transmission systems, it is optimistic that the Shannon limit can be an asymptotic approach to infinity. However, more complex processing at both the transmitter and receiver must be confronted. The limitation of the processing power and the latency required in a real-time system may limit this limit level. Furthermore, the noises due to an electric wideband subsystem will play a major part in such limit approaches. These are the main topics that are addressed in this book.

The transmission at ultrahigh capacity over long-haul fiber core networks and access networks has reached multi-Tbps to Pbps and will soon reach exa-bps levels, especially inside data centers and inter-data center connections in the photonic domain. The nonlinearity and noises set limits in such networks must be overcome in both electronic and photonic domain as well as in the digital processing subsystems.

REFERENCES

1. M. G. Taylor, Coherent detection method using DSP for demodulation of signal and subsequent equalization of propagation impairments, *IEEE Photon. Technol. Lett.*, 16(2), 674–676, 2004.
2. M. G. Taylor, Accurate digital phase estimation process for coherent detection using a parallel digital processor. In *Proc. ECOC 2005 Conf.*, IEEE , New York, 2005.
3. M. Taylor, Phase estimation methods for optical coherent detection using digital signal processing, *IEEE J. Lightwave Technol.*, 27(7), 901–914, 2009.
4. D.-S. Ly-Gagnon, S. Tsukamoto, K. Katoh, and K. Kikuchi, Coherent detection of optical quadrature phase-shift keying signals with carrier phase estimation, *J. Lightwave Technol.*, 24(1), 12–21, 2006.

5. B. J. Puttnam, R. S. Luís, W. Klaus, J. Sakaguchi, J.-M. Delgado Mendinueta, Y. Awaji, N. Wada et al., 2.15 Pb/s transmission using a 22 core homogeneous single-mode multi-core fiber and wideband optical comb. In *Proc. ECOC 2015, PDP.3.1.*

6. C. E. Shannon, A mathematical theory of communication, *Bell Syst. Tech. J.*, 27, 379–423; 623–656, 1948.

7. J. M. Wonzencroft and I. M. Jacobs, *Principles of Communication Engineering*, John Wiley & Sons, New York, 1965, p. 321.

8. G. P. Agrawal, *Nonlinear Fiber Optics*, Academic Press, San Diego, 1995.

9. L. N. Binh, *Digital Optical Communications Systems*, CRC Press, Boca Raton, FL, 2008.

10. L. N. Binh, *Digital Optical Communications Systems*, CRC Press, Boca Raton, FL, 2013.

11. L. N. Binh, *Digital Optical Communications (Optics and Photonics Series)*, CRC Press, Boca Raton, FL, Nov 2008.

12. E. Agrell, A. Alvarado, G. Durisi, and M. Karlsson, Capacity of a nonlinear channel with finite memory, *IEEE J. Lightwave Technol.*, 32(6), 2862–2876, Aug 16, 2014.

13. A. Spellt, C. Kurtzke, and K. Petermann, Ultimate transmission capacity of amplified optical fiber communication systems taking into account of fiber nonlinearities. In *Proc. ECOC'93*, Montreux, Switzerland, Sept 1993.

14. E. E. Narimanov and P. Mitra, The channel capacity of a fiber optics communication system: perturbation theory, *J. of Lightwave Technol.*, 20(3), 530–537, 2002.

15. S. Verdi and T. Sun Han, A general formula for channel capacity, *IEEE Trans. Inform. Theory*, 40(4), 1147–1158, July 1994.

16. K. C. Kao and G. A. Hockham, Dielectric-fibre surface waveguides for optical frequencies, *Proc. IEE*, 113, 1151, 1966.

17. H. Zbinden, H. Bechmann-Pasquinucci, N. Gisin, and G. Ribordy, Quantum cryptography, *Appl. Phys. B*, 67, 743–748, 1998.

18. J. M. Arrazola and N. Lütkenhaus, Quantum communication with coherent states and linear optics, *Phys. Rev. A*, 90(4), 042335, Oct 2014.

19. A. Klimek, M. Jachura, W. Wasilewski, and K. Banaszek, Quantum memory receiver for superadditive communication using binary coherent states, arXiv:1512.06561v1 [quant-ph], Dec 2015.

20. P. Gallion and F. J. Mendieta, Minimum energy per bit in high rate optical communications and quantum communications, *Proc. SPIE*, 8065, 80650F-1, 2011.

21. D. Qiu, Minimum-error discrimination between mixed quantum states, arXiv:0707.3970 [quant-ph], 2007.

22. J. Dressel, T. A. Brun, and A. N. Korotkov, Violating the modified Helstrom bound with nonprojective measurements, arXiv:1410.0096 [quant-ph], 2014.

23. A. Montanaro, *A Lower Bound on the Probability of Error in Quantum State Discrimination*, www.damtp.cam.ac.uk/user/am994/presentations/upper.pdf, University of Bristol, UK, 2008.

24. A. D. Ellis, J. Zhao, and D. Cotter, Approaching the non-linear Shannon limit, *IEEE J. Lightwave Technol.*, 28(4), 423–434, Feb 15, 2010.

25. SHF AG, Berlin Germany, 6-bit ADCs, www.SHF.de.

26. https://www.fujitsu.com/uk/Images/c64.pdf.

27. L. N. Binh, *Advanced in Digital Communications Systems*, 2nd edn., CRC Press Boca Raton, FL, 2015.

28. M. I. Yousef, *Information Transmission Using the Nonlinear Fourier Transform*, Doctor of Philosophy Dissertation, University of Toronto, Toronto, Canada, 2013.

29. M. I. Yousefi, E. S. Rogers Sr, and F. R. Kschischang, Information transmission using the nonlinear Fourier transform, part I: Mathematical tools, *IEEE Trans. Inform. Theory*, 60(7), 4312–4328, 2014.

30. H. Bulow, Experimental demonstration of optical signal detection using nonlinear Fourier transform, *IEEE J. Lightwave Technol.*, 33(7), 1433–1439, 2015.

31. E. B. Desurvire, Capacity demand and technology challenges for lightwave systems in the next two decades, *IEEE J. Lightwave Technol.*, 24(12), 4697–4710, 2006.

32. J. M. Kahn and K.-P. Ho, Spectral efficiency limits and modulation/detection techniques for DWDM systems, *IEEE J. Sel. Topics Quantum Electron.*, 10(2), 259–272, Mar/Apr 2004.

33. C. E. Shannon, A mathematical theory of communication, *Bell Syst. Tech. J.*, 27, 379–423; 623–656, 1948.

3 Optical Coherent Reception and Noise Processes

Detection of optical signals can be carried out at the optical receiver by direct conversion of optical signal power to electronic current in the photodiode (PD) and then electronic amplification. This chapter provides the fundamental understanding of coherent detection (CoD) of optical signals, which requires the mixing of the optical fields of the optical signals and that of the local oscillator (LO), a high-power laser so that its beating product would result in the modulated signals preserving both its phase and amplitude characteristics in the electronic domain. Optical preamplification in CoD can also be integrated at the front end of the optical receiver.

3.1 INTRODUCTION

With the exponential increase in data traffic, especially due to the demand for ultrabroad bandwidth driven by multimedia applications, cost-effective ultrahigh-speed optical networks have become highly desired. It is expected that Ethernet technology will not only dominate in access networks but also will become the key transport technology of next generation metro/core networks. 100 Gigabit Ethernet (100 GbE) is currently considered to be the next logical evolution step after 10 GbE. Based on the anticipated 100 GbE requirements, 100 Gbit/s data rate of serial data transmission per wavelength is required. To achieve this data rate while complying with current system design specifications such as channel spacing, chromatic dispersion (CD), and polarization mode dispersion (PMD) tolerance, coherent optical communication systems with multilevel modulation formats will be desired, because it can provide high spectral efficiency, high receiver sensitivity, and potentially high tolerance to fiber dispersion effects [1–6].* Compared to conventional direct detection in intensity-modulation/direct-detection (IMDD) systems that only detects the intensity of the light of the signal, CoD can retrieve the phase information of the light, and therefore, can tremendously improve the receiver sensitivity.

Coherent optical receivers are important components in long-haul optical fiber communication systems and networks to improve the receiver sensitivity and thus extra transmission distance. Coherent techniques were considered for optical transmission systems in the 1980s when the extension of repeater distance between spans is pushed to 60 km instead of 40 km for single mode optical fiber at bit rate of 140 Gb/s. However, in the late 1980s the invention of optical fiber amplifiers has overcome this attempt? Recently, interests in coherent optical communications have attracted significant research activities for ultrabit rate dense wavelength division multiplexing (DWDM) optical systems and networks. The motivation has been possible due to the fact that (i) the uses of optical amplifiers in cascade fiber spans have added significant noises and thus limit the transmission distance, (ii) the advances of digital signal processors (DSP) whose sampling rate can reach few tens of giga-samples/s allowing the processing of beating signals to recover the phase or phase estimation (PE), (iii) the availability of advanced signal processing algorithms such as Viterbi and Turbo algorithms, and (iv) that the differential coding and modulation and detection of such signals may not require an optical phase-locked loop (OPLL), and hence self-coherent and DSP to recover transmitted signals. These technological advances, especially in the digital processors

---

* Synchronous detection is implemented by mixing the signals and a strong local oscillator in association with the phase locking of the local oscillator to that of the carrier.

at ultra-sampling rate, allow overcoming several difficulties in homodyne coherent reception conducted in the first coherent system generation in the 1980s.

As is well known, a typical arrangement of an optical receiver is that the optical signals are detected by a PD (a PIN diode or avalanche photodiode [APD] or a photon-counting device); electrons generated in the photodetector are then electronically amplified through a front end electronic amplifier. The electronic signals are then decoded for recovery of original format. However, when the fields of incoming optical signals are mixed with those of a LO whose frequency can be identical or different to that of the carrier, the phase, and frequency property of the resultant signals reflect those of the original signals. Coherent optical communication systems have also been reviving dramatically due to electronic processing and availability of stable narrow linewidth lasers.

This chapter deals with the analysis and design of coherent receivers with OPLL and the mixing of optical signals and that of the LO in the optical domain and thence detected by the optoelectronic receivers following this mixing. Thus, both the optical mixing and photodetection devices act as the fundamental elements of a coherent optical receiver. Depending on the frequency difference between the lightwave carrier of the optical signals and that of the LO, the CoD can be termed as heterodyne or homodyne detection. For heterodyne detection, there is a difference in the frequency and thus the beating signal is fallen in a passband region in the electronic domain. Thus, all the electronic processing at the front end must be in this passband region. On the other hand, in homodyne detection, there is no frequency difference and thus the detection is in the baseband of the electronic signal. Both cases would require a locking of the LO and carrier of the signals. An OPLL is thus treated in this chapter.

This chapter is organized as follows: Section 3.2 gives an account of the components of coherent receivers, Section 3.3 outlines the principles of optical coherent detection under heterodyne, homodyne, or intradyne techniques, and Section 3.4 gives details of the OPLL which is a very important development for modern optical coherent detection.

3.2 COHERENT RECEIVER COMPONENTS

The design of an optical receiver depends on the modulation format of the signals and thus transmitted through the transmitter. The modulation of the optical carrier can be in the form of amplitude, phase, and frequency. Furthermore, the phase shaping also plays a critical role in the detection and the bit-error rate (BER) of the receiver and thence the transmission systems. In particular, it is dependent on the modulation in analog or digital, Gaussian or exponential pulse shape, on-off keying or multiple levels, and so on.

Figure 3.1 shows the schematic diagram of a digital coherent optical receiver, which is similar to the direct detection receiver but with an optical mixer at the front end. Figure 3.2 shows the small signal equivalent circuits of such a receiver's front end. However, the phase of the signals at

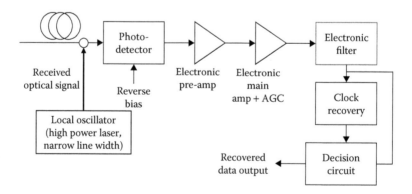

FIGURE 3.1 Schematic diagram of a digital optical coherent receiver with an additional LO mixing with the received optical signals before detected by an optical receiver.

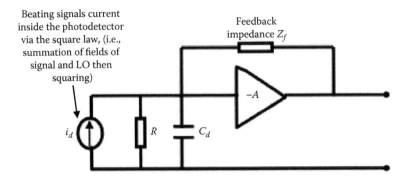

FIGURE 3.2 Schematic diagram of an electronic preamplifier in an optical receiver of a transimpedance electronic amplifier at the front end. The current source represents the electronic current generated in the photodetector due to the beating of the local oscillator and the optical signals. C_d = photodiode capacitance.

base or passband of the detected signals in the electrical domain would remain in the generated electronic current and voltages at the output of the electronic preamplifier. An optical front end is an optical mixer combining the fields of the optical waves of the local laser and the optical signals so that the envelope of the optical signals can be beating with each other to a product with the summation of the frequencies and the difference of the frequencies of the lightwaves. Only the lower frequency term, which falls within the absorption range of the photodetector, is converted into the electronic current preserving both the phase and amplitude of the modulated signals.

Thus, an optical receiver front end, very much the same as that of the direct detection, is connected following the optical processing front end consisting of a photodetector for converting lightwave energy into electronic currents; an electronic preamplifier for further amplification of the generated electronic current followed by an electronic equalizer for bandwidth extension, usually in voltage form; a main amplifier for further voltage amplification; a clock recovery circuitry for regenerating the timing sequence; and a voltage-level decision circuit for sampling the waveform for the final recovery of the transmitted and received digital sequence. Therefore, the optoelectronic preamplifier is followed by a main amplifier with an automatic control to regulate the electronic signal voltage to be filtered and then sampled by a decision circuit with synchronization by a clock recovery circuitry.

An inline fiber optical amplifier can be incorporated in front of the photodetector to form an optical receiver with an optical amplifier front end to improve its receiving sensitivity. This optical amplification at the front end of an optical receiver will be treated in this chapter dealing with optical amplification processes.

The structure of the receiver is thus consisted of four parts: the optical mixing front, the front end section, the linear channel of the main amplifier and automatic gain control (AGC) if necessary, and the data recovery section. The optical mixing front end sums the optical fields of the LO and that of the optical signals. Polarization orientation between these lightwaves is very critical to maximize the beating of the additive field in the PD. Depending on whether the frequency difference is finite or null between these fields of the resulting electronic signals derived from the detector, the electronic signals can be in the baseband or the passband, and the detection technique is termed as a heterodyne or homodyne technique, respectively.

3.3 CoD

Optical CoD can be distinguished by the "demodulation" scheme in communications techniques in association with following definitions: (i) CoD is the mixing between two lightwaves or optical carriers, one is information-bearing lightwaves and the other a LO with an average energy much larger than that of the signals and (ii) demodulation refers to the recovery of baseband signals from the electrical signals.

A typical schematic diagram of a coherent optical communications employing guided wave medium and components is shown in Figure 3.1, in which a narrow band laser incorporating an optical isolator cascaded with an external modulator is usually the optical transmitter. Information is fed via a microwave power amplifier to an integrated optic modulator, commonly used $LiNbO_3$ or EA types. The CoD is a principal feature of coherent optical communications, which can be further distinguished with heterodyne and homodyne techniques depending whether there is a difference or not between the frequencies of the LO and that of the carrier of the signals. A LO is a laser source whose frequency can be tuned and approximately equivalent to a monochromatic source; a polarization controller would also be used to match its polarization with that of the information-bearing carrier. The LO and the transmitted signal are mixed via a polarization maintaining coupler and then detected by a coherent optical receiver. Most of the previous CoD schemes are implemented in a mixture of photonic domain and electronic/microwave domain.

Coherent optical transmission has become the focus of research. One significant advantage is the preservation of all the information of the optical field during detection, leading to enhanced possibilities for optical multilevel modulation. This section investigates the generation of optical multilevel modulation signals. Several possible structures of optical M-ary phase-shift keying (M-ary-PSK) and M-ary quadrature amplitude modulation (M-ary-QAM) transmitters are shown and theoretically analyzed. Differences in the optical transmitter configuration and the electrical driving lead to different properties of the optical multilevel modulation signals. This is shown by deriving general expressions applicable to every M-ary-PSK and M-ary-QAM modulation format and exemplarily clarified for Square-16-QAM modulation.

Coherent receivers are distinguished between synchronous and asynchronous. Synchronous detection requires an OPLL that recovers the phase and frequency of the received signals to lock the LO to that of the signal so as to measure the absolute phase and frequency of the signals relative to that of the LO. Thus, synchronous receivers allow direct mixing of the bandpass signals and the baseband, so this technique is termed as homodyne reception. For asynchronous receivers, the frequency of the LO is approximately the same as that of the receiving signals and no OPLL is required. In general, the optical signals are first mixed with an intermediate frequency (IF) oscillator which is about two to three times that of the 3 dB passband. The electronic signals can then be recovered using electrical PLL at lower carrier frequency in the electrical domain. The mixing of the signals and an LO of an IF is referred to as heterodyne detection.

If no LO is used for demodulating the digital optical signals, then differential or self-homodyne reception may be utilized, which is classically termed as autocorrelation reception process or self-heterodyne detection.

Coherent communications have been an important technique in the 1980s and the early 1990s, but then their research was interrupted with the advent of optical amplifiers in the late 1990s that offer up to 20 dB gain without difficulty. Nowadays, however, coherent systems have once again become the focus of interest, due to the availability of DSP and low-priced components, the partly relaxed receiver requirements at high data rates and several advantages that CoD provides. The preservation of the temporal phase of the CoD enables new methods for adaptive electronic compensation of CD. With regard to WDM systems, coherent receivers offer tunability and allow channel separation via steep electrical filtering. Furthermore, only the use of CoD permits convergence to the ultimate limits of spectral efficiency. To reach higher spectral efficiencies, the use of multilevel modulation is required. For this, too, coherent systems are also beneficial, because all the information of the optical field is available in the electrical domain. This way complex optical demodulation with interferometric detection—which has to be used in direct detection systems—can be avoided and the complexity is transferred from the optical to the electrical domain. Several different modulation formats based on the modulation of all four quadratures of the optical field were proposed in the early 1990s, describing the possible transmitter and receiver structures and calculating the theoretical BER performance. However, a more detailed and practical investigation of multilevel modulation coherent optical systems for today's networks and data rates is missing so far.

Currently, coherent reception has attracted significant interests due to the following reasons: (i) The received signals of the coherent optical receivers are in the electrical domain, which is proportional to that in the optical domain. This, in contrast to the direct detection receivers, allows exact electrical equalization or exact PE of the optical signals. (ii) Using heterodyne receivers, DWDM channels can be separated in the electrical domain by using electrical filters with sharp roll of the passband to the cut-off band. Presently, the availability of ultrahigh sampling rate DSP allows users to conduct filtering in the DSP in which the filtering can be changed with ease.

However, there are disadvantages that coherent receivers would suffer: (i) coherent receivers are polarization sensitive that requires polarization tracking at the front end of the receiver; (ii) homodyne receivers require OPLL and electrical PLL for heterodyne that would need control and feedback circuitry, optical, or electrical, which may be complicated; and (iii) for differential detection the compensation may be complicated due to the differentiation receiving nature.

In a later chapter when some advanced modulation formats are presented for optically amplified transmission systems, the use of photonic components is extensively exploited to take advantage of the advanced technology of integrated optics, planar lightwave circuits. The modulation formats of signals depend on whether the amplitude, the phase, or the frequency of the carrier is manipulated as mentioned in Chapter 2. In this chapter, the detection is coherently converted to the IF range in the electrical domain and the signal envelop. The down-converted carrier signals are detected and then recovered. Both binary-level and multilevel modulations schemes employing amplitude, phase, and frequency shift keying (FSK) modulation are described in this chapter.

Thus, CoD can be distinguished by the difference between the central frequency of the optical channel and that of the LO. Three types can be classified as follows: (i) heterodyne, when the difference is higher than the 3 dB bandwidth of the baseband signal; (ii) homodyne, when the difference is nil; and (iii) intradyne, when the frequency difference falls within the baseband of the signal.

It is noted that to maximize the beating signals at the output of the photodetector, the polarizations of the LO and the signals must be aligned. In practice, this can be best implemented by the polarization diversity technique.

3.3.1 Optical Heterodyne Detection

A typical schematic of a coherent optical reception sub-system is shown in Figure 3.3. Figure 3.4 shows also more details of a heterodyne reception sub-system in which the frequency of the channel carrier is different with that of the local oscillator. The LO, whose frequency can be higher or lower than that of the carrier, is mixed with the information-bearing carrier thus allowing down- or

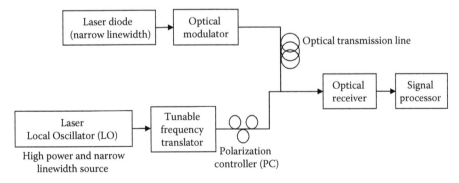

FIGURE 3.3 Typical arrangement of coherent optical communications systems. LD/LC is a very narrow linewidth laser diode as an LO without any phase locking to the signal optical carrier. PM coupler is polarization maintaining fiber coupled device, PC = polarization controller. (Adapted from R. C. Alferness. *IEEE J. Quantum. Elect.*, QE-17, 946–959, 1981; W. A. Stallard et al., *IEEE J. Lightwave Technol.*, LT-4(7), 852–857, July 1986. With permission.)

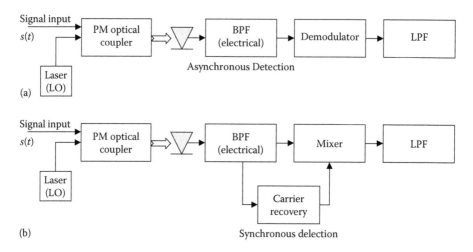

FIGURE 3.4 Schematic diagram of optical heterodyne detection: (a) asynchronous and (b) synchronous, receiver structures. LPF = low pass filter, BPF = bandpass filter, PD = photodiode.

upconversion of the information signals to the IF range. The down-converted electrical carrier and signal envelop is received by the photodetector. This combined lightwave is converted by the PD into electronic current signals, which are then filtered by an electrical bandpass filter (BPF) and then demodulated by a demodulator. A low pass filter (LPF) is also used to remove higher order harmonics of the nonlinear detection photodetection process, the square-law detection. With an envelope detector, the process is asynchronous, hence the name term asynchronous detection. If the down-converted carrier is recovered and then mixed with IF signals, then this is synchronous detection. It is noted that the detection is conducted at the IF range in electrical domain, hence there is a need of controlling the stability of the frequency spacing between the signal carrier and the LO. This means the mixing of these carriers would result into an IF carrier in the electrical domain prior to the mixing process or envelop detection to recover the signals.

The CoD thus relies on the electric field component of the signal and the LO. The polarization alignment of these fields is critical for the optimum detection. The electric field of the optical signals and the LO can be expressed as

$$E_s(t) = \sqrt{2P_s(t)} \cos\{\omega_s t + \phi_s + \varphi(t)\} \tag{3.1}$$

$$E_{LO} = \sqrt{2P_L} \cos\{\omega_{LO}t + \phi_{LO}\} \tag{3.2}$$

where $P_s(t)$ and P_{LO} are the instantaneous signal power and average power of the signals and LO, respectively, $\omega_s(t)$ and ω_{LO} are the signal and LO angular frequencies, ϕ_s and ϕ_{LO} are the phase including any phase noise of the signal and the LO, and $\psi(t)$ is the modulation phase. The modulation can be amplitude with the switching on and off (amplitude shift keying—ASK) of the optical power or phase or frequency with the discrete or continuous variation of the time-dependent phase term. For discrete phase, it can be PSK, differential PSK (DPSK), or differential quadrature PSK (DQPSK), and when the variation of the phase is continuous, we have FSK if the rate of variation is different for the bit "1" and bit "0."

Under an ideal alignment of the two fields, the photodetection current can be expressed by

$$i(t) = \frac{\eta q}{h\upsilon}\left[P_s + P_{LO} + 2\sqrt{P_s P_{LO}} \cos\{(\omega_s - \omega_{LO})t + \phi_s - \phi_{LO} + \varphi(t)\} \right] \tag{3.3}$$

where the higher frequency term (the sum) is eliminated by the photodetector frequency response, η is the quantum efficiency, q is the electronic charge, h is Plank's constant, and υ is the optical frequency.

Thus, the power of the LO dominates the shot-noise process and at the same time boosts the signal level, hence enhancing the signal-to-noise ratio (SNR). The oscillating term is resulted from the beating between the LO and the signal inside the PD, which is proportional to the amplitude and is the square root of the product of the power of the LO and the signal.

The electronic signal power S and shot noise N_s can be expressed as

$$S = 2\Re^2 P_s P_{LO}$$

$$N_s = 2q\Re(P_s + P_{LO})B \qquad (3.4)$$

$$\Re = \frac{\eta q}{h\nu} = \text{responsivity}$$

where B is the 3 dB bandwidth of the electronic receiver. Thus, the optical signal-to-noise ratio (OSNR) can be written as

$$\text{OSNR} = \frac{2\Re^2 P_s P_{LO}}{2q\Re(P_s + P_{LO})B + N_{eq}} \qquad (3.5)$$

where N_{eq} is the total electronic noise equivalent power at the input to the electronic preamplifier of the receiver. From this equation, we can observe that if the power of the LO is significantly increased so that the shot noise dominates over the equivalent noise, at the same time increasing the SNR, the sensitivity of the coherent receiver can only be limited by the quantum noise inherent in the photodetection process. Under this quantum limit, the OSNR_{QL} is given by

$$\text{OSNR}_{QL} = \frac{\Re P_s}{qB} \qquad (3.6)$$

3.3.1.1 ASK Coherent System

Under the ASK modulation scheme, the demodulator of Figure 3.4 is an envelope detector (in lieu of the demodulator) followed by a decision circuitry. That is, the eye diagram is obtained, and a sampling instant is established with a clock recovery circuit. While the synchronous detection would require a locking between the frequencies of the carrier and the LO. The LO frequency is tuned to that the carrier according to the tracking of the frequency component of the beating signal. The amplitude demodulated envelope can then be expressed as

$$r(t) = 2\Re\sqrt{P_s P_{LO}}\cos(\omega_{IF})t + n_x\cos(\omega_{IF})t + n_y\sin(\omega_{IF})t \qquad (3.7)$$

$$\omega_{IF} = \omega_s - \omega_{LO}$$

The IF ω_{IF} is the difference between the frequencies of the LO and the signal carrier, and n_x and n_y are the expected values of the orthogonal noise power components, which are random variables.

$$r(t) = \sqrt{[2\Re P_s P_{LO} + n_x]^2 + n_y^2}\cos(\omega_{IF}t + \Phi)t \text{ with } \Phi = \tan^{-1}\frac{n_y}{2\Re P_s P_{LO} + n_x} \qquad (3.8)$$

3.3.1.1.1 Envelop Detection

The noise power terms can be assumed to follow a Gaussian probability distribution and are independent of each other with a zero mean and a variance σ; the probability density function (PDF) can thus be given as

$$p(n_x, n_y) = \frac{1}{2\pi\sigma^2}e^{-(n_x^2 + n_y^2)/2\sigma^2} \qquad (3.9)$$

With respect to the phase and amplitude, this equation can be written as [3]

$$p(\rho,\phi) = \frac{\rho}{2\pi\sigma^2} e^{-(\rho^2 + A^2 - 2A\rho\cos\phi)/2\sigma^2} \tag{3.10}$$

where

$$\rho = \sqrt{[2\Re\sqrt{P_s(t)P_{LO}} + n_s(t)]^2 + n_y^2(t)}$$

$$A = 2\Re\sqrt{P_s(t)P_{LO}} \tag{3.11}$$

The PDF of the amplitude can be obtained by integrating the phase amplitude PDF over the range of 0 to 2π and given as

$$p(\rho) = \frac{\rho}{\sigma^2} e^{-(\rho^2 + A^2)/2\sigma^2} I_0\left\{\frac{A\rho}{\sigma^2}\right\} \tag{3.12}$$

where I_0 is the modified Bessel function. If a decision level is set to determine the "1" and "0" level, then the probability of error and the BER can be obtained assuming an equal probability of error between the "1s" and "0s":

$$\text{BER} = \frac{1}{2}P_e^1 + \frac{1}{2}P_e^0 = \frac{1}{2}\left[1 - Q\left(\sqrt{2\delta}, d\right) + e^{-(d^2/2)}\right] \tag{3.13}$$

where Q is the Magnum function and δ is given by

$$\delta = \frac{A^2}{2\sigma^2} = \frac{2\Re^2 P_s P_{LO}}{2q\Re\left(P_s + P_{LO}\right)B + i_{N_{eq}}^2} \tag{3.14}$$

When the power of the LO is much larger than that of the signal and the equivalent noise current power, this SNR becomes

$$\delta = \frac{\Re P_s}{qB} \tag{3.15}$$

The physical representation of the detected current and the noise current due to the quantum shot noise and noise equivalent of the electronic preamplification is shown in Figure 3.5, in which the signal current can be general and derived from the output of the detection scheme, that from a photodetector or a back-to-back (B2B) pair of photodetectors of a balanced receiver for detecting the phase difference of DPSK or DQPSK or continuous-phase frequency-shift keying (CPFSK) signals and converting to amplitudes.

The BER is optimum when setting its differentiation with respect to the decision level δ, and an approximate value of the decision level can be obtained as

$$d_{opt} \cong \sqrt{2 + \frac{\delta}{2}} \Rightarrow \text{BER}_{ASK-e} \cong \frac{1}{2}e^{-(\delta/4)} \tag{3.16}$$

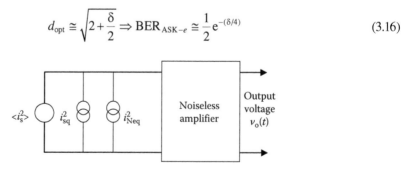

FIGURE 3.5 Equivalent current model at the input of the optical receiver, average signal current and equivalent noise current of the electronic preamplifier as seen from its input port.

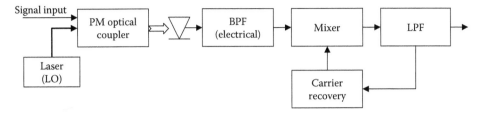

FIGURE 3.6   Schematic diagram of optical heterodyne detection for PSK format.

3.3.1.1.2   Synchronous Detection

ASK can be detected using synchronous detection[*] and the BER is given by

$$\mathrm{BER}_{ASK-S} \cong \frac{1}{2} erfc^{\sqrt{\delta}/2} \tag{3.17}$$

3.3.1.2   PSK Coherent System

Under the PSK modulation format, the detection is similar to that of Figure 3.4 for heterodyne detection (see Figure 3.6) but after the BPF, an electrical mixer is used to track the phase of the detected signal. The received signal is given by

$$r(t) = 2\Re\sqrt{P_s P_{LO}}\cos[(\omega_{IF})t + \varphi(t)] + n_x \cos(\omega_{IF})t + n_y \sin(\omega_{IF})t \tag{3.18}$$

The information is contained in the time-dependent phase term $\varphi(t)$.

When the phase and frequency of the voltage-controlled oscillator (VCO) are matched with those of the signal carrier, then the received electrical signal can be simplified as

$$r(t) = 2\Re\sqrt{P_s P_{LO}}\,a_n(t) + n_x$$
$$a_n(t) = \pm 1 \tag{3.19}$$

Under the Gaussian statistical assumption, the probability of the received signal of a "1" is given by

$$p(r) = \frac{1}{\sqrt{2\pi\sigma^2}}e^{-((r-u)^2/2\sigma^2)} \tag{3.20}$$

Furthermore, the probability of the "0" and that of the "1" are assumed to be equal. We can obtain the BER as the total probability of the received "1" and "0":

$$\mathrm{BER}_{PSK} = \frac{1}{2} erfc(\delta) \tag{3.21}$$

3.3.1.3   Differential Detection

As observed in the synchronous detection, a carrier recovery circuitry is required, usually implemented using a PLL, which complicates the overall receiver structure. It is possible to detect the signal by a self-homodyne process by beating the carrier of one-bit period to that of the next consecutive bit; this is called the differential detection. The detection process can be modified as shown in Figure 3.7, in which the phase of the IF carrier of one bit is compared with that of the next bit, and a difference is recovered to represent the bit "1" or "0." This requires a differential coding at the

---

[*] Synchronous detection is implemented by mixing the signals and a strong local oscillator in association with the phase locking of the local oscillator to that of the carrier.

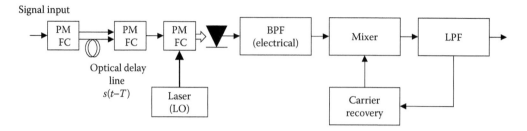

Signal input

FIGURE 3.7 Schematic diagram of optical heterodyne and differential detection for PSK format.

transmitter and an additional phase comparator for the recovery process. In later chapters on DPSK, the differential decoding is implemented in photonic domain via a photonic phase comparator in form of an MZ delay interferometer (MZDI) with a thermal section for tuning the delay time of the optical delay line. The BER can be expressed as

$$\text{BER}_{\text{DPSK}-e} \cong \frac{1}{2} e^{-\delta} \tag{3.22}$$

$$r(t) = 2\Re \sqrt{P_s P_{\text{LO}}} \cos[\pi A_k s(t)] \tag{3.23}$$

where $s(t)$ is the modulating waveform and A_k represents the bit "1" or "0." This is equivalent to the baseband signal, and the ultimate limit is the BER of the baseband signal.

The noise is dominated by the quantum shot noise of the LO, with its square noise current given by

$$i_{N-\text{sh}}^2 = 2q\Re(P_s + P_{\text{LO}}) \int_0^\infty \left|H(j\omega)\right|^2 d\omega \tag{3.24}$$

where $H(j\omega)$ is the transfer function of the receiver system, normally a transimpedance of the electronic preamp and that of a matched filter. As the power of the LO is much larger than the signal, integrating over the dB bandwidth of the transfer function, this current can be approximated by

$$i_{N-\text{sh}}^2 \simeq 2q\Re P_{\text{LO}} B \tag{3.25}$$

Hence, the SNR (power) is given by

$$\text{SNR} \equiv \delta \simeq \frac{2\Re P_s}{qB} \tag{3.26}$$

The BER is the same as that of a synchronous detection and is given by

$$\text{BER}_{\text{homodyne}} \cong \frac{1}{2} erfc\sqrt{\delta} \tag{3.27}$$

The sensitivity of the homodyne process is at least 3 dB better than that of the heterodyne, and the bandwidth of the detection is half of its counterpart due to the double-sideband nature of the heterodyne detection.

3.3.1.4 FSK Coherent System

The nature of FSK is based on the two frequency components that determine the bits "1" and "0." There are a number of formats related to FSK depending on whether the change of the frequencies

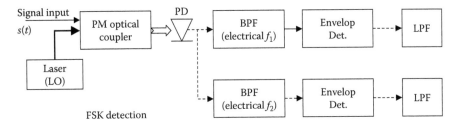

FIGURE 3.8 Schematic diagram of optical homodyne detection of FSK format.

representing the bits is continuous or noncontinuous, the FSK or CPFSK modulation formats. For noncontinuous FSK, the detection is usually performed by a structure of dual frequency discrimination as shown in Figure 3.8, in which two narrow band filters are used to extract the signals. For CPFSK, both the frequency discriminator and balanced receiver for PSK detection can be used. The frequency discrimination is indeed preferred as compared with the balanced receiving structures because it would eliminate the phase contribution by the LO or optical amplifiers, which may be used as an optical preamp.

When the frequency difference between the "1" and "0" equals a quarter of the bit rate, the FSK can be termed as the minimum-shift keying modulation scheme. At this frequency spacing, the phase is continuous between these states.

3.3.2 OPTICAL HOMODYNE DETECTION

Optical homodyne detection matches the transmitted signal phases to that of the LO phase signal. A schematic of the optical receiver is shown in Figure 3.9. The field of the incoming optical signals is mixed with the LO, whose frequency and phase are locked with that of the signal carrier waves via a PLL. The resultant electrical signal is then filtered and therefore a decision circuitry is formed.

3.3.2.1 Detection and OPLL

Optical homodyne detection requires the phase matching of the frequency of the signal carrier and that of the LO. This type of detection would give a very high sensitivity, in principle, of 9 photons/bit. Implementation of such a system would normally require an OPLL, whose structure of a recent development [4] is shown in Figure 3.10. The LO frequency is locked into the carrier frequency of the signals by shifting it to the modulated sideband component via the use of the optical modulator. A single-sideband optical modulator is preferred. However, a double sideband may also be used. This modulator is excited by the output signal of a VCO whose frequency is

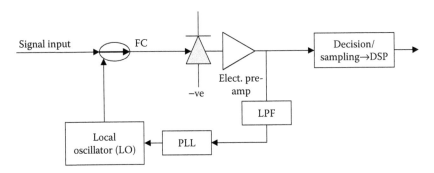

FIGURE 3.9 General structure of an optical homodyne detection system. FC = fiber coupler, LPF = low pass filter, PLL = phase lock loop.

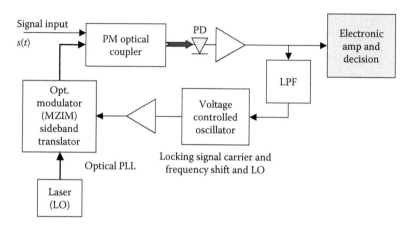

FIGURE 3.10 Schematic diagram of optical homodyne detection—electrical line (dashed) and optical line (continuous and solid) using an OPLL.

determined by the voltage level of the output of an electronic BPF condition to meet the required voltage level for driving the electrode of the modulator. The frequency of the LO is normally tuned to the region such that the frequency difference with respect to the signal carrier falls within the passband of the electronic filter. When the frequency difference is zero, there is no voltage level at the output of the filter and thus the OPLL has reached the final stage of locking. The bandwidth of the optical modulator is important so that it can extend the locking range between the two optical carriers.

Any frequency offset between the LO and the carrier is detected, and noise is filtered by the LPF. This voltage level is then fed to a VCO to generate a sinusoidal wave that is then used to modulate an intensity modulator modulating the lightwaves of the LO. The output spectrum of the modulator would exhibit two sidebands and the LO lightwave. One of these components would then be locked to the carrier. A close loop would ensure a stable locking. If the intensity modulator is biased at the minimum transmission point and the voltage level at the output of the VCO is adjusted to $2V_\pi$ with driven signals of $\pi/2$ phase shift with each other, then we would have carrier suppression and sideband suppression. This ease the confusion of the close loop locking.

Under a perfect phase matching, the received signal is given by

$$i_s(t) = 2\Re\sqrt{p_s P_{LO}}\,\cos\left\{\pi\alpha_k s(t)\right\} \tag{3.28}$$

where a_k takes the value ± 1, and $s(t)$ is the modulating waveform. This is a baseband signal, and thus the error rate is the same as that of the baseband system.

The shot-noise power induced by the LO and the signal power can be expressed as

$$i_{NS}^2 = 2q\Re(p_s + P_{LO})\int_0^\infty |H(j\omega)|\,d\omega \tag{3.29}$$

where $|H(j\omega)|$ is the transfer function of the receiver whose expression, if under a matched filtering, can be

$$|H(j\omega)|^2 = \left[\frac{\sin(\omega T/2)}{\omega T/2}\right]^2 \tag{3.30}$$

where T is the bit period. Then the noise power becomes

$$i_{NS}^2 = q\Re(p_s + P_{LO})\frac{1}{T} \simeq \frac{q\Re P_{LO}}{T} \quad \text{when} \quad p_s \ll P_{LO} \tag{3.31}$$

Thus, the SNR is

$$\text{SNR} = \frac{2\Re p_s P_{LO}}{q\Re P_{LO}/T} = \frac{2p_s T}{q} \tag{3.32}$$

and the BER is

$$P_E = \frac{1}{2}erfc(\sqrt{\text{SNR}}) \rightarrow \text{BER} = erfc(\sqrt{\text{SNR}}) \tag{3.33}$$

3.3.2.2 Quantum Limit Detection

For homodyne detection, a super quantum limit can be achieved. In this case, the LO is used in a very special way that matches the incoming signal field in polarization, amplitude, and frequency and is assumed to be phase locked to the signal. Assuming that the phase signal is perfectly modulated such that it acts in-phase or counter-phase with the LO, the homodyne detection would give a normalized signal current of

$$i_{sC} = \frac{1}{2T}\left[\mp\sqrt{2n_p} + \sqrt{2n_{LO}}\right]^2 \quad \text{for} \quad 0 \le t \le T \tag{3.34}$$

Assuming further that $n_p = n_{LO}$, the number of photon for the LO for generation of detected signals, then the current can be replaced with $4n_p$ for the detection of a "1" and nothing for a "0" symbol.

3.3.2.3 Linewidth Influences

3.3.2.3.1 Heterodyne Phase Detection

When the linewidth of the light sources is significant, the IF deviates due to a phase fluctuation, and the PDF is related to this linewidth conditioned on the deviation $\delta\omega$ of the IF. For a signal power of p_s, the total probability of error is given as

$$P_E = \int_{-\infty}^{\infty} P_C(p_s, \partial\omega)p_{IF}(\partial\omega)\partial\omega \tag{3.35}$$

The PDF of the IF under a frequency deviation can be written as [5]

$$p_{IF}(\partial\omega) = \frac{1}{\sqrt{\Delta\upsilon BT}}e^{-(\partial\omega^2/4\pi\Delta\upsilon B)} \tag{3.36}$$

where $\Delta\upsilon$ is full linewidth at the full width half maximum (FWHM) of the power spectral density and T is the bit period.

3.3.2.3.2 Differential Phase Detection with LO

3.3.2.3.2.1 DPSK Systems
The DPSK detection requires an MZDI and a balanced receiver either in the optical domain or in the electrical domain. If in the electrical domain, then the beating signals in the PD between the incoming signals and the LO would give the beating electronic

current, which is then split. One branch is delay by one-bit period and then summed up. The heterodyne signal current can be expressed as [6]

$$i_s(t) = 2\Re\sqrt{P_{LO}p_s}\,\cos(\omega_{IF}t + \phi_s(t)) + n_x(t)\cos\omega_{IF}t - n_y(t)\sin\omega_{IF}t \tag{3.37}$$

The phase $\phi_s(t)$ is expressed by

$$\phi_s(t) = \varphi_s(t) + \{\varphi_N(t) - \varphi_N(t+T)\} - \{\varphi_{pS}(t) - \varphi_{pS}(t+T)\} - \{\varphi_{pL}(t) - \varphi_{pL}(t+T)\} \tag{3.38}$$

The first term is the phase of the data and takes the value 0 or π. The second term represents the phase noise due to shot noise of the generated current and the third and fourth terms are the quantum shot noise due to the LO and the signals. The probability of error is given by

$$P_E = \int\limits_{-\pi/2}^{\pi/2}\ \int\limits_{-\infty}^{\infty} p_n(\phi_1 - \phi_2)\,p_q(\phi_1)\,\partial\phi_1\partial\phi \tag{3.39}$$

where $p_n(.)$ is the PDF of the phase noise due to the shot noise and $p_q(.)$ is for the quantum phase noise generated from the transmitter and the LO [7].

The probability of error can be written as

$$p_N(\phi_1 - \phi_2) = \frac{1}{2\pi} + \frac{\rho e^{-\rho}}{\pi}\sum_{m=1}^{\infty} a_m \cos\left(m(\phi_1 - \phi_2)\right)$$

$$a_m \sim \left\{\frac{2^{m-1}\Gamma\left[\dfrac{m+1}{2}\right]\Gamma\left[\dfrac{m}{2}+1\right]}{\Gamma[m+1]}\left[I_{m-1/2}\frac{\rho}{2} + I_{(m+1)/2}\frac{\rho}{2}\right]\right\}^2 \tag{3.40}$$

where $\Gamma(.)$ is the gamma function and is the modified Bessel function of the first kind. The PDF of the quantum phase noise can be given as [8]

$$p_q(\phi_1) = \frac{1}{\sqrt{2\pi D\tau}}e^{\phi_1^2/2D\tau} \tag{3.41}$$

where D is the phase diffusion constant, and the standard deviation from the central frequency

$$\Delta\upsilon = \Delta\upsilon_R + \Delta\upsilon_L = \frac{D}{2\pi} \tag{3.42}$$

is the sum of the transmitter and the LO FWHM linewidth. Substituting Equations 3.40 and 3.41 into Equation 3.39, we obtain

$$P_E = \frac{1}{2} + \frac{\rho e^{-\rho}}{2}\sum_{n=0}^{\infty} \frac{(-1)^n}{2n+1}e^{-(2n+1)^2\pi\Delta\upsilon T}\left\{I_{n-1/2}\frac{\rho}{2} + I_{(n+1)/2}\frac{\rho}{2}\right\}^2 \tag{3.43}$$

This equation gives the probability of error as a function of the received power. The probability of error is plotted against the receiver sensitivity, and the product of the linewidth and the bit rate (or the relative bandwidth of the laser linewidth and the bit rate) is shown in Figure 3.11 for DPSK modulation format at 140 Mbps bit rate and the variation of the laser linewidth from 0 to 2 MHz.

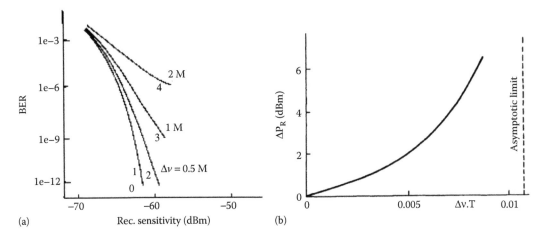

(a)

(b)

FIGURE 3.11 (a) Probability of error versus receiver sensitivity with linewidth as a parameter in MHz. (b) Degradation of optical receiver sensitivity at BER = 10^{-9} for DPSK systems as a function of the linewidth and bit period—bit rate = 140 Mb/s. (Extracted from Nicholson, G., *Elect. Lett.*, 20/24, 1005–1007, 1984. With permission.)

3.3.2.3.3 *Differential Phase Detection Under Self-Coherence*

Recently, the laser linewidth requirement for DQPSK modulation and differential detection for DQPSK has also been studied. No LO is used, which means self-coherent detection. It has been shown that the linewidth of up to 3 MHz of the transmitter laser would not significantly influence the probability of error as shown in Figure 3.12 [8]. Figure 3.13 shows the maximum linewidth of a laser source in a 10 GSymbols/s system. The loose bound is to neglect linewidth if the impact is to double the BER with the tighter bound being to neglect linewidth if the impact is a 0.1-dB SNR penalty.

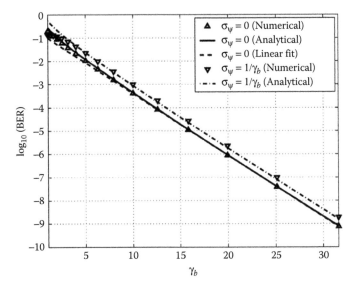

FIGURE 3.12 Analytical approximation (solid line) and numerical evaluation (triangles) of the BER for the cases of zero linewidth and that required to double the BER. The dashed line is the linear fit for zero linewidth. Bit rate 10 Gb/s per channel. (Extracted from S. Savory and T. Hadjifotiou, *IEEE Photonic Technol. Lett.*, 16, 930–932, March 2004. With permission.)

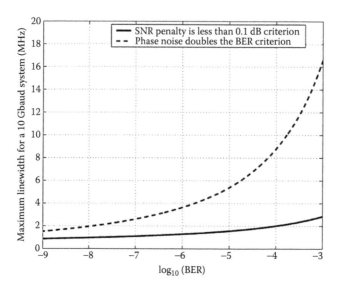

FIGURE 3.13 Criteria for neglecting linewidth in a 10 GSymbols/s system. The loose bound is to neglect linewidth if the impact is to double the BER with the tighter bound being to neglect linewidth if the impact is a 0.1 dB SNR penalty. Bit rate 10 GSymbols/s. (Extracted from Y. Yamamoto and T. Kimura, *IEEE J. Quantum Electron.*, QE-17, 919–934, 1981. With permission.)

3.3.2.3.4 Differential Phase Coherent Detection of Continuous Phase FSK Modulation Format
The probability of error of CPFSK can be derived by taking into consideration the delay line of the differential detector, the frequency deviation, and phase noise [10]. Similar to Figure 3.8, the differential detector configuration is shown in Figure 3.14a, and the conversion of frequency to voltage relationship in Figure 3.14b. If heterodyne detection is employed, then a BPF is used to bring the signals back to the electrical domain.

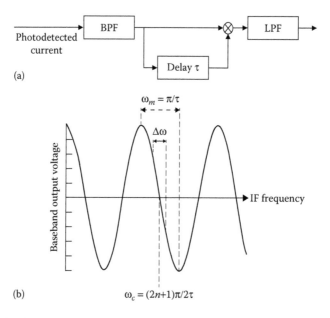

FIGURE 3.14 (a) Configuration of a CPFSK differential detection and (b) frequency to voltage conversion relationship of FSK differential detection. (Extracted from Iwashita, K. and Masumoto, T., *IEEE J. Lightwave Technol.*, LT-5/4, 452–462, 1987, Figure 1. With permission.)

The detected signal phase at the shot-noise limit at the output of the LPF can be expressed as

$$\phi(t) = \omega_c t + a_n \frac{\Delta\omega}{2}\tau + \varphi(t) + \varphi_n(t)$$

(3.44)

$$\text{with} \quad \omega_c = 2\pi f_c = (2n+1)\frac{\pi}{2\tau}$$

where τ is the differential detection delay time, $\Delta\omega$ is the deviation of the angular frequency of the carrier for the "1" or "0" symbol, $\varphi(t)$ is the phase noise due to the shot noise, $n(t)$ is the phase noise due to the transmitter and the LO quantum shot noise and takes the values of ± 1, the binary data symbol.

Thus, by integrating the detected phase from $-\frac{\Delta\omega}{2}\tau \longrightarrow \pi - \frac{\Delta\omega}{2}\tau$, we obtain the probability of error as

$$P_E = \int\limits_{-\frac{\Delta\omega}{2}\tau}^{\pi-\frac{\Delta\omega}{2}\tau/2}\int\limits_{-\infty}^{\infty} p_n(\phi_1-\phi_2)p_q(\phi_1)\partial\phi_1\partial\phi_1$$

(3.45)

Similar to the case of DPSK system, substituting Equations 3.40 and 3.41 into Equation 3.45, we obtain

$$P_E = \frac{1}{2}\frac{\rho e^{-\rho}}{2}\sum_{n=0}^{\infty}\frac{(-1)^n}{2n=1}e^{-(2n+1)^2\pi\Delta\upsilon\tau}\left\{I_{n-1/2}\frac{\rho}{2}+I_{(n+1)/2}\frac{\rho}{2}\right\}^2 e^{-(2n+1)^2\pi\Delta\upsilon\tau}\cos\{(2n+1)\alpha\}$$

(3.46)

$$\alpha = \frac{\pi(1-\beta)}{2} \quad \text{and} \quad \beta = \frac{\Delta\omega}{\omega_m} = 2\pi\tau/T_0$$

where ω_m is the deviation of the angular frequency with m the modulation index, and T_0 is the pulse period or bit period. The modulation index parameter b is defined as the ratio of the actual frequency deviation to the maximum frequency deviation. Figure 3.15 shows the dependence of degradation of the power penalty to achieve the same BER as a function of the linewidth factor $\Delta\upsilon\tau$ and the modulation index β.

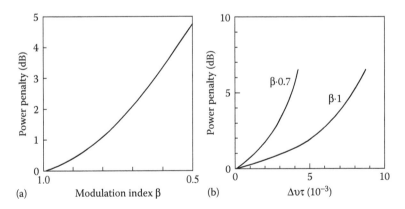

FIGURE 3.15 (a) Dependence of receiver power penalty at a BER of 10^{-9} on modulation index β (ratio between frequency deviation and maximum frequency spacing between f1 and f2). (b) Receiver power penalty at a BER of 10^{-9} as a function of the product of the beat bandwidth and the bit delay time—effects excluding LD phase noise. (Extracted from Iwashita, K. and Masumoto, T., *IEEE J. Lightwave Technol.*, LT-5/4, 452–462, 1987, Figures 3 and 4. With permission.)

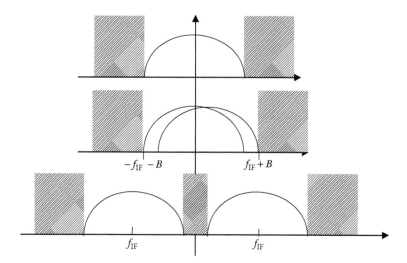

FIGURE 3.16   Spectrum of CoD (a) homodyne, (b) intradyne, and (c) heterodyne.

3.3.3   OPTICAL INTRADYNE DETECTION

Optical phase diversity receivers combine the advantages of the homodyne with minimum signal processing bandwidth and heterodyne reception with no optical phase locking required. The term diversity is well known in radio transmission links that describes the transmission over more than one path. In optical receivers, the optical path is considered as due to different polarization and phase paths. In intradyne detection, the frequency difference, the IF, or the LOFO ((LO) frequency offset) between the LO and the central carrier is nonzero, and lies within the signal bandwidth of the baseband signal as illustrated in Figure 3.16 [11]. Naturally, the control and locking of the carrier and the LO cannot be exact, sometimes due to jittering of the source. Most of the time, the laser frequency is locked stably by oscillating the reflection mirror, and hence the central frequency is varied by a few hundreds of KHz. Thus, intradyne CoD is more realistic. Furthermore, the DSP in modern coherent reception system would be able to extract this difference without much difficulty in the digital domain [12]. Obviously, the heterodyne detection would require a large frequency range of operation of electronic devices, whereas homodyne and intradyne reception require simpler electronics. Either differential or nondifferential format can be used in DSP-based coherent reception. For differential-based reception, the differential decoding would gain an advantage when there are slips in the cycles of bits due to walk-off of the pulse sequences over very long transmission noncompensating fiber lines.

The diversity in phase and polarization can be achieved by using a $\pi/2$ hybrid coupler that splits the polarization of the LO and the received channels and mixing with a $\pi/2$ optical phase shift, and then the mixed signals are detected by a balanced photodetectors. This diversity detection is described in the next few sections (see also Figure 3.21).

3.4   SELF-COHERENT DETECTION AND ELECTRONIC DSP

Coherent techniques described above would offer significant improvement but face a setback due to the availability of stable LO and an OPLL for locking the frequency of the LO and that of the signal carrier.

DSP have been widely used in wireless communications and play key roles in the implementation of DSP-based coherent optical communication systems. DSP techniques have been applied to coherent optical communication systems to overcome the difficulties of OPLL, and also to improve the performance of the transmission systems in the presence of fiber-degrading effects including CD, PMD, and fiber nonlinearities.

Coherent optical receivers have the following advantages: (1) the shot-noise-limited receiver sensitivity can be achieved with a sufficient LO power; (2) closely spaced WDM channels can be separated with electrical filters having sharp roll-off characteristics; and (3) the ability of phase detection can improve the receiver sensitivity compared with the IMDD system [13]. In addition, any kind of multilevel phase modulation formats can be introduced by using the coherent receiver. While the spectral efficiency of binary modulation formats is limited to 1 bit/s/Hz/polarization (which is called the Nyquist limit), multilevel modulation formats with N bits of information per symbol can achieve up to the spectral efficiency of N bits/s/Hz/polarization. Recent research has focused on M-ary PSK and even QAM with CoD, which can increase the spectral efficiency by a factor of $\log_2 M$ [14–16]. Moreover, for the same bit rate, because the symbol rate is reduced, the system can have a higher tolerance to CD and PMD.

However, one of the major challenges in CoD is to overcome the carrier phase noise when using an LO to beat with the received signals to retrieve the modulated phase information. Phase noise can result from lasers, which will cause a power penalty to the receiver sensitivity. A self-coherent multisymbol detection of optical differential M-ary PSK is introduced to improve the system performance; however, higher analog-to-digital conversion resolution and more DSP power are required as compared to a digital coherent receiver [17]. Further, differential encoding is also necessary in this scheme. As for the coherent receiver, initially, an OPLL is an option to track the carrier phase with respect to the LO carrier in homodyne detection. However, an OPLL operating at optical wavelengths in combination with distributed feedback lasers may be quite difficult to be implemented because the product of laser linewidth and loop delay is too large [18]. Another option is to use electrical PLL to track the carrier phase after downconverting the optical signal to an IF electrical signal in a heterodyne detection receiver as mentioned above. Compared to heterodyne detection, homodyne detection offers better sensitivity and requires a smaller receiver bandwidth [19]. On the other hand, coherent receivers employing high-speed analog-to-digital converters (ADCs) and high-speed baseband DSP units are becoming increasingly attractive rather than using an OPLL for demodulation. Conventional block Mth power PE scheme is proposed in [13,18] to raise the received M-ary PSK signals to the Mth power to estimate the phase reference in conjunction with a coherent optical receiver. However, this scheme requires nonlinear operations, such as taking the Mth power, and resolving the $\pm 2\pi/M$ phase ambiguity, which incurs a large latency to the system. Such nonlinear operations would limit further potential for real-time processing of the scheme. In addition, nonlinear phase noises always exist in long-haul systems due to the Gordon–Mollenauer effect [20], which severely affect the performance of a phase-modulated optical system [21]. The results in [22] show that such Mth power PE techniques may not effectively deal with nonlinear phase noise.

The maximum-likelihood (ML) carrier phase estimator derived in [23] can be used to approximate the ideal synchronous CoD in optical PSK systems. The ML phase estimator requires only linear computations, and thus it is more feasible for online processing of real systems. Intuitively one can show that the ML estimation receiver outperforms the Mth power block phase estimator and conventional differential detection, especially when the nonlinear phase noise is dominant, thus significantly improving the receiver sensitivity and tolerance to the nonlinear phase noise. The algorithm of ML phase estimator is expected to improve the performance of coherent optical communication systems using different M-ary PSK and QAM formats. The improvement by DSP at the receiver end can be significant for the transmission systems in the presence of fiber-degrading effects, including CD, PMD, and nonlinearities for both single channel and DWDM systems.

3.5 ELECTRONIC AMPLIFIERS: RESPONSES AND NOISES

3.5.1 INTRODUCTION

The electronic amplifier as a preamplification stage of an optical receiver plays a major role in the detection of optical signals so that optimum SNR and therefore the OSNR can be derived based on the photodetector responsivity. Under CoD, the amplifier noises must be much less than that of the quantum shot noises contributed by the high power level of the LO, which is normally about 10 dB above that of the signal average power.

Thus, this section gives an introduction of electronic amplifiers for wide band signals applicable to ultrahigh-speed, high-gain, and low-noise transimpedance amplifiers (TIAs). We concentrate on differential input TIAs, but address the detailed design of a single input single output with noise suppression technique in Section 3.7 with the design strategy for achieving stability in the feedback amplifier as well as low noise and wide bandwidth. We define the electronic noise of the preamplifier stage as the total equivalent input noise spectral density, that is, all the noise sources (current and voltage sources) of all elements of the amplifier are referred to the input port of the amplifier and thus an equivalent current source is found, from which the current density is derived. Once this current density is found, the total equivalent at the input can be found when the overall bandwidth of the receiver is determined. When this current is known, and with the average signal power, we can obtain without difficulty the SNR at the input stage of the optical receiver, and then the OSNR. On the other hand, if the OSNR required at the receiver is determined for any specific modulation format, then with the assumed optical power of the signal available at the front of the optical receiver and the responsivity of the photodetector we can determine the maximum electronic noise spectral density allowable by the preamplification stage and hence the design of the amplifier electronic circuit.

The principal function of an optoelectronic receiver is to convert the received optical signals into electronic equivalent signals, followed by amplification and sampling and processing to recover properties of the original shapes and sequence. So, at first, the optical domain signals must be converted to electronic current in the photodetection device, the photodetector of either p-i-n or APD, in which the optical power is absorbed in the active region and both electrons and holes generated are attracted to the positive- and negative-biased electrodes, respectively. Thus, the generated current is proportional to the power of the optical signals, hence the name "square law" detection. The p-i-n detector is structured with a $p+$ and $n+$:doped regions sandwiched by the intrinsic layer in which the absorption of optical signal occurs. A high electric field is established in this region by reverse biasing the diode, and thus electrons and holes are attracted to either sides of the diode, resulting in generation of current. Similarly, an APD works with the reverse-biasing level close to the reverse breakdown level of the pn junction (no intrinsic layer) so that electronic carriers can be multiplied in the avalanche flow when the optical signals are absorbed.

This photogenerated current is then fed into an electronic amplifier whose transimpedance must be sufficiently high and generates low noise so that a sufficient voltage signal can be obtained and then further amplified by a main amplifier, a voltage gain type. For high-speed and wide band signals, transimpedance amplification type is preferred as it offers wide band, much wider than high impedance type, though the noise level might be higher. With TIAs, there are two types, the single input single output port and two differential inputs and single output. The output ports can be differential with a complementary port. The differential input TIA offers much higher transimpedance gain (Z_T) and wider bandwidth as well. This is contributed to the use of a long-tail pair at the input and hence reasonable high input impedance that would ease the feedback stability [24–26].

In Section 3.3, a case study of coherent optical receiver is described from the design to implementation, including the feedback control and noise reduction. Although the corner frequency is only a few hundreds of MHz, with limited transition frequency of the transistors, this bandwidth is remarkable. The design is scalable to ultra-wide-band reception subsystems.

3.5.2 Wide Band TIAs

Two types of TIAs are described and distinguished by the term single input TIA and differential input TIA. They are distinguished by whether the amplifiers provide at the input a single port or a differential two-port. The later type is normally designed using a differential transistor pair termed as "a long-tail pair" instead of a single transistor stage for the former type TIA.

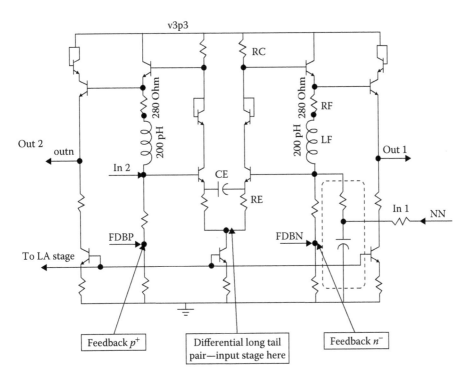

FIGURE 3.17 A typical structure of a differential TIA [27] with differential feedback paths. (Adapted from H. Tran, et al., *IEEE J. Solid St. Circ.*, 39(10), 1680–1689, 2004. With permission.)

3.5.2.1 Single Input, Single Output

We prefer to treat this section as a design example and describe the experimental demonstration of a wide band and low-noise amplifier. In the next section, the differential input TIA is treated with large transimpedance and reasonably low noise.

3.5.2.2 Differential Inputs, Single/Differential Output

An example circuit of the differential input TIA is shown in Figure 3.17, in which a long tail pair or differential pair is employed at the input stage. Two matched transistors are used to ensure the minimum common mode rejection and maximum different mode operation. This pair has a very high input impedance and thus the feedback from the output stage can be stable. Thus, the feedback resistance can be increased up to the limit of the stability locus of the network pole. This thus offers the high transimpedance Z_T and wide bandwidth. A typical Z_T of 3000–6000 Ω can be achieved with 30 GHz 3 dB bandwidth (see Figure 3.19), as shown in Figures 3.18 and 3.19. Also the chip image of the TIA can be seen in Figure 3.18a. Such a TIA can be implemented in either InP or SiGe material. The advantage of SiGe is that the circuit can be integrated with a high-speed Ge-APD detector and ADC and DSP. On the other hand, if implemented in InP then high-speed p-i-n or APD can be integrated and then radio frequency (RF) interconnected with ADC and DSP. The differential group delay may be serious and must be compensated in the digital processing domain.

3.5.3 Amplifier Noise Referred to Input

There are several noise sources in any electronic systems, which include thermal noises, shot noises, and quantum shot noises, especially in optoelectronic detection. Thermal noises result when the operating temperature is well above the absolute temperature at which no random movement of electrons and the resistance of electronic element occur. This type of noise depends on the ion temperature.

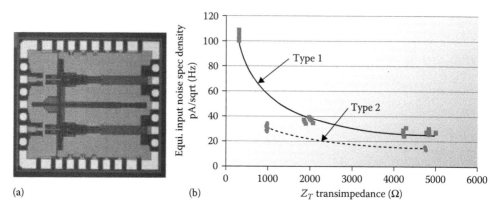

FIGURE 3.18 Differential amplifiers: (a) chip level image and (b) referred input noise equivalent spectral noise density. Inphi TIA 3205 (type 1) and 2850 (type 2). (Courtesy of Inphi Inc., Technical information on 3205 and 2850 TIA device, 2012.)

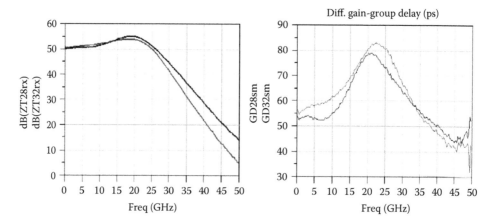

FIGURE 3.19 Differential amplifier: frequency response and differential group delay.

Shot noises are due to the current flowing and random scattering of electrons, and thus this type of noise depends on the strength of the flowing currents such as biasing current in electronic devices. Quantum shot noises are generated due to the current emitted from optoelectronic detection processes, which are dependent on the strength of the intensity of the optical signals or sources imposed on the detectors. Thus, this type of noise depends on signals. In the case of CoD, the mixing of the LO laser and signals normally occurs with the strength of the LO being much larger than that of signal average power. Thus, the quantum shot noises are dominated by that from the LO.

In practice, an equivalent electronic noise source is the total noise as referred to the input of electronic amplifiers that can be measured by measuring the total spectral density of the noise distribution over the whole bandwidth of the amplification devices. Thus, the total noise spectral density can be evaluated and referred to the input port. For example, if the amplifier is a transimpedance type, then the transimpedance of the device is measured first, and then the measure voltage spectral density at the output port can be referred to the input. In this case, it is the total equivalent noise spectral density. The common term employed and specified for TIAs is the total equivalent spectral noise density over the midband region of the amplifying device. The midband region of any amplifier is defined as the flat gain region from DC to the corner 3 dB point of the frequency response of the electronic device.

Figure 3.20 illustrates the meaning of the total equivalent noise sources as referred to the input port of a two-port electronic amplifying device. A noisy amplifier with an input excitation current

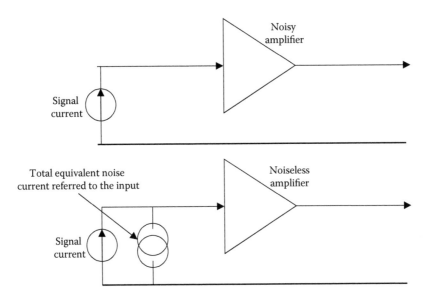

FIGURE 3.20   Equivalent noise spectral density current sources.

source, typically a signal current generated from the PD after the optical to electrical conversion, can be represented with a noiseless amplifier and the current source in parallel with a noise sources whose strength is equal to the total equivalent noise current referred to the input. Thus, the total equivalent current can be obtained by taking the product between this total equivalent current noise spectral density and the 3 dB bandwidth of the amplifying device. Thus, the SNR at the iutput of the electronic amplifier is given by

$$SNR = \frac{square\_of\_current\_generated}{square\_of\_current\_generated + total\_equivalent\_noise\_current\_power} \quad (3.47)$$

From this SNR referred at the input of the electronic front end, one can estimate the eye opening of the voltage signals at the output of the amplifying stage which is normally required by the ADC for sampling and conversion to digital signals for processing. Thus, one can then estimate the require OSNR at the input of the photodetector and hence the launched power required at the transmitter over several span links with certain attenuation factors.

Detailed analyses of amplifier noises and their equivalent noise sources as referred to input ports are given in Annex 2. It is noted that noises have no direction of flow as they always add and do not substract, and thus the noises are measured as noise power and not as a current. Thus, electrical spectrum analyzers are commonly used to measure the total noise spectral density, or the distribution of noise voltages over the spectral range under consideration, which is thus defined as the noise power spectral density distribution.

3.6   DSP SYSTEMS AND COHERENT OPTICAL RECEPTION

3.6.1   DSP-Assisted Coherent Detection

Over the years since the introduction of optical coherent communications in the mid-1980s, the invention of optical amplifiers has left coherent reception behind until recently, when long-haul transmission suffered from nonlinearity of dispersion compensating fibers and standard single mode fiber transmission line due to its small effective area. Furthermore, the advancement of DSP in wireless communication has also contributed to the application of DSP in modern coherent communication systems. Thus, the name "DSP-assisted coherent detection," that is, when a real-time

DSP is incorporated after the optoelectronic conversion of the total field of the LO and that of the signals, the analog received signals are sampled by a high-speed ADC and then the digitalized signals are processed in a DSP. Currently, real-time DSP are intensively researched for practical implementation. The main difference between real-time and offline processing is that the real-time processing algorithm must be effective due to limited time available for processing.

When polarization division multiplexed (PDM) QAM channels are transmitted and received, polarization and phase diversity receivers are employed. The schematics of such receiver are shown in Figure 3.21a. Further, the structures of such reception systems incorporating DSP with the diversity hybrid coupler in optical domain are shown in Figure 3.21b–d. The polarization diversity section with the polarization beam splitters at the signal and LO inputs facilitate the demultiplexing of polarized modes in the optical waveguides. The phase diversity using a 90° optical phase shifter allows the separation of the inphase (I-) and quadrature (Q-) phase components of QAM channels. Using a 2 × 2 coupler also enables the balanced reception using photo-detector pair connected B2B and hence a 3 dB gain in the sensitivity. Section 2.7 of Chapter 2 has described the modulation scheme QAM using I-Q modulators for single polarization or dual polarization multiplexed channels.

3.6.1.1 DSP-Based Reception Systems

The schematic of synchronous coherent receiver based on DSP is shown in Figure 3.22. Once the polarization and the I- and Q-optical components are separated by the hybrid coupler, the positive and negative parts of the I- and Q- are coupled into a balanced optoelectronic receiver as shown in Figure 3.21b. Two PDs are connected B2B so that push–pull operation can be achieved, hence a 3 dB betterment as compared to a single PD detection. The current generated from the B2B connected PDs is fed into a TIA so that a voltage signal can be derived at the output. Further, a voltage-gain amplifier is used to boost these signals to the right level of the ADC so that sampling can be conducted and the analog signals converted to digital domain. These digitalized signals are then fetched into DSP and processing in the "soft domain" can be conducted. Thus, a number of processing algorithms can be employed in this stage to compensate for linear and nonlinear distortion effects due to optical signal propagation through the optical guided medium, and to recover the carrier phase and the clock rate for resampling of the data sequence and so on. Chapter 6 will describe in detail the fundamental aspects of these processing algorithms. Figure 3.22 shows a schematic of possible processing phases in the DSP incorporated in the DSP-based coherent receiver. Besides the soft processing of the optical phase locking as described in Chapter 5, it is necessary to lock the frequencies of the LO and that of the signal carrier to a certain limit within which the algorithms for clock recovery can function, for example within ±2 GHz.

3.6.2   COHERENT RECEPTION ANALYSIS

3.6.2.1   Sensitivity

At ultrahigh bit rate, the laser must be externally modulated; thus, the phase of the lightwave conserves along the fiber transmission line. The detection can be direct, self-coherent, or homodyne and heterodyne. The sensitivity of coherent receiver is also important for the transmission system, especially the PSK scheme under both homodyne and heterodyne transmission techniques. This section gives the analysis of receiver for synchronous coherent optical fiber transmission systems. Consider that the optical fields of the signals and LO are coupled via a fiber coupler with two output ports 1 and 2. The output fields are then launched into two photodetectors connected B2B and then the electronic current is amplified using a transimpedance type and further equalized to extend the bandwidth of the receiver. Our objective is to obtain the receiver penalty and its degradation due to imperfect polarization mixing and unbalancing effects in the balanced receiver. A case study of the design, implementation, and measurements of an optical balanced receiver electronic circuit and noise suppression techniques is given in Section 3.7 (Figure 3.23).

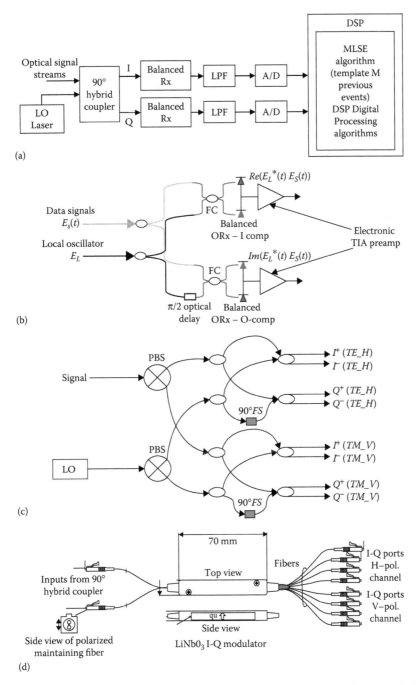

FIGURE 3.21 Scheme of a synchronous coherent receiver using DSP for PE for coherent optical communications. (a) Generic scheme, (b) detailed optical receiver using only one polarization phase diversity coupler, (c) hybrid 90° coupler for polarization and phase diversity, (d) typical view of a hybrid coupler with two input ports and eight output ports of structure in (c). TE_V, TE_H = transverse electric mode with vertical (V) or horizontal (H) polarized mode, TM = transverse magnetic mode with polarization orthogonal to that of the TE mode. FS = phase shifter; PBS = polarization beam splitter; and MLSE = maximum-likelihood phase estimation. (Adapted from S. Zhang et al., A comparison of phase estimation in coherent optical PSK system. In Photonics Global '08, Paper C3-4A-03, Singapore, December 2008; S. Zhang et al., Adaptive decision-aided maximum likelihood phase estimation in coherent optical DQPSK system. In OptoElectronics and Communications Conference (OECC) '08, Paper TuA-4, pp. 1–2, Sydney, Australia, July 2008.)

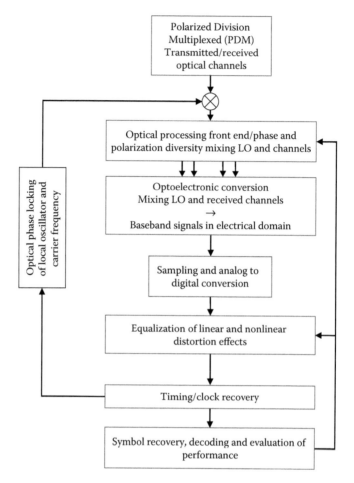

FIGURE 3.22 Flow of functionalities of DSP processing in a QAM-coherent optical receiver with possible feedback control.

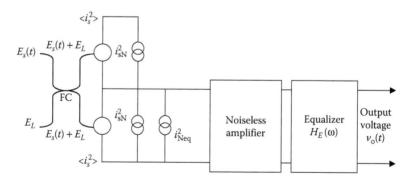

FIGURE 3.23 Equivalent current model at the input of the optical balanced receiver under CoD, average signal current, and equivalent noise current of the electronic preamplifier as seen from its input port and equalizer. FC = fiber coupler.

The following parameters are commonly used in analysis:

E_s	Amplitude of signal optical field at the receiver
E_L	Amplitude of local oscillator optical field
P_s, P_L	Optical power of signal and local oscillator at the input of the photodetector
$s(t)$	The modulated pulse
$\langle i_{NS}^2(t) \rangle$	Mean square noise current (power) produced by the total optical intensity on the photodetector
$\langle i_s^2(t) \rangle$	Mean square current produced by the photodetector by $s(t)$
$S_{NS}(t)$	Shot noise spectral density of $\langle i_s^2(t) \rangle$ and local oscillator power
$i_{Neq}^2(t)$	Equivalent noise current of the electronic preamplifier at its input
$Z_T(\omega)$	Transfer impedance of the electronic preamplifier
$H_E(\omega)$	Voltage transfer characteristic of the electronic equalizer followed the electronic preamplifier

The combined field of the signal and LO via a directional coupler can be written with their separate polarized field components as

$$E_{sX} = \sqrt{K_{sX}}\, E_S \cos(\omega_s t - \phi_{m(t)})$$

$$E_{sY} = \sqrt{K_{sY}}\, E_S \cos(\omega_s t - \phi_{m(t)} + \delta_s)$$

$$E_{LX} = \sqrt{K_{LX}}\, E_L \cos(\omega_L t) \tag{3.48}$$

$$E_{LY} = \sqrt{K_{LY}}\, E_L \cos(\omega_L t + \delta_L)$$

$$\phi_{m(t)} = \frac{\pi}{2} K_m s(t)$$

where $\phi_{m(t)}$ represents the phase modulation, K_m is the modulation depth, and K_{sX}, K_{sY}, K_{LX}, and K_{LY} are the intensity fraction coefficients in the X and Y directions of the signal and LO fields, respectively.

Thus, the output fields at ports 1 and 2 of the FC in the X-plane can be obtained using the transfer matrix as

$$\begin{bmatrix} E_{R1X} \\ E_{R2X} \end{bmatrix} = \begin{bmatrix} \sqrt{K_{sX}(1-\alpha)} \cos(\omega_s t - \phi_{m(t)}) & \sqrt{K_{LX}\alpha} \sin(\omega_L t) \\ \sqrt{K_{sX}\alpha} \sin(\omega_s t - \phi_{m(t)}) & \sqrt{K_{LX}(1-\alpha)} \cos(\omega_L t) \end{bmatrix} \begin{bmatrix} E_s \\ E_L \end{bmatrix} \tag{3.49}$$

$$\begin{bmatrix} E_{R1Y} \\ E_{R2Y} \end{bmatrix} = \begin{bmatrix} \sqrt{K_{sY}(1-\alpha)} \cos(\omega_s t - \phi_{m(t)}) & \sqrt{K_{LY}\alpha} \sin(\omega_L t + \delta_L) \\ \sqrt{K_{sY}\alpha} \sin(\omega_s t - \phi_{m(t)}) & \sqrt{K_{LY}(1-\alpha)} \cos(\omega_L t) + \delta_L \end{bmatrix} \begin{bmatrix} E_s \\ E_L \end{bmatrix} \tag{3.50}$$

with α defined as the intensity coupling ratio of the coupler. Thus, the field components at ports 1 and 2 can be derived by combining the X and Y components from Equations 3.49 and 3.50; thus, the total powers at ports 1 and 2 are given as

$$P_{R1} = P_s(1-\alpha) + P_L\alpha + 2\sqrt{P_s P_L \alpha(1-\alpha)K_p} \sin(\omega_{IF} t + \phi_{m(t)} + \phi_p - \phi_e)$$

$$P_{R2} = P_s\alpha + P_L(1-\alpha) + 2\sqrt{P_s P_L \alpha(1-\alpha)K_p} \sin(\omega_{IF} t + \phi_{m(t)} + \phi_p - \phi_e + \pi)$$

with $\quad K_p = K_{sX}K_{LX} + K_{sY}K_{LY} + 2\sqrt{K_{sX}K_{LX}K_{sY}K_{LY}} \cos(\delta_L - \delta_s)$ $\tag{3.51}$

$$\phi_p = \tan^{-1}\left[\frac{\sqrt{K_{sX}K_{LY}} \sin(\delta_L - \delta_s)}{\sqrt{K_{sX}K_{LX}} + \sqrt{K_{sY}K_{LY}} \cos(\delta_L - \delta_s)} \right]$$

where ω_{IF} is the intermediate angular frequency, which is equal to the difference between the frequencies of the LO and the carrier of the signals. ϕ_e is the phase offset, and $\phi_p - \phi_e$ is the demodulation reference phase error.

In Equation 3.51, the total field of the signal and the LO are added and then the product of the field vector and its conjugate is taken to obtain the power. Only the term with frequency that falls within the range of the sensitive of the photodetector would produce the electronic current. Thus, the term with the sum of the frequency of the wavelength of the signal and LO would not be detected and only the product of the two terms would be detected as given.

Now assuming a binary PSK (BPSK) modulation scheme, the pulse has a square shape with amplitude +1 or −1, the PD is a p-i-n type, and the PD bandwidth is wider than the signal 3 dB bandwidth followed by an equalized electronic preamplifier. The signal at the output of the electronic equalizer or the input signal to the decision circuit is given by

$$\hat{v}_D(t) = 2K_H K_p \sqrt{P_s P_L \alpha(1-\alpha)K_p} \int_{-\infty}^{\infty} H_E(f)df \int_{-\infty}^{\infty} (t)dt \sin\left(\frac{\pi}{2}K_m\right)\cos(\phi_p - \phi_e)$$

$$\rightarrow \hat{v}_D(t) = 2K_H K_p \sqrt{P_s P_L \alpha(1-\alpha)K_p} \sin\left(\frac{\pi}{2}K_m\right)\cos(\phi_p - \phi_e)$$

(3.52)

$$K_H = 1 \text{ for homodyne}; K_H = 1/\sqrt{2} \text{ for heterodyne}$$

For a perfectly balanced receiver, $K_B = 2$ and $\alpha = 0.5$; otherwise $K_B = 1$. The integrals of the first line in Equation 3.52 are given by

$$\int_{-\infty}^{\infty} H_E(f)df = \frac{1}{T_B} \qquad \because H_E(f) = \sin c(\pi T_B f)$$

$$\int_{-\infty}^{\infty} s(t)dt = 2T_B$$

(3.53)

$V_D(f)$ is the transfer function of the matched filter for equalization, and T_B is the bit period. The total noise voltage as a sum of the quantum shot noise generated by the signal and the LO and the total equivalent noise of the electronic preamplifier at the input of the preamplifier and at the output of the equalizer is given by

$$\left\langle v_N^2(t) \right\rangle = \frac{[K_B \alpha S'_{IS} + (2-K_B)S_{Ix} + S_{IE}] \int_{-\infty}^{\infty} |H_4(f)|^2 df}{K_{IS}^2}$$

or

(3.54)

$$\left\langle v_N^2(t) \right\rangle = \frac{[K_B \alpha S'_{IS} + (2-K_B)S_{Ix} + S_{IE}]}{K_{IS}^2 T_B}$$

For homodyne and heterodyne detection, we have

$$\left\langle v_N^2(t) \right\rangle = \Re q \frac{P_L}{\lambda T_B}[K_B \alpha S'_{IS} + (2-K_B)S_{Ix} + S_{IE}]$$

(3.55)

where the spectral densities S'_{IX}, S'_{IE} are given by

$$S'_{IX} = \frac{S_{IX}}{S'_{IS}}$$

$$(3.56)$$

$$S'_{IE} = \frac{S_{IE}}{S'_{IS}}$$

Thus, the receiver sensitivity for BPSK and equiprobable detection and Gaussian density distribution is given by

$$P_e = \frac{1}{2} erfc\left(\frac{\delta}{\sqrt{2}}\right)$$

$$(3.57)$$

with δ given by

$$P_e = \frac{1}{2} erfc\left(\frac{\delta}{\sqrt{2}}\right) \quad \text{with } \delta = \frac{\hat{v}_D}{2\sqrt{\langle v_N^2 \rangle}}$$

$$(3.58)$$

Thus, using Equations 3.52, 3.55, and 3.58 we obtain the receiver sensitivity in the linear power scale as

$$P_s = \langle P_s(t) \rangle = \frac{\Re q\delta^2}{4\lambda T_B K_H^2} \frac{[K_B \alpha S'_{IS} + (2 - K_B)S_{Ix} + S_{IE}]}{\eta K_p(1-\alpha)\alpha K_B^2 \sin^2\left(\frac{\pi}{2}K_m\right)\cos^2(\phi_p - \phi_e)}$$

$$(3.59)$$

3.6.2.2 Shot Noise-Limited Receiver Sensitivity

In the case when the power of the LO dominates the noise of the electronic preamplifier and the equalizer, the receiver sensitivity (in linear scale) is given as

$$P_s = \langle P_{sL} \rangle = \frac{\Re q\delta^2}{4\lambda T_B K_H^2}$$

$$(3.60)$$

This shot-noise-limited receiver sensitivity can be plotted as shown in Figure 3.24.

3.6.2.3 Receiver Sensitivity under Nonideal Conditions

Under a nonideal condition, the receiver sensitivity departs from the shot-noise-limited sensitivity and is characterized by the receiver sensitivity penalty PD_T as

$$PD_T = 10Log\frac{\langle P_s \rangle}{\langle P_{sL} \rangle} dB$$

$$(3.61)$$

$$PD_T = 10Log_{10}\left[\frac{K_B \alpha S'_{IS} + (2 - K_B)S_{Ix} + S_{IE}}{K_B \alpha}\right]$$

$$-10Log_{10}\left[K_B(1-\alpha)\right]$$

$$-10Log_{10}\left([\eta][K_p]\sin^2\left(\frac{\pi}{2}K_m\right)\cos^2(\phi_p - \phi_e)\right)$$

$$(3.62)$$

where η is the LO excess noise factor.

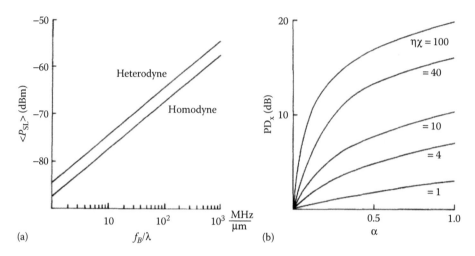

(a)   (b)

FIGURE 3.24 (a) Receiver sensitivity of coherent homodyne and heterodyne detection, signal power versus bandwidth over the wavelength. (b) Power penalty of the receiver sensitivity from the shot-noise-limited level as a function of the excess noise of the LO. (Extracted from Hodgkinson, I., *IEEE J. Lightwave Technol.*, 5, 573–587, 1987, Figures 1 and 2. With permission.)

The receiver sensitivity is plotted against the ratio f_B/λ for the case of homodyne and heterodyne detection, as shown in Figure 3.24a, and the power penalty of the receiver sensitivity against the excess noise factor of the LO, as shown in Figure 3.24b. Receiver power penalty can be deduced as a function of the total electronic equivalent noise spectral density, and as a function of the rotation of the polarization of the LO that can be found in [30]. Furthermore, in [31], the receiver power penalty and the normalized heterodyne center frequency can vary as a function of the modulation parameter and as a function of the optical power ratio at the same polarization angle.

3.6.3 DIGITAL PROCESSING SYSTEMS

A generic structure of the coherent reception and DSP system is shown in Figure 3.25, in which the DSP system is placed after the sampling and conversion from analog state to digital form. Obviously, the optical signal fields are beating with the LO laser whose frequency would be approximately identical to the signal channel carrier. The beating occurs in the square law photodetectors, that is, the summation of the two fields are squared and the product term is decomposed into the difference

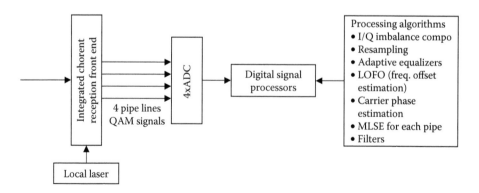

FIGURE 3.25 Coherent reception and the DSP system.

and summation term; thus, only the difference term is fallen back into the baseband region and amplified by the electronic preamplifier, which is a balanced differential transimpedance type.

If the signals are complex, then there are the real and imaginary components that form a pair. Another pair comes from the other polarization mode channel. The digitized signals of both the real and imaginary parts are processed in real time or offline. The processors contain the algorithms to combat a number of transmission impairments such as the imbalance between the inphase and the quadrature components created at the transmitter, the recovery of the clock rate and timing for resampling, the carrier PE for estimation of the signal phases, adaptive equalization for compensation of propagation dispersion effects using ML phase estimation, and so on. These algorithms are built into the hardware processors or memory and loaded to processing subsystems.

The sampling rate must normally be twice that of the signal bandwidth to ensure that the Nyquist criteria are satisfied. Although this rate is very high for 25 G to 32 GSy/s optical channels, Fujitsu ADC has reached this requirement with a sampling rate of 56 G to 64 GSa/s, as depicted in Figure 3.37.

The linewidth resolution of the processing for semiconductor device fabrication has been progressed tremendously over the year in an exponential trend as shown in Table 3.1. These progresses could be made due to the successes in the lithographic techniques using optical at short wavelength such as the UV, the electronic optical beam, and X-ray lithographic with appropriate photoresist such as SU-80, which would allow the line resolution to reach 5 nm in 2020. So, if we plot the trend in a log-linear scale as shown in Figure 3.26, a linear line is obtained, meaning that the resolution is reduced exponentially. When the gate width is reduced, the electronic speed would increase tremendously; at 5 mm, the speed of the electronic CMOS device in SiGe would reach several tens of GHz. Regarding the high-speed ADC and digital-to-analog converter (DAC), the clock speed is increased by paralleling, delaying of extracted outputs of the registers, and taking the summation of all the digitized digital lines to form a very high speed operation. For example, for Fujitsu 64 GSa/s DAC or ADC, the applied clock sinusoidal waveform is only 2 GHz. Figure 3.27 shows the progresses in the speed development of Fujitsu ADC and DAC.

3.6.3.1 Effective Number of Bits

3.6.3.1.1 Definition

Effective number of bits (ENOBs) is a measure of the quality of a digitized signal. The resolution of a DAC or ADC is commonly specified by the number of bits used to represent the analog value, in principle, giving 2^N signal levels for an N-bit signal. However, all real signals contain a certain amount of noise. If the converter is able to represent signal levels below the system noise floor, the lower bits of the digitized signal only represent system noise and do not contain useful information. ENOB specifies the number of bits in the digitized signal above the noise floor. Often, ENOB is also used as a quality measure for other blocks such as sample-and-hold amplifiers. This way

---

TABLE 3.1

Milestones of Progresses of Linewidth Resolution

Semiconductor Manufacturing Processes and Spatial Resolution (Gate Width)			
10 μm—1971	800 nm (0.80 μm)—1989: UV lithography	90 nm—2002: electron lithography	14 nm—approx. 2014: X-ray lithography
3 μm—1975	600 nm (0.60 μm)—1994	65 nm—2006	10 nm—approx. 2016: X-ray lithography
1.5 μm—1982	350 nm (0.35 μm)—1995	45 nm—2008	7 nm—approx. 2018: X-ray lithography
1 μm—1985	250 nm (0.25 μm)—1998	32 nm—2010	5 nm—approx. 2020: X-ray lithography
	180 nm (0.18 μm)—1999	22 nm—2012	
	130 nm (0.13 μm)—2000		

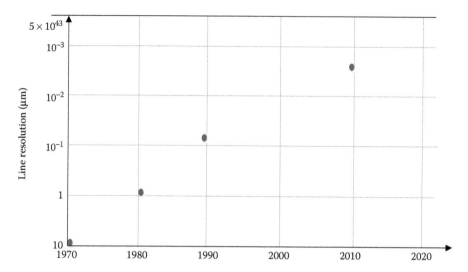

FIGURE 3.26   Semiconductor manufacturing with resolution of line resolution.

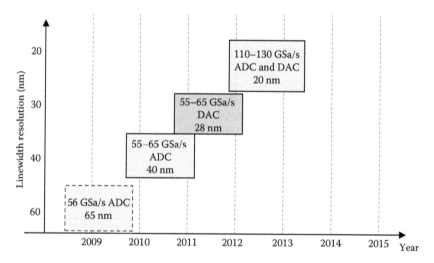

FIGURE 3.27   Evolution of ADC and DAC operating speed with corresponding linewidth resolution.

analog blocks can also be easily included to signal-chain calculations as the total ENOB of a chain of blocks is usually below the ENOB of the worst block.

Thus, we can represent the ENOB of a digitalized system by

$$\text{ENOB} = \frac{\text{SINAD} - 1.76}{6.02} \tag{3.63}$$

where all values are given in dB, and the signal-to-noise and distortion ratio (SINAD) is the ratio of the total signal including distortion and noise to the wanted signal; the 6.02 term in the divisor converts decibels (a $\log_{10}$ representation) to bits (a $\log_2$ representation); and the 1.76 term comes from quantization error in an ideal ADC[*].

---

[*] http://en.wikipedia.org/wiki/ENOB - cite_note-3 (access date: Sept. 2011).

This definition compares the SINAD of an ideal ADC or DAC with a word length of ENOB bits with the SINAD of the ADC or DAC being tested. Indeed the SINAD is a measure of the quality of a signal from a communications device, often defined as

$$\text{SINAD} = \frac{P_{\text{sig}} + P_{\text{noise}} + P_{\text{distorion}}}{P_{\text{noise}} + P_{\text{distorion}}} \tag{3.64}$$

where P is the average power of the signal, noise, and distortion components. SINAD is usually expressed in dB and is quoted alongside the receiver sensitivity, to give a quantitative evaluation of the receiver sensitivity. Note that with this definition, unlike SNR, a SINAD reading can never be less than 1 (i.e., it is always positive when quoted in dB).

When calculating the distortion, it is common to exclude the DC components. Because of the widespread use, SINAD has collected a few different definitions. SINAD is calculated as one of the following: (i) The ratio of (a) total received power, that is, the signal to (b) the noise-plus-distortion power. This is modeled by the equation above. (ii) The ratio of (a) the power of original modulating audio signal, that is, from a modulated radio frequency carrier to (b) the residual audio power, that is, noise-plus-distortion powers remaining after the original modulating audio signal is removed. With this definition, it is now possible for SINAD to be less than 1. This definition is used when SINAD is used in the calculation of ENOB for an ADC.

Example: Consider the following measurements of a 3-bit unipolar DAC with reference voltage $V_{\text{ref}} = 8$ V:

Digital input	000	001	010	011	100	101	110	111
Analog output (V)	−0.01	1.03	2.02	2.96	3.95	5.02	6.00	7.08

The offset error in this case is −0.01 V or −0.01 LSB as 1 V = 1 LSB (lower sideband) in this example. The gain error is $(7.08 + 0.01)/(7/1)/1 = 0.09$ LSB, where LSB stands for the least significant bits. Correcting the offset and gain error, we obtain the following list of measurements: (0, 1.03, 2.00, 2.93, 3.91, 4.96, 5.93, 7) LSB. This allows the INL and DNL to be calculated: INL = (0, 0.03, 0, −0.07, −0.09, −0.04, −0.07, 0) LSB, and DNL = (0.03, −0.03, −0.07, −0.02, 0.05, −0.03, 0.07, 0) LSB.

Differential nonlinearity (DNL): For an ideal ADC, the output is divided into $2N$ uniform steps, each with Δ width as shown in Figure 3.28. Any deviation from the ideal step width is called DNL and is measured in number of counts (lower sidebands). For an ideal ADC, the DNL is 0 LSB. In a practical ADC, DNL error comes from its architecture. For example, in an successive-approximation-register ADC, DNL error may be caused near the mid-range due to mismatching of its DAC.

Integral nonlinearity (INL) is a measure of how closely the ADC output matches its ideal response. INL can be defined as the deviation in LSB of the actual transfer function of the ADC from the ideal transfer curve. INL can be estimated using DNL at each step by calculating the cumulative sum of DNL errors up to that point. In reality, INL is measured by plotting the ADC transfer characteristics. INL is popularly measured using either (i) best-fit (best straight line) method or (ii) end point method.

Best-fit INL: The best-fit method of INL measurement considers offset and gain error. One can see in Figure 3.29 that the ideal transfer curve considered for calculating the best-fit INL does not go through the origin. The ideal transfer curve drawn here depicts the nearest first-order approximation to the actual transfer curve of the ADC.

The intercept and slope of this ideal curve can lend us the values of the offset and gain error of the ADC. Quite intuitively, the best-fit method yields better results for INL. For this reason, many times, this is the number present on ADC datasheets.

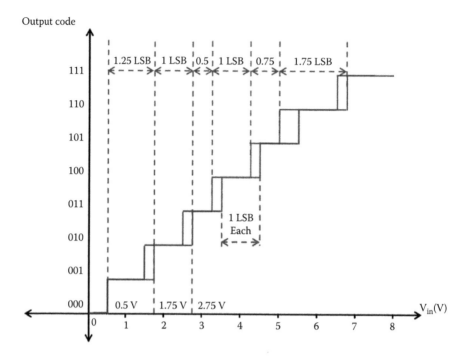

FIGURE 3.28 Representation of DNL in a transfer curve of an ADC.

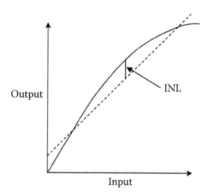

FIGURE 3.29 Best-fit INL.

The only real use of the best-fit INL number is to predict distortion in time-variant signal applications. This number would be equivalent to the maximum deviation for an AC application. However, it is always better to use the distortion numbers than INL numbers. To calculate the error budget, end-point INL numbers provide a better estimation. Also, this is the specification that is generally provided in datasheets. So, one has to use this instead of end-point INL.

End-point INL: The end-point method provides the worst-case INL. This measurement passes the straight line through the origin and maximum output code of the ADC (Figure 3.6). As this method provides the worst-case INL, it is more useful to use this number as compared to the one measured using best fit for DC applications. This INL number would be typically useful for error budget calculation. This parameter must be considered for applications involving precision measurements and control (Figure 3.30).

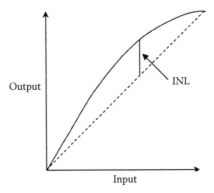

FIGURE 3.30 End-point INL.

The absolute and relative accuracy can now be calculated. In this case, the ENOB absolute accuracy is calculated using the largest absolute deviation D, in this case, 0.08 V:

$$D = \frac{V_{ref}}{2^{ENOB}} \rightarrow ENOB = 6.64 \text{ bits} \tag{3.65}$$

The ENOB relative accuracy is calculated using the largest relative (INL) deviation d, in this case, 0.09 V.

$$d = \frac{V_{ref}}{2^{ENOB}} \rightarrow ENOB_{rel} = 6.47 \text{ bits} \tag{3.66}$$

For this kind of ENOB calculation, note that the ENOB can be larger or smaller than the actual number of bits (ANOBs). When the ENOB is smaller than the ANOB, this means that some of the LSBs of the result are inaccurate. However, one can also argue that the ENOB can never be larger than the ANOB, because you always have to add the quantization error of an ideal converter, which is ±0.5 LSB. Different designers may use different definitions of ENOB!

3.6.3.1.2 High-Speed ADC and DAC Evaluation Incorporating Statistical Property

The ENOB of an ADC is considered as the number of bits that an analog signal can convert to its digital equivalent by the number of levels represented by the modulo-2 levels, which are reduced due to noises contributed by electronic components in such a convertor. Thus, only an effective number of equivalent bits can be accounted for. Hence, the term ENOB is proposed.

As shown in Figure 3.31b, a real ADC can be modeled as a cascade of two ideal ADCs and additive noise sources and an AGC amplifier [31]. The quantized levels are thus equivalent to a specific ENOB as far as the ADC is operating in the linear nonsaturated region. If the normalized signal amplitude/power surpasses unity, the saturated region, then the signals are clipped. The decision level of the quantization in an ADC normally varies following a normalized Gaussian PDF; thus, we can estimate the RMS noise introduced by the ADC as

$$RMS\_noise = \sqrt{\int_{-\frac{LSB}{2}}^{\frac{LSB}{2}} \int_{-\infty}^{\infty} x^2 \frac{\frac{1}{\sqrt{2\pi\sigma^2}} \exp\left(-\frac{(x-y)^2}{2\sigma^2}\right)}{LSB} dxdy} = \sqrt{LSB^2/12 + \sigma^2} \tag{3.67}$$

where:
σ is the variance
x is the variable related to the integration of decision voltage
similarly, y for integration inside one LSB

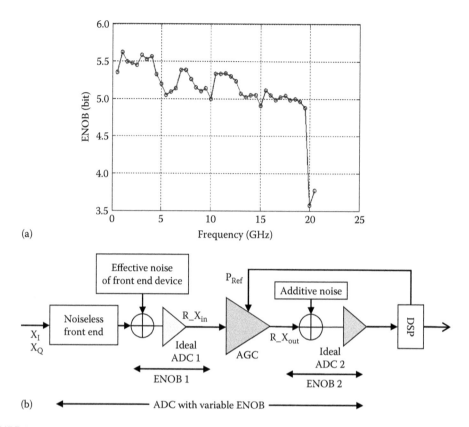

FIGURE 3.31 (a) Measured ENOB frequency response of a commercial real-time DSA of 20 GHz bandwidth and sampling rate of 50 GSa/s. (b) Deduced ADC model of variable ENOB based on the experimental frequency response of (a) and spectrum of broadband signals.

Given the known quantity $LSB^2/12$ by the introduction of the ideal quantization error, σ^2 can be determined via the Gaussian noise distribution. We can thus deduce the ENOB values corresponding to the levels of Gaussian noise as

$$\text{ENOB} = N - \log_2\left(\frac{\sqrt{LSB^2/12 + (A\sigma)^2}}{LSB/\sqrt{12}}\right) \tag{3.68}$$

where A is the RMS amplitude derived from the noise power. According to the ENOB model, the frequency response of ENOB of the digital sampling analyzer (DSA) is shown in Figure 3.31a, with the excitation of the DSA by sinusoidal waves of different frequencies. As observed, the ENOB varies with respect to the excitation frequency, in the range from 5 to 5.5. Having known the frequency response of the sampling device, what is the ENOB of the device when excited with broadband signals? This indicates the different resolution of the ADC of the receiver of the transmission operating under different noisy and dispersive conditions; thus, an equivalent model of ENOB for performance evaluation is essential. We note that the amplitudes of the optical fields arrived at the receiver vary depending on the conditions of the optical transmission line. The AGC has a nonlinear gain characteristic in which the input-sampled signal power level is normalized with respect to the saturated (clipping) level. The gain is significantly high in the linear region and saturated in the high level. The received signal R_X_{in} is scaled with a gain coefficient according to $R\_X_{out} = R\_X_{in}/\sqrt{P_{in\_av}/P_{Ref}}$, where the signal-averaged power P_{in_av} is estimated and the gain is scaled relative to the reference power level P_{Ref} of the AGC; then a linear scaling factor is used to obtain the output sampled value R_X_{out}. The gain of the AGC is also adjusted according to the signal energy, via the feedback control path from the DSP (see Figure 3.31b). Thus, new values of ENOB can be evaluated with noise distributed across the frequency spectrum of the signals, by an averaging process. This signal-dependent ENOB is now denoted as ENOBs.

3.6.3.1.3 Impact of ENOB on Transmission Performance

Figure 3.32a shows the BER variation with respect to the OSNR under B2B transmission using the simulated samples at the output of the 8-bit ADC with ENOBs and full ADC resolution as parameters. The difference is due to the noise distribution (Gaussian or uniform). Figure 3.32b depicts the variation of BER versus OSNR, with ENOBs as the variation parameter in case of offline data with ENOB of DSA shown in Figure 3.1a. Several more tests were conducted to ensure the effectiveness of our ENOB model. When the sampled signals presented to the ADC are of different amplitudes, controlled, and gain nonlinearly adjusted by the AGC, different degrees of clipping effect would be introduced. Thus, the clipping effect can be examined for the ADC of different quantization levels but with identical ENOBs, as shown in Figure 3.33a, for the B2B experiment. Figure 3.33b–e shows, with BER as a parameter, the contour plots of the variation of the adjusted reference power level of the AGC and ENOBs for the cases of 1500 km long-haul transmission of full CD compensation and non-CD compensation operating in the linear (0 dBm launch power in both links) and nonlinear regimes of the fibers with the launch power of 4 and 5 dBm, respectively. When the link is fully CD compensated, the nonlinear effects further contribute to the ineffectiveness of the ADC resolution and hence moderate AGC freedom in the performance is achieved. On the other hand, in the case of non-CD compensation link (Figure 3.33d and e), the dispersive pulse sampled amplitudes are lower with less noise allowing the resolution of the ADC to be higher via the nonlinear gain of the AGC; thus, effective PE and equalization can be achieved. We note that the offline data sets, employed prior to the processing using ENOBs to obtain the contours of Figure 3.33, produce the same BER contours of 2×10^{-3} for all cases. Hence, a fair comparison can be made when the ENOBs model is used. The opening distance of the BER contours indicates the dynamic range of the ENOBs model, especially the AGC. It is obvious from Figure 3.33a–e that the dynamic range of the model is higher for noncompensating than for full CD compensated transmission and even for the case of B2B. However, for nonlinear scenario for both cases, the requirement for ENOBs is higher for the

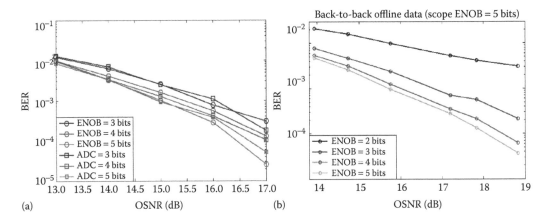

FIGURE 3.32 (a) B2B performance with different ENOBs values of the ADC model with simulated data (8-bit ADC) and (b) OSNR versus BER under different simulated ENOBs of offline data obtained from an experimental digital coherent receiver. (From Mao, B. N. et al., Investigation on the ENOB and clipping effect of real ADC and AGC in coherent QAM transmission system. In *Proceedings of ECOC*, Geneva, 2011.)

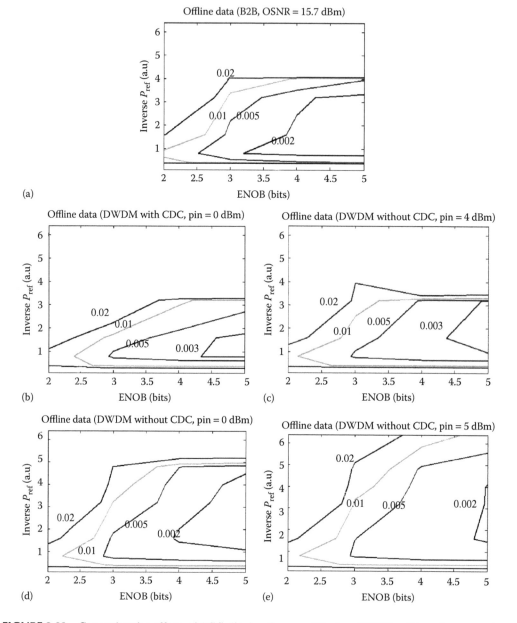

FIGURE 3.33 Comprehensive effects of AGC clipping (inverse of P_{Ref}) and ENOBs of the coherent receiver, experimental transmission under (a) B2B, (b) DWDM linear operation with full CD compensation (CDC) and (b) linear, (c) nonlinear regions with 4 dBm launch power; and (d) non-CD compensation and (d) linear, (e) nonlinear region with launch power of 5 dBm.

dispersive channel (Figure 3.33c and e). This may be due to the cross-phase modulation effects of adjacent channels, and hence more noise.

3.6.3.2   Digital Processors

The structures of the DAC and ADC are shown in Figures 3.34 and 3.35, respectively. Normally, there would be four DACs in an IC, in which each DAC section is clocked with a clock sequence which is derived from a lower frequency sinusoidal wave injected externally into the DAC. Four units are required for the in-phase and quadrature phase components of QAM-modulated polarized channels;

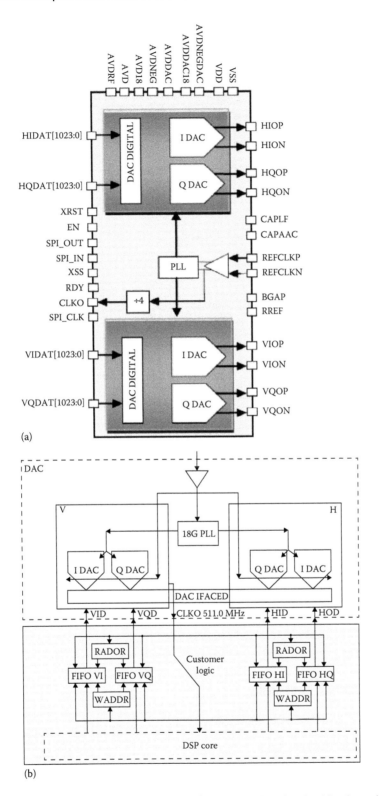

FIGURE 3.34 Fijitsu DAC structures for four channel PDM_QPSK signals: (a) schematic diagram and (b) processing function.

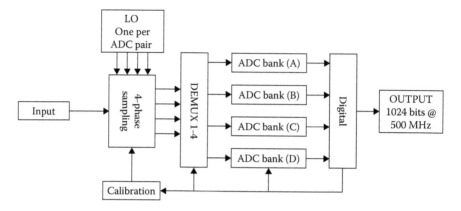

FIGURE 3.35   ADC principles of operations (CHAIS).

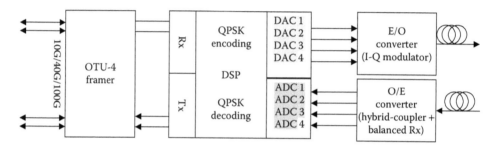

FIGURE 3.36   Schematic of a typical structure of ADC and ADC transceiver subsystems for PDM-QPSK modulation channels.

thus the notations of I_{DAC} and Q_{DAC} are shown in the diagram. Similarly, the optical received signals of PDM-QAM would be sampled by a four-phase sampler and then converted to digital form into four groups of I and Q lanes for processing in the DSP subsystem. Because of the interleaving of the sampling clock waveform, the digitalized bits appear simultaneously at the end of a clock period that is sufficiently long, so that the sampling number is sufficiently large to achieve all samples. For example, as shown in Figure 3.35, 1024 samples are achieved at a periodicity corresponding to 500 MHz cycle clock for 8-bit ADC. Thus, the clock has been slowed down by a factor of 128, or alternatively, the sampling interval is $1/(128 \times 500\ \text{MHz}) = 1/64\ \text{GHz}$. The sampling is implemented using a CHArged mode Interleaved Sampler (CHAIS).

Figure 3.36 shows a generic diagram of an optical DSP-based transceiver employing both DAC and ADC under QPSK modulated or QAM signals. The current maximum sampling rate of 64 GSa/s is available commercially. An IC image of the ADC chip is shown in Figure 3.37.

3.7   CONCLUDING REMARKS

This chapter has described the principles of coherent reception and associated techniques with noise considerations and main functions of the DSP. The DSP algorithms will be described, not in a separate chapter, but between lines in the chapters.

Furthermore, the matching of the LO laser and that of the carrier of the transmitted channel is very important for effective CoD, if not degradation of the sensitivity of results. The International Telecommunication Union standard requires that for DSP- based coherent receiver the frequency offset between the LO and the carrier must be within the limit of ±2.5 GHz. Furthermore, in practice it is expected that in network and system management the tuning of the LO is to be done

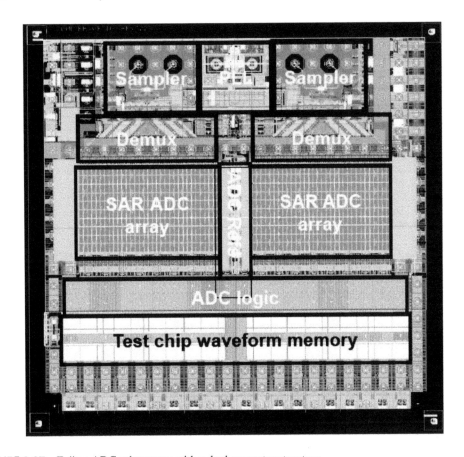

FIGURE 3.37 Fujitsu ADC subsystems with a dual convertor structure.

remotely and automatic locking of the LO with some prior knowledge of the frequency region to set the LO initial frequency. Thus, this action briefly describes the optical phase locking the LO source for an intradyne coherent reception subsystem.

REFERENCES

1. A. H. Gnauck and P. J. Winzer, Optical phase-shift-keyed transmission, *J. Lightwave Technol.*, 23, 115–130, 2005.
2. R. C. Alferness, Guided wave devices for optical communication, *IEEE J. Quantum. Elect.*, QE-17, 946–959, 1981.
3. W. A. Stallard, A. R. Beaumont, and R. C. Booth, Integrated optic devices for coherent transmission, *IEEE J. Lightwave Technol.*, LT-4(7), 852–857, July 1986.
4. V. Ferrero and S. Camatel, Optical phase locking techniques: An overview and a novel method based on single sideband sub-carrier modulation, *Opt. Express*, 16(2), 818–828, 21 January 2008.
5. I. Garrett and G. Jacobsen, Theoretical analysis of heterodyne optical receivers for transmission systems using (semiconductor) lasers with nonnegligible linewidth, *IEEE J. Lightwave Technol.*, LT-3/4, 323–334, 1986.
6. G. Nicholson, Probability of error for optical heterodyne DPSK systems with quantum phase noise, *Electron. Lett.*, 20/24, 1005–1007, 1984.
7. S. Shimada, *Coherent Lightwave Communications Technology.* Chapman and Hall, London, 1995, p. 27.
8. Y. Yamamoto and T. Kimura, Coherent optical fiber transmission system, *IEEE J. Quantum Electron.*, QE-17, 919–934, 1981.

9. S. Savory and T. Hadjifotiou, Laser linewidth requirements for optical DQPSK optical systems, *IEEE Photonic Technol. Lett.*, 16(3), 930–932, March 2004.

10. K. Iwashita and T. Masumoto, Modulation and detection characteristics of optical continuous phase FSK transmission system, *IEEE J. Lightwave Technol.*, LT-5/4, 452–462, 1987.

11. F. Derr, Coherent optical QPSK intradyne system: Concept and digital receiver realization, *IEEE J. Lightwave Technol.*, 10(9), 1290–1296, 1992.

12. G. Bosco, I. N. Cano, P. Poggiolini, L. Li, and M. Chen,MLSE-based DQPSK transmission in 43Gb/s DWDM long-haul dispersion managed optical systems, *IEEE J. Lightwave Technol.*, 28(10), May 15, 2010.

13. D.-S. Ly-Gagnon, S. Tsukamoto, K. Katoh, and K. Kikuchi, Coherent detection of optical quadrature phase-shift keying signals with carrier phase estimation, *J. Lightwave Technol.*, 24, 12–21, 2006.

14. E. Ip and J. M. Kahn, Feedforward carrier recovery for coherent optical communications, *J. Lightwave Technol.*, 25, 2675–2692, 2007.

15. L. N. Binh, Dual-ring 16-Star QAM direct and coherent detection in 100 Gb/s optically amplified fiber transmission: Simulation, *Opt. Quant. Electron.*, August 5, 2008. Accepted December 2008. Published on line Dec 2008.

16. L. N. Binh, Generation of multi-level amplitude-differential phase shift keying modulation formats using only one dual-drive Mach-Zehnder interferometric optical modulator, *Opt. Eng.*, 48(4), 2009.

17. M. Nazarathy, X. Liu, L. Christen, Y. K. Lize, and A. Willner, Self-coherent multisymbol detection of optical differential phase-shift-keying, *J. Lightwave Technol.*, 26, 1921–1934, 2008.

18. R. Noe, PLL-free synchronous QPSK polarization multiplex/diversity receiver concept with digital I&Q baseband processing, *IEEE Photonics Technol. Lett.*, 17, 887–889, 2005.

19. L. G. Kazovsky, G. Kalogerakis, and W.-T. Shaw, Homodyne phase-shift-keying systems: Past challenges and future opportunities, *J. Lightwave Technol.*, 24, 4876–4884, 2006.

20. J. P. Gordon and L. F. Mollenauer, Phase noise in photonic communications systems using linear amplifiers, *Opt. Lett.*, 15, 1351–1353, 1990.

21. H. Kim and A. H. Gnauck, Experimental investigation of the performance limitation of DPSK systems due to nonlinear phase noise, *IEEE Photonics Technol. Lett.*, 15, 320–322, 2003.

22. S. Zhang, P. Y. Kam, J. Chen, and C. Yu, Receiver sensitivity improvement using decision-aided maximum likelihood phase estimation in coherent optical DQPSK system. In *Conference on Lasers and Electro-Optics/Quantum Electronics and Laser Science and Photonic Applications Systems Technologies*, Technical Digest (CD) (Optical Society of America, 2008), paper CThJJ2.

23. P. Y. Kam, Maximum-likelihood carrier phase recovery for linear suppressed-carrier digital data modulations, *IEEE Trans. Commun.*, COM-34, 522–527, June 1986.

24. E. M. Cherry and D. A. Hooper, *Amplifying Devices and Amplifiers*, John Wiley & Sons, New York, 1965.

25. E. Cherry and D. Hooper, The design of wide-band transistor feedback amplifiers, *Proc. IEE*, 110(2), 375–389, 1963.

26. N. M. S. Costa and A. V. T. Cartaxo, Optical DQPSK system performance evaluation using equivalent differential phase in presence of receiver imperfections, *IEEE J. Lightwave Technol.*, 28(12), 1735–1744, June 2010.

27. H. Tran, F. Pera, D. S. McPherson, D. Viorel, and S. P. Voinigescu, 6-kΩ, 43-Gb/s differential transimpedance-limiting amplifier with auto-zero feedback and high dynamic range, *IEEE J. Solid St. Circ.*, 39(10), 1680–1689, 2004.

28. S. Zhang, P. Y. Kam, J. Chen, and C. Yu, A comparison of phase estimation in coherent optical PSK system. In *Photonics Global '08*, Paper C3-4A-03, Singapore, December 2008.

29. S. Zhang, P. Y. Kam, J. Chen, and C. Yu, Adaptive decision-aided maximum likelihood phase estimation in coherent optical DQPSK system. In *OptoElectronics and Communications Conference (OECC) '08*, Paper TuA-4, pp. 1–2, Sydney, Australia, July 2008.

30. I. Hodgkinson, Receiver analysis for optical fiber communications systems, *IEEE J. Lightwave Technol.*, 5(4), 573–587, 1987.

31. N. Stojanovic, An algorithm for AGC optimization in MLSE dispersion compensation optical receivers, *IEEE Trans. Circ. Syst. I*, 55, 2841–2847, 2008.

32. B. N. Mao et al., Investigation on the ENOB and clipping effect of real ADC and AGC in coherent QAM transmission system. In *Proceedings of ECOC*, Geneva, 2011.

4 Optical Noncoherent Reception and Noises Processes

This chapter describes the fundamental understanding of the detection of optical signals using direct techniques. Coherent detection has been presented in Chapter 3. Optical preamplification at the input of the optical receivers is not treated here but considered from parts by parts inserted in other chapters. Noncoherent or direct detection (DD) techniques can now be found in several low-cost optical systems in access networks to deliver multi-Tbps to end users for the coming data center-centric (DCC) networks and ultrabroadband optical networks for the fifth generation (5G) wireless communications cloud-based networking. Further, these techniques can be found very popular in interdata center communications.

4.1 INTRODUCTION

Optical receivers are important components in optical fiber communication systems and networks. Optical receivers are normally placed at the far end of the transmission links or they are used in the front end of an optical repeater in terrestrial optical systems and at terminal front ends in optical networks. Their principal function is conversion of optical signals into electronic forms for further electronic amplification and signal processing such as clock recovery, retiming, digital level detection, and sampling.

Typical arrangement of an optical receiver is that the optical signals are detected by a photodiode (a PIN diode or avalanche photodiode [APD] or a photon counting device), electrons generated in the photodetector (PD) are then electronically amplified through a front end electronic amplifier. The electronic signals are then decoded for recovery of original format.

Optical coherent communication systems have also been reviving dramatically due to electronic processing and availability of stable narrow linewidth lasers. Thus, we also describe the coherent detection (CoD) techniques in the next chapter so as to update readers with this renew emerging trend in the twenty-first century.

This chapter deals with the analysis and design of electronic amplifiers incorporated with the photodetection device as the fundamental elements of an optical receiver. In recent years the development of optical amplifiers, particularly the rare-earth doped optical fiber amplifiers, has advanced to a particle stage that these fiber amplifiers are coming to be a standard photonic component in most terrestrial optical fiber communications systems. With this in mind, we need to include the design of optical amplifiers as a standalone system or inline optical receivers acting as an optical repeater or as a preamplifier front end placed before the PD and followed by an electronic amplifier.

Noises generated in the electronic preamplifier lower the signal-to-noise ratio or reduce the eye opening of the received digital signals at its output. The equivalent noise power of the electronic preamplifier increases as a square function of the frequency. Thus, it is necessary to reduce this frequency-dependent noise. A noise matching network can be placed at the front end that would act as a noise filter. An example for high-frequency electronic preamplifier is given in Section 4.5.

Optical amplifiers, the rare-earth doped optical fiber amplifiers, have advanced and would provide a boosting in the optoelectronic detection. The integration of an OA would further improve the CoD and a treatment of this optically preamplified detection process is also given. The

invention of optical amplifiers in the 1990s has thus revolutionized the design of the optical transmission over long haul as they offer significant optical gain that would compensate for fiber losses
and other losses of the optical components.

This chapter is organized as follows: Section 4.2 gives an overview of the roles of an optical
receiver in various optical transmission systems and their applications in networking. Section 4.3
then gives an account of the optical components of an optical receiver for DD. Noises in PD and
integration with the electronic preamplifier are described in this section. Methods for calculations
of receiver noises are also described. The performance evaluation of the binary optical receiver
given in terms of noise and signal power at the output of the receiver is described. Two examples of
the design and noise estimation when a field effect transistor (FET) or a bipolar junction transistor
(BJT) is used as the front end electronic amplification device are illustrated. Section 4.5 gives an
ample of noise matching network integrated at the front end of the electronic preamplifier. Finally,
Section 4.6 gives concluding remarks.

4.2   OPTICAL RECEIVERS IN VARIOUS SYSTEMS

In modern optical and photonic communications systems, the optical amplifiers can be incorporated
in front of a front end optoelectronic receiver to form an optical amplified preamplifier. This type of
optical receivers would be treated in Chapter 10, which deals with optical amplification.

The ultimate goals of the design of optical receiver is to determine the minimum optical energy
in terms of the number of photons per bit period required at the input of the PD so that it would
satisfy a certain optical signal-to-noise ratio for an analog optical system or the sensitivity of a digital optical communications satisfying a certain bit-error rate (BER). In other words, the minimum
optical power required at a certain bit rate in a digital communication system so that the decision
circuitry can detected with a specified BER (e.g., BER = 10^{-9} or 10^{-12}).

Table 4.1 shows the important requirement for optical receivers in various communications systems from a terrestrial to local and wide area networks to undersea submarine systems.

The electronic amplifier is considered as a block diagram rather than the detail circuit configuration. However for completeness, a section is dedicated to the design of the electronic front end

TABLE 4.1
Optical Receiver Characteristics and Their Relative Importance in Typical Optical-Fiber Communications Systems

	Features	Undersea Optical Communications	Terrestrial Communications and WAN	Point-to-Point	LAN/DATA Center
High receiver sensitivity	Maximum repeater spacing	Critical	Moderate critical	Moderate	Moderate
Wide dynamic range	Flexible and convenient systems configurations	Moderate	Moderate critical	Moderate critical	Critical
Bit rate transparency	Variable bit rate operation	Not required	Not required	Desirable	Desirable
Bit rate dependency	Flexible	Accommodated by the use of appropriate line codes and scrambling/descrambling		Desirable	Desirable
Fast acquisition time	Short preamble bit sequence	Not required	Not required	Moderate critical	Moderate critical

amplifier. It is essential that there be a generic approach for the design of optical receivers, which is the ultimate objective of this chapter.

4.3   RECEIVER COMPONENTS

The design of an optical receiver depends on the modulation format of the signals and thus transmitted through the transmitter. In particular, it is dependent on the modulation in analog or digital, Gaussian or exponential pulse shape, on-off keying (OOK) or multiple levels and, so on.

Optical signals treated in this chapter are involved with an intensity modulator (IM) and DD, OOK or amplitude shift keying (ASK) systems. They are thus treated with an assumption that light-faces arrived at the receiver are assumed to be polarization independent and their polarization plays no part in the degradation of the optical signals, except due to the polarization mode dispersion that would be independently given in the characteristics of the optical fiber (Appendix of Chapter 12). The coherent optical systems have been treated in many other textbooks on optical communications where the polarization and the coherence of the optical source play a very significant role in the design and implementation of the optical fiber communications systems. These coherent optical communications are of great interest to the research and development community in the early to mid-1980s but no longer now due to the invention of the Erbium-doped optical amplifiers (EDFAs) in the late 1980s.

Figure 4.1 shows the schematic diagram of a digital optical receiver. An optical receiver front end consists of a PD for converting lightwave energy into electronic currents, an electronic pre-amplifier for further amplification of the generated electronic current, usually in voltage form, a main amplifier for further voltage amplification, a clock recovery circuitry for regenerating the timing sequence and a voltage-level decision circuit for sampling the waveform for the final recovery of the transmitted and received digital sequence. Therefore, the optoelectronic preamplifier is followed by a main amplifier with an automatic control to regulate the electronic signal voltage to be filtered and then sampled by a decision circuit with synchronization by a clock recovery circuitry.

The inline fiber optical amplifier can be incorporated in front of the PD to form an optical receiver with an optical amplifier front end to improve its receiving sensitivity. This optical amplification at the front end of an optical receiver will be treated in [1] dealing with optical amplification processes.

The structure of the receiver is thus consisted of three parts: the front end section, the linear channel of the main amplifier and automatic gain control (AGC), and the data recovery section. These sections are described in order below.

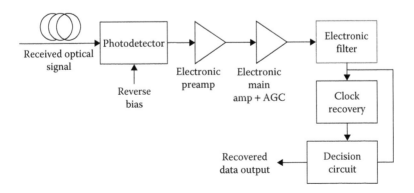

FIGURE 4.1   Schematic diagram of the digital optical receiver IM/DD.

4.3.1 PHOTODIODES

A PD detects and converts the optical input power into an electric current output. The ideal PD would be highly quantum efficient; ideally it would add no noises to the received signals, respond uniformly to all signals with different wavelengths within the transparent windows of optical fibers around 1300 and 1550 nm, and finally it would not be saturated and behave linearly as a function of signal amplitude. There are several different types of PDs that are commercially available. Of the semiconductor-based types, the photodiode is used almost exclusively for high-speed fiber optic systems due to its compactness and thus extremely high bandwidth. The two most commonly used types of photodiodes are the p-type intrinsic n-type (PIN) and APD. Detailed reviews of these photodiodes have been presented in the well-known published literature [2–4]. The fundamental characteristics of these two device types are briefly described for completeness.

4.3.1.1 PIN Photodiode

A PIN PD consists of three regions of semiconductors, the heavily doped p r, the intrinsic, and the n doped sections (Figure 4.3). It is essential that the p−n junction is reversed-biased to cause mobile electrons and holes moving away from the junction, thus increases the width of the depletion layer and hence a high electric field is produced by the immobile charges at both ends. If lightwaves are incident on the PIN surface and absorbed in the *high-field depleted* region, the electron–hole pairs generated will move at saturation-limited velocity. If the incoming photons fall on the other region, no significant electronic current would be generated. Thus, because the width of the p region, W_p, is inversely proportional to its doping concentration, decreases in the doping level to intrinsic behavior would widen W_p; hence the depleted region is now wider and compatible with the absorption region to produce large photon-generated current. A p^+ region must be added to make good ohmic contact. Wider intrinsic region gives higher quantum efficiency, η but lower response rate.

4.3.1.2 Avalanche Photodiodes

High-field multiplication region is added adjacent to the lightly doped depletion region–intrinsic as in contrast with the PIN structure. Photons absorbed in the intrinsic region produces electron–hole pairs. These pairs drift to the high-field region whereby they gain sufficient energy (E, field must be above the threshold for impact ionization to occur) so that to ionize bound electrons in the valence band on colliding with them. Thus, the avalanche effect occurs only when the electric field existed in the high-field multiplication region above the threshold for impact ionization. A guard-ring is usually added to prevent leakage of the surface current, I_{surf} (Figure 4.4).

4.3.1.3 Quantum Efficiency and Responsivity

Two most important characteristics of a PD are its quantum efficiency and its response speed. These parameters depend on the material band gap, the operating wavelength, the doping and thickness of the p, i, and n regions of the device. The quantum efficiency, η is the number of electron–hole carrier pairs generated per incident photon of energy $h\upsilon$ and is given by

$$\eta = \frac{\text{number of electron pairs generated}}{\text{number of incident photons}} = \frac{I_p/q}{P_o/h\upsilon} = \frac{I_p h\upsilon}{P_o q} \tag{4.1}$$

where I_p is the photocurrent and P_o is the incident optical power. The performance of a photodiode is often characterized by its responsivity $\Re$, which is related to the quantum efficiency by

$$\Re = \left(\frac{I_p}{P_o}\right)G = \left(\frac{\eta q}{h\upsilon}\right)G = \left(\frac{\eta q\lambda}{hc}\right)G \tag{4.2}$$

where G is the APD average multiplication factor and $G = 1$ for the non-APD case. The optical power is given by

$$P_o = \frac{i_s}{G\Re} \tag{4.3}$$

where it means that for the same incident optical power, APD could produce higher photocurrent than PIN. Thus, on the basis of Equations 4.1 and 4.2, we could model PIN and APD.

4.3.1.4 High-Speed PDs

When the speed of the PDs is increased, for example, to be used in high-speed systems above 10 Gb/s, PIN detectors are normally used and the effective receiving area is smaller to reduce the junction capacitance. This decreased area then forces the detector to have lower breakdown voltage.

4.4 DETECTION AND NOISES

The front end subsystem is the most important component because noise contributed by this part severely affects the overall performance of the system. The design of an amplifier front end depends on the sensitivity required and the bandwidth of the system. For example, an optical front end amplifier for an optical time domain reflectometer (OTDR) would require a very highly sensitive front end while its bandwidth is not critical. On the other hand, a front end optical amplifier for a high bit rate optical communications requires a wide bandwidth amplifier as well as a reasonably high sensitivity and low noise contribution. As observed in Table 4.1, optical front end amplifiers and demands on their characteristics with their limit are given.

In general, optically amplified front ends are required for two types of optical systems: the short link in local area or metropolitan area networks, and the terrestrial link systems. The former requires high data accumulation rate, and the latter requires a fast acquisition rate and reasonable sensitivity.

Thus, electronic preamplifiers can take two most typical structures: the high impedance (HI) and the transimpedance (TI) front end. The difference is that in TI configuration there is a shunt feedback from the output to the input. This TI configuration offers a wide bandwidth and the HI provides a high sensitivity due its noise that is almost negligible compared with the TI model. The TI and HI configurations are shown in Figure 4.2.

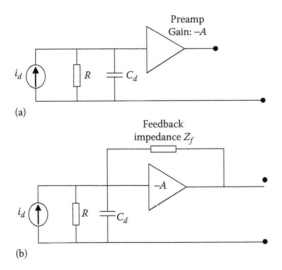

FIGURE 4.2 Schematic diagram of the electronic preamplifier in an optical receiver. (a) High impedance (HI) optical front end receiver and (b) transimpedance (TI) optical front end receiver. The current source represents the electronic current generated in the photodetector (PD). C_d = photodiode capacitance.

A PD can be represented by a current source in parallel with a capacitor and a shot noise current as shown in Figure 4.2. The noise current (quantum shot noise) is not shown. A diode capacitance due the capacitance dropped across the depletion region of the pn junction is also included. In fact, this diode capacitance C_d is a critical component for the design of the PD.

The HI front end would have very HI Z_i as the input impedance of the electronic amplifier. This HI would limit the bandwidth of the overall amplifier. Usually Z_i would be a few hundred of ohms. While a TI front end offers lower receiver sensitivity, the bandwidth is increased by a factor of A, the amplifier linear gain.

4.4.1 Linear Channel

The linear channel in optical receivers consists of a main electronic amplifier and a low pass filter or an equalizer whose objectives are to reduce the noise or distortion in the channel without incurring further intersymbol interference (ISI). The preamplifier and the main amplifier have the bandwidth normally wider than that of the system. Thus in calculating the total noise contribution, the noise spectra density must be integrated over the electronic amplifier bandwidth.

4.4.2 Data Recovery

Once the optical signals are detected and amplified, the data streams are sampled with respect to the time interval recovered by a clock recovery circuit, normally by using a surface acoustic wave filter if the bit rate is high, and then amplified and fed into a triggering circuit. High-speed optical receivers whose bandwidth are normally wider than 2.5 GHz, are usually integrated using CMOS hybrid circuit technology with digital signal processor and analog to digital converter.

4.4.3 Noises in PDs

The schematics of the photodectors of type p-i-n and APD are shown in Figures 4.3 and 4.4 respectively in which the strength of the field distribution across their structure is also depicted.

The most important noise source in the PD is the fundamental quantum or shot noise caused by the intrinsic fluctuations in the photoexcitation of electronic carriers. In any interval of T seconds, the probability of exactly N primary electrons generated is given by the time-varying Poisson distribution

$$P[N,(t_0,t_0+T)] = \frac{\Lambda^N e^{-\Lambda}}{N!} \tag{4.4}$$

where

$$\Lambda = \int_{t_0}^{t_0+T} \lambda(t)dt \xrightarrow{\text{where}} \lambda(t) = \frac{\eta}{\hbar v}p(t)+\lambda_d$$

with λ_d the number of electrons generated under the dark condition, or the dark current electrons of the PD; thus Λ is the average number of primary electrons produced during the interval.

For APD, the internal avalanche multiplication factor is also a random process (Figure 4.4). The variation of the avalanche gain gives rise to excess noise. The probability of distribution of the avalanche gain depends on the types of APD. In particular, it is a function of the ratio of the hole ionization probability to that of the electron, the factor k. The extra noise is represented by an excess noise factor $F(g)$ given by

$$\langle g^2 \rangle = F(g)G^2 \text{ with } G = \langle g \rangle \tag{4.5}$$

The term $F(g)$ is usually determined by experimental works and can be approximated as

$$F(G) \simeq kG + \left(2 - \frac{1}{G}\right)(1-k) \tag{4.6}$$

and usually $F(G)$ can be estimated by

$$F(G) \simeq G^x \tag{4.7}$$

For Ge APD, $x \sim 1$, while for well-designed Si APD, $x \sim 0.4$ or 0.5.

4.4.4 Receiver Noises

Before proceeding to the systems calculations to determine the performance of optical receivers, it is necessary that the noises generated in the PD and the preamplifier front end are considered.

Readers can in fact consider investigating the system calculation and returning back to the noise calculation provided that they are taking either the total equivalent noise spectral density at the input of the detector or its noise (Figure 4.2). This section describes all noise mechanisms related to the photodetection process including the electronic noise associated with the receiver.

Shot noises and thermal noises are the two most significant noises in optical detection systems. Shot noises are generated by either quantum process or electronic biasing. The noises are specified in noise spectral density that is square of noise current per unit frequency (in Hz). Thus, the noise spectral density is to be integrated over the total amplifier bandwidth to obtain the equivalent noise currents.

4.4.4.1 Shot Noises

Electrical shot noises are generated by the random generation of streams of electrons (current). In optical detection, shot noises are generated by (i) biasing currents in electronic devices, and (ii) photocurrents generated by the photodiode.

For a bias current I, the spectral current noise density S_I given by

$$S_I = \frac{d(i_I^2)}{df} = 2qI \quad \text{in A}^2/\text{Hz} \tag{4.8}$$

where q is the electronic charge. The current i_I represents the noise current due to the biasing current I.

4.4.4.2 Quantum Shot Noise

The average current $\langle i_s^2 \rangle$ generated by the PD by an optical signal with an average optical power P_{in} is given by

$$S_Q = \frac{d\langle i_s^2 \rangle}{df} = 2q\langle i_s^2 \rangle \quad \text{in A}^2/\text{Hz} \tag{4.9}$$

In the case when the APD is used, the noise spectral density is given by

$$S_Q = \frac{d\langle i_s^2 \rangle}{df} = 2q\langle i_s^2 \rangle\langle G_n^2 \rangle \quad \text{in A}^2/\text{Hz} \tag{4.10}$$

where $\langle G_n^2 \rangle$ is the average avalanche gain of the detector. It is noted here again that the dark currents generated by the PD must be included to the total equivalent noise current at the input after it is

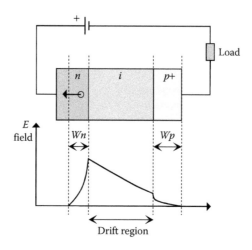

FIGURE 4.3   Schematic diagram of a p-i-n photodiode.

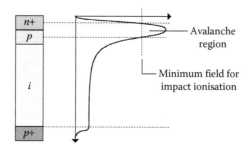

FIGURE 4.4   Schematic diagram of an avalanche photodiode (APD).

evaluated. These currents are generated even in the absence of the optical signal. The dark currents can be eliminated by cooling the PD to at least below the temperature of liquid nitrogen (77°K).

4.4.4.3   Thermal Noise

At a certain temperature, the conductivity of a conductor varies randomly. This random movement of electrons generates a fluctuating current even in the absence of an applied voltage. The thermal noise of a resistor R is given by

$$S_R = \frac{d\left(i_R^2\right)}{df} = \frac{2k_B T}{R} \quad \text{in A}^2/\text{Hz} \tag{4.11}$$

where:
  k_B is Boltzmann's constant
  T is the absolute temperature (in K)
  R is the resistance (in ohms)
  i_R denotes the noise current due to resistor R

4.4.5   Noise Calculations

In this section, the design of a receiver for use in a digital communication system and the methods for noise calculations are described. Binary and multilevel operations are also given.

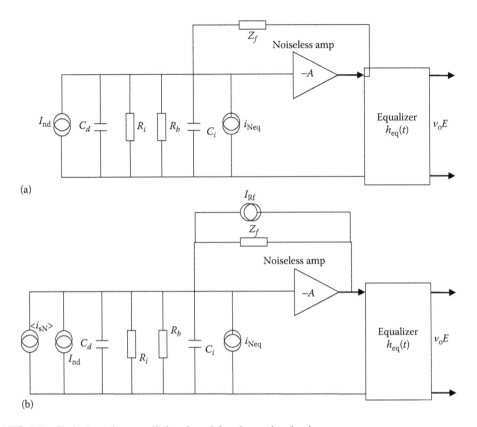

FIGURE 4.5 Optical receiver small signal model and associated noise sources.

A schematic of an optical preamplifier of the receiver and an electronic analog equalizer for the detection of optical digital modulated signals is shown in Figure 4.5. An equalizer is considered to extend the bandwidth of the receiver to the range of several GHz for ultrahigh-speed operations of these digital optical receivers. This is similar to earlier design considerations of optical receiver when multimode fibers were used in the first generation of optical fiber communications systems [5–7]. The noise sources and the small amplification circuit model can be simplified as shown in Figure 4.6, in which all the noise sources of the electronic amplifier and the PD are presented and grouped into total noise current sources at the input and output of the amplifier. These sources are then transferred to total equivalent current sources as seen into the input of the amplifier. Thus, it is very straightforward to find out the signal-to-noise ratio and contribution of noises from the electronic amplification process and the quantum shot noise process in the detection of optical signals.

Our goal is to obtain an analytical expression of the noise spectral density equivalent to a source looking into the electronic amplifier including the quantum shot noises of the PD. A general method for deriving the equivalent noise current at the input is by representing the electronic device by a Y-equivalent linear network as shown in Figure 4.6. The two current noise sources $\mathrm{d}\left(i_N'^2\right)$ and $\mathrm{d}\left(i_N''^2\right)$ represent the summation of all noise currents at the input and at the output of the Y-network. This can be transformed into a Y-circuit with the noise current at the input as follows:

The output voltages V_o in Figure 4.6a can be written as

$$V_o = \frac{i_N'(Y_f - Y_m) + i_N''(Y_i + Y_f)}{Y_f(Y_m + Y_i + Y_o + Y_L) + Y_i(Y_o + Y_L)} \qquad (4.12)$$

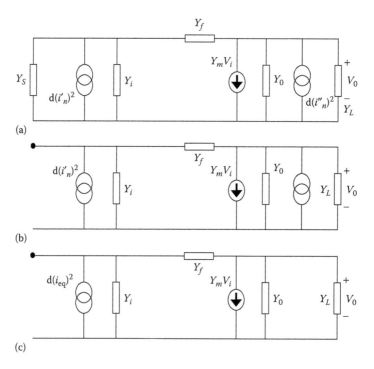

FIGURE 4.6   Small equivalent circuits including noise sources: (a) Y-parameter model representing the ideal current model and all current noise sources at the input and output ports, (b) with noise sources at input and output ports, and (c) with a total equivalent noise source at the input.

and that in Figure 4.6b as

$$V_o = \frac{(i'_N)_{eq}(Y_f - Y_m)}{Y_f(Y_m + Y_i + Y_o + Y_L) + Y_i(Y_o + Y_L)} \tag{4.13}$$

Thus comparing these two equations we can deduce the equivalent noise current at the input of the detector:

$$i_{Neq} = i'_N + i''_N \frac{Y_i + Y_f}{Y_f - Y_m} \tag{4.14}$$

Then reverting to the mean square generators for a noise source, we have

$$d(i_{Neq})^2 = d(i'_N)^2 + d(i''_N)^2 \left| \frac{Y_i + Y_f}{Y_f - Y_m} \right|^2 \tag{4.15}$$

It is therefore expected that if the Y-matrix of the front end low-noise amplifier is known, the equivalent noise at the input of the amplifier can be obtained using Equation 4.15.

4.5   PERFORMANCE CALCULATIONS FOR BINARY DIGITAL OPTICAL SYSTEMS

As described above, the noises of the optical receivers consist of the thermal noises and quantum shot noises due to the bias currents and the photocurrent generated by the PD with and without the optical signals. Thus this quantum shot noise is strongly signal dependent.

These noises degrade the sensitivity of the receiver and thus a penalty can be estimated. Another source of interference that would also result in signal penalty is the ISI. The goal of this section is to obtain an analytical expression for the receiver sensitivity of the optical receiver in an IM/DD pulse-code-modulated system.

4.5.1 Signals Received

Assume that the received signal power is

$$p(t) = \sum_j a_j h_p(t - jT_B)$$

(4.16)

where T_B is the bit period. The average output voltage is thus given by

$$\langle v_o \rangle = \Re \langle G_n \rangle \left[\sum_j a_j \frac{1}{T_B} \int_{-T_b/2}^{T_b/2} h_p(t - jT_B) dt \right] R_I A$$

(4.17)

$h_p(t - jT_B)$ is the impulse response of the system evaluated at each time interval. R_I is the channel amplifier. It is assumed that the overall amplifier has a flat gain response A over the bandwidth of the system,

$$a_j \big|_{t=0} = \begin{cases} b_0 \approx 0 \\ b_1 \end{cases}$$

(4.18)

with b_0 the energy when a transmitted "0" is received and b_1 the energy when a transmitted "1" is received. The summation over a number of periods is necessary to take into account the contribution of adjacent optical pulses.

We now have to distinguish between two cases when a "0" or a "1" is transmitted and received.

4.5.1.1 Case (a): OFF or a Transmitted "0" is Received

Using Equations 4.17 and 4.18, we have

$$\langle v_o \rangle_0 = v_{o0} = \Re \frac{b_0}{T_B} G R_I A \cong 0$$

(4.19)

with the total equivalent noise voltage at the output, v_{NTo}^2,

$$v_{\text{NTo}}^2 = v_{\text{NA}}^2 = i_{\text{Neq}}^2 R_I^2 A^2$$

(4.20)

4.5.1.2 Case (b): ON Transmitted "1" is Received

In this case, the average signal voltage at the output is received as

$$\langle v_o \rangle_1 = v_{o1} = \Re \frac{b_1}{T_b} \langle G_n \rangle R_I A$$

(4.21)

with a total noise equivalent mean voltage at the output given by

$$v_{\text{NT1}}^2 = v_{o\text{SN}}^2 + v_{\text{NA}}^2$$

(4.22)

where $v_{o\text{SN}}^2$ is the signal-dependent shot noise. v_{NA}^2 is the amplifier noise at the output and given by

$$v_{\text{NA}}^2 = i_{\text{Neq}}^2 R_I^2 A^2 B$$

(4.23)

The signal-dependent noise is in fact the quantum shot noise and is given by

$$v_{o\text{SN}}^2 = \int_0^B 2q \langle i_s \rangle_1 \langle G_n^2 \rangle R_I^2 A^2 df$$

(4.24)

where B is the 3 dB bandwidth of the overall amplifier, and $\langle i_s \rangle_1$ is the average photocurrent current received when a "1" was transmitted. This current is estimated as follows:

$$\langle i_s \rangle_1 = \sum_{-\infty}^{\infty} \Re \frac{b_1}{T_B} \int_{-T_b/2}^{T_b/2} h_p(t - jT_B)dt \tag{4.25}$$

or

$$\langle i_s \rangle_1 = \Re \frac{b_1}{T_B} \int_{-\infty}^{\infty} h_p(t)dt \tag{4.26}$$

Using normalization with $\int_{-\infty}^{\infty} h_p(t)dt = 1$, Equation 4.26 becomes

$$\langle i_s \rangle_1 = \Re \frac{b_1}{T_B} \tag{4.27}$$

4.5.2   Probability Distribution

The optical systems under consideration are a typical IM/DD system in which the optical energy of each pulse period is equivalent to that of at least a few hundred photons. This number is large enough so that a Gaussian distribution of the probability density is warranted.

The probability density function of a "0" transmitted and received by the optical receiver is thus given by (Figure 4.7)

$$p\left[v_o \big| \text{"0"}\right] = \frac{1}{\left(2\pi v_{\text{NT0}}^2\right)^{1/2}} \exp\left[\frac{-(v_0 - v_{00})^2}{2v_{\text{NT0}}^2}\right] \tag{4.28}$$

and similarly for a "1" transmitted,

$$p\left[v_o \big| \text{"1"}\right] = \frac{1}{\left(2\pi v_{\text{NT1}}\right)^{1/2}} \exp\left[\frac{-(v_0 - v_{01})^2}{2v_{\text{NT1}}^2}\right] \tag{4.29}$$

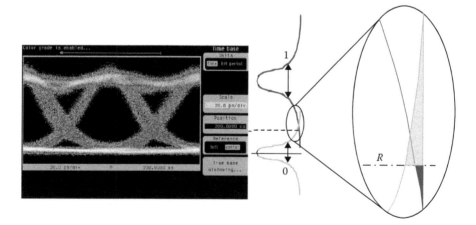

FIGURE 4.7 Time-dependent fluctuating signal in the receiver in response to the random digital bit stream and the probability density distribution as a function of the output voltage level and the decision level for recovery of the transmitted signals.

The total probability of error or a BER is defined [8,9] as

$$BER = p(1)p(0/1) + p(0)p(1/0) \qquad (4.30)$$

where $p(1)$ and $p(0)$ are the probabilities of receiving "1" and "0," respectively, and $p(1/0)$ and $p(0/1)$ are the probabilities of deciding "1" when "0" is transmitted and vice versa.

In a binary bit stream "1" and "0" are likely to occur equally, that is, $p(1) = p(0) = 0.5$; then Equation 4.30 becomes

$$BER = \frac{1}{2}[p(0/1) + p(1/0)] \qquad (4.31)$$

Thus as equal probability of transmitting "0" and "1" are assumed then for a decision voltage level of d as indicated in Figure 4.5, the total probability of error P_E is the summation of the errors of deciding "0" or "1" by integrating the probability density function over the shaded regions, thus given by

$$BER = P_E = \frac{1}{2}\int_d^\infty p\left[v_0\left|\text{"0"}\right.\right]dv_0 + \frac{1}{2}\int_{-\infty}^d p\left[v_0\left|\text{"1"}\right.\right]dv_0 \qquad (4.32)$$

Substituting for the probability distribution using Equation 4.30 leads to

$$BER = \frac{1}{2\sqrt{\pi}}\int_{\frac{d-v_{00}}{v_{NT0}}}^\infty e^{-\frac{x^2}{2}}dx + \frac{1}{2\sqrt{\pi}}\int_{\frac{v_{01}-d}{v_{NT1}}}^d e^{-\frac{x^2}{2}}dx \qquad (4.33)$$

The functions in Equation 4.33 have the standard form of the complementary error function $Q(\alpha)$, which is defined as

$$Q(\alpha) = \frac{1}{\sqrt{2}}\int_\alpha^\infty e^{-x^2/2}dx \qquad (4.34)$$

Then Equation 4.33 becomes

$$BER = \frac{1}{2}\left[Q\left(\frac{d}{v_{NT0}}\right) + Q\left(\frac{v_{01}-d}{v_{NT1}}\right)\right] \qquad (4.35)$$

The $Q(\delta)$ function is a standard function, and this curve is shown in Figure 4.8. It is noted that for a BER = 10^{-9}, the value of δ is about 6, which is the normal standard for communications at a bit rate of 155 Mbps to 40 Gb/s. However in laboratory demonstration, it requires a BER of 10^{-12}, which corresponds to $a\delta$ of 7.

4.5.3 MINIMUM AVERAGE OPTICAL RECEIVED POWER

Again using the condition $p[v_o/\text{"0"}] = p[v_o/\text{"1"}]$, we have from Equation 4.35

$$\frac{d}{v_{NT0}} = \frac{v_{01}-d}{v_{NT1}} \equiv \delta \qquad (4.36)$$

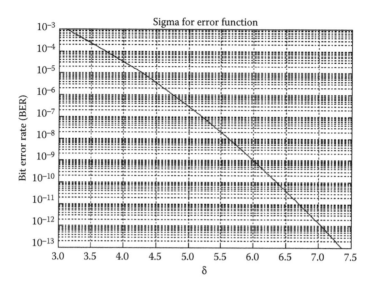

FIGURE 4.8 BER as a function of the δ parameter. Sometimes δ is assigned as the system Q magnum function.

Now assuming $v_{00} = 0$,

$$\text{BER} = Q(\delta) \tag{4.37}$$

Thus by eliminating the variable d from Equation 4.36, we obtain

$$\delta = \frac{v_{01} - \delta v_{NT0}}{v_{NT1}} \tag{4.38}$$

Or alternatively

$$v_{01} = \delta(v_{NT0} + v_{NT1}) \tag{4.39}$$

Substituting v_{01}, v_{NT1}, and v_{NT0}, we have

$$\Re \frac{b_1}{T_b} \langle G_n \rangle R_I A = \partial \left\{ \left[2\Re q \frac{b_1}{T_B} \langle G_n^2 \rangle B R_I^2 A^2 + v_{NA}^2 \right]^{1/2} + v_{NA} \right\} \tag{4.40}$$

However, the amplifier noise voltage v_{NA} can be given by $v_{NA} = i_{Neq} R_I A B^{1/22}$. Thus by substituting this noise voltage and eliminating $R_I A$ and solving for the energy essential for the "1" transmitted and received at the PD, the energy b_1 can be written as

$$b_1 = \frac{q}{\Re} \left[\delta^2 G^x + \frac{2\delta i_{Neq} T_B}{qG} \right] \tag{4.41}$$

where we have used the approximation $\langle G_n \rangle = G$ and $\langle G_n^2 \rangle = \langle G \rangle^{2+x}$ with x the factor dependent on the ionization ration in an APD. For PIN PD, $G = 1$.

The optical receiver sensitivity, denoted as RS, can thus be defined as

$$\text{Receiver sensitivity} = \text{RS} = 10 \text{Log}_{10} \frac{P_{av}}{P_0} \quad \text{in dBm} \tag{4.42}$$

where $P_{av} = b_1/T_b$ and $P_0 = 1.0$ mW (the reference power level for evaluating power in dBm). This is the optical receiver sensitivity that is defined as the minimum optical power required for the receiver to operate reliably with a BER below a specified value.

In Equation 4.42, there are two terms clearly specifying the dependence of the signal dependence δ and the amplifier noise contribution. It is recalled that the term b_1 represents the optical energy of "1" required for the optical receiver to detect with a certain BER. The term q/R on the RHS of Equation 4.41 is equivalent to optical power required to generate one electron, or the number of photons required at the optical receiver to generate one electron. Typical measure for the optical receiver is the number of photons required for it to operate with a specified BER. Thus, the term $[\delta^2 G^x]$ indicates the number of photons required to overcome the opening of the eye at a particular BER and the term $2\delta i_{Neq}T_b/qG$ indicates the number of photons required to overcome the noises of the electronic preamp at a specific BER. With APD, the gain of the avalanche gain does also contribute to an increase of the noise level and hence higher power. However, the current avalanche gain would compensate for this.

We note that the analysis here has not taken into account the effects of pulse shape and the transfer function characteristics of the receiver. Thus in this case, an equivalent current noise power as seen from the input port of the electronic preamplifier would simplify the analysis of the optical receiver sensitivity or SNR. The effects of the pulse shape and transfer function of the receiver are considered in the next few sections.

4.5.3.1 Fundamental Limit: Direct Detection

Referring to Equation 4.41, we could see that in the case when there are no electronic noises, the first term is the number of photons in the quantum limit of the DD receiver. Thus, we have

$$n_p = \left[\delta^2 G^x\right] \tag{4.43}$$

For a PIN detector and a BER of 10^{-9}, the number of photons of the quantum limit is 36.

Example:

For an ASK modulated optical signal at a bit rate of 10 Gb/s, estimate the number of photon energy required for the incoming symbol "1" so that a BER of 10^{-9} can be achieved.

4.5.3.2 Equalized Signal Output

Recall that the optical signal $p(t)$ fallen on the input of the PD is given as

$$p(t) = \sum_{k=-\infty}^{\infty} a_k h_p(t' - kT_B) \tag{4.44}$$

where T_B is signal bit period, and $h_p(t)$ is the input optical shape constrained by

$$\int_{-\infty}^{+\infty} h_p(t)\,dt = 1 \tag{4.45}$$

It is noted that this function represents the optical power variation as a function of time and thus it is always positive.

Now neglecting the DC component due to dark current, the signal at the output of the equalizer can be written as

$$\langle v_{0E}(t)\rangle = \Re G p(t) * h_{fe}(t) * h_{eq}(t) \tag{4.46}$$

where $h_{fe}(t)$ is the impulse response of the optical receiving subsystem and $h_{eq}(t)$ is the impulse response of the electronic equalizer and $*$ indicates the convolution.

The optical receiver is usually dominated by the pole contributed by the diode capacitance C and associated resistance of the network. It can be shown without difficulty that (refer to Figure 4.5) the frequency domain transfer function is given by

$$H_{fe}(\omega) = -\frac{R_f}{1 + j\omega(R_f C / A)} \tag{4.47}$$

with A the open loop voltage gain of the electronic preamplifier, R_f the feedback resistance, and C the total input equivalent capacitance at the input port of the amplifier. Thus knowing the input optical shape, the equalizer transfer function for a given pulse shape can be determined by multiplying the transfer function of the receiver and that of the equalizer. It is noted that the 3 dB bandwidth of a feedback receiver is given by $A/R_f C$, which is about A times the bandwidth of the receiver without feedback. Thus the demand on the equalizer is much less for this case.

4.5.3.3 Photodiode Shot Noise

The output noise power due to the random fluctuation of the multiplied Poisson nature of the current (t) produced by the PD can be written [10] as

$$\left\langle i_{SN}^2 \right\rangle = \int_{-\infty}^{+\infty} qG^{2+x}\Re \left\{ \sum_{k=-\infty}^{\infty} a_k h_p(t' - kT_B) + \lambda_d \right\} h_I^2(t - t') dt' \tag{4.48}$$

where $h_I(t) = h_{fe}(t) * h_{eq}(t)$ is the overall *current* impulse response of the optical preamplifier and the equalizer in cascade. λ_d is the electrons/second generated by the PD when no lightwave is shined on the PD. It is important to note that the quantum shot noise power depends on the signal levels $\{a_k\}$ and the time t, that is, the noise is signal dependent and nonstationary.

Without the loss of generality, we can consider the output quantum shot noise at the instant $t = 0$; then by replacing $t = 0$ and $t' = \tau$, Equation 4.48 becomes

$$\left\langle i_{SN}^2(0) \right\rangle = \int_{-\infty}^{+\infty} qG^{2+x}\Re \left\{ \sum_{k=-\infty}^{\infty} a_k h_p(\tau - kT) h_I^2(-\tau) \right\} d\tau + \int_{-\infty}^{+\infty} qG^{2+x}\Re \lambda_d h_I^2(\tau) d\tau \tag{4.49}$$

The first term is contributed by the signal amplitude, the signal-dependent shot noise, and the second term indicates the contribution of the dark current of the PD.

The shot noise power due to the signal sequence $\{a_k\}$ at the time $t = 0$ can thus be rewritten as

$$\left\langle i_{SN}^2(0) \right\rangle = qG^{2+x}\Re \left\{ \sum_{k=-\infty}^{\infty} a_k (h_p * h_I^2)(-kT_B) \right\} \tag{4.50}$$

$$\left\langle i_{SN}^2(0) \right\rangle = qG^{2+x}\Re \left\{ \sum_{k=-\infty}^{\infty} a_k (h_p * h_I^2)(-kT_B) \right\} = qG^{2+x}\Re a_0 g(0) + qG^{2+x}\Re \left\{ \sum_{\substack{k=-\infty \\ k \neq 0}}^{\infty} a_k g(-kT_B) \right\} \tag{4.51}$$

with $(h_p * h_I^2)(-kT_B) = g(-kT_B)$

The first term is the quantum shot noise due the pulse under consideration and the second term is due to the preceding and succeeding signal pulses. The effect of the succeeding pulses on the noise at $t = 0$ is likely because there are some delay due to signal traveling through the system.

This dependence of noise power on preceding and succeeding pulses is similar to ISI. However, it is important to note that this effect cannot be eliminated by equalizing the pulse to obtain zero ISI, because $h_0(kT) = 0$ does not imply that $g(kT) = 0$. The exact estimation of the noise power contributed to the considered pulse $\langle i_{SN}^2(0) \rangle$ requires statistical knowledge of the distribution of the pulse sequence. This is commonly unknown and thus some criteria must be used. The worst case that would contribute the largest noise effect is when the sequence $\{a_k\}$ has the adjacent symbols of maximum values b_M. In this case, we have

$$\langle i_{SN}^2(0) \rangle = qG^{2+x} \Re a_0 g(0) + qG^{2+x} \Re \left\{ \sum_{\substack{k=-\infty \\ k \neq 0}}^{\infty} b_M g(-kT_B) \right\} \tag{4.52}$$

The frequency domain of the noise power can be estimated as

$$\left\langle \left[i_{SN}^F(0) \right]^2 \right\rangle = qG^{2+x} \Re \left\{ \frac{b_M}{T_B} \sum_{\substack{k=-\infty \\ k \neq 0}}^{\infty} b_M G\left(\frac{2\pi N}{T_B} \right) - \frac{b_M - a_0}{2\pi} \int_{-\infty}^{\infty} G(\omega) d\omega \right\} \tag{4.53}$$

where

$$G(\omega) = \Im[g(t)]$$

$$G(\omega) = \frac{1}{2\pi} H_p(\omega)[H_I(\omega) * H_I(\omega)] \tag{4.54}$$

$$\sum_{k=-\infty}^{\infty} g(-kT_B) = \frac{1}{T_B} \sum_{n=-\infty}^{\infty} b_M G\left(\frac{2\pi N}{T_B} \right)$$

The worst case total noise power is therefore given by

$$\left\langle \left[i_{SN}^F(0) \right]^2 \right\rangle = \left\langle i_{sd}^2 \right\rangle + \left\langle i_{SN}^2(0) \right\rangle \tag{4.55}$$

with $\langle i_{sd}^2 \rangle$ the shot noise due to the dark current of the PD.

4.5.4   TOTAL OUTPUT NOISES AND PULSE SHAPE PARAMETERS

It is now advantageous to normalize the input optical pulse and the output equalized pulse as follows:

$$h_p'(t) = T_B h_p(tT_B)$$
$$h_o'(t) = T_B h_o(tT_B) \tag{4.56}$$

Or equivalently in the frequency domain

$$H_p'(f) = H_p\left(\frac{2\pi f}{T_B} \right)$$
$$H_o'(f) = \frac{1}{T_B} H_o\left(\frac{2\pi f}{T_B} \right) \tag{4.57}$$

These temporal and spectral functions depend only on the shapes of the input and output pulses and not on the epoch T. The normalization can be achieved by setting

$$H'_p(0) = 1 \qquad (4.58)$$

The requirement on $H'_o(f)$ for zero ISI is

$$\sum_{-\infty}^{\infty} H'_0(f + k) = 1 \ \text{ for } |f| < 0.5 \qquad (4.59)$$

Thus, all the output noises can be expressed in terms of these normalized functions. We can thus obtain from Equation 4.54

$$H_I\left(\frac{2\pi f}{T_B}\right) = \frac{T_B}{\Re G} H'(f)$$

$$\text{with} \ \ H'(f) = \frac{H'_o(f)}{H'_p(f)} \qquad (4.60)$$

The normalized effects of the pulse shaping are given as:

$$I_1 = \int_{-\infty}^{\infty} H'_p[H'*H'(f)]df$$

$$I_2 = \int_{-\infty}^{\infty} |H'(f)|^2 \, df \qquad (4.61)$$

$$I_3 = \int_{-\infty}^{\infty} |H'(f)|^2 \, f^2 df$$

The term I_1 influences the quantum shot noise of the PD from the optical signal received at the input, the term I_2 influences the dark current shot noises and the thermal noises due to the input resistance of the preamplifier, and the term I_3 influences the frequency-dependent parts of the transfer function of the electronic preamp.

4.5.4.1 FET Front End Optical Receiver

As an example of the estimation of noises at the output of the optical receiver, we illustrate an optical receiver with a PD followed by an FET front end. As shown in Figure 4.5, for an FET front end we have the following parameters:

$$C = C_a + C_d \qquad (4.62)$$

$$R_p = \frac{R_b R_i}{R_b + R_i} \qquad (4.63)$$

where:
   R_b is the bias resistance
   R_i is the input resistance of the amplifier at the midband
   C_d is the diode capacitance under reverse bias
   C_a is the input capacitance of the amplifier

The spectral density of the thermal noises due to the bias resistance and the feedback resistance R_f is given by

$$S_R = \frac{2k_BT}{R_f}$$

$$S_{R_b} = \frac{2k_BT}{R_b}$$

(4.64)

with T the temperature and k_B Boltzmann's constant. Then the noise power at the equalizer output is given by

$$\langle i_{RN}^2 \rangle = 2k_BT\left(\frac{1}{R_f}+\frac{1}{R_b}\right)\cdot\int_{-\infty}^{\infty}|H_I(2\pi f)|^2\,\mathrm{d}f = 2k_BT\left(\frac{1}{R_f}+\frac{1}{R_b}\right)\cdot\left(\frac{1}{\Re G}\right)^2 I_2T_b$$

(4.65)

where T_b is the bit period and T is the absolute temperature. The noise due to the transconductance of the FET is given by

$$\langle i_{g_mN}^2 \rangle = \frac{2k_BT\Gamma P}{g_m}\cdot\frac{1}{2\pi}\int_{-\infty}^{\infty}|H_I(2\pi f)|^2\left(\frac{1}{R_f^2}+\omega^2C^2\right)\mathrm{d}\omega$$

$$= \frac{2k_BT\Gamma P}{g_m}\cdot\left(\frac{1}{\Re G}\right)^2\left(\frac{T_bI_2}{R_f^2}+\frac{(2\pi C)^2}{T_b}I_3\right)$$

(4.66)

with g_m the transconductance of the FET, $\Gamma \sim 0.7$ the factor related to materials (0.7 for GaAs), and P the imperfection factor of the FET device.

The quantum shot noise of the PD is given by

$$\langle [i_{SN}^F(0)]^2 \rangle = qG^{2+x}\Re\left\{ \frac{b_M}{T}\sum_{\substack{k=-\infty \\ k\neq 0}}^{\infty}b_MG\left(\frac{2\pi N}{T}\right) - \frac{b_M-a_0}{2}\int_{-\infty}^{\infty}G(\omega)\mathrm{d}\omega \right\}$$

$$= \frac{q}{\Re}G^x\left[a_0I_1+b_M\left(\sum\nolimits_1-I_1\right) \right]$$

(4.67)

with $\displaystyle\sum\nolimits_1 = \sum_{-\infty}^{\infty}H_p'(k)\left[H'(k)*H'(k)\right]$

4.5.4.2 BJT Front End Optical Receiver

BJT can also be employed as the front end amplification stage of the optical receivers. A typical circuit design of such a receiver is shown in Figure 4.9, in which three BJTs have been used with a direct coupled pair and a peaking stage to extend the bandwidth of the overall gain and then a shunt feedback to form a TI configuration.

4.5.4.2.1 Noise Generators

Electrical shot noises are generated by the random generation of streams of electrons (current). In optical detection, shot noises are generated by (i) biasing currents in electronic devices and (ii) photocurrents generated by the photodiode.

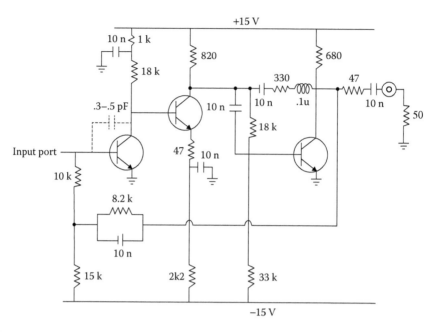

FIGURE 4.9 Design circuit of the electronic preamplifier for the balanced optical amplifier. All bipolar junction transistors (BJTs) are either of types Phillips BFR90A or BFT24 (see the detailed design in Chapter 5).

A biasing current I generates a spectral current density S_I given by

$$S_I = \frac{d(i_I^2)}{df} = 2qI \quad \text{in A}^2/\text{Hz} \tag{4.68}$$

where q is the electronic charge. The quantum shot noise $\langle i_s^2 \rangle$ generated by the PD by an optical signal with an average optical power P_in is given by

$$S_Q = \frac{d\langle i_s^2 \rangle}{df} = 2q\langle i_s^2 \rangle \quad \text{in A}^2/\text{Hz} \tag{4.69}$$

In the case when the APD is used, the noise spectral density is given by

$$S_Q = \frac{d\langle i_s^2 \rangle}{df} = 2q\langle i_s^2 \rangle\langle G_n^2 \rangle \quad \text{in A}^2/\text{Hz} \tag{4.70}$$

It is noted here again that the dark currents generated by the PD must be included to the total equivalent noise current at the input after it is evaluated. These currents are generated even in the absence of the optical signal. These dark currents can be eliminated by cooling the PD to at least below the temperature of liquid nitrogen (77 K).

At a certain temperature, the conductivity of a conductor varies randomly. The random movement of electrons generates a fluctuating current even in the absence of an applied voltage. The thermal noise spectral density of a resistor R is given by

$$S_R = \frac{d\left(i_R^2\right)}{df} = \frac{4k_BT}{R} \quad \text{in A}^2/\text{Hz} \tag{4.71}$$

where:

k_B is Boltzmann's constant

T is the absolute temperature (in K)

R is the resistance (in ohms)

i_R denotes the noise current due to resistor R

4.5.4.2.2 Equivalent Input Noise Current

Our goal is to obtain an analytical expression of the noise spectral density equivalent to a source looking into the electronic amplifier including the quantum shot noises of the PD. A general method for deriving the equivalent noise current at the input is by representing the electronic device by a Y-equivalent linear network as shown in Figure 4.3. The two current noise sources $di_n'^2$ and $di_n''^2$ represent the summation of all noise currents at the input and at the output of the Y-network. This can be transformed into a Y-network circuit with the noise current referred to the input as follows:

The output voltages V_0, as referred to in Figure 4.6, can be written as

$$V_0 = \frac{i_N'(Y_f - Y_m) + i_N''(Y_i + Y_f)}{Y_f(Y_m + Y_i + Y_o + Y_L) + Y_i(Y_o + Y_L)} \tag{4.72}$$

and then using the equivalent model, we have

$$V_0 = \frac{(i_N')_{eq}(Y_f - Y_m)}{Y_f(Y_m + Y_i + Y_o + Y_L) + Y_i(Y_o + Y_L)} \tag{4.73}$$

Thus comparing these two equations, we can deduce the equivalent noise current at the input of the detector as

$$i_{Neq} = i_N' + i_N'' \frac{Y_i + Y_f}{Y_f - Y_m} \tag{4.74}$$

Then reverting to mean square generators for a noise source, we have

$$d(i_{Neq})^2 = d(i_N')^2 + d(i_N'')^2 \left| \frac{Y_i + Y_f}{Y_f - Y_m} \right|^2 \tag{4.75}$$

It is therefore expected that if the Y-matrix of the front end low-noise amplifier is known as the equivalent noise at the input of the amplifier, which can be obtained by using Equation 4.75.

For a given source, the input noise current power of a BJT front end can be found by (see also Chapter 5 and Appendix A)

$$i^2{}_{Neq} = \int_0^B d(i_{Neq})^2 = a + \frac{b}{r_E} + c r_E \tag{4.76}$$

where:

B is the bandwidth of the electronic preamplifier

r_E is the emitter resistance of the front end transistor of the preamplifier

a, b, and c are the parameters depending on the circuit elements and amplifier bandwidth

Hence, an optimum value of r_E can be found. An optimum biasing current is to be set for the collector current of the BJT such that i^2_{Neq} is at minimum,

$$r_{Eopt} = \sqrt{\frac{b}{c}} \quad \text{hence} \quad \rightarrow i^2_{Neq}\Big|_{r_E = r_{Eopt}} = a + 2\sqrt{bc} \tag{4.77}$$

If two types of BJT are considered as Phillips BFR90A and BFT24, then a good approximation of the equivalent noise power can be found as

$$a = \frac{8\pi B^3}{3}\left\{r_B C_s^2 + \left(C_s + C_f + C_{tE}\right)\tau_T\right\}$$

$$b = \frac{B}{\beta_N} \tag{4.78}$$

$$c = \frac{4\pi B^3}{3}\left(C_s + C_f + C_{tE}\right)^2$$

The theoretical estimation of the transistors can be derived from the measured scattering parameters as given by the manufacturer as

BFR90A

$r_{Eopt} = 59 \ \Omega$

$I_{Eopt} = 0.44 \ \text{mA}$ \hfill (4.79)

$I_{eq}^2 = 7.3 \times 10^{-16} \ \text{A}^2 \longrightarrow \dfrac{I_{eq}^2}{B} = 4.9 \times 10^{-24} \ \text{A}^2/\text{Hz}$

$I_{eq} = 27 \ \text{nA} \longrightarrow \dfrac{I_{eq}}{\sqrt{B}} = 2.21 \ \text{pA}/\sqrt{\text{Hz}}$

BFT24

$r_{Eopt} = 104 \ \Omega$

$I_{Eopt} = 0.24 \ \text{mA}$ \hfill (4.80)

$I_{eq}^2 = 79.2 \times 10^{-16} \ \text{A}^2 \longrightarrow \dfrac{I_{eq}^2}{B} = 6.1 \times 10^{-24} \ \text{A}^2/\text{Hz}$

$I_{eq} = 30.2 \ \text{nA} \longrightarrow \dfrac{I_{eq}}{\sqrt{B}} = 2.47 \ \text{pA}/\sqrt{\text{Hz}}$

Note that the equivalent noise current depends largely on some not well-defined values such as the capacitance, the transit times and the base spreading resistance, and the short circuit current gain β_N. The term $I_{eq}/\sqrt{B}$ is usually specified as the noise spectral density equivalently referred to the input of the electronic preamplifier.

4.6   AN HEMT MATCHED NOISE NETWORK PREAMPLIFIER

A third-order noise matching network (shown in Figure 4.10) has been obtained by Park [11] to tailor for 10 GHz low-noise optical receiver. This network is inserted in our preamplifier modeling and the electronic circuit is connected as shown in Figure 4.11.

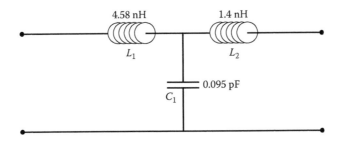

FIGURE 4.10 Third-order noise matching network.

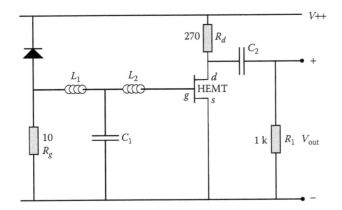

FIGURE 4.11 Electronic circuit of an HEMT noise matched network preamplifier.

For high receiving bit rate, we need to design a preamplifier with high-frequency response. In electronics, the drain-to-source high electron mobility transistor (HEMT) parasitic capacitance, Cds, is negligibly small at high frequency. Furthermore, we could reduce the complexity of solving this circuit by considering the *Miller effect* capacitance. Thus, we are able to construct the small signal high-frequency response equivalent circuit for the electronic circuit and the simplified equivalent circuit is shown in Figure 4.11.

The above equivalent circuit shown in Figure 4.12 can be further simplified, as shown in Figure 4.13.

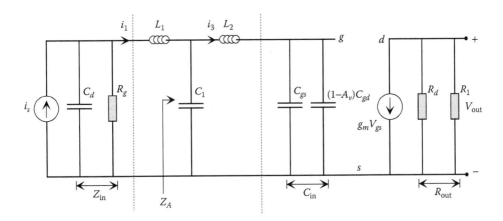

FIGURE 4.12 Equivalent circuit of an HEMT noise matched network preamplifier.

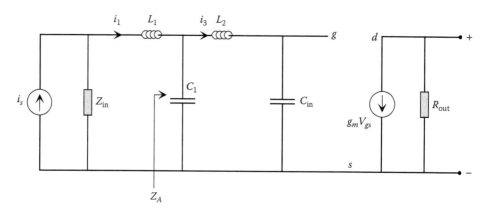

FIGURE 4.13   Simplified equivalent circuit of an HEMT noise matched network preamplifier.

Thus, the required preamplifier's transfer function can be obtained as follows:

$$i_3 = \left(\frac{sC_{in}}{s^3 L_2 C_1 C_{in} + sC_1 + sC_{in}} \right) \cdot i_1 \tag{4.81}$$

$$i_1 = \left(\frac{Z_{in}}{Z_{in} + Z_A} \right) \cdot i_s \tag{4.82}$$

By substituting Equations 4.82 into 4.81, we obtain the transfer function of the HEMT matched noise network preamplifier,

$$H_{HEMT} = \frac{V_{out}}{i_s} = -g_m R_{out} \left(\frac{1}{s^3 L_2 C_1 C_{in} + sC_1 + sC_{in}} \right) \cdot \left(\frac{Z_{in}}{Z_{in} + Z_A} \right) \tag{4.83}$$

where:

$$R_{out} = \frac{R_d R_l}{R_d + R_l} \tag{4.84}$$

$$Z_{in} = \frac{R_g}{sR_g C_d + 1} \tag{4.85}$$

$$Z_A = sL_1 + \frac{\left(\dfrac{1}{sC_1} \right) \cdot \left(sL_2 + \dfrac{1}{sC_{in}} \right)}{\dfrac{1}{sC_1} + \dfrac{1}{sC_{in}} + sL_2} = \frac{s^4 L_1 L_2 C_1 C_{in} + s^2 L_1 C_1 + s^2 L_1 C_{in} + s^2 L_2 C_{in} + 1}{s^3 L_2 C_1 C_{in} + s\left(C_1 + C_{in} \right)} \tag{4.86}$$

4.6.1   Noise Theory and Equivalent Input Noise Current

The circuit diagram in Figure 4.20 can be represented with reflection coefficients shown in Figure 4.16. We denote the reflection coefficient of Z_{ph}, Z_s, and Z^*_{opt} by Γ_{ph}, Γ_s, and Γ^*_{opt}, which are normalized to Z_0 and defined by

$$\Gamma_x = \frac{Z_x - Z_0}{Z_x + Z_0} \tag{4.87}$$

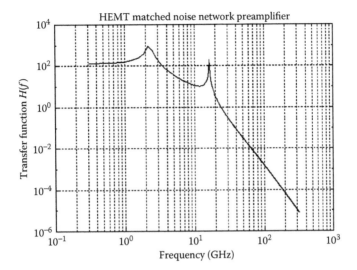

FIGURE 4.14 Frequency response of an HEMT noise matched network preamplifier.

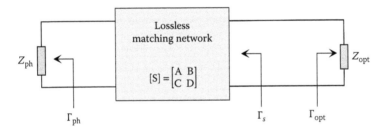

FIGURE 4.15 Lossless matching network with source and load impedances.

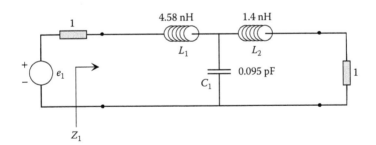

FIGURE 4.16 Matched third-order noise matching network to find the S-parameters.

where x = ph, s, and opt. It is useful to obtain the S-parameters of the third-order noise matching network defined in the matrix form (Figures 4.14 and 4.15) as

$$S = \begin{bmatrix} S_{11} & S_{12} \\ S_{21} & S_{22} \end{bmatrix} \tag{4.88}$$

From Figure 4.16, we find Z_1, which is given by

$$Z_1 = sL_1 + \left(sC_1 + \frac{1}{1+sL_2} \right)^{-1} = \frac{s^3 C_1 L_1 L_2 + s^2 C_1 L_1 + s(L_1 + L_2) + 1}{s^2 C_1 L_2 + sC_1 + 1} \tag{4.89}$$

Applying the circuit theory of the T-network and by substituting Equation 4.88, we obtain

$$S_{11} = \frac{Z_1 - 1}{Z_1 + 1} = \frac{s^3 C_1 L_1 L_2 + s^2 (C_1 L_1 - C_1 L_2) + s(L_1 + L_2 - C_1)}{s^3 C_1 L_1 L_2 + s^2 (C_1 L_1 + C_1 L_2) + s(L_1 + L_2 + C_1) + 2} \tag{4.90}$$

From circuit theory, S_{21} is given by

$$S_{21} = 2\sqrt{\frac{r_1}{r_2}} \left(\frac{V_2}{e_1} \right) \tag{4.91}$$

From Figure 4.16, we can obtain

$$\frac{V_2}{e_1} = \frac{1}{s^3 C_1 L_1 L_2 + s^2 (C_1 L_1 + C_1 L_2) + s(L_2 + L_1 + C_1) + 2} \tag{4.92}$$

Hence, by substituting Equation 4.92 into 4.90, we have

$$S_{21} = \frac{2}{s^3 C_1 L_1 L_2 + s^2 (C_1 L_1 + C_1 L_2) + s(L_2 + L_1 + C_1) + 2} \tag{4.93}$$

Thus using the symmetrical property, the other two S-parameters can be obtained as follows:

$$S_{22} = S_{11} = \frac{s^3 C_1 L_1 L_2 + s^2 (C_1 L_1 - C_1 L_2) + s(L_1 + L_2 - C_1)}{s^3 C_1 L_1 L_2 + s^2 (C_1 L_1 + C_1 L_2) + s(L_1 + L_2 + C_1) + 2} \tag{4.94}$$

$$S_{12} = S_{21} = \frac{2}{s^3 C_1 L_1 L_2 + s^2 (C_1 L_1 + C_1 L_2) + s(L_2 + L_1 + C_1) + 2} \tag{4.95}$$

By using Equation 4.87, we can derive all the required reflection coefficients. First of all, consider the equivalent photodiode circuit shown in Figure 4.17.

Then, Z_{ph} can be found as

$$Z_{ph} = R_s + \frac{1}{sC_d} \tag{4.96}$$

From Equation 4.96, we have

$$\Gamma_{ph} = \frac{Z_{ph} - Z_0}{Z_{ph} + Z_0} \tag{4.97}$$

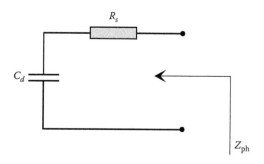

FIGURE 4.17   Equivalent circuit of a photodiode detector.

Then, we substitute Z_{ph} from Equation 4.96

$$\Gamma_{ph} = \frac{sC_d(R_s - 1) + 1}{sC_d(R_s + 1) + 1} \tag{4.98}$$

Knowing the S-parameters and Γ_{ph}, we can obtain Γ_s as

$$\Gamma_s = S_{22} + \frac{S_{12}S_{21}\Gamma_{ph}}{1 - \Gamma_{ph}S_{11}} \tag{4.99}$$

and

$$\Gamma_{opt} = \Gamma_s = S_{22} + \frac{S_{12}S_{21}\Gamma_{ph}}{1 - \Gamma_{ph}S_{11}} \tag{4.100}$$

The equivalent input noise current density appearing across the photodiode junction capacitance can be expressed in terms of reflection coefficients [11] as

$$i_{eq}^2 = 4kTR_s(\omega C_d)^2 \frac{\left(1 - \left|\Gamma_{opt}^*\right|^2\right)}{G_M\left|1 - \Gamma_s\Gamma_{opt}^*\right|^2}\left\{F_{min}\left(1 - \left|\Gamma_s\right|^2\right) + \frac{4R_n}{Z_0}\frac{\left|\Gamma_s - \Gamma_{opt}\right|^2}{\left|1 + \Gamma_{opt}\right|^2}\right\}\Delta f \tag{4.101}$$

where F_{min} and R_n can be obtained from the given data in Table 4.2, and G_M is the transducer power gain of the lossless matching network. Our goal is to minimize the equivalent input noise current density. This can be done by minimizing the photodiode junction capacitance C_d, the series resistance R_s of photodiode, the minimum noise figure F_{min}, and the noise resistance R_n, and by maximizing the transducer power gain. Among these parameters in Equation 4.101, only the transducer power gain G_M and the output reflection coefficient Γ_s are related to the noise matching network and these may be optimized by design. The transducer power gain can be expressed in terms of reflection coefficients [11] and is defined by

$$G_M = 1 - \left|\frac{(1 + \Gamma_s)(1 - \Gamma_{opt}) - (1 - \Gamma_s)(1 + \Gamma_{opt})}{(1 + \Gamma_s)(1 - \Gamma_{opt}^*) + (1 - \Gamma_s)(1 + \Gamma_{opt}^*)}\right|^2 \tag{4.102}$$

TABLE 4.2
Noise Parameters of the Packaged HEMT with
0.3 μm Gate Length

Frequency (GHz)	Γ_{opt} (Mag)	Γ_{opt} (Ang)	F_{min} (dB)	R_n (Ω)
2	0.79	30	0.33	29
4	0.73	59	0.35	21
6	0.68	87	0.44	14.5
8	0.63	119	0.55	9.5
10	0.59	139	0.66	6
12	0.55	164	0.75	4

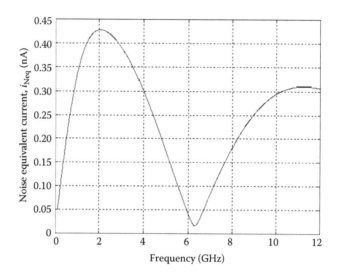

FIGURE 4.18   Equivalent input noise current density of an HEMT noise matched front end receiver.

The transducer power gain has its maximum value of unity when $\Gamma_s = \Gamma_{opt}$. This condition is equivalent to the output admittance of the noise matching network being conjugately matched to the complex conjugate of the optimum source admittance (Figure 4.18).

4.7   REMARKS

This chapter has addressed the considerations of optical receivers in which the electronic currents are generated when the optical energy/power from the modulated optical signals is absorbed by the photosensitive regions directly. The noise generation process is described and related small signal models of electronic preamplifier. An equivalent noise current as seen at the input port of the electronic preamplifier is presented as well as method to derive it for the front end amplifier. Both FET and BJT types are given. Further, the effects of the pulse shape on the output signals are also given related to the transfer function of electronic preamplifier. A noise matching network at the front end of the optical preamplifier would reduce the noise effects at high frequency and this is crucial in the design of an ultra-wide-band optical receiver.

These noise models can be integrated into coherent receivers that will be described in the chapter. For a coherent optical receiver, a local oscillator, a powerful laser mixes the optical signals with its output field and gives a beating signal in the PD due to its square law property. The phase and amplitude of the optical signals are preserved in the electronic domain and thus processing in the electronic domain can be performed to extract the digital and analog property of the original signals.

APPENDIX A: NOISE EQUATIONS IN SMALL SIGNAL AMPLIFIERS

Referring to the small signal and noise model given in Figures 4.19 and 4.20, the noises generated in a transistor can be expressed as

$$di_1^2 = 4k_B T g_B df$$

$$di_2^2 = 2qI_c df \simeq 2k_B T g_m df$$

$$di_3^2 = 2qI_B df \simeq 2k_B T (1 - \alpha_N) g_m df$$

$$(4.103)$$

$$di_4^2 = \text{shot noise of diodes and thermal noise of bias in } g_B \text{ resistors}$$

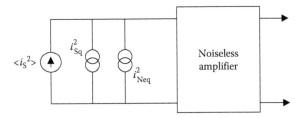

FIGURE 4.19 Equivalent noise current at the input and noiseless amplifier model; i_{Sq}^2 is the quantum shot noise, which is signal dependent, and i_{Neq}^2 is the total equivalent noise current referred to the input of the electronic amplifier.

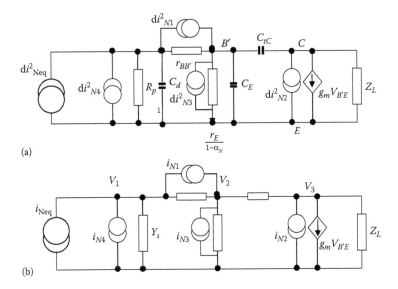

FIGURE 4.20 (a) Approximated noise equivalent and small signal model of a BJT front end; (b) generalized noise and small signal model circuit. Note that $r_B = r_{sd} + r_{BB'}$ and $C_d = C_p + C_i$, with $r_{BB'}$ the base resistance, r_{sd} the diode resistance, C_d the photodiode capacitance, and C_i the input capacitance.

with g_B the base conductance, I_C the collector bias current, and I_B the base bias current. From nodal analysis of the small signal equivalent circuit given in Figure 4.20, we can obtain the relationship

$$\begin{bmatrix} Y_s + g_B & -g_B & 0 \\ -g_B & g_B + y_1 + y_f & -y_f \\ 0 & g_m - y_f & y_f + y_2 \end{bmatrix} \begin{bmatrix} V_1 \\ V_2 \\ V_3 \end{bmatrix} = \begin{bmatrix} i_{eq} - i_{N1} \\ i_{N1} + i_{N3} \\ i_{N2} \end{bmatrix} \tag{4.104}$$

Hence, V_3, V_2, and V_1 can be found by using Euler's rule for the matrix relationship

$$V_3 = \frac{\Delta_{13}}{\Delta}(i_{eq} - i_{N1}) + \frac{\Delta_{23}}{\Delta}(i_{N1} + i_{N3}) + \frac{\Delta_{33}}{\Delta}(i_{N2}) \tag{4.105}$$

Hence, the noise currents as referred to the input are

$$d(i_1'^2) = \left| \frac{Y_s}{g_B} \right|^2 dI_1^2 = \omega_2 C_s^2 r_B 4kT df \tag{4.106}$$

$$d(i_2'^2) = \left| (Y + y_1 + y_f) + \frac{Y_s}{g_B}(y_1 + y_f) \right|^2 \left| \frac{1}{y_f - g_m} \right|^2 = \left| \frac{1}{y_f - g_m} \right|^2 dI_2^2$$

$$= \left(\frac{1}{\beta_N r_E^2} + \omega^2 \left(C_0^2 - \frac{2C_B r_B}{\beta_N r_E} \right) + \omega^4 2C_s r_B C_B^2 \right) 2kTr_E df \qquad (4.107)$$

$$d(i_3'^2) = \left| \frac{Y_s}{g_B} + 1 \right|^2 dI_3^2 = \left(\omega C_s^2 r_B^2 + 1 \right) \frac{2kT}{\beta r_E} df \qquad (4.108)$$

where we have assumed that $\omega C_f \ll g_m$.

REFERENCES

1. Binh, L.N., Advanced Digital Optical Communications, 2ed, CRC Press Taylors and Francis Group, Boca Raton, FL, 2015.
2. Alferness, R.C., Waveguide electrooptic modulators, *IEEE J. Microwave Theory and Tech.*, 30, 1982, 1121–1137.
3. Booth, R.C., LiNbO$_3$, integrated optic devices for coherent optical fiber systems, *Thin Solid Films*, 126, 1985, 167–176.
4. Ritchie, S. and A.G. Steventon, The potential of semiconductors for optical-integrated circuits, *Proc. of Conference Digital Optical Circuit Technology*, NATO, France, vol. 362, 1985, pp. 1111–1120.
5. Personick, S.D., Receiver design for digital fiber optics communications Systems Part I, *Bell Syst. Tech. J.*, 52, 1973, 843–86.
6. Personick, S.D., Optical detectors and receivers, *IEEE J. Lightwave Technol.*, 26(9), 1976, 1005–1020.
7. Hullet, J.L. and T.V. Muoi, A feedback receive amplifier for optical transmission systems, *IEEE Trans. Com.*, 24(10), 1976, 1180–1185.
8. Dogliotti, R., G. Luvison, and G. Pirani, Error probability in optical fiber transmission systems, *IEEE Trans. Inf. Theory*, 25(2), 1979, 170–178.
9. Personick, S.D., *Optical Fiber Transmission Systems*. Plenum Press, New York, 1980.
10. Personick, S.D., Receiver design for digital fiber optic communications systems, *Bell Syst. Tech. J.*, 52, 1973, 843–886.
11. Yamamoto, Y., Receiver performance evaluation of various digital optical modulation-demodulation systems in the 0.5–1.0 μm wavelength region, *IEEE J. Quant. Electron.*, 16, 1980, 1251–1259.

5 Noise Suppression Techniques in Balanced Receiver

It has been shown that a balanced optical receiver can suppress the excess noise intensity generated from the local oscillator (LO) [1–3]. In a balanced receiver, an optical coupler is employed to mix a weak optical signal with a LO field of average powers of E_S^2 and E_L^2, respectively. Figure 5.1 shows the generic diagram of a balanced optical receiver under a coherent detection scheme in which two photodetectors and amplifiers are operating in a push–pull mode.

Usually the magnitude of the optical field of the LO E_L is much greater than E_S, the signal field. The fields at the output of a 2×2 coupler can thus be written as

$$
\begin{bmatrix} \tilde{E}_1 \\ \tilde{E}_2 \end{bmatrix} = \begin{bmatrix} \sqrt{1-k} & \sqrt{k}\,e^{j\pi/2} \\ \sqrt{k}\,e^{j\pi/2} & \sqrt{1-k} \end{bmatrix} \begin{bmatrix} \tilde{E}_S \\ \tilde{E}_L \end{bmatrix}
\tag{5.1}
$$

where:
E_S is the field of the received optical signal
E_L is the field of the LO (possibly a distributed feedback laser)
k is the intensity coupling coefficient of the coupler

For a 3 dB coupler, we have $k = 0.5$. The E_S and E_L fields can then be written as

$$
\begin{aligned}
E_S &= \sqrt{2}\,\tilde{E}_S\,e^{j(\omega_1 t + \phi_1)} \\
E_L &= \sqrt{2}\,\tilde{E}_L\,e^{j(\omega_2 t + \phi_2)}
\end{aligned}
\tag{5.2}
$$

where $\tilde{E}_S$ and $\tilde{E}_L$ are the magnitude of the optical fields of the signal and the LO laser. The electronic currents generated from the photodetectors (PDs) corresponding to the average optical power and the signals are given by

$$
i_1(t) = \frac{\Re}{2}|E_1|^2 = \frac{\Re_1}{2}\left\{ \begin{array}{l} \dfrac{1}{2}\left(E_S^2 + E_L^2\right) + \\[2mm] E_S E_L \cos\left[\left(\omega_1 + \omega_2\right)t + \phi_1 + \phi_2 - \dfrac{\pi}{2}\right] \end{array} \right\} + N_1(t)
\tag{5.3}
$$

$$
i_2(t) = \frac{\Re}{2}|E_2|^2 = \frac{\Re_2}{2}\left\{ \begin{array}{l} \dfrac{1}{2}\left(E_S^2 + E_L^2\right) \\[2mm] + E_S E_L \cos\left[\begin{array}{l} \left(\omega_1 + \omega_2\right)t \\[1mm] + \phi_1 + \phi_2 - \dfrac{\pi}{2} \end{array} \right] \end{array} \right\} + N_2(t)
\tag{5.4}
$$

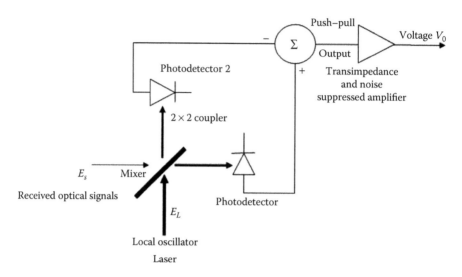

FIGURE 5.1 Generic block diagram of a coherent balanced detector optical receiver.

where:

ϕ_1, ϕ_2 and ω_1, ω_2 are the phase and frequency of the signal and LO, respectively
$N_1(t)$ and $N_2(t)$ are the noises resulting from the photodetection process
$\mathfrak{R}_1, \mathfrak{R}_2$ are the responsivities of the PDs 1 and 2, respectively

It is assumed that the two photodiodes have the same quantum efficiency, thence their responsivity. Thus, the total current of a back-to-back PD pair referred to the input of the electronic amplifier is

$$i_{Teq}(t) = \left\{ \begin{array}{l} I_{dc1} - I_{dc2} \\ \\ + (\mathfrak{R}_1 + \mathfrak{R}_2) E_S E_L \cos \left[\begin{array}{l} (\omega_1 - \omega_2)t \\ \\ + \phi_1 - \phi_2 - \dfrac{\pi}{2} \end{array} \right] \end{array} \right\} + N_1(t) + N_2(t) \qquad (5.5)$$

where the first two terms I_{dc1} and I_{dc2} are the detector currents generated by the detector pair due to the reception of the power of the LO, which is a continuous wave source; thus these terms appear as the DC constant currents that are normally termed as the shot noises due to the optical power of the LO. The third term is the beating between the LO and the signal carrier, the signal envelop. The noise currents $N_{1eq}(t)$ and $N_{2eq}(t)$ are seen as the equivalence at the input of the electronic amplifier from the noise processes that are generated by the PDs and the electronic amplifiers. Noise components are the shot noise due to bias currents, the quantum shot noise that is dependent on the strength of the signal and the LO, the thermal noise due to the input resistance of the amplifier and the equivalent noise current referred to the input of the electronic amplifier contributed by all the noise sources at the output port of the electronic amplifier. We denote by $N_{lep}(t)$ the quantum shot noise generated due to the average current produced by the PD pair in reception of the average optical power of the signal sequence.

The difference of the produced electronic currents can be derived using the following techniques: (i) the generated electronic currents of the PDs can be coupled through a 180° microwave coupler and then fed to an electronic amplifier; (ii) the currents are fed to a differential electronic amplifier; or (iii) a balanced PD pair connected back to back and fed to a small signal electronic amplifier [4]. The first two techniques require stringent components and normally not preferred as contrasted by the high performance of the balanced receiver structure of (iii).

Regarding the electronic amplifier, the electronic preamplifier, a transimpedance (TI) configuration is selected due to its wide band and high dynamics. However, it suffers high noises due to the equivalent input impedance of the shunt feedback impedance, normally around a few hundred ohms. In the next section, the theoretical analysis of noises of the optical balanced receiver is described.

5.1 ANALYTICAL NOISE EXPRESSIONS

In the electronic preamplifier, the selection of the transistor as the first-stage amplifying device is very critical. Either a field effect transistor (FET) or bipolar junction transistor (BJT) could be used. However, the BJT is preferred for wide band application due to its robustness to noises. The disadvantages of using the BJT as compared with the FET are due to its small base resistance that leads to high thermal noise. However, for a shunt feedback amplifier, the resistance of Millers' equivalent resistance as referred to the input of the amplifier is much smaller and thus dominates over that of the base resistance. The advantage of the BJT over the FET is that its small signal gain follows an exponential trend with respect to the small variation of the driving current derived from the photodetection as compared with parabolic for FET. The FET may also offer high input impedance between the gate to source of the input port, but may not offer much improvement in a feedback amplification configuration in terms of noises. This section focuses on the use and design of BJT multistage shunt feedback electronic amplifiers.

Noises of electronic amplifying devices can be represented by superimposing all the noise current generators to the small signal equivalent circuit. These noise generators represent the noises introduced into the circuit by physical sources/processes at different nodes. Each noise generator can be expressed by a noise spectral density current square or power as shown in Figure 5.2. This figure gives a general model of small signal equivalent circuit including noise current sources of any transistor, which can be represented by the transfer $[y]$ matrix parameters. Indeed, the contribution of the noise sources of the first stage of the electronic preamplifier is dominant plus that of the shunt feedback resistance.

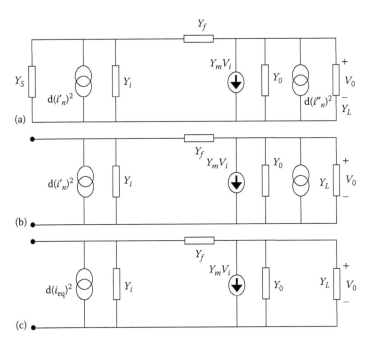

FIGURE 5.2 Small equivalent circuits including noise sources: (a) Y-parameter model representing the ideal current model and all current noise sources at the input and output ports, (b) with noise sources at input and output ports, and (c) with a total equivalent noise source at the input.

The current generator $di_n'^2/df$ is the total equivalent noise generator contributed by all the noise sources to node 1 including the thermal noises of all resistors connected to the node. Similarly, for the noise source, $di_n''^2/df$ is referred to the output node.

5.2   NOISE GENERATORS

Electrical shot noises are generated by the random generation of streams of electrons (current). In optical detection, shot noises are generated by (i) biasing currents in electronic devices and (ii) photo currents generated by the photodiode.

A biasing current I generates a shot noise power spectral density S_I given by

$$S_I = \frac{d(i_I^2)}{df} = 2qI \qquad A^2/Hz \tag{5.6}$$

where q is the electronic charge. The quantum shot noise $\langle i_s^2 \rangle$ generated by the PD by an optical signal with an average optical power P_{in} is given by

$$S_Q = \frac{d\langle i_s^2 \rangle}{df} = 2q\langle i_s^2 \rangle \qquad A^2/Hz \tag{5.7}$$

If the PD is an APD type, then the noise spectral density is given by

$$S_Q = \frac{d\langle i_s^2 \rangle}{df} = 2q\langle i_s^2 \rangle\langle G_n^2 \rangle \qquad A^2/Hz \tag{5.8}$$

It is noted here again that the dark currents generated by the PD must be included to the total equivalent noise current at the input after it is evaluated. These currents are generated even in the absence of the optical signal. These dark currents can be eliminated by cooling the PD to at least below the temperature of liquid nitrogen (77 K).

At a certain temperature, the conductivity of a conductor varies randomly. The random movement of electrons generates a fluctuating current even in the absence of an applied voltage. The thermal noise spectral density of a resistor R is given by

$$S_R = \frac{d\left(i_R^2\right)}{df} = \frac{4k_BT}{R} \qquad \text{in } A^2/Hz \tag{5.9}$$

where:
  k_B is Boltzmann's constant
  T is the absolute temperature (in K)
  R is the resistance in ohms
  i_R denotes the noise current due to resistor R

5.3   EQUIVALENT INPUT NOISE CURRENT

Our goal is to obtain an analytical expression of the noise spectral density equivalent to a source looking into the electronic amplifier including the quantum shot noises of the PD. A general method for deriving the equivalent noise current at the input is given by representing the electronic device by a Y-equivalent linear network as shown in Figure 5.3. The two current noise sources $di_n'^2$ and $di_n''^2$ represent the summation of all noise currents at the input and at the output of the Y-network. This can be transformed into a Y-network circuit with the noise current referred to the input as follows.

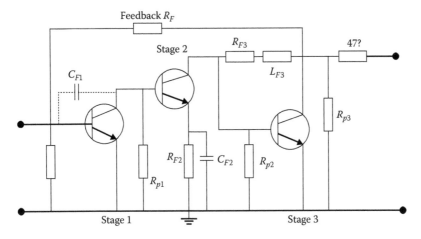

FIGURE 5.3 AC circuit model of a three-stage feedback electronic preamplifier.

The output voltages V_0 of Figure 5.3a can be written as

$$V_0 = \frac{i'_N(Y_f - Y_m) + i''_N(Y_i + Y_f)}{Y_f(Y_m + Y_i + Y_o + Y_L) + Y_i(Y_o + Y_L)} \tag{5.10}$$

and for Figure 5.3b

$$V_0 = \frac{(i'_N)_{eq}(Y_f - Y_m)}{Y_f(Y_m + Y_i + Y_o + Y_L) + Y_i(Y_o + Y_L)} \tag{5.11}$$

Thus comparing the above two equations, we can deduce the equivalent noise current at the input of the detector as

$$i_{Neq} = i'_N + i''_N \frac{Y_i + Y_f}{Y_f - Y_m} \tag{5.12}$$

Then reverting to mean square generators for a noise source, we have

$$d(i_{Neq})^2 = d(i'_N)^2 + d(i''_N)^2 \left| \frac{Y_i + Y_f}{Y_f - Y_m} \right|^2 \tag{5.13}$$

It is therefore expected that if the Y-matrix of the front end section of the amplifier is known, the equivalent noise at the input of the amplifier can be obtained by using Equation 5.13.

We propose a three-stage electronic preamplifier in AC configuration as shown in Figure 5.3. The details of the design of this amplifier are given in the next section. Small-signal and associated noise sources of this amplifier are given in Section 5.8. As can be seen, this general configuration is a forward path of shut, series, and shut stages that reduces the interaction between stages due to the impedance levels [7]. Shunt–shunt feedback is placed around the forward path; hence stable transfer function is the transfer impedance that is important for transferring the generated electronic photo-detected current to the voltage output for further amplification and data recovery.

For a given source, the input noise current power of a BJT front end can be found by

$$i^2{}_{Neq} = \int_0^B d(i_{Neq})^2 = a + \frac{b}{r_E} + cr_E \tag{5.14}$$

where:

B is the bandwidth of the electronic preamplifier

r_E is the emitter resistance of the front end transistor of the preamplifier

The parameters $a, b,$ and c are dependent on the circuit elements and amplifier bandwidth

Hence, an optimum value of r_E can be found and an optimum biasing current can be set for the collector current of the BJT such that i_{Neq}^2 is at its minimum, as

$$r_{Eopt} = \sqrt{\frac{b}{c}} \quad \text{hence} \quad \rightarrow i^2_{Neq}\Big|_{r_E = r_{Eopt}} = a + 2\sqrt{bc} \qquad (5.15)$$

If two types of BJT are considered, Phillips BFR90A and BFT24,[*] then a good approximation of the equivalent noise power can be found as

$$a = \frac{8\pi B^3}{3}\left\{ r_B C_s^2 + \left(C_s + C_f + C_{tE}\right)\tau_T \right\}$$

$$b = \frac{B}{\beta_N} \qquad (5.16)$$

$$c = \frac{4\pi B^3}{3}\left(C_s + C_f + C_{tE}\right)^2$$

The theoretical estimation of the parameters of the transistors can be derived from the measured scattering parameters as given by the manufacturer [5] as

For transistor type BFR90A

$$r_{Eopt} = 59\ \Omega \quad \text{for } I_{Eopt} = 0.44\,\text{mA}$$

$$I_{eq}^2 = 7.3 \times 10^{-16}\,\text{A}^2 \quad \text{hence} \quad \frac{I_{eq}^2}{B} = 4.9 \times 10^{-24}\,\text{A}^2/\text{Hz} \qquad (5.17)$$

$$I_{eq} = 27\,\text{nA} \quad \text{thus} \quad \frac{I_{eq}}{\sqrt{B}} = 2.21\,\text{pA}/\sqrt{\text{Hz}}$$

For transistor type BFT24

$$r_{Eopt} = 104\ \Omega \quad \text{for } I_{Eopt} = 0.24\,\text{mA}$$

$$I_{eq}^2 = 79.2 \times 10^{-16}\,\text{A}^2 \quad \text{hence} \quad \frac{I_{eq}^2}{B} = 6.1 \times 10^{-24}\,\text{A}^2/\text{Hz} \qquad (5.18)$$

$$I_{eq} = 30.2\,\text{nA} \quad \text{thus} \quad \frac{I_{eq}}{\sqrt{B}} = 2.47\,\text{pA}/\sqrt{\text{Hz}}$$

Note that the equivalent noise current depends largely on some not well-defined values such as the capacitance, the transit times, the base spreading resistance, and the short circuit current gain β_N. The term $I_{eq}/\sqrt{B}$ is usually specified as the noise spectral density equivalent referred to the input of the electronic preamplifier.

---

[*] These BJT types were most popular for less than 1 GHz bandwidth amplifier design and can be considered as a standard transistor for the high-frequency design. The design model can be scaled to 100s GHz using a modern HEMT or SiGe CMOS transistor.

5.4 POLE-ZERO PATTERN AND DYNAMICS

An AC model of a three-stage electronic preamplifier is shown in Figure 5.3, and the design circuit is shown in Figure 5.4. As briefly mentioned above, there are three stages with feedback impedance from the output to the input. The subscripts of the resistors, capacitors, and inductors indicate the order of the stages. The first stage is a special structure of shunt feedback amplification in which the shunt resistance is increased to infinity. The shunt resistance is in order of hundreds of ohms for the required bandwidth, thus contributing to noises of the amplifier. This is not acceptable. The shunt resistance increases the pole of this stage and approaches the origin. The magnitude of this pole is reduced by the same amount of that of the forward path gain. Thus, the poles of the close loop amplifier remain virtually unchanged. As R_F is increased, the pole p_1 decreases but G_1 is increased. Hence, $G_1 p_1/s - p_1$ remains constant. Thus, the position of the root locus is almost unchanged.

A compensating technique for reducing the bandwidth of the amplifier is to add capacitance across the base collector of the first stage. This may be necessary if oscillation occurs due to the phase shift becoming unacceptable at $GH = 1$ with G the open loop gain and H the feedback transfer function.

The second stage is a series feedback stage with feedback peaking. The capacitance C_{F2} is chosen such that at high frequencies it begins to bypass the feedback resistor. Thence the feedback admittance partially compensates for the normal high-frequency drop in gain associated with the base and stray capacitances. Furthermore, if the capacitance is chosen such that $R_{F2}C_{F2} = \tau_r$, the transfer admittance and input impedance become single-pole, which would be desirable [6]. The first and second stages are direct coupled. Effect of the first stage pole on the root locus of a shunt feedback electronic amplifier is shown in Figure 5.5. The locus of the pole position due to the change of component values indicates the stability of the amplifier design.

The third stage uses an inductive peaking technique. For a shunt stage with a resistive load, the forward path gain has only one pole and hence there is only one real pole in the close loop response. A complex pole pair can be obtained by placing a zero in Z_{F2} and hence a pole in the feedback transfer function $H(\omega)$. Figure 5.6a–c shows the high-frequency singularity pattern of individual stages. Figure 5.6d shows the root-locus diagram. It can be calculated that the poles take up the positions in Figure 5.6e when the loop gain GH is 220.

In the low-frequency region, the loop gain has five poles and three zeroes on the negative axis, and two zeroes at the origin. The largest low-frequency pole is set at approximately the input coupling capacitor and the input resistance leading to a low-frequency cutoff of around tens of kilohertz.

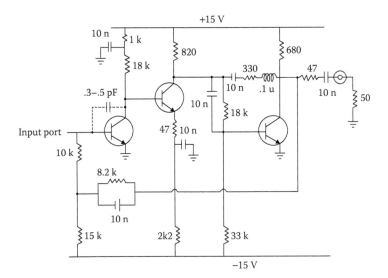

FIGURE 5.4 Design circuit of the electronic preamplifier for the balanced optical amplifier.

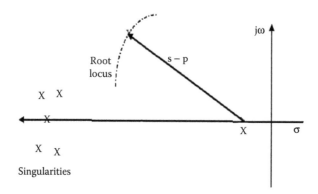

FIGURE 5.5 Effect of the first stage pole on the root locus of a shunt feedback electronic amplifier. The root locus is given by $1 = (GH)_{\text{midband}} \sum s - z_i / z_i \sum p_i / s - p_i$. "X" = poles, "o" zeroes positions of the transfer function.

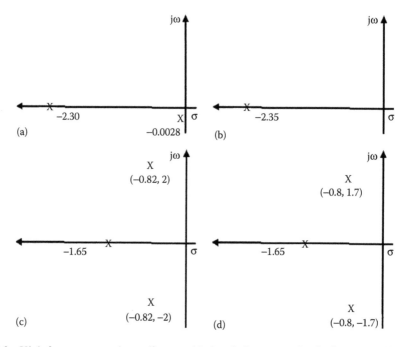

FIGURE 5.6 High-frequency root-locus diagram: (a) singularity pattern for the first stage, (b) second stage, (c) third stage, and (d) root locus for the close loop pattern for $GH = 220$.

The singularities of individual stages can be approximated by:

Stage 1: The dominant pole at

$$p = -\frac{1}{\beta_N R_{L1} C_{F1}} \tag{5.19}$$

and the other pole located at

$$p = -\frac{C_{F1}}{C_{F1} + C_{L1}} \frac{1}{\tau_T + C_{F1}/g_m} \tag{5.20}$$

and the midband gain $= \beta_N R_{L1}$ with the load of stage 1 of R_{L1}

Stage 2: The dominant pole at

$$p = -\frac{1}{r_B G_{T2} \tau_T}; \text{ and the midband gain} = G_{T2} = 1/R_{F2} \tag{5.21}$$

Stage 3: Complex pole pair at

$$|p|^2 = -\frac{R_L}{\tau_T^2 R_{F3}}; \sigma = -\frac{(R_L + R_{F3})}{2\tau_F R_{F3}}; \tau_F = \frac{L_{F3}}{R_{F3}}; \text{ and a zero at} -\frac{1}{\tau_F}; \text{ with a gain} = -R_{F3} \tag{5.22}$$

The feedback configuration of the circuit of the TI electronic preamplifier has a 10 K or 15 K resistor whose noises would contribute to the total noise current of the amplifier. The first and second stages are direct-coupled hence eliminating the level shifter and a significant amount of stray capacitance. The peaking capacitor required for the second stage is in order of 0.5 pF; hence no discrete component can be used. This may not require any component as the stay capacitance may suffice.

The equivalent noise current at the input of the TI amplifier (TIA) is approximately proportional to the square of the capacitance at the base of the transistor. Therefore, minimization of the stray capacitance at this point must be conducted by shortening the connection lead as much as possible. As a rough guide, critical points on the circuit where stray capacitance must be minimized are not at signal ground and have impedance and small capacitance to ground. For instance, at the base of the output stage, the capacitance to the ground is more tolerable at this point and therefore at the collector to the ground of the second transistor, than say the collector of the last stage, which should ideally have no capacitance to ground.

An acceptable step response can be achieved by manipulating the values of changing C_{F1}, R_{F3}, and L_{F3}. Since C_{F1} contributes to the capacitance at the base of the first stage, C_{F1} is minimized. The final value of C_{F1} is 0.5 pF (about half twist of two wires), R_{F1} is 330 Ω, and L_{F1} is about 0.1 μF (2 cm of wire). The amplifier is sensitive to parasitic capacitance between the feedback resistor and the first two stages. Thus, a grounded shield is placed between the 10 K resistor and the first and second stages.

The expressions for the singularities for the first two stages are similar as described above. The base collector capacitance of the third stage affected the position of the poles considerably. The poles of the singularities of this stage are as follows: the midband transresistance is $-R_{F3}$, with a zero at $z = 1/\tau_F$; and a complex pole pair given by

$$|p_p|^2 = \frac{1}{L_F C_F + R_F C_F \tau_T + \tau_T \tau_F \dfrac{R_F}{R_L}}; \quad \sigma = \frac{1}{2}|p_p|^2 \left(R_F C_F + \tau_T \left(1 + \frac{R_F}{R_L} \right) \right) \tag{5.23}$$

In addition, a large pole given approximately at

$$p = -\frac{1}{L_F C_F \tau_T |p_p|^2} \tag{5.24}$$

Figure 5.7 shows the open loop singularities (a), the root-locus diagram (b), and the close loop (c). The two large poles can be ignored without any significant difference. Similarly, for the root-locus diagram, the movement of the large poles is negligible, the pole and zero pair can be ignored, and the movement of the remaining three poles can be calculated by considering just these three poles as the open loop singularities.

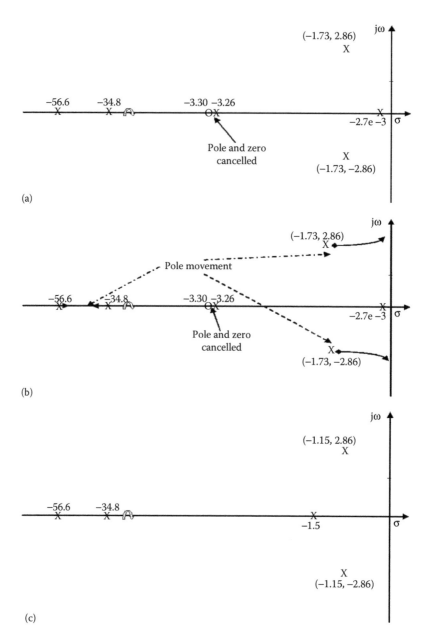

FIGURE 5.7 Pole-zero patterns in the s-plane and their dynamics of (a) open loop, (b) root locus, and (c) close loop.

5.5 RESPONSES AND NOISES MEASUREMENTS

5.5.1 RISE-TIME AND 3 DB BANDWIDTH

Based on the pole-zero patterns given in Section 5.4, the step responses can be estimated and contrasted with the measured curve as shown in Figure 5.8. The experimental setup for the rise time measurement is shown in Figure 5.9 in the electrical domain without using a photodetector. An artificial current source is implemented using a series resistor with minimum stray capacitance. This testing in the electrical domain is preferred over the optical technique as the rise time of an

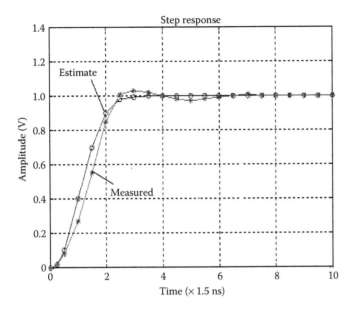

FIGURE 5.8   Step response of the amplifier: o—estimated; *—measured.

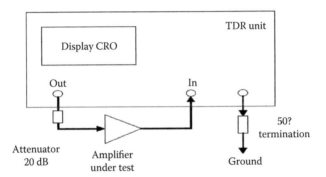

FIGURE 5.9   Experimental setup for the measurement of the rise time in the electrical domain using a time domain reflectometer (TDR).

electrical time domain reflectometer (TDR) is sufficiently short so that it does not influence the measurement of the rise time.* If the optical method is used, the principal problem is that we must able to modulate the intensity of the lightwave with a very sharp step function. This is not possible when the Mach–Zehnder intensity modulator (MZIM) is used as its transfer characteristics following a square of a cosine function. A very short pulse sequence laser such as a mode-locked fiber laser can be used. However, this is not employed in this work as the bandwidth of our amplifier is not that very wide and the TDR measurement is sufficient to give us a reasonable value of the rise time.

The passband of the TIA is also confirmed with the measurement of the frequency response by using a scattering parameter test set HP 85046 A, the scattering parameter S_{21} frequency response is measured, and the 3 dB passband is confirmed at 190 MHz.

---

* TDR unit is currently available in several ultra-wideband digital real-time oscilloscopes provided by Agilent Keysight or Tektronix.

5.5.2 Noise Measurement and Suppression

Two pig-tailed PDs are used and mounted with back-to-back configuration at the input of the transistor of stage 1. A spectrum analyzer is used to measure the noise of the amplifier shown in Figure 5.10. The background noise of the analyzer was measured to be -88 dBm. The expected noise referred to the input is 9.87×10^{-24} A^2/Hz. When the amplifier is connected to the spectrum analyzer, this noise level is increased to -85 dBm, which indicates that the noise at the output of the amplifier is around -88 dBm.

Since this power is measured into 50 Ω, using a spectral bandwidth of 300 kHz, the input current noise can be estimated as

$$\int di^2_{\text{Neq}} = \frac{v^2_N}{R^2_T} = \frac{10^{-11} \times 50}{(5k)^2} = 2 \times 10^{-17.5} \text{A}^2 \rightarrow \frac{di^2_{\text{Neq}}}{df} = \frac{2e^{-17.5} \text{A}^2}{3 \times 10^5} = 1.06 \times 10^{-23} \text{A}^2/\text{Hz} \qquad (5.25)$$

Thus, the measured noise is very close to the expected value.

5.5.3 Requirement for Quantum Limit

The noise required for near quantum limit operation can be estimated. The total shot noise referred to the input is

$$i^2_{\text{T-shot}} = \left(\Re_1 + \Re_2\right)\left|E_L\right|^2 \qquad (5.26)$$

with $\Re_{1,2}$ the responsivity of the photodetectors 1 and 2, respectively. The total excess noise referred to the input is

$$i^2_{\text{NT-shot}} = \left(\Re_1 + \Re_2\right)\left|E_L\right|^2; \quad i^2_{\text{N-excess}} = 2q\gamma(\omega)\left(\Re_1 + \Re_2\right)^2\left|E_L\right|^4 \qquad (5.27)$$

The excess noise from each detector is correlated so γ is a function of frequency, typically between 104 and 1010 A^{-1}. A receiver would operate within 1 dB of the quantum limit if the shot noise is about 6 dB above the excess noise and the amplifier noise. The amplifier noise power spectral density is 6.09×10^{-14} A^2/Hz. Assuming that a complete cancellation of the excess noise can be achieved, the LO required power is given as

$$i^2_{\text{NT-shot}} > 4i^2_{\text{NTeq}} \rightarrow \left|E_L\right|^2 > 0.24 \text{ mW} \approx -6.2 \text{ dBm} \qquad (5.28)$$

Since there is bound to excess noise due imperfection of the balancing of the two detectors, the power of the LO is expected to be slightly greater than this value.

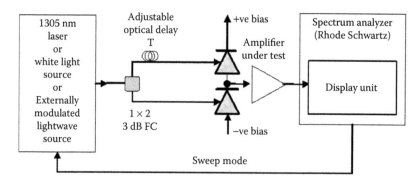

FIGURE 5.10 Experimental setup to measure excess noise impact and cancellation of a balanced receiver. FC = fiber coupler.

The output voltage of the amplifier can be monitored so noises contributed by different processes can be measured. When the detectors are not illuminated, we have

$$v_0^2 = Z_T^2 i_{Neq}^2 \qquad (5.29)$$

with Z_T the TI and v_0 the output voltage of the amplifier. It is known that incoherent white light is free from excess intensity noise, so when the detectors are illuminated with white light sources, we have

$$v_0^2 = Z_T^2 \left(i_{Neq}^2 + i_{N\text{-shot}}^2 \right) \qquad (5.30)$$

When the LO is turned on illuminating the detector pair, then

$$v_0^2 = Z_T^2 \left(i_{Neq}^2 + i_{N\text{-shot}}^2 + i_{N\text{-excess}}^2 \right) \qquad (5.31)$$

5.5.4   EXCESS NOISE CANCELLATION TECHNIQUE

The presence of uncanceled excess noise can be shown by observing the output voltage of the electronic preamplifier using a spectrum analyzer. First, the detectors are illuminated with incoherent white light and measure the output noise. Then the detectors are illuminated with the laser LO at a power level equal to that of the white light source; thus the shot noise would be at the same magnitude. Thus, the increase in the output noise will be due to the uncanceled intensity noise. This, however, does not demonstrate the cancellation of the LO intensity noise.

A well-known method for this cancellation is to bias the LO just above its threshold [3]. This causes the relaxation resonance frequency to move over the passband. Thus, when the outputs of the single or balanced receivers are compared, the resonance peak is observed for the former case and not in the later case in which the peak is suppressed. Since the current is only in the μA scale in each detector for a LO operating in the region near the threshold, the dominant noise is that of the electronic amplifier. This method is thus not suitable here. However, the relaxation resonance frequency is typically between 1 and 10 GHz. A better method can be developed by modulating the LO with a sinusoidal source and measuring the difference between single and dual photodetection receivers. The modulation can be implemented at either one of the frequencies to observe the cancellation of the noise peak power or by sweeping the LO over the bandwidth of the amplifier so that the cancellation of the amplifier can be measured. For fair comparison, the LO power would be twice that of the single detector case so the same received power of both cases can be almost identical and hence the signal-to-noise ratio (SNR) at the output of the amplifier can be derived and compared.

5.5.5   EXCESS NOISE MEASUREMENT

An indication of excess noise and its cancellation within the 0–200 MHz frequency range by the experimental setup can be observed, as shown in Figure 5.10. A laser of 1305 nm wavelength is used with an output power of −20 dBm employed as the LO so that the dominant noise is that of the electronic transimpedance preamplifier as explained above. The LO is intensity modulator by the sweep generator from the spectrum analyzer of 3 GHz frequency range (Rhode Schwartz model). The noise suppression pattern of the same trend can be observed with suppression about every 28 MHz interval over 200 MHz, the entire range of the TIA.

Figure 5.11 shows the noise spectral density for the case of single and dual detectors. Note that the optical paths have different lengths. This path difference is used for cancellation of the excess noise as described in [7,8]. It can be shown that the optical delay line leads to a cancellation of noise following a relationship of $\{1 - \cos(2\pi fT)\}$ with T the delay time of the optical path of one of the fibers from the fiber coupler to the detector. Hence the maxima of the cancellation occur at $f = m/T$

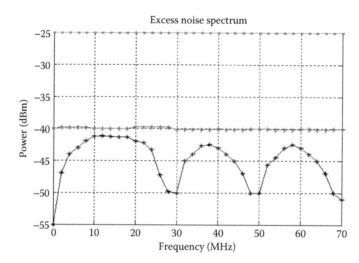

FIGURE 5.11 Excess noise for a single detector and dual detectors with noise cancellation using delay and filter at the input port of the preamplifier.

where m is an integer. From Figure 5.10, we can deduce the optical path length as $f = 30$ MHz or $T = 31.7$ ns; thus $d = cT/n_{\text{eff}} = 6.8$ m with n_{eff} (~1.51 at 1550 nm) the effective refractive index of the guided mode in the fiber. The total length of the fiber delay is equivalent to about 6.95 m in which the effective refractive index of the propagation mode is estimated to be 1.482 at the measured wavelength. Thus, there is a small discrepancy, which can be accounted for the uncertainty of the exact value of the effective refractive index of the fiber.

Using this method, Abbas et al. [1] achieved a cancellation of 3.5 dB, while 8 dB is obtained in our setup. This discrepancy can be understood to come from the non-indential characteristics of the PDs of the pair of the photodetectors of the balanced reception system.

In our initial measurements, we have observed some discrepancies that are due to (i) unmatched properties of the photodetection of the balanced detector pair; (ii) the delay path has three extra optical connections and therefore more loss, so even when there is no delay the excess noise would not be completely cancelled. Thus, at some stage the longer path was implemented with a tunable optical delay path so as to match the delay path to the null frequency of the RF wave to achieve cancellation of the excess noise. This noise cancellation can lead to an improvement of 16 dB in the optical signal-to-noise ratio (OSNR) due to the dual detector configuration and thus an improvement of near 100 km length of standard single mode fiber.

The noise cancellation is periodic following a relationship with respect to the delay time T imposed by the optical path of the interferometer T as $(1 - \cos 2\pi fT)$. The optical delay path difference is estimated by adjusting the optical delay line, which is implemented by a fiber path coupled out of a fiber end to open air path and back to another fiber path via a pair of Selfoc lenses.

It is noted that the LO light source can be directly modulated with a sweep sinusoidal signal derived from the electrical spectrum analyzer. This light source has also been externally modulated using a MZIM of bandwidth much larger than that of the electronic amplifier (about 25 GHz). The excess noise characteristics obtained using both types of sources are almost identical. Furthermore, with the excess noise reduction of 8 dB and the signal power for the balanced receiver operating in a push–pull mode then the received electrical current would be double that of a single detector receiver. That means a gain of 3 dB in the SNR by this balanced receiver.

5.6 REMARKS

Here, we consider the design and implementation of a balanced receiver using a dual detector configuration and a BJT front end preamplifier. In view of recent growing interests in a coherent optical transmission system with electronic digital signal processing [8–10], a noise-suppressed balanced receiver operating in the multi-GHz bandwidth range would be essential. Noise cancellation using optical delay line in one of the detection paths leads to suppression of excess noise in the receiver.

In this section, we demonstrate, as an example, by design analysis and implementation, a discrete wide band optical amplifier with a bandwidth of around 190 MHz and a total input noise equivalent spectral density of $10^{-23}\text{A}^2/\text{Hz}$. This agrees well with the predicted value using a noise model analysis. It is shown that if sensible construction of the discrete amplifier, minimizing the effects of stay capacitances and appropriate application of compensation techniques, a bandwidth of 190 MHz can be easily obtained for BJTs whose transition frequencies can only be of a few GHz. Furthermore, by using an optimum emitter current for the first stage and minimizing the biasing and feedback resistance and capacitance at the input, a low noise amplifier with an equivalent noise current of about $1.0\,\text{pA}/\sqrt{\text{Hz}}$ can be achieved. Furthermore, the excess noise cancellation property of the receiver is found to give the maximum SNR of 8 dB with a matching of the two photodetectors.

Although the bandwidth of the electronic amplifier reported here is only 190 MHz, which is about 1% of the 100 Gb/s target bit rate for modern optical Ethernet. The bandwidth achieved by our amplifier has reached around the maximum region that can be achieved with discrete transistor stages whose transition frequency is only 5 GHz. The amplifier configuration reported here can be scaled up to the multi-GHz region for 100 Gb/s receiver using an integrated electronic amplification device without much difficulty. We thus believe that the design procedures for determining of the pole and zero patterns on the s-plane and their dynamics for stability consideration are essential so that any readers who are interested in its implementation using Monolithic Microwave Integrated Circuit technology can use. If the amplifier is implemented in the multi-GHz range, then the microwave design technique employing a noise figure method may be most useful when incorporating the design methodology reported in this section.

5.7 BROADBAND LOW NOISE HIGH TRANSFER IMPEDANCE BALANCED DIFFERENTIAL TIA OPTICAL RECEIVER

Under tremendous rapid growth of numerous multimedia applications in Internet, high-speed transceivers operating at tens of gigabits per second are required to fulfill the data transfer volume. As the speed is getting higher, the bandwidth of the electronic receiver must be wider; hence the noises contributed to the detected signals, as seen in the above analyses and demonstration. There is a need for low noise and wide band TIA with large TI over the mid-band region. One of such TIAs is the differential input and different output TIA. Furthermore, higher order modulation formats, especially the M-QAM, require linearity in the highly sensitive TIA, thus an automatic gain control stage must be incorporated in such differential TIA. This section gives an overview of this type of TIA_AGC circuit where simplified noise considerations are given following those described in [6,11,12].

The typical architecture of a 40 Gb/s receiver is shown in Figure 5.12, which consists of a front end amplifier and a analog-to-digital converter (ADC) plus digital signal processors (DSP) for signal processing including clock data recovery (CDR) via the phase recovery, dispersion compensation of both linear and nonlinear effects, and so on. The front end amplifier must enhance the weak input signals to several hundreds of mVs for the subsequent clock and data recovery or sampled by the ADC and then DSP processing. The CDR circuit should extract the

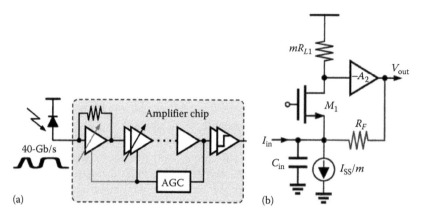

(a)   (b)

FIGURE 5.12 (a) Generic structure of TIA with AGC incorporated and (b) feedback TIA circuit employing common gate CMOS circuit.

clock from the jittery incoming data while automatically retiming and demultiplexing the data into lower-speed streams.

APPENIDX A: NOISE EQUATIONS

Refer to the small signal and noise model given in Figures A.1 and A.2; the spectral density of noises due to collector bias current I_C and small signal transconductance g_m and base conductance g_B generated in a BJT can be expressed as

$$di_1^2 = 4k_BTg_Bdf; \quad di_2^2 = 2qI_Cdf + 2k_BTg_mdf$$

$$di_3^2 = 2qI_Bdf + 2k_BT(1-\alpha_N)g_mdf; \text{ and} \tag{A.1}$$

$$di_4^2 = \sum \text{shot noise of diodes and thermal noise of bias in } g_B \text{ resistors}$$

From nodal analysis of the small signal equivalent circuit given in Figure A.2, we can obtain the relationship

$$\begin{bmatrix} Y_s + g_B & -g_B & 0 \\ -g_B & g_B + y_1 + y_f & -y_f \\ 0 & g_m - y_f & y_f + y_2 \end{bmatrix} \begin{bmatrix} V_1 \\ V_2 \\ V_3 \end{bmatrix} = \begin{bmatrix} i_{eq} - i_{N1} \\ i_{N1} + i_{N3} \\ i_{N2} \end{bmatrix} \tag{A.2}$$

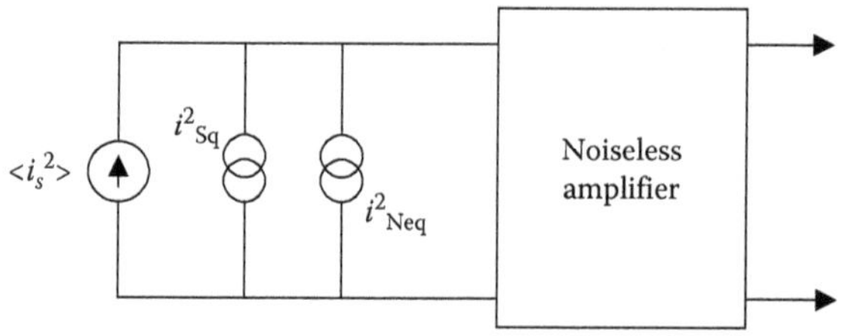

FIGURE A.1 Equivalent noise current at the input and noiseless amplifier model. i_{Sq}^2 is the quantum shot noise, which is signal dependent, and i_{Neq}^2 is the total equivalent noise current referred to the input of the electronic amplifier.

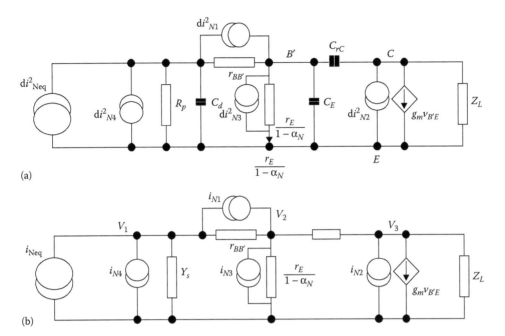

FIGURE A.2 (a) Approximated noise equivalent and small signal model of a BJT front end. (b) Generalized noise and small signal model circuit. Note that $r_B = r_{sd} + r_{BB'}$ and $C_d = C_p + C_i$ with $r_{BB'}$ the base resistance, r_{sd} the diode resistance, C_d the photodiode capacitance, and C_i the input capacitance.

Hence, V_3, V_2, V_1 can be found by using Euler's rule for the matrix relationship

$$V_3 = \frac{\Delta_{13}}{\Delta}(i_{eq} - i_{N1}) + \frac{\Delta_{23}}{\Delta}(i_{N1} + i_{N3}) + \frac{\Delta_{33}}{\Delta}(i_{N2}) \tag{A.3}$$

Hence the noise currents referred to the input are

$$d(i_1'^2) = \left|\frac{Y_s}{g_B}\right|^2 dI_1^2 = \omega_2 C_s^2 r_B 4kT df \tag{A.4}$$

$$d(i_2'^2) = \left|(Y + y_1 + y_f) + \frac{Y_s}{g_B}(y_1 + y_f)\right|^2 \left|\frac{1}{y_f - g_m}\right|^2 = \left|\frac{1}{y_f - g_m}\right|^2 dI_2^2$$

$$= \left(\frac{1}{\beta_N r_E^2} + \omega^2 \left(C_0^2 - \frac{2C_B r_B}{\beta_N r_E}\right) + \omega^4 2C_s r_B C_B^2\right) 2kT r_E df \tag{A.5}$$

and

$$d(i_3'^2) = \left|\frac{Y_s}{g_B} + 1\right|^2 dI_3^2 = \left(\omega C_s^2 r_B^2 + 1\right)\frac{2kT}{\beta r_E} df \tag{A.6}$$

Here we have assumed that

$$\omega C_f \ll g_m$$

$$C_0 = C_s + C_E + C_f + \frac{C_s r_s}{\beta r_E} \approx C_s + C_{tE} + C_f + \frac{\tau_t}{r_E} = C_x + \frac{\tau_t}{r_E} \tag{A.7}$$

with $\qquad C_x = C_E + C_f; C_E = C_{tE} + \dfrac{\tau_t}{r_E}$

The total noise spectral density referred to the input of the amplifier, as shown in Figures A.1 and A.2, is thus given as

$$\frac{di_{Neq}^2}{df^2} = \frac{di_{1eq}^2}{df^2} + \frac{di_{2eq}^2}{df^2} + \frac{di_{3eq}^2}{df^2} = 2k_BT \left[\frac{1}{\beta r_E} \frac{\beta+1}{\beta} + \omega^2 \left(\frac{2}{r_B} + \frac{1}{\beta r_E} \right) + C_0 r_E - \frac{2C_s r_B C_x}{\beta} + \omega^2 C_s^2 r_B^2 C_x^2 r_E \right] \tag{A.8}$$

Thence the total noise power referred to the input is given as

$$i_{Neq}^2 = \int_0^B di_{Neq}^2 = a + \frac{b}{r_E} + c r_E;$$

with

$$a \approx 2k_BT \left[\frac{8\pi^2 B^3}{3} \left(C_s r_B + C_x \tau_t + \frac{C_s^2 r_B^2}{\beta} \right) + \frac{(2\pi)^4}{5} B^5 C_s^2 r_B^2 2C_y \tau_t \right]$$

$$b \approx 2k_BT \left[\frac{B}{\beta} + \frac{4\pi^2 B^3}{3} \left(\frac{C_s r_B^2}{\beta} + \tau_t^2 \right) + \frac{16\pi^4 B^5}{5} C_s^2 r_B^2 \tau_t^2 \right] \tag{A.9}$$

$$c \approx 2k_BT \left[\frac{4\pi^2 B^3 C_x^2}{3} + \frac{16\pi^4 B^5}{5} C_s^2 r_B^2 C_y^2 \right] \quad \text{with } C_y = C_{tE} + \frac{\tau_t}{r_E}$$

where the first term in each coefficient $a, b,$ and c is dominant for a bandwidth of 190 MHz of the two transistor types Phillip BFR90A and BFT24 under considerations.

If the dependence of C_E on I_E is ignored, then Equation A.9 becomes

$$i_{Neq}^2 = 2k_BT \left[\frac{1}{r_E} \frac{B}{\beta} + r_E \left(\frac{4\pi^2 B^3}{3} C_0^2 \right) + \frac{16\pi^4 B^5}{5} C_s^2 r_B^2 C_x^2 + \frac{8\pi^2 B^3 C_s^2 r_B}{3} \right] \tag{A.10}$$

For the transistor BFR90A, the optimum emitter resistance is $r_{Eopt} = 49\ \Omega$ and the total equivalent noise power referred to the input is $4.3 \times 10^{-16} A^2$, which is moderately high.

Clearly, the noise power is a cubic dependence on the bandwidth of the amplifier. This is thus very critical for an ultra-wide-band optical receiver. Thus, it is very important to suppress the excess noise of the optical receiver due to the quantum shot noise of the LO.

APPENDIX B: NOISES IN DIFFERENTIAL TIA

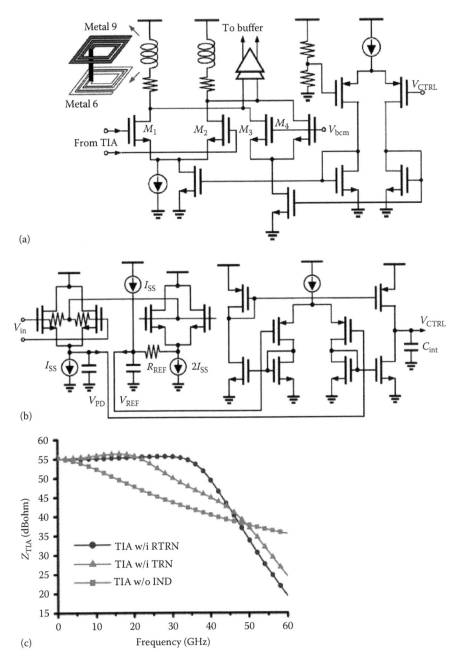

FIGURE B.1 A TIA with (a) the TRN (b) a TIA without inductors are designed with the same gain and power dissipation for comparison and (c) simulated frequency response of the TIA with the RTRN.

REFERENCES

1. Abbas, G.L., V. Chan, and T.K. Yee, A dual-detector optical heterodyne receiver for local oscillator noise suppression, *IEEE J. Lightwave Technol.*, 3(5), 1985, 1110–1122.
2. Kasper, B., C. Burns, J. Talman, and K. Hall, Balanced dual detector receiver for optical heterodyne communication at Gb/s rate, *Elect. Lett.*, 22(8), 10, 1986, 413–414.
3. Alexander, S., Design of wide-band optical heterodyne balanced mixer receivers, *IEEE J. Lightwave Technol.*, 5(4), 1987, 523–537.
4. Binh, L.N., *Digital Optical Communications*, CRC Press, Boca Raton, FL, 2008, Chapter 4.
5. Phillips, *Handbook of Semiconductor*, Vol. S10, Eindoven, the Netherlands, 1987.
6. Cherry, E.M. and D.E. Hooper, *Amplifying and Low Pass Amplifier Design*. John Wiley and Sons, New York, 1968, Chapters 4 and 8.
7. Sterlin, R., R. Battiig, P. Henchoz, and H. Weber, Excess noise suppression in a fiber optic balanced heterodyne detection system, *Opt. Quant. Electron.*, 18, 1986, 445–454.
8. Kahn, J. and E. Ip, Principles of digital coherent receivers for optical communications, paper OTuG5, *Proc. OFC 2009*, San Diego, USA, March 2009.
9. Zhang, C., Y. Mori, K Igarashi, K. Katoh, and K. Kikuchi, Demodulation of 1.28-Tbit/s polarization-multiplexed 16-QAM signals on a single carrier with digital coherent receiver, Paper OTuG3, *Proc. OFC*, San Diego, USA, March 2009.
10. Binh, L.N., *Digital Optical Communications*. CRC Press, Taylors & Francis Group, Boca Raton, FL, 2008, Chapter 11.
11. Liao, C.-F. and S.-L. Liu, 40 Gb/s trans-impedance-AGC amplifier and CDR circuit for broadband data receivers in 90 nm CMOS, *IEEE J. Solid-State Circ.*, 43(3), 2008, 648.
12. Cherry, E.M., Loop gain, input impedance, and output impedance of feedback amplifiers, *IEEE Cir. Syst. Mag.*, 8(1), 2008, 55–71.

6 DSP-Based Coherent Optical Transmission Systems

6.1 INTRODUCTION

A generic flow chart of the digital signal processing (DSP) optical reception systems is introduced in Figure 6.1. It is noted that the clock/timing recovered signals is fed back into the sampling unit of the analog-to-digital converter (ADC) so as to obtain the best correct timing for sampling the incoming data sequence for processing in the DSP. Any errors made at this stage of timing will result into high deviation of the bit-error rate (BER) in the symbol decoder shown in Figure 6.1. It is also noted that the vertical polarized channel (V-pol) and the horizontal polarized (H-pol) channel are detected and their in-phase (I-) and quadrature (Q-) components are produced in the electrical domain with signal voltage conditioned for the conversion to the digital domain by the ADC.

The processing of the sampled sequence from the received optical data and photodetected electronic signals passing through the ADC relies on the timing recovery from the sampled events of the sequence. The flowing stages of the blocks given in Figure 6.1 may be changed or altered accordingly depending on the modulation formats and pulse shaping, for example, the Nyquist pulse shapes in Nyquist superchannel transmission systems.

This chapter attempts to illustrate the performance of the processing algorithms in optical transmission systems employing coherent reception techniques over a highly dispersive optical transmission line, especially the multispan optically amplified non-DCM long-haul distance. First, the quadrature phase-shift keying (QPSK) homodyne scheme is examined, and then the 16 quadrature amplitude modulation (16-QAM) incorporating both polarized channels multiplexed in the optical domain, hence the term polarized division multiplexed (PDM)-QPSK or PDM-16-QAM. We then expand the study for superchannel transmission systems in which several subchannels are closely spaced in the spectral region so as to increase the spectral efficiency so that the total effective bit rate must reach at least 1 Tb/s (Tbps). Because of the overlapping of adjacent channels there are possibilities that modifications of the processing algorithms are to be made.

Furthermore, the nonlinearity impairments on transmitted subchannels would be degrading the system performance, and the application of backpropagation techniques described in Chapter 3 must be combined with linear and nonlinear equalization schemes so as to effectively combat the performance degrading.

6.2 QUADRATURE PHASE-SHIFT KEYING SYSTEMS

6.2.1 CARRIER PHASE RECOVERY

Homodyne coherent reception requires a perfect match of the frequency of the signal carrier and the local oscillator (LO). Any frequency difference will lead to the phase noise of the detected signals. This is the largest hurdle for the first optical coherent system initiated in the mid-1980s. In DSP-based coherent reception systems, the recovery of the carrier phases and hence the frequency is critical to achieve the most sensitive reception with maximum performance in the BER, or evaluation of the probability of error. This section illustrates the recovery of the carrier phase for QPSK and 16-QAM optical transmission systems whose constellations of a single circle feature or multilevel circular distribution. How would the DSP algorithms perform under the physical impairment effects on the recovery of the phase of the carrier?

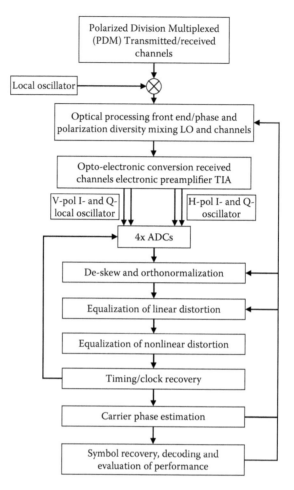

FIGURE 6.1 Flow of functionalities of DSP processing in a coherent optical receiver of a coherent transmission system. A feedback path diverted from the timing and phase recovery to ADC, hence a close-loop control feedback of the processing.

6.2.2   112G QPSK COHERENT TRANSMISSION SYSTEMS

Currently, several equipment manufacturers are striving to provide commercial advanced optical transmission systems at 100 Gbps employing coherent detection (CoD) techniques for long-haul backbone networks and metro networks as well. Since the 1980s, it has been well known that single-mode optical fibers can support such transmission due to the preservation of the guided modes and its polarized modes of the weakly guiding linearly polarized (LP) electromagnetic waves [1]. Naturally, both transmitters and receivers must satisfy the coherency conditions of narrow linewidth sources and coherent mixing with a LO, an external cavity laser (ECL), to recover both the phase and amplitude of the detected lightwaves. Both polarized modes of the LP modes can be stable over a long distance so far to provide the PDM channels, even with polarization mode dispersion (PMD) effects. All linear distortion due to PMD and chromatic dispersion (CD) can be equalized in the DSP domain employing algorithms in real-time processors provided in Section 6.2.6.

It is thus very important to ensure that these subsystems perform the coherent detection and transmitting functions. This section thus presents a summary of the tests conducted with a back-to-back (B2B) transmission of QPSK PDM channels. The symbol rate of the transmission system is 28 GSy/s under the modulation format PDM-CSRZ-QPSK. It is noted that the differential QPSK (DQPSK) encoder and the bit pattern generator are provided.

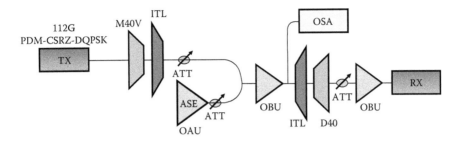

FIGURE 6.2 Setup of the PDM QPSK optical transmission system.

The transmission system is arranged as shown in Figure 6.2. The carrier suppressed return-to-zero (CSRZ) QPSK transmitter consists of a CSRZ optical modulator that is biased at the minimum transmission point of the transfer characteristics of the Mach–Zehnder interferometric modulator (MZIM) driven by a sinusoidal signal whose frequency is half of the symbol rate or 14 GHz for 28 GSy/s. A wavelength-division multiplexing (WDM) mux (multiplexer) is employed to multiplex other wavelength channels located within the C-band (1530–1565 nm). An optical amplifier (EDFA [Erbium:doped fiber amplifier] type) is employed at the front end of the receiver so that noises can be superimposed on the optical signals to obtain the optical signal-to-noise ratio (OSNR). The DSP-processed signals in the digital domain are carried out offline and the BER is obtained.

The transmitter consists of an ECL, an encore type, a polarization splitter coupled with a 45° aligned ECL beam, two separate CS-RZ external LiNbO₃ modulators, and then two I-Q optical modulators. The linewidth of the ECL is specified at about 100 kHz, and with external modulators we can see that the spectrum of the output modulated lightwaves is dominated by the spectrum of the baseband modulation signals. However, we observed that the laser frequency is oscillating about 300 MHz due to the integration of a vibrating grating so as to achieve stability of the optical frequency.

The receivers employed in this system are of two types. One is a commercialized type, Agilent N 4391 in association with an Agilent external LO, and the other one is a transimpedance amplifier (TIA) type including a photodetector pair (PDP) connected B2B in a push–pull manner and then a broadband TIA. In addition to the electronic reception part, a $\pi/2$ hybrid coupler including a polarization splitter, $\pi/2$ phase shift and polarization combiner that mixed the signal polarized beams and those of the LO (an ECL type identical to the one used in the transmitter), is employed as the optical mixing subsystem at the front end of the receiver. The mixed polarized beams (I-Q signals in the optical domain) are then detected by balanced receivers. I-Q signals in the electrical domain are the sampled and stored in the real-time oscilloscope (Tektronix 7200). The sampled I-Q signals are then processed offline using the algorithms provided in the scope or own developed algorithms such as the evaluation of error vector magnitude (EVM) described above for Q factor and thence BER.

Both the transmitter and receivers function with the required OSNR for the B2B of about 15 dB at a BER of 2×10^{-3}. It is noted that the estimation of the amplitude and phase of the received constellation is quite close to the received signal power and the noise contributed by the balanced receiver with a small difference, due to the contribution of the quantum shot noise contributed by the power of the LO. The estimation technique was described in Chapter 4. Figure 6.3 shows the BER versus OSNR for B2B QPSK PDM channels. As shown in Figure 6.4 (the variation of BER as a function of the signal energy over noise for AWGN noise), for 4QAM or QPSK coherent, the SNR is expected at about 8 dB for BER of 1×10^{-3}. Experimental processing of such a scheme in B2B configuration shows an OSNR of about 15.6 dB. This is due to 3 dB split by the polarized channels and additional noises contributed by the receiver; hence about 15.6 dB OSNR is required. The forward error coding (FEC) is set at 1×10^{-3}. The receiver is the Agilent type as mentioned above.

A brief analysis of the noises at the receiver is as follows. The noise is dominated by the quantum shot noise generated by the power of the LO, which is about at least 10 times greater than that of the signals. Thus, the quantum shot noises generated at the output of the photodetector (PD) is

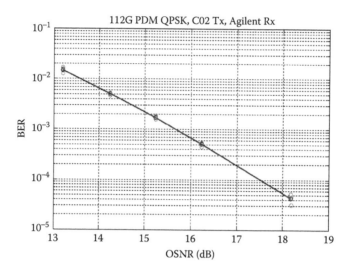

FIGURE 6.3   B2B OSNR versus BER performance with Agilent Rx in a 112 Gbps PDM-CSRZ-DQPSK system.

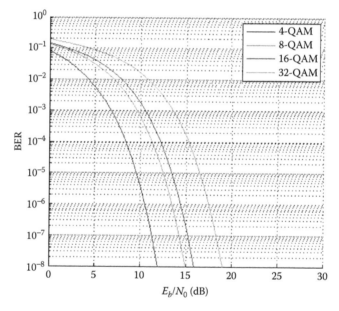

FIGURE 6.4   Theoretical BER versus SNR (energy per bit over the noises contained in that bit period) for a different level QAM scheme obtained by bertool.m of MATLAB® under Gaussian noise probabilistic distribution.

$$i^2_{N-LO} = 2qI_{LO}B$$

$$I_{LO} = 1.8 \text{ mA at } 0.9 \text{ PD quantum efficiency}$$

$$B = 31 \text{ GHz BW U2t Agilent Rx}$$

$$i^2_{N-LO} = 2 \times 10^{-19} \times 1.8 \times 10^{-3} \times 30 \times 10^9 \qquad (6.1)$$

$$= 10.8 \times 10^{-12}$$

$$\longrightarrow i_{N-LO} = 3.286 \text{ μA}$$

The bandwidth of the electronic preamplifier of 31 GHz is taken into account. This shot noise current due to the LO imposed on the PD pair is almost compatible with that of the electronic noise of the electronic receiver given that the noise spectral density equivalent at the input of the electronic amplifier of the U2t balanced receiver is specified at 80 pA/$\sqrt{\text{Hz}}$, that is,

$$\left(80\times10^{-12}\right)^2\times30\times10^9 = 19.2\times10^{-12}\,\text{A}^2 \xrightarrow[\text{noise\_current}]{} i_{\text{Neq.}} = 4.38\ \mu\text{A} \qquad (6.2)$$

Thus, any variation in the LO would affect this shot noise in the receiver. It is thus noted that with the transimpedance of the electronic preamplifier estimated at 150 Ω, a dBm difference in the LO would be contributed to a change of the voltage noise level of about 0.9 mV in the signal constellation obtained at the output of the ADC. A further note is that the noise contributed by the electronic front end of the ADC has not been taken into account. We note that differential TIA offer at least 10 times higher transimpedance of around 3000 Ω over a 30 GHz midband. These TIAs offer much higher sensitivities as compared to single input TIA type [2,3].

6.2.3 I-Q Imbalance Estimation Results

There are imbalance due to the propagation of the polarized channels and the I- and Q-components. They must be compensated to minimize the error. The I-Q imbalance of the Agilent BalRx and U²t BalRx is less than 2°, which might be negligible for the system as shown in Figure 6.5. This imbalance must be compensated for in the DSP domain.

6.2.4 Skew Estimation

In addition to the imbalance of the I- and Q-components due to optical coupling and electronic propagation in high-frequency cables, there are propagation delay time differences between these components that must be compensated. The skew estimation is shown in Figure 6.6, obtained over a number of data sets.

Abnormal skew variation from time to time was also observed, which should not happen if there is no modification on the hardware. Considering the skew variation happened with the Agilent receiver, which only has very short radio frequency (RF) cable and tight connection,

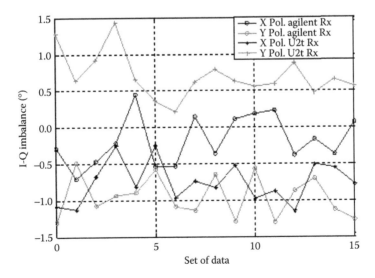

FIGURE 6.5   I-Q imbalance estimation results for both Rx. Note maximum imbalance phase of ±1.5°.

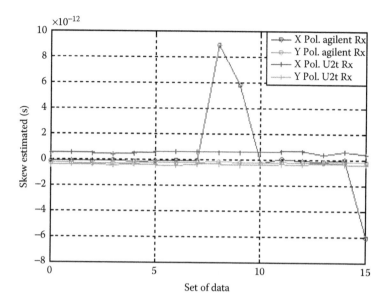

FIGURE 6.6   Skew estimation results for both types of Rx.

there is a high probability that the skew happened inside the Tektronix oscilloscope more than at the optical or electrical connection outside.

Figure 6.7 shows the BER versus the OSNR when the skew and the imbalance between I- and Q-components of the QPSK transmitter are compensated. The OSNR of DQPSK is improved by about 0.3–0.4 dB at 1×10^{-3} BER compared with the result without I-Q imbalance and skew compensation. In the time domain, compared with the result that we obtained in 2009, the required OSNR of DQPSK at a BER of 1×10^{-3} is about 14.7 dB, which is an improvement of 0.1 dB. The required OSNR at 1×10^{-3} BER of QPSK is about 14.7 dB, which is about the common performance of the state of the art for QPSK. For a BER $= 1 \times 10^{-3}$, an imbalanced CMRR $= -10$ dB would create

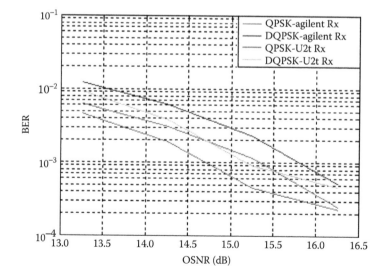

FIGURE 6.7   OSNR versus BER for two types of integrated coherent receiver after compensating I-Q imbalance and skew.

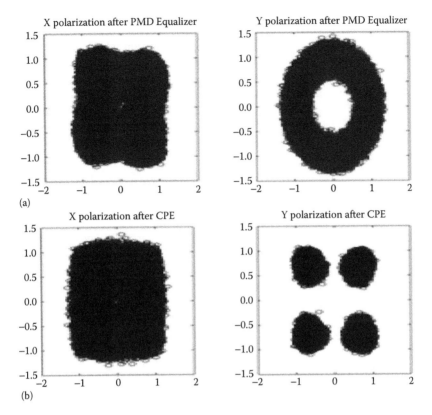

FIGURE 6.8 Constellation (a) after PMD module and (b) after CPE algorithm module.

a penalty of 0.2 dB in the OSNR for the Agilent receiver and an improvement of 0.7 dB for a commercial balanced receiver employed in the Rx subsystems as shown in Figures 6.7 and 6.8.

Figure 6.9 shows the structure of the ECL incorporating a reflection mirror, which is vibrating at a slow frequency of around 300 MHz. A control circuit would be included to indicate the electronic control of the vibration and cooling of the laser so as to achieve stability and eliminating of Brillouin scattering effects.

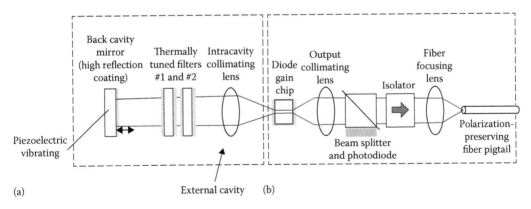

FIGURE 6.9 External cavity structure of the laser with the back mirror vibration in the external cavity structure. (a) Tuning optics and (b) DFB-like optics.

6.2.5 Fractionally Spaced Equalization of CD and PMD

Ip and Kahn [4] have employed the fractional spaced equalization scheme with mean square error to evaluate its effectiveness of PDM amplitude shift keying (ASK) with a nonreturn-to-zero (NRZ) or return to zero (RZ) pulse-shaping transmission system. Their simulation results are displayed in Figure 6.10 for the maximum allowable CD normalized in ratio with respect to the dispersion parameter of the SSMF the versus the number of equalizer taps N for a 2-dB power penalty at a launched OSNR of 20 dB per symbol for ASK RZ and NRZ pulse shapes, using a Bessel antialiasing filter at a sampling rate $1/T = M/KT_s$; with T_s = symbol period, and M/K the fractional ratio, as shown in Figure 6.28 for a fractional ratio of (a) $M/K = 1$, (b) 1.5 and (c) 2, and (d), (e), and (f) using a Butterworth antialiasing filter. The sampling antialiasing filter is employed to ensure that artificial fold back to the spectrum is avoided. The filter structures Bessel and Butterworth give much similar performance for fractional spaced equalizers but less tap for the Bessel filtering case when an equally spaced equalizer is used (see Figure 6.28a and d).

The effects of the crosstalk coupling between the two polarized modes can be minimized by representing the linear fiber channel with the model shown in Figure 6.11 with $h_{11}, h_{12}, h_{22}, h_{21}$ the discrete impulse response of the filtering effects (delay due to PMD of first order) throughput and crosscoupling of the two polarized sequences [4,5].

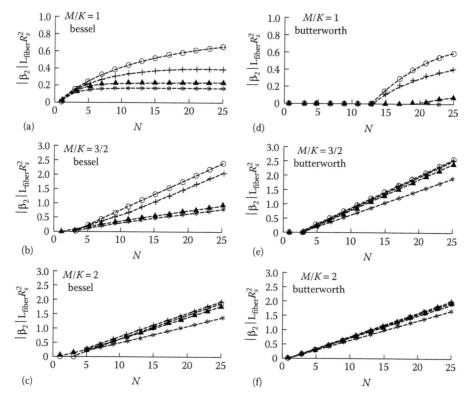

FIGURE 6.10 Maximum allowable CD versus the number of equalizer taps N for a 2-dB power penalty at an input SNR of 20 dB per symbol for RZ and NRZ pulse shapes, using a Bessel antialiasing filter. "○" denotes transmission using NRZ pulses, "×" denotes 33% RZ, "Δ" denotes 50% RZ, and "+" denotes 67% RZ, by fractional spaced equalizer (FSE) and sampling rate R_s $1/T = M/KT_s$; with T_s the symbol period, M/K is the fractional ratio (a) $M/K = 1$, (b) 1.5, (c) 2, (d), (e), and (f) using a Butterworth antialiasing filter. The modulation scheme is ASK with NRZ and RZ pulse shaping. N = number of taps of the equalizer. (Extracted from E. Ip and J.M. Kahn, *IEEE J. Lightwave Technol.*, 25(8), Aug. 2007, 2033. With permission.)

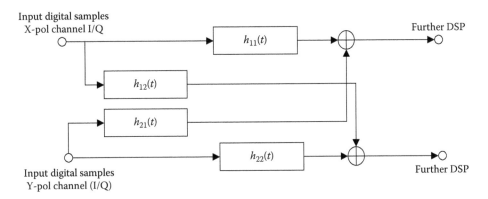

FIGURE 6.11 Baseband crosscoupling model in the digital domain to equalize PMD and CD effects.

6.2.6 Linear, Nonlinear Equalization, and Backpropagation Compensation of Linear and Nonlinear Phase Distortion

Ip and Kahn [6] have fist developed and applied the backpropagation, as given in Chapter 2, to equalize the distortion due to nonlinear impairment of optical channel transmission through the single-mode optical fibers. The backpropagation algorithm is simply a reverse phase rotation at the end of each span of the multispan link. The rotating phase is equivalent to the phase exerted on the signals in the frequency domain with a square of the frequency dependence. Thus, this backpropagation is efficient in the aspect that the whole span can be compensated so as to minimize the numerical processes, hence less processing time and central processing unit time of the digital signal processor.

Figure 6.12 shows the equalized constellations of a 21.4 GSy/s QPSK modulation scheme system after transmission through 25 × 80 km non-DCF spans under the equalization using (a) linear compensation only, (b) nonlinear equalization, and (c) combined backpropagation and linear equalization. Obviously, the backpropagation contributes to the improvement of the performance of the system.

Figure 6.13 shows the phase errors of the constellation states at the receiver versus launched power of a 25 × 80 km multispan QPSK 21.4 GSy/s transmission system. The results extracted from [5] shows the performance of backpropagation phase rotation per span for 21.4 Gbps 50% RZ-QPSK transmitted over 25 × 80 km spans of single-mode fiber (SMF) with five reconfigurable optical add-drop modules (ROADMs), with 10% CD undercompensation. The algorithm is processed offline of received sampled data after 25 × 80 km standard single-mode fiber (SSMF) propagation via the use of the nonlinear Schrödinger equation and the coherent reception technique described in Chapter 4. It is desired that the higher the launched power the better the OSNR that can be employed for longer distance transmission. So for fractional space, ratios of 3 and 4 offer higher launched power and thus be the preferred equalization scheme as compared with equal space or sampling rate equal to that of the symbol rate. The ROADM is used to equalize the power of the channel under consideration as compared with other dense wavelength-division multiplexing (DWDM) channels.

6.3 16-QAM SYSTEMS

Consider that the 16-QAM received a symbol signal whose phase Φ denotes the phase offset. The symbol d_k denotes the magnitude of the QAM symbols and n_k the noises superimposed on the symbol at the sampled instant. The received symbols can be written as

$$r_k = d_k e^{j\phi} + n_k; \qquad k = 1, 2, \ldots, L \qquad (6.3)$$

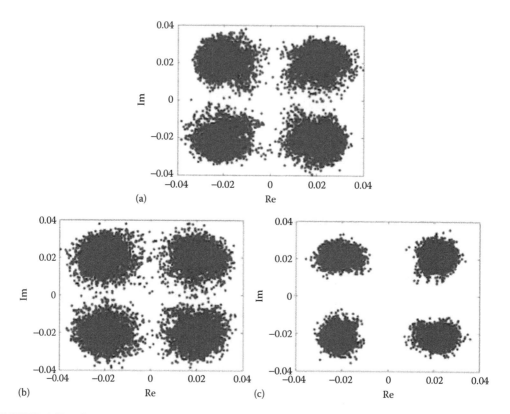

FIGURE 6.12 Constellation of a QPSK scheme as monitored at (a) with linear CD equalization only (b) nonlinear phase noise compensation and (c) after backpropagation processing combining with linear equalization. (Simulation results extracted from Spinner, B., *IEEE J. Sel. Top. Quant. Elect.*, 16(5), 1180–1192, 2010. With permission.)

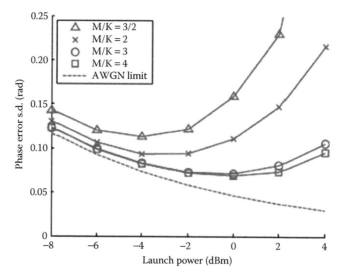

FIGURE 6.13 Phase errors at the receiver versus launched power or 25 × 80 km multispan QPSK at 21.4 GSy/s. Performance of backpropagation phase rotation per span for 21.4 Gbps 50% RZ-QPSK transmitted over 25 × 80 km spans of SMF with 5 ROADMs, with 10% CD undercompensation using fractionally spaced equalizer with *M/K* factor of 1.5–4. (Simulation results extracted from Spinner, B., *IEEE J. Sel. Topics Quant Elect.*, 16(5), 1180–1192, 2010. With permission.)

Using the maximum likelihood sequence estimator, the phase of the symbol can be estimated as

$$l(\phi) = \sum_{k=1}^{L} \ln \left\{ \sum_{d} e^{-\frac{1}{2\sigma^2}\left|r_k - de^{j\phi}\right|^2} \right\} \tag{6.4}$$

or effectively one would take the summation of the contribution of all states of the 16-QAM on the considered symbol measured as the geometrical distance in a natural logarithmic scale with the noise contribution of a standard deviation σ.

The frequency offset estimation for 16-QAM can be conducted by partitioning the 16-QAM constellation into a number of basic QPSK constellations, as shown in Figure 6.14. There two QPSK constellations in the 16-QAM whose symbols can be extracted from the received sampled data set. They are then employed to estimate the phase of the carrier as described in Section 6.2.1 on carrier phase estimation for QPSK modulated transmission systems. At first a selection of the innermost QPSK constellation, classified as Class I symbols can be carried out, thence an estimation of the frequency offset of the 16-QAM transmitted symbols can be obtained. Thence, an FO compensation algorithm is conducted, and the phase recovery of all 16-QAM symbols can be derived. Further confirmation of the difference of carrier phase recovery or estimation can be conducted with the constellation of Class I as indicated in Figure 6.14a.

Carrier phase recovery based on the Viterbi–Viterbi algorithm on the Class I QPSK subconstellation of the 16-QAM may not be sufficient and so a modified scheme has been reported by Fatadin et al. [8]. Further refining this, estimations of the carrier phase for 16-QAM by partition and rotating are made so as to match certain symbol points to those of Class I constellation of the 16-QAM. The procedures are as shown in the flow diagram of Figure 6.15.

Conduct the partition into different classes of constellation and then rotate Class II symbols with an angle in either clockwise or anticlockwise of $\pm\theta_{rot} = \pi/4 - \tan^{-1}(1/3)$. To avoid opposite rotation with respect to the real direction, estimate the error in the rate of changes of the phase variation or frequency by the use the fourth power of the argument of the angles of two consecutive symbols, given by

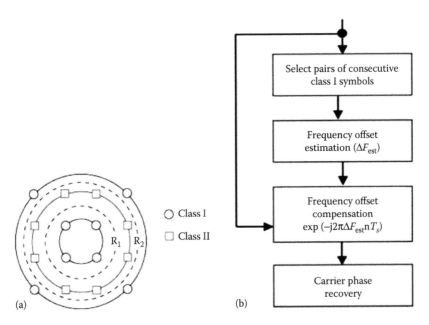

FIGURE 6.14 Processing of 16-QAM for carrier phase estimation (a) constellation of 16-QAM and (b) processing for carrier phase recovery with classes I and II of circulator subconstellation. (From I. Fatadin and S.J. Savory, *IEEE Photonic Technol. Lett.*, 23(17), 2001, 1246–1248. With permission.)

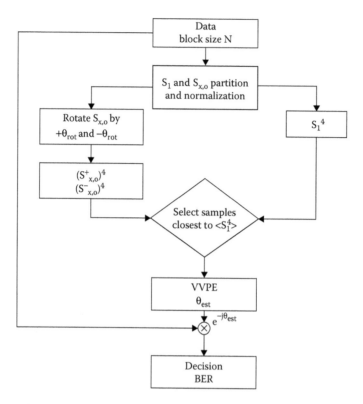

FIGURE 6.15 Refine carrier phase recovery of 16-QAM by rotation of Class I and Class II subconstellations.

$$\Delta F_{\text{est.}} = \frac{1}{8\pi T_s} \arg \left(\sum_{k=0}^{N} S_{k+1} S_k \right)^4 \tag{6.5}$$

to check their quadratic mean, then select the closer symbol, and fin apply the standard Viterbi–Viterbi procedure. Louchet et al. did also employ a similar method [9] and confirm the effectiveness of such a scheme. The effects on the constellation of the 16-QAM due to different physical phenomena are shown in Figure 6.16. Clearly, the FO would generate the phase noises in Figure 6.16a and influencing both the I- and Q-components by the CD of small amount (so as to see the constellation noises) and the delay of the polarized components on the I- and Q-components. These distortions of the constellation allow practical engineers to assess the validity of algorithms that are normally separate and independent and implemented in serial mode. This is in contrast to the constellations illustrated for QPSK as shown in Figure 6.8. Figure 6.17 also shows the real-time signals that result from the beating of the two sinusoidal waves of FO beating in a real-time oscilloscope.

Noé et al. [10] have simulated the carrier phase recovery for QPSK with PDM channels under constant modulus algorithm (CMA) and decision-directed with and without modification in which the error detected in each stage would be updated. The transmission system under consideration is B2B with white Gaussian noises and phase noises superimposed on the signals. For an OSNR of 11 dB, the CMA with modification offers a BER of 1×10^{-3} while it is 4×10^{-2} for a CMA without modification. This indicates that updating the matrix coefficients is very critical to recover the original data sequence. The modified CMA was also recognized to be valid for 16-QAM.

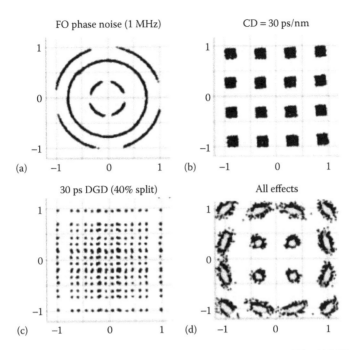

FIGURE 6.16 16-QAM constellation under influence (a) phase rotation due to FO of 1 MHz and no amplitude distortion; (b) residual CD impairment; (c) DGD of PMD effect; and (d) total phase noises effect. (E. Ip and J.M. Kahn, *IEEE J. Lightwave Technol.*, 25(8), Aug. 2007, 2033. With permission.)

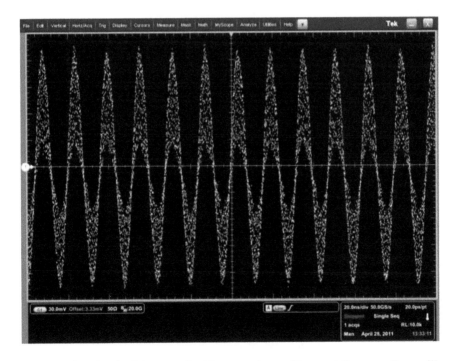

FIGURE 6.17 Beating signals of the two mixed lasers as observed by a real-time sampling oscilloscope.

6.4 TERA-BITS/S SUPERCHANNEL TRANSMISSION SYSTEMS

6.4.1 Overview

PDM-QPSK has been exploited as the 100 Gbps long-haul transmission commercial systems, the optimum technologies for 400 GE/1 TE transmission for next generation optical networking have now attracted significant interests for deploying ultrahigh capacity information over the global inter- net backbone networks. Further intense research on Tbps transmission systems have also attracted several research groups as the logical rate to increase from 100 Gbps. The development of hardware platforms for 1 to N Tbps is critical for proving the design concept. The Tbps can be considered as a superchannel that is defined an optical channel comprising a number of subrate subchannels whose spectra would be the narrowest allowable. Thus, to achieve efficient spectral efficiency, phase shap- ing is required, and one of the most efficient technique is the Nyquist pulse shaping. Thus, Nyquist QPSK can be considered as the most effective formats for delivery of high spectral efficiency and effective in coherent transmission and reception as well as equalization at both transmitting and reception ends.

Thus, in this section, we describe detailed design and experimental platform for delivery of Tbps using Nyquist QPSK at a symbol rate of 28–32 GSy/s and 10 subcarriers. The generation of subcarriers has been demonstrated using either recirculating frequency shifting (RFS) or nonlinear driving of an I-Q modulator to create five subcarriers per main carrier, thus two main carriers are required. Nyquist pulse shaping is used to effectively pack multiplexed channels whose carriers are generated by the comb generation technique. A digital-to-analog converter (DAC) with a sampling rate varied from 56 to 64 GSy/s is used for generating a Nyquist pulse shape including the equaliza- tion of the transfer functions of the DAC and optical modulators.

6.4.2 Nyquist Pulse and Spectra

The raised-cosine filter is an implementation of a low-pass Nyquist filter, that is, one that has the property of vestigial symmetry. This means that its spectrum exhibits an odd *symmetry* about $1/2T_s$, where T_s is the symbol period. Its frequency domain representation is a "brick-wall-like" function, given by

$$H(f) = \begin{cases} T_s & |f| \leq \dfrac{1-\beta}{2T_s} \\[2ex] \dfrac{T_s}{2}\left[1+\cos\left(\dfrac{\pi T_s}{\beta}\left\{|f|-\dfrac{1-\beta}{2T_s}\right\}\right)\right] & \dfrac{1-\beta}{2T_s} < |f| \leq \dfrac{1+\beta}{2T_s} \\[2ex] 0 & \text{otherwise} \end{cases} \tag{6.6}$$

with $0 \leq \beta \leq 1$

This frequency response is characterized by two values: β, the *roll-off factor* (*ROF*), and T_s, the reciprocal of the symbol rate in Sy/s, that is, $1/2T_s$ the half bandwidth of the filter. The impulse response of such a filter can be obtained by analytically taking the inverse Fourier transformation of Equation 6.6, in terms of the normalized sinc function, as

$$h(t) = \text{sinc}\left(\frac{t}{T_s}\right)\frac{\cos\left(\dfrac{\pi\beta t}{T_s}\right)}{1-\left(2\dfrac{\pi\beta t}{T_s}\right)^2} \tag{6.7}$$

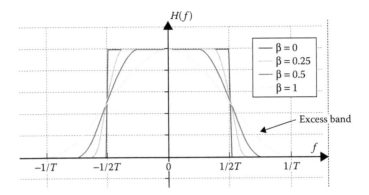

FIGURE 6.18 Frequency response of raised-cosine filter with various values of the ROF β.

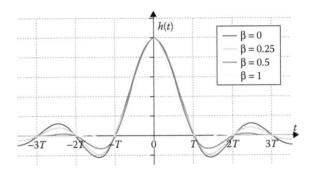

FIGURE 6.19 Impulse response of raised-cosine filter with the ROF β as a parameter.

where the *ROF*, β, is a measure of the *excess bandwidth* of the filter, that is, the bandwidth occupied beyond the Nyquist bandwidth as from the amplitude at $1/2T$. Figure 6.18 depicts the frequency spectra of raised-cosine pulse with various ROF. Their corresponding time domain pulse shapes are given in Figure 6.19.

When used to filter a symbol stream, a Nyquist filter has the property of eliminating intersymbol interference (ISI), as its impulse response is zero at all nT (where n is an integer), except when $n = 0$. Therefore, if the transmitted waveform is correctly sampled at the receiver, the original symbol values can be recovered completely. However, in many practical communications systems, a matched filter is used at the receiver, so as to minimize the effects of noises. For zero ISI, the net response of the product of the transmitting and receiving filters must equate to $H(f)$, thus we can write

$$H_R(f)H_T(f) = H(f) \tag{6.8}$$

Or alternatively, we can rewrite that

$$|H_R(f)| = |H_T(f)| = \sqrt{|H(f)|} \tag{6.9}$$

The filters which can satisfy the conditions of Equation 6.9 are the *root-raised-cosine* filters. The main problem with root-raised-cosine filters is that they occupy a larger frequency band than that of the Nyquist sinc pulse sequence. Thus, for the transmission system, we can split the overall raised-cosine filter with a root-raised-cosine filter at both the transmitting and receiving ends, provided the system is linear. This linearity is to be specified accordingly. An optical fiber transmission system can be considered to be linear if the total power of all channels is under the nonlinear SPM threshold limit. When it is over this threshold, a weakly linear approximation can be used.

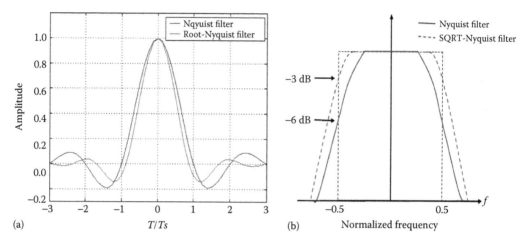

FIGURE 6.20 (a) Impulse and (b) corresponding frequency response of *sinc* Nyquist pulse shape or root-raised-cosine Nyquist filters.

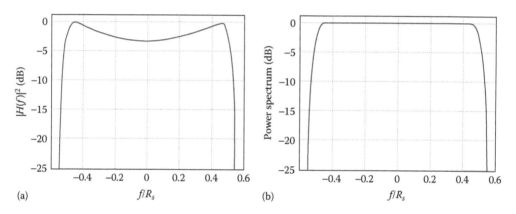

FIGURE 6.21 (a) Desired Nyquist filter for spectral equalization and (b) output spectrum of the Nyquist filtered QPSK signal.

The design of a Nyquist filter influences the performance of the overall transmission system. Oversampling factor, selection of ROF for different modulation formats, and finite impulse response (FIR) Nyquist filter design are key parameters to be determined. If taking into account the transfer functions of the overall transmission channel including fiber, wavelength selective switch (WSS), and the cascade of the transfer functions of all O/E-components, the total channel transfer function is more Gaussian-like. To compensate this effect in the Tx-DSP, one would thus need a special Nyquist filter to achieve the overall frequency response equivalent to that of the rectangular or raised cosine with ROF shown in Figures 6.20 and 6.21.

6.4.3 SUPERCHANNEL SYSTEM REQUIREMENTS

Transmission distance: As next generation of backbone transport, the transmission distance should be comparable to the previous generation, namely a 100 Gbps transmission system. As the most important requirement, we require that the 1 Tbps transmission for long haul should be 1500–2000 km, and for metro application ~300 km.

CD tolerance: As the SSMF fiber CD factor/coefficient 16.8 ps/nm is the largest among the current deployed fibers, CD tolerance should be up to 30,000 ps/nm at the central channel whose wavelength is approximated at 1550 nm. At the edge of C-band, this factor is expected to increase by about 0.092 ps/(nm^2 km) or about 32,760 ps/nm at 1560 nm and 26,400 ps/nm at 1530 nm.[*]

PMD tolerance: The worse case of deployed fiber with 2000 km would have a differential group delay (DGD) of 75 ps or about 3 symbol period for 25 GSy/s per subchannel. So the PMD (mean all-order DGD) tolerance is 25 ps.

SOP rotation speed: According to the 100 Gbps experiments, state-of-polarization (SOP) rotation can be up to 10 kHz, we take the same spec as the 100G system.

Modulation format: PDM-QPSK for long-haul transmission; PDM-16-QAM for metro application.

Spectral efficiency: Compared to a 100G system with an increase of factor 2. Both Nyquist-WDM and CO-OFDM can fulfill this. However, it depends on technological and economical requirements that would determine the suitability of the technology for optical network deployment.

Table 6.1 tabulates the system specifications of various transmission structures with parameters of subsystems, especially when the comb generators employed using either recirculating or nonlinear generation techniques. The DSP reception and offline DSP are integrated in these systems.

6.4.4 System Structure

6.4.4.1 DSP-Based Coherent Receiver

A possible structure of a superchannel transmission system can be depicted in Figure 6.22. At the transmitter, the data inputs can be inserted into the pulse-shaping and individual data streams can be formed. A DAC can be used to shape the pulse with a Nyquist equivalent shape, the raised-cosine form whose spectra also follow a raised cosine with the ROF β varying from 0.1 to 0.5. If this off factor takes the value of 0.1, the spectra would follow an approximate shape of a rectangular. A comb generator can be used to generate equally spaced subcarriers for the superchannel from a single carrier laser source, commonly an ECL of a very narrow band of linewidth of about 100 kHz. These comb-generated subcarriers (see Figure 6.23) are demultiplexed into subcarriers and fed into a bank of I-Q optical modulators as described in Chapter 3, and Nyquist pulse shape pulse sequence as the output of the DAC is then employed to modulate these subcarriers to form the superchannels at the output of an optical multiplexer shown in the block on the left side of Figure 6.22. More details of the transmitter for superchannels are shown in Figure 6.24. It is noted that the generation of a comb source can be recirculating of the shifting of the original carrier around a closed loop. The frequency shift is the spacing frequency between the subcarriers. So the Nth subcarrier would be the Nth time circulation of the original carrier. There would be superimposing of noises due to the amplification simulated emission incorporated in the loop, and this would be minimized by inserting into the loop an optical filter whose bandwidth would be the same as or wider than that of the superchannel.

The fiber transmission line is optically amplified optical fiber multispan without incorporating any dispersion compensating fibers. Thus, the transmission is very dispersive. The broadening of a 40-ps width pulse would spread across at least 80–100 symbol period after propagating over 3000 km of SSMF. Thus, one can assume that the pulse launched into the fiber of the first span would be considered to be an impulse as compared to that after 3000 km SSMF propagation.

After the propagation over the multispan non-DCF line, the transmitted subchannels are demuxed via a wavelength splitter into individual subchannels, with minimum crosstalk. Each subchannel is then coherently mixed with an LO, which is generated from another comb source incorporating an optical phase-locked loop (OPLL) to lock the comb into that of the subcarriers of a superchannel. Thus, a comb generator is indicated on the right side, the reception system of Figure 6.22. The coherently mixed subchannels are then detected by a balanced receiver (as described in Chapter 4

---

[*] Technical specification of Corning fiber G.652 SSMF given in Ref. [11].

TABLE 6.1
1 Tbps Offline System Specifications

Parameter	Superchannel RCFS Comb Gen	Superchannel Nonlinear Comb Gen	Some Specs	Remarks
Bit rate	1, 2, …, N Tbps (whole C-band)	1, 2, …, N Tbps	~1.28 Tbps at 28–32 GB	20% OH for OTN, FEC
Number of ECLs	1	$N \times 2$		
Nyquist roll-off	0.1 or less	0.1 or less		DAC pre-equalization required
Baud rate (Gbaud)	28–32	28–32	28, 30, or 31.5 Gbaud	Pending on FEC coding allowance
Transmission distance	2500	2500	1200 (16 span)–2000 km (25 spans) 2500 km (30 spans) 500 km	20%FEC required for long-haul application Metro application
Modulation format	QPSK/16-QAM	QPSK/16-QAM	Multicarrier Nyquist WDM PDM-DQPSK/QAM Multicarrier Nyquist WDM PDM-16-QAM	For long haul For long haul For metro
Channel spacing			4 × 50 GHz 2 × 50 GHz	For long haul For long haul
Launch power	≪0 dBm if 20 Tbps is used		~−3 to 1 dBm lower if $N > 2$	Depending on QPSK/16-QAM and long haul/metro can be different
B2B ROSNR @ 2×10^{-2} (BOL) (dB)	14.5	14.5	15 dB for DQPSK 22 dB for 16-QAM	1 dB hardware penalty 1 dB narrow filtering penalty
Fiber type	SSMF G.652 Or G.655)	SSMF G.652 (or 655)	G.652 SSMF	
Span loss	22	22	22 dB (80 km)	
Amplifier	EDFA ($G > 22$ dB); NF < 5 dB		EDFA (OAU or OBU)	
BER	2×10^{-3}	2×10^{-3}	Pre-FEC 2×10^{-2} (20%) or 1×10^{-3} classic FEC (7%)	
CD penalty (dB)			0 dB at ±3000 ps/nm <0.3 dB at ±30,000 ps/nm;	16.8 ps/nm/km and 0.092 ps/(nm² km)
PMD penalty (DGD)			0.5 dB at 75 ps, 2.5 symbol periods	
SOP rotation speed	10 kHz	10 kHz	10 kHz	OPLL may require due to oscillation of the LO carrier
Filters cascaded penalty			<1dB at 12 pcs WSS	
Driver linearity	Required	Required	THD < 3%	16-QAM even more strict

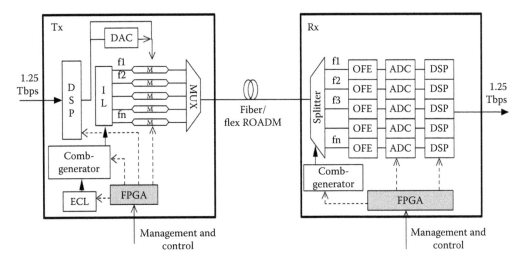

FIGURE 6.22 A possible structure of a superchannel Tbps transmission system.

and Section 6.2 of this chapter), and then electronically amplified and fed into the sampler and ADC. The digital signals are then processed in the DSP of each subchannel system or a parallel and interconnected DSP system. In these DSP, the sequence of processing algorithms is employed to recover the carrier phase, and hence the clock recovery, compensating for the linear and nonlinear dispersion, and the evaluation of the BER versus different parameters such as OSNR. Figure 6.25 shows the modulated spectra of five channels whose subcarriers are selected from the multiple sub-carrier sources shown in Figure 6.23. The modulation is QPSK with Nyquist pulse shaping.

6.4.4.2 Optical Fourier Transform-Based Structure

The superchannel transmission system can also be structured using optical fast Fourier transform (FFT) as demonstrated in [12] and shown in Figure 6.26, in which the components of Mach–Zehnder delay interferometer (MZDI) (see Figure 6.26a) act as spectral filter and splitter (see Figure 6.26b), the optical FFT. The outputs of these MZDI are then fed into coherent receivers and processed digi-tally, as in Figure 6.26c, with the electro-absorption modulator (EAM) performing the switching function so as to time demultiplexing the ultrafat signal speed to lower speed sequence so that the detection system can decode and convert to the digital domain for further processing.

The spectra of superchannels at different positions in the transmission system can be seen in Figure 6.27a through c. It is noted that the pulse shape is Nyquist, and the subcahnnels are placed close and satisfying the orthogonal condition; thus the name optical orthogonal frequency division multiplexing (OFDM) is used to indicate this superchannel arrangement.

6.4.4.3 Processing

The processing of superchannels can be considered as similar to the digital processing of individual subchannels except when there may be crosstalk between subchannels due to overlapping of certain spectral regions between the considered channel and its adjacent channels.

Thus, for the Nyquist QPSK subchannel, the DSP processing would be much the same as for QPSK DWDM for 112G described above with care taken for the overlapping either at the transmit-ter or at the receiver. The EVM is a parameter that indicates the scattering of the vector formed by I- and Q-components departing from the center of the constellation point. The variance of this EVM in the constellation plane is used to evaluate the noises of the detected states, thence the Q factor can be evaluated with ease and thus the BER by using the probability density function and the mag-nitude of the vector of a state on the constellation plane. The BER of the subchannels of the OFDM superchannel is shown in Figure 6.28b and c [9] for different percentage overloading due to FEC.

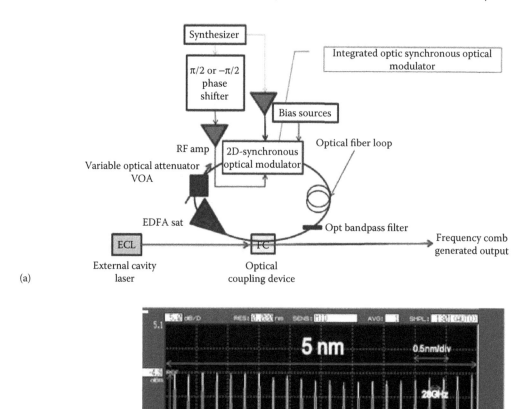

(a)

(b)

FIGURE 6.23 Block diagram of (a) an RFS comb generator and (b) a typical generated spectrum of the comb generator with 28 GHz spacing between channels over more than 5 nm in spectral region and about 30 dB carrier-to-noise ratio (CNR).

The loading factor is important as this will increase the speed or symbol rate of the subchannel one has to offer. The higher this percentage, the higher the increase in the symbol rate is and thus requiring high-speed devices and components.

In the receiver of the optical OFDM superchannel system of [8], to judge the effectiveness of the optical FFT receiver, three alternative receiver concepts and tested them for a QPSK signal. A QPSK signal is chosen because it was not possible to receive a 16-QAM signal with the alternative receivers owing to their inferior performance [8]. First, a subcarrier with a narrow bandpass filter is used to extract a subchannel. The filter passband is adjusted in for best performance of the received signal (Figures 6.28 and 6.27a). The selected filter bandwidth is 25 GHz. The constellation diagram shows severe distortion. When using narrow optical filtering, one has to accept a compromise between

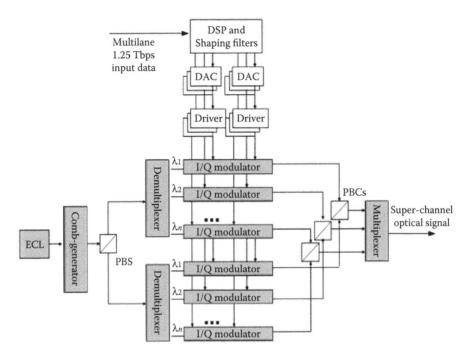

FIGURE 6.24 Generic detailed architecture of a superchannel transmitter. PBC = polarization beam combiner, PBS = polarization beam splitter. I/Q = in-phase–quadrature phase. DAC = digital-to-analog converter. ECL = external cavity laser.

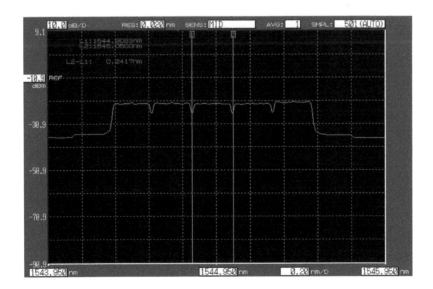

FIGURE 6.25 Selected five subcarriers with modulation.

crosstalk from neighboring channels (as modulated OFDM subcarriers necessarily overlap) and ISI owing to the increasing length of the impulse response when narrow filters are used. Narrow filters can be used, however, if the ringing from ISI is mitigated by additional time gating. The reception of a subcarrier using a coherent receiver is then performed. In the coherent receiver, the signal is down converted in a hybrid coupler as described in Chapter 5, and detected using balanced detectors and

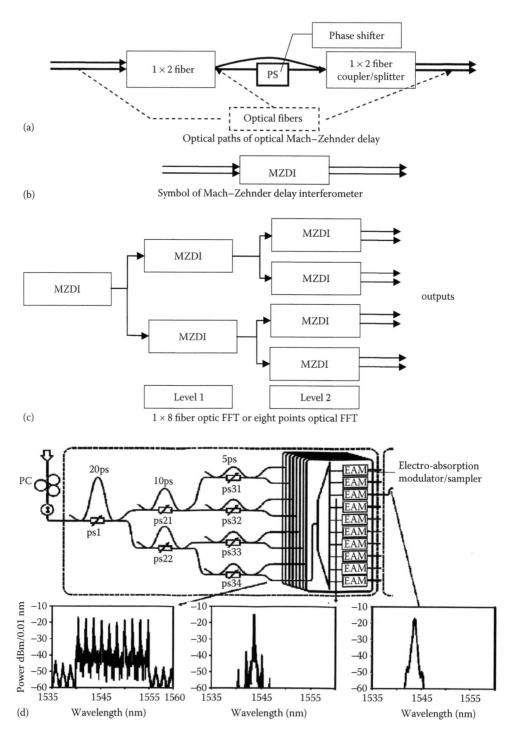

FIGURE 6.26 Operations by guided wave components using fiber optics (a) guided wave optical path of an MZDI or asymmetric interferometer with phase delay tunable with by thermal or electro-optic effects (b) block diagram representation (c) implementation of optical FFT using cascade stages of a fiber optical MZDI structure (extracted from [13]). EAM = electro-absorption modulator used for demultiplexing in the time domain. Note also phase shifters employed in MZDIs between stages. Insets are spectra of optical signals at different stages as indicated of the optical FFT (serial type). EAM = electro-absorption modulator; PC = polarization controller, and (d) optical spectrum at various locations as shown in the schematics.

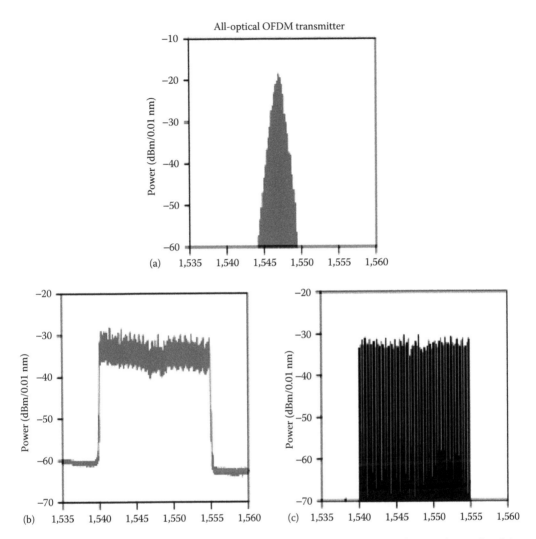

FIGURE 6.27 Spectrum of superchannels and demuxed channels. All subchannels are orthogonal and thus the name orthogonal frequency-division multiplexing (OFDM) (a) At output of transmitter, (b) after fiber propagation and (c) after polarization demux. (Extracted from Fatadin, I. et al., *IEEE Photonic Technol. Lett.*, 2010. With permission.)

sampled in a real-time oscilloscope. Using a combination of error low-pass filtering due to the limited electrical bandwidth of the oscilloscope and DSP, the subcarrier is extracted from the received signal. This receiver performs better than the filtering approach, but a larger electrical bandwidth and sampling rate of the ADC and additional DSP would be needed to eliminate the crosstalk from other subcarriers and then to achieve a performance similar to that of the optical FFT.

Thus, optical OFDM may offer significant advantages for a superchannel but additional processing time would be required, while for Nyquist QPSK superchannels allow better performance and less complexity in the receiver DSP subsystem structure.

6.4.5 TIMING RECOVERY IN NYQUIST QAM CHANNEL

Nyquist pulse shaping is one of the efficient methods to pack adjacent subchannels into a superchannel. The timing recovery of such a Nyquist subchannel is critical for sampling the data received and improving the transmission performance. Timing recovery can be done either before or after the

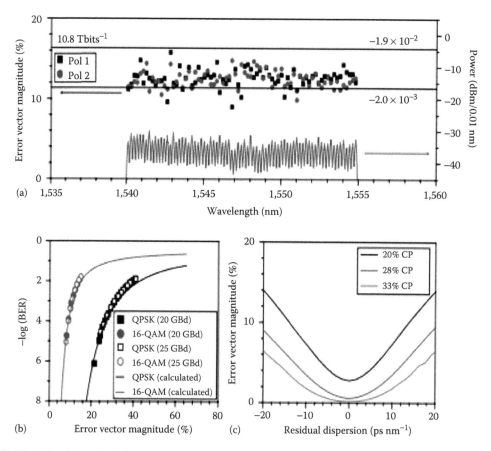

FIGURE 6.28 All-optical 10.8 Tbps OFDM results. (a) Measured EVM for both polarizations (symbols) and for all subcarriers of the OFDM signal decoded with the all-optical FFT. The estimated BER for all subcarriers is below the third-generation FEC limit of 1.9×10^{-2}. The optical spectrum far left race is drawn beneath. (b) Relationship between BER and EVM. Measured points (symbols) and calculation of BER as a function of EVM for QPSK and 16-QAM. (c) Tolerance toward residual CD of the implemented system decoded with the eight-point FFT for a cyclic prefix of 20%, 28%, and 33%. (Extracted from Fatadin, I. et al., *IEEE Photonic Technol. Lett.*, 2010. With permission.)

PMD compensator. The phase detector scheme is shown in Figure 6.29, a Godard type [14], which is a first-order linear scheme. After CD compensation (CD^{-1} blocks), the signal is sent to a SOP modifier to improve the clock extraction. The clock performance of the NRZ QPSK signal in the presence of a first-order PMD, characterized by a DGD and azimuth, is presented in Figure 6.3. The azimuth of 45° and DGD of a half symbol/unit interval completely destroys the clock tone. Therefore, the SOP modifier is required for enabling the clock extraction. In practical systems, a raised-cosine filter is used to generate Nyquist pulses. A filter pulse response is defined by two parameters, the ROF β and the symbol period T_s, and described by taking the inverse Fourier transform of Equation 6.6. The Godard phase detector cannot recover the carrier phase even with small β, thus the channel spectra is close to rectangular. A higher order phase detector must be used to effectively recover the timing clock period as shown in [15]. A fourth power-law PD (PLPD) with prefiltering presented in [16] as shown in Figure 6.29b can deal with small β values.

The PLPD operates by first splitting and forming the combination of these components X- and Y-polarized channels, then conducting the frequency domain detection and regenerating through the voltage-controlled oscillator (VCO), the frequency shift required for the ADC to ensure the sampling timing is correct with the received sampled for processing.

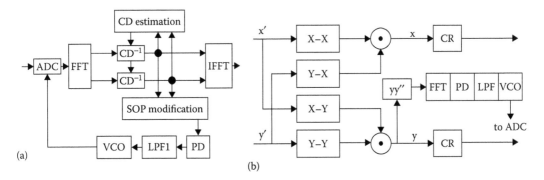

FIGURE 6.29 (a) Godard phase detector algorithm. (b) Forth-order PLPD. VCO = voltage-controlled oscillator, FFT = fast Fourier transform, IFFT = inverse FFT, CD = chromatic dispersion, SOP = state of polarization.

The use of such a phase detector in coherent optical receivers requires large hardware effort. Because of PMD effects the direct implementation of this method before PMD compensation is almost impossible. Therefore, the PLPD implementation in the frequency domain after the PMD compensation is proven to be the most effective and performing well even in the most extreme cases with ROF equal to zero.

6.4.6 128 Gbps 16-QAM Superchannel Transmission

An experimental setup by Dong et al. [17] for the generation and transmission of six channels carrying 128-Gbps under modulation format and polarization multiplexing PDM-16-QAM signal. The two 16-Gbaud electrical 16-QAM signals are generated from the two arbitrary waveform generators. Laser sources at 1550.10 nm and the second source with a frequency spacing of 0.384 nm (48 GHz) are generated from two ECLs, each with the linewidth less than 100 kHz and the output power of 14.5 dBm, respectively. Two I/Q MODs are used to modulate the two optical carriers with I- and Q-components of the 64 Gbps (16-Gbaud) electrical 16-QAM signals after the power amplification using four broadband electrical amplifiers/drivers, respectively. Two phase shifters with the bandwidth of 5 kHz to 22.5 GHz provide 2-symbol extra delay to decorrelate the identical patterns. For the operation to generate 16-QAM, the two parallel Mach–Zehnder modulators for I/Q modulation are both biased at the null point and driven at full swing to achieve zero-chirp and phase modulation. The phase difference between the upper and the lower branch of I/Q MZIM is also controlled at the null point. The data input is shaped so that about 0.99 ROF of raised-cosine pulse shape could be generated.

The power of the signal is boosted using polarization-maintaining EDFAs. The transmitted optical channels are the mixed with an LO and polarization demultiplexed via a $\pi/2$ hybrid coupler, then I- and Q-components are detected by four pairs of balanced photodetectors (PDP). They are then transimpedance amplified and resampled.

The sampled data are then processed in the following sequence: CD compensation, clock recovery, resampling and going through a classical CMA, a three-stage CMA, frequency offset compensation, a feed-forward (FF) phase equalization, least-mean-square (LMS) equalizer, and finally differentially detected to avoid cycle slip effects.

Furthermore, the detailed processing [18] for the electrical polarization recovery is achieved using a three-stage blind equalization scheme: (i) First, the clock is extracted using the "square and filter" method, and then the digital signal is resampled at twice of the baud rate based on the recovery clock. (ii) Second, a T/2-spaced time-domain FIR filter is used for the compensation of CD, where the filter coefficients are calculated from the known fiber CD transfer function using the frequency-domain truncation method. (iii) Third, adaptive filters employing two complex-valued, 13 tap coefficients, and partial T/2-space are employed to retrieve the modulus of the 16-QAM signal.

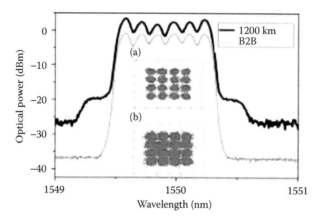

FIGURE 6.30   Spectra of Nyquist channels, 6 × 128G PDM 16-QAM B2B and 1200 km transmission.

The two adaptive FIR filters are based on the classic CMA and followed by a three-stage CMA, to realize multimodulus recovery and polarization demultiplexing. The carrier recovery is performed in the subsequent step, where the fourth power is used to estimate the frequency offset between the LO and the received optical signal. The phase recovery is obtained by FF and the LMS algorithms for LO frequency offset compensation. Finally, differential decoding is used for BER calculating after decision.

The spectra of the six Nyquist channels under back-to-back (B2B) and 1200 km SSMF transmission are shown in Figure 6.30. It is noted that even for 0.01 rolloff of Nyquist channels, we do not observe the flatness of individual channels in the diagram. The constellation as expected would require significant amount of equalization and processing.

On average the achieved BER of 2×10^{-3} with a launched power of −1 dBm over a 1200-km SSMF nondispersion compensation link was demonstrated by the authors of Ref. [17]. The link is determined by a recirculating loop consisting of four spans with each span length of 80 km SSMF and an inline EDFA. WSS is employed wherever necessary to equalize the average power of the subchannels. The optimum launched power is about −1 dBm for the six-subchannel superchannel transmission.

6.4.7   450 Gbps 32QAM Nyquist Transmission Systems

Further spectral packing of subchannels in a superchannel can be done with Nyquist pulse shaping and predistortion or pre-equalization at the transmitting side. Zhou et al. [19] have recently demonstrated the generation and transmission of 450 Gbps WDM channels over the standard 50 GHz ITU-T grid optical network at a net spectral efficiency of 8.4 bps/Hz. This result is accomplished by the use of Nyquist-shaped, PDM 32QAM, or 5 bits/symbol × 2 (polarized modes) × 45 GSy/s to give a total of 450 Gbps. Both pre- and posttransmission digital equalization techniques are employed to overcome the limitation of the DAC bandwidth. Nearly ideal Nyquist pulse shaping with an ROF of 0.01 allows guard bands of only 200 MHz between subcarriers. To mitigate the narrow optical filtering effects from the 50 GHz-grid ROADM, a broadband optical pulse-shaping method is employed. By combining electrical and optical shaping techniques, the transmission of five 450 Gbps PDM-Nyquist 32QAM on the 50 GHz grid over 800 km and one 50 GHz-grid ROADM was proven with soft DSP equalization and processing. The symbol rate is set at 28 GSy/s.

It is noted that the transmission SSMF length is limited to 800 km due to the reduced Euclidean geometrical distance between constellation points of 32QAM and by avoiding the accumulated ASE noises contributed by EDFA in each span. Raman optical amplifiers with distributed gain are used in a recirculating loop of 100 km ultralarge area fibers. The BER performance for all five subchannels is shown in Figure 6.31a with the insertion of the spectra of all subchannels. Note the near-flat spectrum of each subchannel that indicates the near-Nyquist pulse shaping. Figure 6.31b and c

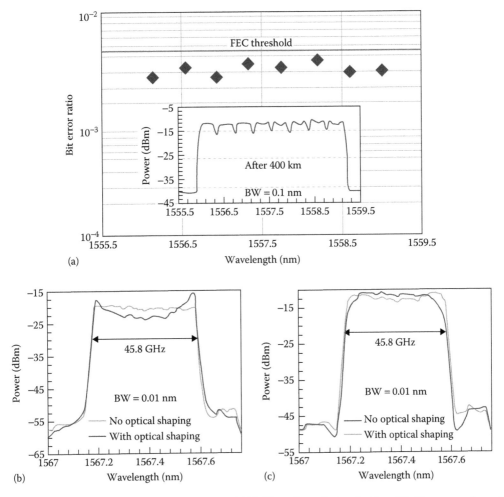

FIGURE 6.31 (a) Spectra of five subchannels of 450 Gbps channel, that is, 5 × 450 Gbps superchannels after 400 km (one loop circulating) of 5 × 80 km plus Raman amplification. BER versus wavelength and FEC threshold at 2.3×10⁻³; (b) spectrum of a single subchannel before and (c) after WSS with and without optical shaping by optical filters. *Note:* Dark line for original spectra, and light line for spectra after spectral shaping. (Extracted from Don, Z. et al., *IEEE J. Lightwave Technol.*, 30, 4000–4006, 2012. With permission.)

show the spectra of a single subchannel before and after WSS with and without optical filtering that performs as spectra shaping. Further to this published work, Zhou et al. [20] have also reported a time-multiplexed 64QAM and transmitting over 1200 km, that is, three circulating around the ring circumstance of 400 km consisting of four 100 km optically amplified span incorporating Raman pumped amplification sections and EDFA. About 5 dB penalty between 32QAM and 64QAM in receiver sensitivity at the same FEC BER of 6×10^{-3} is obtained. Note that electrical time-division multiplexing of digital sequences from the arbitrary waveform generators was implemented by interleaving so that higher symbol rate can be achieved. Furthermore, 20% soft decision FEC using a quasicyclic LDPC code is employed to achieve a BER threshold of 2.4×10^{-2}; thus the 20% extra overhead is required on the symbol rate.

Simulated results by Bosco et al. [21] without using soft FEC also show the variation of the maximum reach distance with a BER of 2×10^{-3} for capacity in the C-band (bottom axis) and spectral efficiency for PM-BPSK, PM-QPSK, PM-8QAM, and PM-16-QAM as shown in Figure 6.32 for SSMF non-DCF optical transmission lines as well as nonzero dispersion-shifted fibers (NZDSFs) indicated by dashed lines. The design of NZ-DSF was described in Chapter 2.

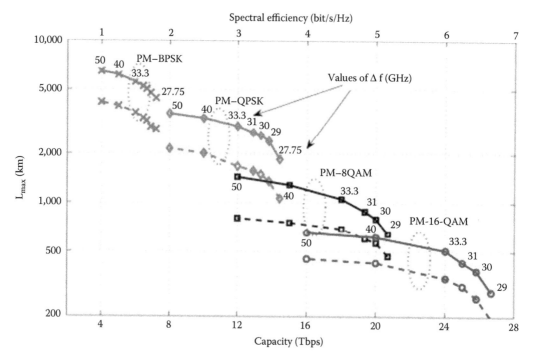

FIGURE 6.32 Transmission reach distance variation with respect to spectral efficiency and total capacity of superchannels. Different QAM schemes are indicated PM = polarization multiplexing. The numbers are for the bandwidth of the subchannel. (Extracted from Zhou, X. et al., 1200 km Transmission of 50 GHz spaced, 5 × 504-Gb/s PDM-32-64 hybrid QAM using Electrical and Optical Spectral Shaping. In *Proceedings of OFC*, Los Angeles, CA, 2012. With permission.)

6.4.8 DSP-Based Heterodyne Coherent Reception Systems

We have so far described optical transmission systems under homodyne coherent reception that is when the frequencies of the lightwave carriers and that of the LO are equal. The original motivation of using homodyne detection is to eliminate the 3 dB degradation as compared with the heterodyne technique, and this is possible under the DSP-based reception algorithm to avoid the difficulties of locking of the LO and the channel carrier. Under the classical heterodyne reception, the "at least" 3 dB loss comes from the splitting of the received signals in the electrical domain and then multiplying by sinusoidal cosine and sine RF oscillator to extract the in-phase and quadrature components.

So far, we have discussed homodyne reception DSP-based optical transmission systems that are considered for extensive deployment in commercial coherent communication systems for 100G, 400G, 1T, or beyond. However, with the development of large-bandwidth and high-speed electronic ADCs and photodetectors (PDs), once again, coherent detection with DSP has been allowed; the mitigation of impairments in optical transmission can be compensated by equalization in the electrical domain. As we have seen in Sections 3.2 and 3.3, for homodyne detection in PDM systems, the I/Q-components of each polarization state should be separated in the optical domain with full information. Thus, four balanced PD pairs incorporating with a photonic dual-hybrid structure and four-channel time-delay synchronized ADCs are required.

By up-converting I- and Q-components to the intermediate frequency (IF) at the same time, not only can heterodyne coherent detection half the number of the balanced PDs and ADCs of the coherent receiver, but also there is no need to consider the delays between I- and Q-components in the PDM signal. Therefore, the four output ports of the optical hybrid can be also halved accordingly. However, this heterodyne technique can possibly be restricted by the bandwidth of the PDs.

Furthermore, the external down conversion of the IF signals may enhance the complexity of the reception system. Currently, the tremendous progresses in increasing the sampling rate and bandwidth of ADCs and PDs give a high possibility of exploiting a simplified heterodyne detection. With large-bandwidth PDs and ADCs, down conversion of the IF, I/Q separation of quadrature signal, and equalization for the PDM and nonlinear effect can all be realized in the digital domain; the DSP followed the ADCs. A heterodyne detection in the transmission system is a limited 5-Gbps 4-ary quadrature amplitude modulation (4QAM) signal over 20 km in [22] and limited 20-Mbaud 64 and 128QAM over 525 km in [23], and then reaching Tbps in Dong et al. [24]. High-order modulation formats, such as PDM-16-QAM and PMD-64QAM, taking advantage of high capacity, can offer to be spectrally effective for 100G or beyond by adopting heterodyne detection.

For the 100G or beyond coherent system with required transmission distance shorter than 1000 km, the inferiority of the SNR sensitivity in heterodyne detection is not so obvious. Conversely, less number of ADCs and easy implementation of the DSP for IF down conversion make heterodyne detection a potential candidate for a 100G or beyond transmission system. Figure 6.33a shows the schematic diagram of the heterodyne reception with digital processing. The quadrature amplitude modulated signals are transmitted and imposed onto the $\pi/2$ hybrid coupler, which is now simplified, and there is no $\pi/2$ phase shifter as compared to the hybrid coupler discussed in Chapter 4 and

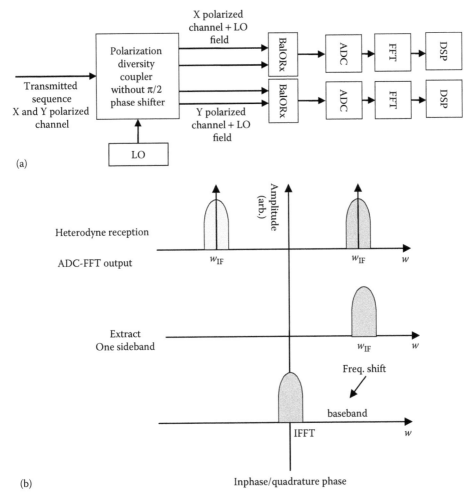

FIGURE 6.33 The principles of heterodyne coherent reception (a) principal blocks and (b) spectra to recover baseband signals. (Adapted from Nakazawa, M. et al., *Electron. Lett.*, 42, 710–712, 2006. With permission.)

Sections 3.3.2.3.3 and 3.3.2.3.4. The coherent mixing of the LO and the modulated channel would result in the time domain signals that are RF enveloped covering the lightwaves and limited within the symbol period. Thus, both the real and imaginary parts appear in the electronic signals produced after the balanced detection in the PDPs in which the beating happens (refer to Figure 6.33b). The electronic currents produced after the PDPs are then amplified via the TIA to produce voltage-level signals that are conditioned to appropriate levels so that they can be sampled by the ADCs. Thence, doing the FFT will produce the two-sided spectrum that exhibits the frequency shifting of the baseband to the RF or IF frequency. Both the in-phase and quadrature components are embedded in the two-sided spectrum. One sideband of the spectrum can be used to extract the I- and Q-parts of the QAM signals for further processing. Ideally, the compensation of the CD should be done in the first stage, then followed by the carrier phase recovery and finally resampling with correct timing.

The following sequence of processing in the digital domain was conducted: (i) sampling in ADC and conversion from the analog voltage level to digital sampled states; (ii) FFT to obtain frequency-domain spectrum; (iii) extract one-sided spectrum and do a frequency shifting to obtain the baseband samples of the spectrum; (iv) resampling with two times the sampling rate; hence (v) compensation of CD; (vi) carrier phase and clock recovery; (vii) using the recovered clock to resample the data sequence; (viii) conducting normal CMA and three-stage CMA to obtain the initial constellation; (ix) frequency equalization of frequency difference of the LO; (x) FF phase equalization, LMS equalization for PMD compensation; and finally, (xi) differential decoding of the samples and symbols to determine the transmission performance BER with respect to certain OSNR measured in the optical domain and launched power at the input of the transmission line.

It is further noted that for the equalization based on DSP, a T/2-spaced time-domain FIR filter is used for the compensation of CD. The two complex-valued, 13-tap, T/2-spaced adaptive FIR filters are based on the classic CMA followed by a three-stage CMA, to realize multimodulus recovery and polarization demultiplexing.

The B2B BER versus the OSNR of the heterodyne reception optical transmission system is shown in Figure 6.34 for PDM-16-QAM 128 Gbps per subchannel with different IFs and compared

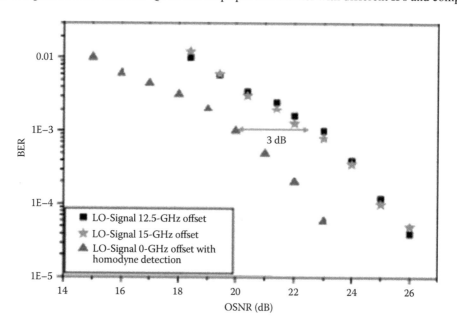

FIGURE 6.34 B2B BER versus OSNR of the heterodyne reception system of PDM-16-QAM 112 Gbps × eight superchannels with different IF frequencies. (Extracted from Nakazawa, M. et al., *Electron. Lett.*, 42, 710–712, 2006. With permission.)

with a homodyne detection scheme. It is noted that 3 dB penalty of heterodyne as compared with homodyne reception due to the fact that only one-sided spectrum is employed. Thus, at BER of 4×10^{-3}, the required OSNR is about 23 dB. The transmission distance is 720 km of SSMF incorporating EDFA and non-DCF.

6.5 CONCLUDING REMARKS

The total information capacity transmitted over a single SMF has been increased tremendously due to the spectral packing of subchannels by pulse-shaping techniques and DSP processing algorithms as well as soft FEC so as to allow higher order of modulation QAM. Thus, digital processing in real time and digitally based coherent optical transmission systems have been proven to the most modern transmission systems for the global internet in the near future. The principal challenges are now lying on the realization of application-specific integrated circuits (using microelectronics technology) or ultrafast field-programmable gate array based systems. The concepts of processing in the digital domain have been proven in offline systems, and thus more efficient algorithms are required for real-time systems. These remain topical research topics and engineering issues. Higher symbol rates may be possible when wider bandwidth optical modulators and DAC and ADC systems are available, for example, grapheme plasmonic silicon modulators [25–27] and their integration with microelectronic DAC and ADC, DSP systems. Ultrahigh speed DACs with a wideband reaching 35 GHZ are now commercially available [28], will enable generation of complex formats and high-level modulation with pre-emphasis that enables the DSP-corporation to push the capacity of optical networks to the multi-exa-bps. via ultrahigh speed multi-Tbps/channel transmission.

The processing algorithms for QAM vary from level to level but essentially they can be based on the number of circles existing in multilevel QAM as compared to a monocycle constellation of the QPSK scheme. The algorithms developed for QPSK can be extended and modified for higher level circular constellation.

REFERENCES

1. T.E. Bell, Communications: Coherent optical communication shows promise, the FCC continues on its path of deregulation, and satellite communications go high-frequency, *IEEE Spectrum*, 23(1), 1986, 49–52.
2. Linear Circuits Inc., From Single ended to different input trans-impedance amplifier, http://circuits.linear.com/267 (accessed July 8, 2015).
3. J.S. Weiner, A. Leven, V. Houtsma, Y. Baeyens, Y.-K. Chen, P. Paschke, Y. Yang et al., SiGe Differential transimpedance amplifier with 50-GHz bandwidth, *IEEE J. Solid-State Circ.*, 38(9), 2003, 1512–1517.
4. E. Ip and J.M. Kahn, Digital equalization of chromatic dispersion and polarization mode dispersion, *IEEE J. Lightwave Technol.*, 25(8), Aug. 2007, 2033.
5. B. Spinner, Equalizer design and complexity for digital coherent receivers. *IEEE J. Sel. Top. Quant. Elect.*, 16(5), 2010, 1180–1192.
6. E. Ip and J.M. Kahn, Compensation of dispersion and nonlinear impairments using digital back propagation. *IEEE J. Lightwave Technol.*, 26(20), Oct. 15, 2008, 3416.
7. I. Fatadin and S.J. Savory, Compensation of frequency offset for 16-QAM optical coherent systems using QPSK partitioning, *IEEE Photonic Technol. Lett.*, 23(17), 2001, 1246–1248.
8. I. Fatadin, D. Ives, and S.J. Savory, Laser linewidth tolerance for 16-QAM coherent optical systems using QPSK partitioning, *IEEE Photonic Technol. Lett.,* 22(9), May 1, 2010, 631–633.
9. H. Louchet, K. Kuzmin, and A. Richter, Improved DSP algorithms for coherent 16-QAM transmission. In *Proceedings of ECOC'08*, paper Tu.1.E6, Belgium, Sept. 2008.
10. R. Noé, T. Pfau, M. El-Darawy, and S. Hoffmann, Electronic polarization control algorithms for coherent optical transmission, *IEEE J. Sel. Top. Quant. Elect.*, 16(5), Sept. 2010, 1193–1199.
11. L.N. Binh, *Digital Optical Communications*. CRC Press, Boca Raton, FL, 2010, Chapter 3, Appendix.

12. D. Hillerkuss, R. Schmogrow, T. Schellinger, M. Jordan, M. Winter, G. Huberl, T. Vallaitis et al., 26 Tbit/s line-rate super-channel transmission utilizing all-optical fast fourier transform processing, *Nat. Photonics*, 5, Jun. 2011, 364.

13. D. Hillerkuss, M. Winter, M. Teschke, A. Marculescu, J. Li, G. Sigurdsson, K. Wormsl et al., Simple all-optical FFT scheme enabling Tbit/s real-time signal processing, *Opt. Express*, 18(9), April 26, 2010, 9324–9340.

14. N. Godard, Passband timing recovery in an all-digital modem receiver, *IEEE Trans. Comm.*, 26, May 1978, 517–523.

15. N. Stojanovic, N.G. Gonzalez, C. Xie, Y. Zhao, B. Mao, J. Qi, L.N. Binh, Timing recovery in Nyquist coherent optical systems. In *International Conference on Telecommunications Systems*, Serbia, 2012.

16. T.T. Fang and C.F. Liu, Fourth-power Law clock recovery with pre-filtering. In *Proceedings of ICC*, Geneva, Switzerland, May 1993, vol. 2, pp. 811–815.

17. Z. Dong, X. Li, J. Yu, and N. Chi, 128-Gb/s Nyquist-WDM PDM-16QAM generation and transmission over 1200-km SMF-28 with SE of 7.47 b/s/Hz, *IEEE J. Lightwave Technol.*, 30(24), Dec. 15, 2012, 4000–4006.

18. Z. Don et al., 128-Gb/s Nyquist-WDM PDM-16QAM generation and transmission over 1200-km SMF-28 with SE of 7.47 b/s/Hz, *IEEE J. Lightwave Technol.*, 30(24), Dec. 15, 2012, 4000–4006.

19. X. Zhou, L.E. Nelson, P. Magill, R. Isaac, B. Zhu, D.W. Peckham, P.I. Borel, and K. Carlson, PDM-Nyquist-32QAM for 450-Gb/s per-channel WDM transmission on the 50 GHz ITU-T grid, *IEEE J. Lightwave Technol.*, 30(4), Feb. 15, 2012, 553.

20. X. Zhou, L.E. Nelson, R. Isaac, P. Magill, B. Zhu, D.W. Peckham, P. Borel et al., 1200 km transmission of 50 GHz spaced, 5×504-Gb/s PDM-32-64 hybrid QAM using electrical and optical spectral shaping. In *Proceedings of OFC*, Los Angeles, CA, 2012.

21. G. Bosco, V. Curri, A. Carena, P. Poggiolini, and F. Forghieri, On the performance of Nyquist-WDM terabit superchannels based on PM-BPSK, PM-QPSK, PM-8QAM or PM-16QAM subcarriers, *IEEE J. Lightwave Technol.*, 29(1), Jan. 2011, 53.

22. R. Zhu, K. Xu, Y. Zhang, Y. Li, J. Wu, X. Hong, and J. Lin, QAM coherent subcarrier multiplexing system based on heterodyne detection using intermediate frequency carrier modulation. In *Microwave Photonics*, IEEE Pub house, New York, 2008, pp. 165–168.

23. M. Nakazawa, M. Yoshida, K. Kasai, and J. Hongou, 20 Msymbol/s, 64 and 128 QAM coherent optical transmission over 525 km using heterodyne detection with frequency-stabilized laser, *Electron. Lett.*, 42(12), 2006, 710–712.

24. Z. Dong, X. Li, J. Yu, and J. Yu, Generation and transmission of 8 × 112-Gb/s WDM PDM-16QAM on a 25-GHz grid with simplified heterodyne detection, *Opt. Express*, 21(2), Jan. 28, 2013, 1773.

25. M. Liu and X. Zhang, Graphene-based optical modulators. In *Proceedings of OFC*, Los Angeles, CA, paper OTu1I.7, 2012.

26. A. Melikyan, L. Alloatti, A. Muslija, D. Hillerkuss, P.C. Schindler, J. Li, R. Palmer et al., High-speed plasmonic phase modulators, *Nat. Photonics*, 8, 2014, 229–233. doi:10.1038/nphoton.2014.9.

27. M. Xu, F. Li, T. Wang, J. Wu, L. Lu, L. Zhou, and Y. Su, Design of an Electro-Optic Modulator Based on a Silicon-Plasmonic Hybrid Phase Shifter, *IEEE J. Lightwave Technol.*, 31(8), 2014, 1170–1177.

28. 6-Bit DAC SHF 614 A, Published: Mar. 19, 2015, http://www.shf.de/new-6-bit-dac-shf-614a/ (accessed July 15, 2016).

7 Optical Modulation and Phasor Vector Representation

A direct modulated photonic transmitter can consist of a single or multiple lightwave sources that can be modulated directly by manipulating the driving current of the laser diode. On the other hand, the laser source can be turned on at all times and its output lightwaves are externally modulated via an integrated optical modulator. This modulation subsystem is now commonly known as external modulated optical transmitter. This externally modulated transmitter preserves the linewidth of the laser and hence its coherence. The idea of external modulation was first proposed in 1969 by P.K. Tien in an article on integrated optics and reviewed in 1977 [1].

This chapter presents the techniques for modulation of lightwaves, externally, not directly manipulating the stimulated emission from inside the laser cavity, via the use of electro-optic effects. Advanced modulation formats have recently attracted much attention for enhancement of the transmission efficiency since the mid-1980s for coherent optical communications. Hence, the preservation of the narrow linewidth of the laser source is critical for operation bit rates in the range of several tens of Gb/s. Thus, external modulation is essential.

Three typical types of optical modulators are presented in this chapter including the lithium niobate ($LiNbO_3$) electro-optic modulators, the electro-absorption (EA) modulators, and polymeric integrated modulators. Their operating principles, device physical structures, device parameters and their applications, and driving condition for generation of different modulation formats as well as their impacts on system performance are also discussed in this chapter.

7.1 OPTICAL MODULATORS

The modulation of lightwaves via an external optical modulator can be classified into three types depending on the special effects that alter the lightwave property, especially the intensity or the phase of the lightwave carrier. In an external modulator, the intensity is normally manipulated by manipulating the phase of the carrier lightwaves guided in one path of an interferometer. A Mach–Zehnder interferometric structure is the most common type, especially the lithium niobate type [2–5].

The EA modulator employs the Franz and Keldysh effect that is observed as lengthening the wavelength of the absorption edge of a semiconductor medium under the influence of an electric field [6,7]. In a quantum structure such as the multiquantum well structure, this effect is called the Stark effect, or the EA effect. The EA modulator can be integrated with a laser structure on the same integrated circuit chip. For the $LiNbO_3$ modulator, the device is externally connected to a laser source via an optical fiber.

The total insertion loss (IL) of a semiconductor intensity modulator is about 8–10 dB including fiber-waveguide coupling loss that is rather high. However, this loss can be compensated by a semiconductor optical amplifier that can be integrated on the same circuit. Compared with $LiNbO_3$ its total IL is about 3–4 dB, which can be affordable as an Er-doped fiber amplifier is now readily available.

The driving voltage for the EA modulator is usually lower than that required for $LiNbO_3$. However, the extension ratio is not as high as that of $LiNbO_3$ type, which is about 25 dB as compared to 10 dB for the EA modulator. This feature contrasts the operating characteristics of the $LiNbO_3$ and EA modulators. Although the driving voltage for the EA modulator is about 2–3 V and 5–7 V for $LiNbO_3$, the former type would be preferred for intensity or phase modulation formats due to the extinction ratio that offers much lower *zero* noise level and hence high quality factor.

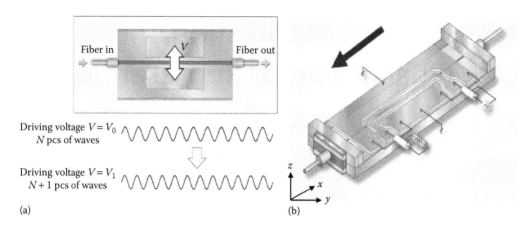

FIGURE 7.1 Electro-optic phase modulation in an integrated modulator using $LiNbO_3$. Electrode imped-ance matching is not shown. (a) Schematic diagram. (b) Integrated optic structure.

7.1.1 PHASE MODULATORS

The phase modulator is a device that manipulates the *phase* of lightwaves considered as the optical carrier signals, under the influence of an electric field created by an applied voltage. When voltage is not applied to the RF electrode, the number of periods of the lightwaves, n, exists in a certain path length. When voltage is applied to the RF electrode, one or a fraction of one period of the wave is added, which now means that $(n + 1)$ waves exist in the same length. In this case, the phase has been changed by 2π and the half voltage of this is called the driving voltage. In case of long-distance optical transmission, waveform is susceptible to degradation due to the nonlinear effect such as self-phase modulation (SPM) and so on. A phase modulator can be used to alter the phase of the carrier to compensate for this degradation. The magnitude of the change of the phase depends on the change of the refractive index created via the electro-optic effect that in turn depends on the orientation of the crystal axis with respect to the direction of the established electric field by the applied signal voltage.

An integrated optic phase modulator operates in a similar manner except that the lightwave carrier is guided via an optical waveguide with diffused or ion-exchanged confined regions for $LiNbO_3$, and rib-waveguide structures for semiconductor type. Two electrodes are deposited so that an electric field can be established across the waveguiding crosssection to produce a change of the refractive index via the electro-optic or EA effect as shown in Figure 7.1. For ultrafast operation, one of the electrodes is a traveling wave type or hot electrode and the other is a ground electrode. The traveling wave electrode must be terminated with matching impedance at the end so as to avoid wave reflection. Usually, a quarter wavelength impedance is used to match the impedance of the traveling wave electrode to that of the 50 Ω transmission line.

A phasor representation of a phase-modulated lightwave can be by the circular rotation at a radial speed of ω_c. Thus, the vector with an angle ϕ represents the magnitude and phase of the lightwave.

7.1.2 INTENSITY MODULATORS

The basic structured LN modulator comprises (i) two waveguides, (ii) two Y-junctions, and (iii) RF/DC traveling wave electrodes. Optical signals coming from the lightwave source are launched into the LN modulator through the polarization-maintaining fiber; it is then equally split into two branches at the first Y-junction on the substrate. When no voltage is applied to the RF electrodes, the two signals are recombined constructively at the second Y-junction and coupled into a single output. In this case, output signals from the LN modulator are recognized as "ONE." When

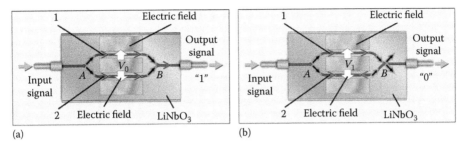

(a) (b)

FIGURE 7.2 Intensity modulation using interferometric principles in waveguide structures in LiNbO$_3$. (a) ON—constructive interference mode. (b) OFF—destructive interference mode. Optical guided wave paths 1 and 2. Electric field is established across the optical waveguide.

voltage is applied to the RF electrode, due to the electro-optic effects of the LN crystal substrate, the waveguide refractive index is changed, and hence the carrier phase in one arm is advanced though retarded in the other arm. Thence the two signals are recombined destructively at the second Y-junction; they are transformed into higher order mode and radiated at the junction. If the phase retarding is in multiple odd factor of π, the two signals are completely out of phase; the combined signals are radiated into the substrate and the output signal from the LN modulator is recognized as a "ZERO." The voltage difference that induces this "ZERO" and "ONE" is called the driving voltage of the modulator, and is one of the important parameters in deciding the modulator's performance (Figure 7.2).

7.1.2.1 Phasor Representation and Transfer Characteristics

Consider an interferometric intensity modulator that consists of an input waveguide that splits into two branches and the recombines to form a single output waveguide. If the two electrodes are initially biased with voltages V_{b1} and V_{b2}, then the initial phases exerted on the lightwaves would be $\phi_1 = \pi V_{b1}/V_\pi = -\phi_2$, which are indicated by the bias vectors shown in Figure 7.3b. From these positions, the phasors are swinging according to the magnitude and sign of the pulse voltages applied to the electrodes. They can be switched to the two positions that can be constructive or destructive. The output field of the lightwave carrier can be represented by

$$E_0 = \frac{1}{2} E_{iRMS} e^{j\omega_c t} \left(e^{j\phi_1(t)} + e^{j\phi_2(t)} \right) \tag{7.1}$$

where ω_c is the carrier radial frequency, E_{iRMS} is the root-mean-square value of the magnitude of the carrier, and $\phi_1(t)$ and $\phi_2(t)$ are the temporal phases generated by the two time-dependent pulse sequences applied to the two electrodes. With the voltage levels varying according to the magnitude of the pulse sequence, one can obtain the transfer curve as shown in Figure 7.3a. This phasor representation can be used to determine exactly the biasing conditions and magnitude of the RF or digital signals required for driving the optical modulators to achieve 50%, 33%, or 67% bit period pulse shapes.

The power transfer function of the Mach–Zehnder modulator can be expressed as[*]

$$P_0(t) = \alpha P_i \cos^2 \frac{\pi V(t)}{V_\pi} \tag{7.2}$$

[*] Note that this equation is represented for single-drive MZIM—it is the same for dual-drive MZIM provided that the bias voltages applied to the two electrodes are equal and opposite in signs. The transfer curve of the field representation would have half the periodic frequency of the transmission curve shown in Figure 7.3.

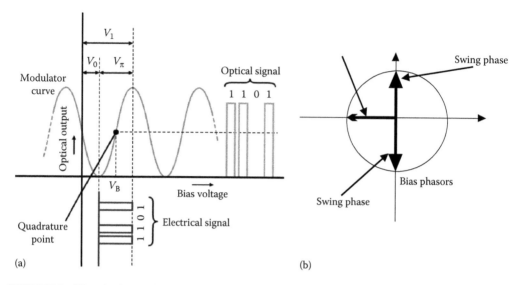

FIGURE 7.3 Electrical-to-optical transfer curve of an interferometric intensity modulator. (a) Optical power versus applied voltage—transfer characteristics. (b) Phasor representations of biased state and the swinging phase states due to the signal amplitude.

where:
   $P_0(t)$ is the output transmitted power
   α is the modulator total IL
   π is the input power (usually from the laser diode)
   $V(t)$ is the time-dependent signal applied voltage
   V_π is the driving voltage so that a π phase shift is exerted on the lightwave carrier

It is necessary to set the static bias on the transmission curve through the bias electrode. It is common practice to set the bias point at 50% transmission point or a $\pi/2$ phase difference between the two optical waveguide branches, the quadrature bias point. As shown in Figure 7.3, electrical digital signals are transformed into optical digital signal by switching voltage to both ends of quadrature points on the positive and negative.

One factor that affects the modulator performance is the drift of the bias voltage. For the Mach–Zehnder interferometric modulator (MZIM) type, it is very critical when it is required to bias at the quadrature point or at minimum or maximum locations on the transfer curve. DC drift is the phenomenon occurring in $LiNbO_3$ due to the build-up of charges on the surface of the crystal substrate. Under this drift, the transmission curve gradually shifts in the long term [8,9]. In the case of the $LiNbO_3$ modulator, the bias point control is vital as the bias point will shift long term. To compensate for the drift, it is necessary to monitor the output signals and feed it back into the bias control circuits to adjust the DC voltage so that operating points stay at the same point, for example, the quadrature point. It is the manufacturer's responsibility to reduce DC drift so that DC voltage is not beyond the limit throughout the lifetime of a device. A feedback control circuit must be used to keep the biasing point stable with respect to the variations of temperature and drift of the optical modulation due to the charge build-up.

7.1.2.2 Chirp-Free Optical Modulators

Due to the symmetry of the crystal refractive index of the uniaxial anisotropy of the class 3m of $LiNbO_3$, the crystal cut and the propagation direction of the electric field affect both modulator efficiency, denoted as driving voltage and modulator chirp. The uniaxial property of $LiNbO_3$ is

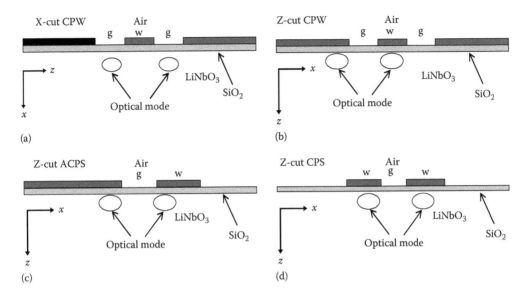

FIGURE 7.4 Commonly used electrode structure and crystal orientation for interferometric modulation to maximize the use of the overlap integral between the optical guided mode and the electric field distribution for largest electro-optic coefficients. ACPS, asymmetric co-planar structure; CPS, coplanar structure (symmetric); g, gap width between electrodes; w, width of active travelling wave electrode.

shown to affect the variation of the change of the refractive indices according to the polarization of lightwaves and the cut of the crystal in which the waveguide and modulator are fabricated. As shown in Figure 7.4, in the case of the Z-cut structure, as a hot electrode is placed on top of the waveguide, RF field flux is more concentrated, and this results in the improvement of overlap between the RF and optical field. However, overlap between RF in the ground electrode and waveguide is reduced in the Z-cut structure so that overall improvement of driving voltage for the Z-cut structure compared to X-cut is approximately 20%. The different overlapping area for the Z-cut structure results in a chirp parameter of 0.7, whereas X-cut and Z-propagation have almost zero chirp due to its symmetric structure. A number of commonly arranged electrode and waveguide structures are shown in Figure 7.4 to maximize the interaction between the traveling electric field and the optical guided waves. Furthermore, a buffer layer, normally SiO_2, is used to match the velocities between these waves so as to optimize the optical modulation bandwidth.

7.1.3 Structures of Photonic Modulators

Figure 7.5a and b shows the structure of an MZ intensity modulator using a single- and dual-electrode configurations, respectively. The thin line electrode is called the *hot* electrode or traveling wave electrode. RF connectors are required for launching the RF data signals to establish the electric field required for electro-optic effects. Impedance termination is also required. Optical fibers' pig tails are also attached to the end faces of the diffused waveguide. The mode spot size of the diffused waveguide is not symmetric, and hence some diffusion parameters are controlled so that maximizing the coupling between the fiber and the diffused or rib waveguide can be achieved. Due to this mismatch between the mode spot sizes of the circular and diffused optical waveguides, there occurs coupling loss. Furthermore, the difference between the refractive indices of fiber and $LiNbO_3$ is quite substantial and thus Fresnel reflection loss would also incur.

Figure 7.5c shows the structure of a polarization modulator that is essential for multiplexing of two polarized data sequences so as to double the transmission capacity, for example, 40–80 Gb/s. Furthermore, this type of polarization modulator can be used as a polarization rotator in a

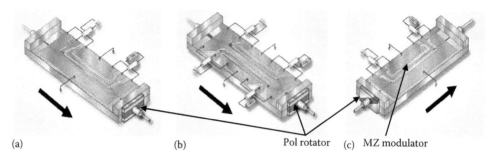

(a)   (b)   Pol rotator   (c)   MZ modulator

FIGURE 7.5 Intensity modulators using LiNbO$_3$. (a) Single-drive electrode. (b) Dual electrode structure. (c) Electro-optic polarization scrambler using LiNbO$_3$.

TABLE 7.1
Typical Operational Parameters of Optical Intensity Modulators

Parameters	Typical Values	Definition/Comments
Modulation speed	10 Gb/s	Capability to transmit digital signals
Insertion loss	Max 5 dB	Defined as the optical power loss within the modulator
Driving voltage	Max 4 V	The RF voltage required to have a full modulation
Optical bandwidth	Min 8 GHz	3 dB roll-off in efficiency at the highest frequency in the modulated signal spectrum
ON/OFF extinction ratio	Min 20 dB	The ratio of maximum optical power (ON) to minimum optical power (OFF)
Polarization extinction ratio	Min 20 dB	The ratio of two polarization states (TM and TE guided modes) at the output

polarization dispersion compensating subsystem [10,11]. Currently, these types of modulators with I-Q modulation have been reported [12].

7.2 RETURN-TO-ZERO OPTICAL PULSES

7.2.1 GENERATION

Figure 7.6 shows the conventional structure of an return-to-zero (RZ)-ASK transmitter in which two external LiNbO$_3$ MZIMs can be used. The MZIM shown in this transmitter can be either a single or dual drive (push pull) type. Operational principles of the MZIM were presented in Section 7.1.2. The optical on-off keying (OOK) transmitter would normally consist of a narrow-linewidth laser source to generate lightwaves whose wavelength satisfies the International Telecommunication Union (ITU) grid standard.

The first MZIM, commonly known as the pulse carver, is used to generate the periodic pulse trains with a required RZ format. The suppression of the lightwave carrier can also be carried out at this stage if necessary, commonly known as the carrier-suppressed RZ (CSRZ). Compared to other RZ types, CSRZ pulse shape is found to have attractive attributes for long-haul wavelength division multiplexing (WDM) transmissions including the π phase difference of adjacent modulated bits, suppression of the optical carrier component in optical spectrum, and narrower spectral width.

Different types of RZ pulses can be generated depending on the driving amplitude of the RF voltage and the biasing schemes of the MZIM. The equations governing the RZ pulses electric field waveforms are

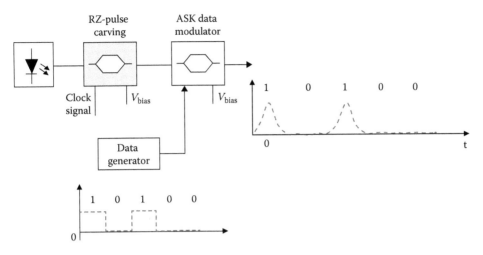

FIGURE 7.6   Conventional structure of an OOK optical transmitter utilizing two MZIMs.

$$E(t) = \begin{cases} \sqrt{\dfrac{E_b}{T}}\sin\left[\dfrac{\pi}{2}\cos\left(\dfrac{\pi t}{T}\right)\right] & \text{67\% duty-ratio RZ pulses or CSRZ} \\[4mm] \sqrt{\dfrac{E_b}{T}}\sin\left[\dfrac{\pi}{2}\left(1+\sin\left(\dfrac{\pi t}{T}\right)\right)\right] & \text{33\% duty-ratio RZ pulses or RZ33} \end{cases}$$

(7.3)

where:
  E_b is the pulse energy per transmitted bit
  T is one bit period

The 33% duty-ratio RZ pulse is denoted as RZ33 pulse, whereas the 67% duty-ratio RZ pulse is known as the CSRZ type. The art in generation of these two RZ pulse types stays at the difference of biasing point on the transfer curve of an MZIM.

The bias voltage conditions and the pulse shape of these two RZ types, the carrier suppression and nonsuppression of maximum carrier, can be implemented with the biasing points at the minimum and maximum transmission point of the transmittance characteristics of the MZIM, respectively. The peak-to-peak amplitude of the RF driving voltage is $2V_\pi$, where V_π is the required driving voltage to obtain a π phase shift of the lightwave carrier. Another important point is that the RF signal is operating at only a half of the transmission bit rate. Hence, pulse carving is actually implementing the frequency doubling. The generations of RZ33 and CSRZ pulse train are demonstrated in Figure 7.7a and b.

The pulse carver can also utilize a dual-drive MZIM that is driven by two complementary sinusoidal RF signals. This pulse carver is biased at $-V_{\pi/2}$ and $+V_{\pi/2}$ with the peak-to-peak amplitude of $V_{\pi/2}$. Thus, a π phase shift is created between the state "1" and "0" of the pulse sequence and hence the RZ with alternating phase 0 and π. If the suppression of the lightwave carrier is required, then the two electrodes are applied with a voltages of magnitude of half of V_π, and thus a total swing voltage amplitude of V_π.

Although RZ modulation offers improved performance, RZ optical systems usually require more complex transmitters than those in the nonreturn-to-zero (NRZ) ones. Compared to only one stage for modulating data on the NRZ optical signals, two modulation stages are required for generation of RZ optical pulses [13–16].

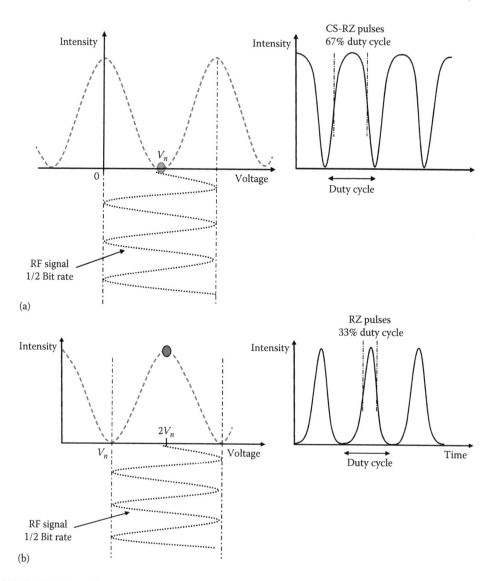

FIGURE 7.7   Bias point and RF driving signals for generation of (a) CSRZ and (b) RZ33 pulses.

7.2.2   PHASOR REPRESENTATION

Phasor representation in brief is the expression of a high-frequency oscillation of waves in terms of a vector from its amplitude (in the peak value) and its phase angle. The oscillation frequency is very high and thus can be considered stationary. Thus, phasor diagram can be constructed to illustrate the evolution of the waves when its phase varies, such as in the case of modulation by RF signals whose RF frequency is very low compared to that of the optical waves. The noise variation such as the phase noises or beating noises between the amplifier random noises and the high-frequency signals can result in the probabilistic noises surrounding the phasor vectors. Thus, now it is very commonly known as the EVM (error vector magnitude), which is used to evaluate the bit-error rate (BER) in coherent reception systems. This section presents the fundamental concepts of phasors. Its applications can be seen throughout this book in the understanding of the evolution of the optical waves under modulation by applied RF signals. Various modulation schemes can be derived from this phasor concept.

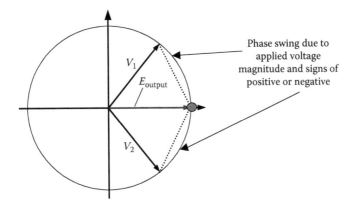

FIGURE 7.8 Phasor representation for generation of output field in dual-drive MZIM.

Recalling Equation 7.1 we have

$$E_o = \frac{E_i}{2}\left[e^{j\varphi_1(t)} + e^{j\varphi_2(t)}\right] = \frac{E_i}{2}\left[e^{j\pi v_1(t)/V_\pi} + e^{j\pi v_2(t)/V_\pi}\right] \tag{7.4}$$

It can be seen that the modulating process for generation of RZ pulses can be represented by a phasor diagram as shown in Figure 7.7, where the optical frequency component $e^{j\omega t}$ has been removed to indicate that the wavevector is rotating at this angular frequency and considered to be stationary. This technique gives a clear understanding of the superposition of the fields at the coupling output of two arms of the MZIM. Here, a dual-drive MZIM is used, that is the data driving signals [$V_1(t)$] and inverse data (data: $V_2(t) = -V_1(t)$) are applied into each arm of the MZIM respectively and the RF voltages swing in inverse directions. Applying the phasor representation, vector addition, and simple trigonometric calculus, the process of generation RZ33 and CSRZ is explained in detail and verified (Figure 7.8).

The widths of these pulses are commonly measured at the position of full-width at half maximum (FWHM). It is noted that the measured pulses are intensity pulses, whereas we are considering the addition of the fields in the MZIM. Thus, the normalized E_o field vector has the value of $\pm 1/\sqrt{2}$ at the FWHM intensity pulse positions and the time interval between these points gives the FWHM values.

7.2.2.1 Phasor Representation of CSRZ Pulses

Key parameters including the V_{bias} and the amplitude of the RF driving signal are shown in Figure 7.9a. Accordingly, its initialized phasor representation is demonstrated in Figure 7.9b.

The values of the key parameters are outlined as follows: (i) V_{bias} is $\pm V_{\pi/2}$. (ii) Swing voltage of driving RF signal on each arm has the amplitude of $V_{\pi/2}$ (i.e., $V_{p-p} = V_\pi$). (iii) RF signal operates at half of bit rate ($B_R/2$). (iv) At the FWHM position of the optical pulse, $E_{out} = \pm 1/\sqrt{2}$ and the component vectors V_1 and V_2 form with vertical axis a phase of $\pi/4$, as shown in Figure 7.10.

Considering the scenario for generation of a 40 Gb/s CSRZ optical signal, the modulating frequency is f_m ($f_m = 20$ GHz $= B_R/2$). At the FWHM positions of the optical pulse, the phase is given by the following expressions:

$$\frac{\pi}{2}\sin(2\pi f_m) = \frac{\pi}{4} \Rightarrow \sin 2\pi f_m = \frac{1}{2} \Rightarrow 2\pi f_m = \left(\frac{\pi}{6},\frac{5\pi}{6}\right) + 2n\pi \tag{7.5}$$

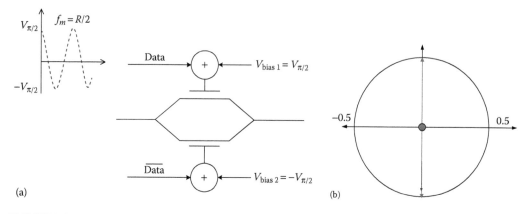

(a)   (b)

FIGURE 7.9 Initialized stage for generation of CSRZ pulse. (a) RF driving signal and the bias voltages. (b) Initial phasor representation.

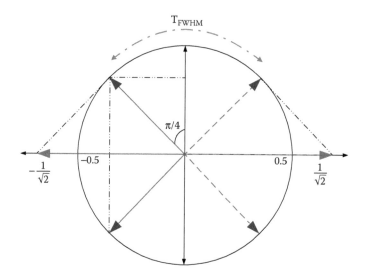

FIGURE 7.10 Phasor representation of CSRZ pulse generation using dual-drive MZIM.

Thus, the calculation of T_{FWHM} can be carried out and hence the duty cycle of the RZ optical pulse can be obtained as given in the following expressions:

$$T_{\mathrm{FWHM}} = \left(\frac{5\pi}{6} - \frac{\pi}{6}\right)\frac{1}{R2\pi} = \frac{1}{3}\pi \times \frac{1}{R} \Rightarrow \frac{T_{\mathrm{FWHM}}}{T_{\mathrm{BIT}}} = \frac{1.66 \times 10^{-4}}{2.5 \times 10^{-11}} = 66.67\% \tag{7.6}$$

The result obtained in Equation 7.6 clearly verifies the generation of CSRZ optical pulses from the phasor representation.

7.2.2.2 Phasor Representation of RZ33 Pulses

Key parameters including the V_{bias}, the amplitude of driving voltage, and its corresponding initialized phasor representation are shown in Figure 7.11a and b, respectively.

The values of the key parameters are as follows: (i) V_{bias} is V_{π} for both arms. (ii) Swing voltage of driving RF signal on each arm has the amplitude of $V_{\pi/2}$ (i.e., $V_{p-p} = V_{\pi}$). (iii) RF signal operates at half of bit rate ($B_R/2$).

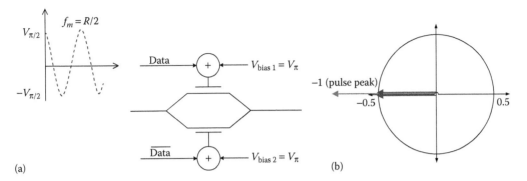

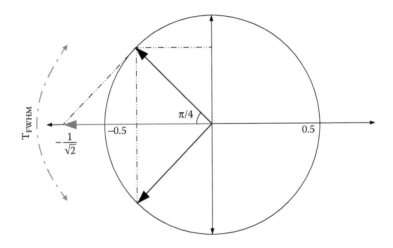

FIGURE 7.11 Initialized stage for generation of RZ33 pulse. (a) RF driving signal and the bias voltage. (b) Initial phasor representation.

FIGURE 7.12 Phasor representation of RZ33 pulse generation using dual-drive MZIM.

At the FWHM position of the optical pulse, the $E_{0output} = \pm(1/\sqrt{2})$ and the component vectors V_1 and V_2 form with horizontal axis a phase of $\pi/4$ as shown in Figure 7.12.

Considering the scenario for generation of a 40 Gb/s CSRZ optical signal, the modulating frequency is f_m ($f_m = 20$ GHz $= B_R/2$). At the FWHM positions of the optical pulse, the phase is given by the following expressions:

$$\frac{\pi}{2}\cos(2\pi f_m t) = \frac{\pi}{4} \Rightarrow t_1 = \frac{1}{6f_m} \tag{7.7}$$

$$\frac{\pi}{2}\cos(2\pi f_m t) = -\frac{\pi}{4} \Rightarrow t_2 = \frac{1}{3f_m} \tag{7.8}$$

Thus, the calculation of T_{FWHM} can be carried out and hence the duty cycle of the RZ optical pulse can be obtained as given in the following expressions:

$$T_{FWHM} = \frac{1}{3f_m} - \frac{1}{6f_m} = \frac{1}{6f_m} \therefore \frac{T_{FWHM}}{T_b} = \frac{1/6f_m}{1/2f_m} = 33\% \tag{7.9}$$

The result obtained in Equation 7.9 clearly verifies the generation of RZ33 optical pulses from the phasor representation.

7.3 DIFFERENTIAL PHASE-SHIFT KEYING

7.3.1 BACKGROUND

Digital encoding of data information by modulating the phase of the lightwave carrier is referred to as optical phase-shift keying (PSK). In early days, optical PSK was studied extensively for coherent photonic transmission systems. This technique requires the manipulation of the absolute phase of the lightwave carrier. Thus, the precise alignment of the transmitter and demodulator center frequencies for the coherent detection (CoD) is required. These coherent optical PSK systems face severe obstacles such as broad linewidth and chirping problems of the laser source. Meanwhile, the differential phase-shift keying (DPSK) scheme overcomes those problems, because the DPSK optically modulated signals can be detected incoherently. This technique only requires the coherence of the lightwave carriers over one bit period for the comparison of the differentially coded phases of the consecutive optical pulses.

A binary "1" is encoded if the present input bit and the past encoded bit are of opposite logic, whereas a binary "0" is encoded if the logics are similar. This operation is equivalent to an XOR logic operation. Hence, an XOR gate is employed as a differential encoder. NOR can also be used to replace XOR operation in differential encoding as shown in Figure 7.13a. In DPSK, the electrical data "1" indicates a π phase change between the consecutive data bits in the optical carrier, while the binary "0" is encoded if there is no phase change between the consecutive data bits. Hence, this encoding scheme gives rise to two points located exactly at π phase difference with respect to each other in signal constellation diagram. For continuous PSK such as the minimum-shift keying (MSK), the phase evolves continuously over a quarter of the section, and thus a phase change of $\pi/2$ between one phase state and the other. This is indicated by the inner bold circle as shown in Figure 7.13b.

7.3.2 OPTICAL DPSK TRANSMITTER

Figure 7.14 shows the structure of a 40 Gb/s DPSK transmitter in which two external LiNbO$_3$ MZIMs are used. Operational principles of an MZIM were presented above. The MZIMs shown in Figure 7.14 can be either of single or dual drive type. The optical DPSK transmitter also consists of a narrow-linewidth laser to generate a lightwave whose wavelength conforms to the ITU standard grid.

The RZ optical pulses are then fed into the second MZIM through which the RZ pulses are modulated by the precoded binary data to generate RZ-DPSK optical signals. Electrical data pulses are differentially precoded in a precoder using the XOR coding scheme. Without pulse carver, the structure shown in Figure 7.14 is an optical NRZ-DPSK transmitter. In data modulation

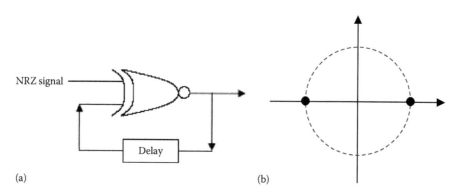

(a)   (b)

FIGURE 7.13   (a) DPSK precoder. (b) Signal constellation diagram of DPSK. (From Tien, P.K., *Rev. Mod. Phys.*, 49, 361–420, 1977. With permission.)

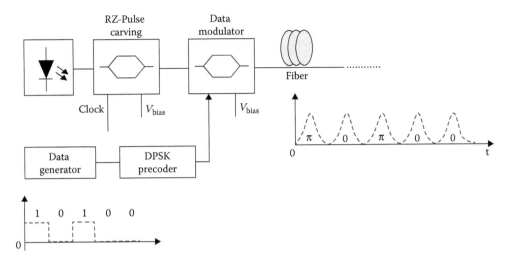

FIGURE 7.14 DPSK optical transmitter with RZ pulse carver.

for DPSK format, the second MZIM is biased at the minimum transmission point. The precoded electrical data has peak-to-peak amplitude equal to $2V_\pi$ and operates at the transmission bit rate. The modulation principles for generation of optical DPSK signals are demonstrated in Figure 7.6.

The electro-optic phase modulator (E-OPM) might also be used for generation of DPSK signals instead of MZIM. Using the optical phase modulator, the transmitted optical signal is chirped, whereas using MZIM, especially the X-cut type with Z-propagation, chirp-free signals can be produced. However, in practice, a small amount of chirp might be useful for transmission [22].

7.4 GENERATION OF MODULATION FORMATS

Modulation is the process facilitating the transfer of information over a medium, for example, a wireless or optical environment. Three basic types of modulation techniques are based on the manipulation of a parameter of the optical carrier to represent the information digital data. These are amplitude shift keying (ASK), PSK, and frequency shift keying (FSK). In addition to the manipulation of the carrier, the occupation of the data pulse over a single period would also determine the amount of energy concentrated and the speed of the system required for

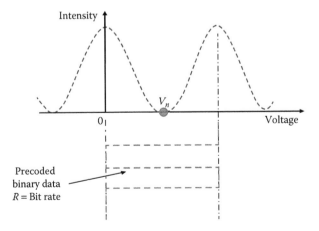

FIGURE 7.15 Bias point and RF driving signals for generation of optical DPSK format.

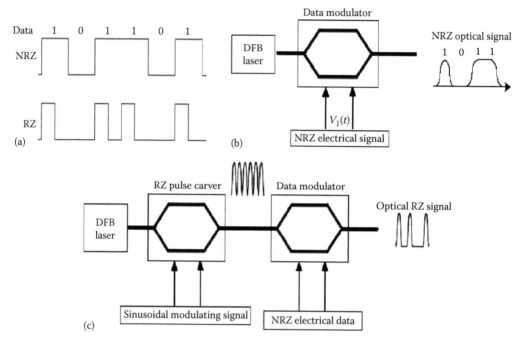

FIGURE 7.16   (a) Baseband NRZ and RZ line coding for the 101101 data sequence. (b) Block diagram of the NRZ photonics transmitter. (c) RZ photonics transmitter incorporating a pulse carver.

transmission. The pulse can remain constant over a bit period or RZ level within a portion of the period. These formats would be named NRZ or RZ. They are combined with the modulation of the carrier to form various modulations formats that are presented in this section. Figure 7.16 shows the base band signals of the NRZ and RZ formats and the corresponding block diagram of a photonic transmitter.

7.4.1 AMPLITUDE MODULATION OOK-RZ FORMATS

There are a number of advanced formats used in advanced optical communications; based on the intensity of the pulse, they may include NRZ, RZ, and duobinary. These ASK formats can also be integrated with the phase modulation to generate discrete or continuous phase NRZ or RZ formats. Currently, the majority of 10 Gb/s installed optical communication systems have been developed with NRZ due to its simple transmitter design and bandwidth efficient characteristic. However, RZ format has higher robustness to fiber nonlinearity and polarization mode dispersion (PMD). In this section, the RZ pulse is generated by MZIM commonly known as *pulse carver* as arranged in Figure 7.18. There are a number of variations in RZ format based on the biasing point in transmission curve shown in Table 7.2. The phasor representation of the biasing and driving signals can be observed in Table 7.2.

CSRZ has been found to have more attractive attributes in long-haul WDM transmissions compared to conventional RZ format due the possibility of reducing the upper level of the power contained in the carrier that serve no purpose in the transmission but only increase the total energy level so approaching the nonlinear threshold level faster. CSRZ pulse has optical phase difference of π in adjacent bits, removing the optical carrier component in optical spectrum and reducing the spectral width. This offers an advantage in compact WDM channel spacing.

TABLE 7.2

Summary of RZ Format Generation and Characteristics of Single-Drive MZIM Based on Biasing Point, Drive Signal Amplitude, and Frequency

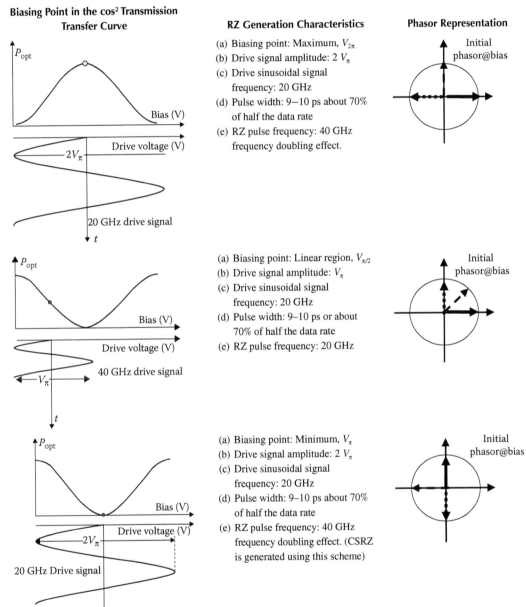

Biasing Point in the $\cos^2$ Transmission Transfer Curve	RZ Generation Characteristics	Phasor Representation
	(a) Biasing point: Maximum, $V_{2\pi}$ (b) Drive signal amplitude: $2\,V_\pi$ (c) Drive sinusoidal signal frequency: 20 GHz (d) Pulse width: 9–10 ps about 70% of half the data rate (e) RZ pulse frequency: 40 GHz frequency doubling effect.	Initial phasor@bias
	(a) Biasing point: Linear region, $V_{\pi/2}$ (b) Drive signal amplitude: V_π (c) Drive sinusoidal signal frequency: 20 GHz (d) Pulse width: 9–10 ps or about 70% of half the data rate (e) RZ pulse frequency: 20 GHz	Initial phasor@bias
	(a) Biasing point: Minimum, V_π (b) Drive signal amplitude: $2\,V_\pi$ (c) Drive sinusoidal signal frequency: 20 GHz (d) Pulse width: 9–10 ps about 70% of half the data rate (e) RZ pulse frequency: 40 GHz frequency doubling effect. (CSRZ is generated using this scheme)	Initial phasor@bias

7.4.2 AMPLITUDE MODULATION CSRZ FORMATS

The suppression of the carrier can be implemented by biasing the MZ interferometer in such a way so that there is a π phase shift between the two arms of the interferometer. The magnitude of the sinusoidal signals applied to an arm or both arms would determine the width of the optical output pulse sequence. The driving conditions and phasor representation are shown in Table 7.2.

7.4.3 DISCRETE PHASE MODULATION NRZ FORMATS

The term discrete phase modulation is referred to differential phase-shift keying (DPSK). The quadrature DSPK (QDPSK) or DPSK are modulation techniques which indicate that the states of the phases of the lightwave carrier are switched from one distinct location on the phasor diagram to the other state, for example from 0 to π or $-\pi/2$ to $-\pi/2$ for binary PSK (BPSK), or even more evenly spaced PSK levels as in the case of M-ary PSK (Figure 7.15).

7.4.3.1 Differential Phase-Shift Keying

Information encoded in the phase of an optical carrier is commonly referred to as optical PSK. In early days, PSK requires precise alignment of the transmitter and demodulator center frequencies [17]. Hence, the PSK system is not widely deployed. With the DPSK scheme introduced, CoD is not critical because DPSK detection only requires source coherence over one-bit period by comparison of two consecutive pulses.

A binary "1" is encoded if the present input bit and the past encoded bit are of opposite logic and a binary "0" is encoded if the logic is similar. This operation is equivalent to XOR logic operation. Hence, an XOR gate is usually employed in differential encoder. NOR can also be used to replace XOR operation in differential encoding as shown in Figure 7.17.

In optical application, electrical data "1" is represented by a π phase change between the consecutive data bits in the optical carrier, while state "0" is encoded with no phase change between the consecutive data bits. Hence, this encoding scheme gives rise to two points located exactly at π phase difference with respect to each other in the signal constellation diagram shown in Figure 7.17b.

A RZ-DPSK transmitter consists of an optical source, pulse carver, data modulator, differential data encoder, and a channel coupler. The channel coupler model is not developed in simulation by assuming no coupling losses when an optical RZ-DPSK modulated signal is launched into the optical fiber. This modulation scheme has combined the functionality of the dual-drive MZIM modulator of pulse carving and phase modulation.

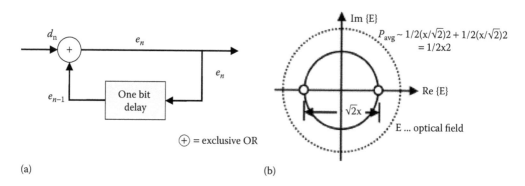

(a)   (b)

FIGURE 7.17 (a) The encoded differential data are generated by $e_n = d_n \oplus e_{n-1}$. (b) Signal constellation diagram of DPSK.

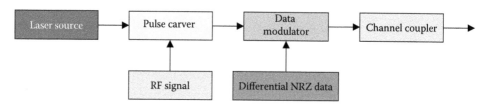

FIGURE 7.18 Block diagrams of the RZ-DPSK transmitter.

The pulse carver, usually an MZ interferometric intensity modulator, is driven by a sinusoidal RF signal for single-drive MZIM and two complementary electrical RF signals for dual-drive MZIM, to carve pulses out from optical carrier signal forming RZ pulses. These optical RZ pulses are fed into second MZ intensity modulator where RZ pulses are modulated by differential NRZ electrical data to generate RZ-DPSK. This data phase modulation can be performed using straight line phase modulator but the MZ waveguide structure has several advantages over phase modulator due to its chirpless property. Electrical data pulses are differentially precoded in a differential precoder as shown in Figure 7.17a. Without the pulse carver and sinusoidal RF signal, the output pulse sequence follows NRZ-DPSK format, that is the pulse would occupy 100% of the pulse period and there is no transition between the consecutive "1s."

7.4.3.2 Differential Quadrature PSK

This differential coding is similar to DPSK except that each symbol consists of two bits that are represented by the two orthogonal axial discrete phases at $(0, \pi)$ and $(-\pi/2, +\pi/2)$, as shown in Figure 7.19, or two additional orthogonal phase positions are located on the imaginary axis of Figure 7.17b.

7.4.3.3 NRZ-DPSK

Figure 7.20 shows the block diagram of a typical NRZ-DPSK transmitter. Differential precoder of electrical data is implemented using the logic explained in the previous section. In phase modulation of an optical carrier, the MZ modulator known as the edata phase modulator is biased at a minimum point and driven by data swing of $2V_\pi$. The modulator showed an excellent behavior that the phase of the optical carrier will be altered by π exactly when the signal transits the minimum point of the transfer characteristic.

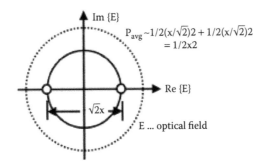

FIGURE 7.19 Signal constellation diagram of DQPSK. Two bold dots are orthogonal to the DPSK constellation.

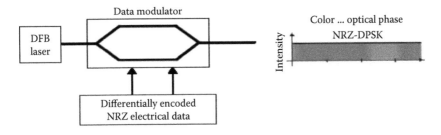

FIGURE 7.20 Block diagram of the NRZ-DPSK photonics transmitter.

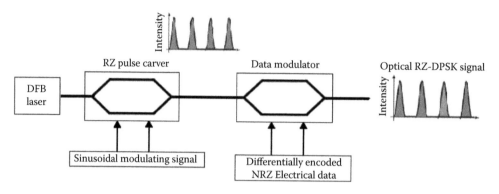

FIGURE 7.21 Block diagram of the RZ-DPSK photonics transmitter.

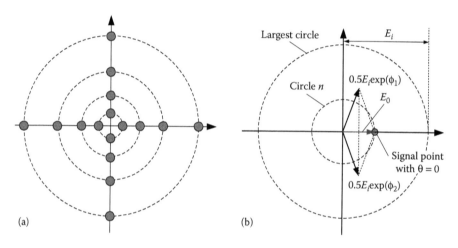

FIGURE 7.22 (a) Signal constellation of 4-ary ADPSK format and (b) phasor representation of a point on the constellation point for driving voltages applied to dual-dive MZIM.

7.4.3.4 RZ-DPSK

The arrangement of the RZ-DPSK transmitter is essentially similar to RZ-ASK, as shown in Figure 7.21, with the data intensity modulator replaced with the data phase modulator. The difference between them is the biasing point and the electrical voltage swing. Different RZ formats can also be generated.

7.4.3.5 Generation of M-ary Amplitude Differential Phase-Shift Keying (M-ary ADPSK) Using One MZIM

As an example, a 16-ary MADPSK signal can be represented by a constellation shown in Figure 7.22. It is, indeed, a combination of a 4-ary ASK and a DQPSK scheme. At the transmitting end, each group of four bits $[D_3D_2D_1D_0]$ of user data are encoded into a symbol; among them two least significant bits $[D_1D_0]$ are encoded into four phase states $[0, \pi/2, \pi, 3\pi/2]$ and the other two most significant bits, $[D_3D_2]$, are encoded into four amplitude levels. At the receiving end, as MZ delay interferometers are used for phase comparison and detection, a diagonal arrangement of the signal constellation shown in Figure 7.22a is preferred. This simplifies the design of a transmitter and receiver, and minimizes the number of phase detectors, hence leading to high receiver sensitivity.

To balance the BER between ASK and DQPSK components, the signal levels corresponding to four circles of the signal space should be adjusted to a reasonable ratio, which depends on the noise

TABLE 7.3

Driving Voltages for 16-ary MADPSK Signal Constellation

Circle 0			Circle 1			Circle 2			Circle 3		
Phase	$V_t(t)$ volt	$V_2(t)$ volt	Phase	$V_t(t)$ volt	$V_2(t)$ volt	Phase	$V_t(t)$ volt	$V_2(t)$ volt	Phase	$V_t(t)$ volt	$V_2(t)$ volt
0	0.38	−0.38	0	0.30	−0.30	0	0.21	−0.21	0	0.0	0.0
$\pi/2$	0.88	0.12	$\pi/2$	0.80	0.20	$\pi/2$	0.71	0.29	$\pi/2$	0.5	0.5
π	−0.62	0.62	π	−0.7	0.70	π	−0.79	−0.79	π	−1.0	1.0
$3\pi/2$	−0.12	−0.88	$3\pi/2$	−0.20	−0.8	$3\pi/2$	−0.29	−0.71	$3\pi/2$	−0.5	−0.5

power at the receiver. As an example, if this ratio is set to $[I_0/I_1/I_2/I_3] = [1/1.4/2/2.5]$, where I_0, I_1, I_2, and I_3 are the intensities of the optical signals corresponding to circle 0, circle 1, circle 2, and circle 3, respectively, then by selecting E_i equal to the amplitude of circle 3 and V_π equal to 1, the driving voltages should have the values given in Table 7.3. Inversely speaking, one can set the outermost level such that its peak value is below the nonlinear SPM threshold, and the voltage level of the outermost circle would be determined. The innermost circle is limited to the condition that the largest signal-to-noise ratio (SNR) should be achieved. This is related to the optical SNR (OSNR) required for a certain BER. Thus, from the largest amplitude level and smallest amplitude level, we can then design the other points of the constellation.

Furthermore, to minimize the effect of intersymbol interference (ISI), 66%-RZ and 50%-RZ pulse formats are also used as alternatives to the traditional NRZ counterpart. These RZ pulse formats can be created by a pulse carver that precedes or follows the dual-drive MZIM modulator. Mathematically, waveforms of NRZ and RZ pulses can be represented by the following equations, where E_{on}, $n = 0, 1, 2, 3$ are peak amplitudes of the signals in circle 0, circle 1, circle 2, and circle 3 of the constellation, respectively:

$$p(t) = \begin{cases} E_{on} & \text{for NRZ} \\[2ex] E_{on}\cos\left(\dfrac{\pi}{2}\cos^2\left(\dfrac{1.5\pi t}{T_s}\right)\right) & \text{for 66\%-RZ} \\[2ex] E_{on}\cos\left(\dfrac{\pi}{2}\cos^2\left(\dfrac{2\pi t}{T_s}\right)\right) & \text{for 50\%-RZ} \end{cases} \qquad (7.10)$$

7.4.4 CONTINUOUS PHASE MODULATION

In Section 7.4.3, the optical transmitters for discrete PSK modulation formats have been described. Obviously, the phase of the carrier has been used to indicate the digital states of the bits or symbols. These phases are allocated in a noncontinuous manner around a circle corresponding to the magnitude of the wave. Alternatively, the phase of the carrier can be continuously modulated and the total phase changes at the transition instants, usually at the end of the bit period, would be the same as those for discrete cases. Because the phase of the carrier continuously varies during the bit period, this can be considered as an FSK modulation technique, except that the transition of the phase at the end of one bit to the beginning of next bit would be continuous. The continuity of the carrier phase at these transitions would reduce the signal bandwidth and hence more tolerable to dispersion effects and higher energy concentration for effective transmission over the optical guided medium. One of the examples of the reduction of the phase at the transition is the offset DQPSK that is a minor but important variation on the QPSK or DQPSK. In OQPSK, the Q-channel is shifted by half a symbol period so that the transition instants of I- and Q-channel signals do not occur at the same time. The result of this simple change is that the phase shifts at any one time are limited and hence the offset QPSK is more constant envelope than the normal QPSK.

The enhancement of the efficiency of the bandwidth of the signals can be further improved if the phase changes at these transitions are continuous. In this case, the change of the phase during the symbol period is continuously changed by using half-cycle sinusoidal driving signals with the total phase transition over a symbol period being a fraction of π, depending on the levels of this PSK modulation. If the change is $\pi/2$, then we have an MSK scheme. The orthogonality of the I- and Q-channels will also reduce further the bandwidth of the carrier-modulated signals. In this section, we describe the basic principles of optical MSK and the photonic transmitters for these modulation formats. Ideally, the driving signal to the phase modulator should be a triangular wave so that a linear phase variation of the carrier in the symbol period is linear. However, when a sinusoidal function is used, there are some nonlinear variations; we thus term this type of MSK a nonlinear MSK format. This nonlinearity contributes to some penalty in the OSNR, which will be explained in a later chapter. Furthermore, the MSK as a special form of offset differential quadrature phase-shift keying (ODQPSK) is also described for optical systems.

7.4.4.1 Linear and Nonlinear MSK

MSK is a special form of continuous-phase FSK (CPFSK) signal in which the two frequencies are spaced in such a way that they are orthogonal and hence have minimum spacing between them, defined by

$$s(t) = \sqrt{\frac{2E_b}{T_b}} \cos\left[2\pi f_1 t + \theta(0)\right] \quad \text{for symbol 1} \tag{7.11}$$

$$s(t) = \sqrt{\frac{2E_b}{T_b}} \cos\left[2\pi f_2 t + \theta(0)\right] \quad \text{for symbol 0} \tag{7.12}$$

As shown in Equations 7.11 and 7.12, the signal frequency change corresponds to higher frequency for data-1 and lower frequency for data-0. Both frequencies, f_1 and f_2, are defined by

$$f_1 = f_c + \frac{1}{4T_b} \tag{7.13}$$

$$f_2 = f_c - \frac{1}{4T_b} \tag{7.14}$$

Depending on the binary data, the phase of signal changes; data-1 increases the phase by $\pi/2$, while data-0 decreases the phase by $\pi/2$. The variation of phase follows paths of sequence of straight lines in phase trellis (Figure 7.23), in which the slopes represent frequency changes. The change in carrier frequency from data-0 to data-1, or vice versa, is equal to half the bit rate of incoming data [5]. This is the minimum frequency spacing that allows the two FSK signals representing symbols 1 and 0 to be coherently orthogonal in the sense that they do not interfere with one another in the process of detection.

An MSK signal consists of both I- and Q-components, which can be written as

$$s(t) = \sqrt{\frac{2E_b}{T_b}} \cos\left[\theta(t)\right] \cos\left[2\pi f_c t\right] - \sqrt{\frac{2E_b}{T_b}} \sin\left[\theta(t)\right] \sin\left[2\pi f_c t\right] \tag{7.15}$$

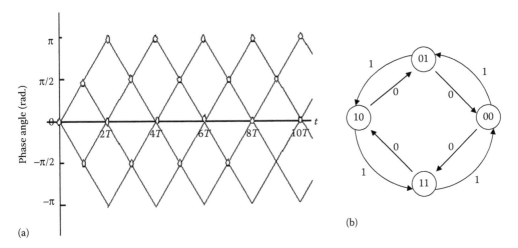

(a)

(b)

FIGURE 7.23   (a) Phase trellis for MSK. (b) State diagram for MSK. (From K.K. Pang, *Digital Transmission*. Melbourne: Mi-Tec Publishing, 2002, p. 58. With permission.)

The in-phase component consists of a half-cycle cosine pulse defined by

$$s_I(t) = \pm\sqrt{\frac{2E_b}{T_b}}\cos\left(\frac{\pi t}{2T_b}\right), \quad -T_b \leq t \leq T_b \tag{7.16}$$

while the quadrature component takes the form

$$s_Q(t) = \pm\sqrt{\frac{2E_b}{T_b}}\sin\left(\frac{\pi t}{2T_b}\right), \quad 0 \leq t \leq 2T_b \tag{7.17}$$

During even bit interval, the I-component consists of a positive cosine waveform for phase of 0, while a negative cosine waveform for phase of π; during odd bit interval, the Q-component consists of a positive sine waveform for phase of $\pi/2$, while a negative sine waveform for phase of $-\pi/2$ (as shown in Figure 7.23). Any of four states can arise: 0, $\pi/2$, $-\pi/2$, π. However, only state 0 or π can occur during any even bit interval and only $\pi/2$ or $-\pi/2$ can occur during any odd bit interval. The transmitted signal is the sum of I- and Q-components and its phase is continuous with time.

Two important characteristics of MSK are as follows: each data bit is held for two-bit period, meaning that the symbol period is equal two-bit period ($h = 1/2$) and the I- and Q-components are interleaved. I- and Q-components are delayed by one-bit period with respect to each other. Therefore, only I- or Q-component can change at a time (when one is at zero crossing, the other is at maximum peak). The precoder can be a combinational logic as shown in Figure 7.24.

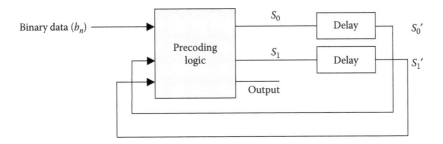

FIGURE 7.24   Combinational logic, the basis of the logic for constructing the precoder.

A truth table can be constructed based on the logic state diagram and combinational logic diagram above. For positive half-cycle cosine wave and positive half-cycle sine wave, the output is 1; for negative half-cycle cosine wave and negative half-cycle sine wave, the output is 0. Then, a K-map can be constructed to derive the logic gates of the precoder, based on the truth table. The following three precoding logic equations are derived:

$$S_0 = \overline{b_n} \, \overline{S_0'} S_1' + b_n \overline{S_0'} S_1' + \overline{b_n} S_0' S_1' + b_n S_0' \overline{S_1'} \tag{7.18}$$

$$S_1 = \overline{S_1'} = \overline{b_n} \overline{S_1'} + b_n \overline{S_1'} \tag{7.19}$$

$$\text{Output} = \overline{S_0} \tag{7.20}$$

The resultant logic gates construction for the precoder is as shown in Figure 7.24.

7.4.4.2 MSK as a Special Case of CPFSK

CPFSK signals are generated by modulating the upper (USB) and lower sideband (LSB) frequency carriers f_1 and f_2 as expressed in Equations 7.18 and 7.19 to give

$$s(t) = \sqrt{\frac{2E_b}{T_b}} \cos\left[2\pi f_1 t + \theta(0)\right] \quad \text{symbol '1'} \tag{7.21}$$

$$s(t) = \sqrt{\frac{2E_b}{T_b}} \cos\left[2\pi f_2 t + \theta(0)\right] \quad \text{symbol '0'} \tag{7.22}$$

Where $f_1 = f_c + (1/4T_b)$ and $f_2 = f_c - (1/4T_b)$ with T_b is the bit period.

The phase slope of lightwave carrier changes linearly or nonlinearly with the modulating binary data. In case of linear MSK, the carrier phase linearly changes by $\pi/2$ at the end of the bit slot with data "1," while it linearly decreases by $\pi/2$ with data "0." The variation of phase follows paths of well-defined phase trellis in which the slopes represent frequency changes. The change in carrier frequency from data-0 to data-1, or vice versa, equals half the bit rate of incoming data [12]. This is the minimum frequency spacing that allows the two FSK signals representing symbols 1 and 0 to be coherently orthogonal in the sense that they do not interfere with one another in the process of

TABLE 7.4

Truth Table Based on MSK State Diagram

$b_n S_0' S_1'$	$S_0' S_1'$	Output
100	01	1
001	00	1
100	01	1
101	10	0
010	01	1
101	10	0
110	11	0
111	00	1
000	11	0
011	10	0

detection. MSK carrier phase is always continuous at bit transitions. The MSK signal in Equations 7.21 and 7.22 can be simplified as

$$s(t) = \sqrt{\frac{2E_b}{T_b}} \cos[2\pi f_c t + d_k \frac{\pi t}{2T_b} + \Phi_k], \quad kT_b \leq t \leq (k+1)T_b \tag{7.23}$$

and the baseband equivalent optical MSK signal is represented as

$$\tilde{s}(t) = \sqrt{\frac{2E_b}{T_b}} \exp\left\{ j\left[d_k \frac{\pi t}{2T} + \Phi(t,k) \right] \right\}, \quad kT \leq t \leq (k+1)T$$

$$= \sqrt{\frac{2E_b}{T_b}} \exp\left\{ j[d_k 2\pi f_d t + \Phi(t,k)] \right\} \tag{7.24}$$

where:
   $d_k = \pm 1$ are the logic levels
   f_d is the frequency deviation from the optical carrier frequency
   $h = 2f_d T$ is defined as the frequency modulation index

In case of optical MSK, $h = 1/2$ or $f_d = 1/(4T_b)$.

7.4.4.3 MSK as an Offset Differential Quadrature Phase-Shift Keying

Equation 7.18 can be rewritten to express MSK signals in form of I-Q components as

$$s(t) = \pm \sqrt{\frac{2E_b}{T_b}} \cos\left(\frac{\pi t}{2T_b} \right) \cos[2\pi f_c t] \pm \sqrt{\frac{2E_b}{T_b}} \sin\left(\frac{\pi t}{2T_b} \right) \sin[2\pi f_c t] \tag{7.25}$$

The I- and Q-components consist of half-cycle sine and cosine pulses defined by

$$s_I(t) = \pm \sqrt{\frac{2E_b}{T_b}} \cos\left(\frac{\pi t}{2T_b} \right) \quad -T_b < t < T_b \tag{7.26}$$

$$s_Q(t) = \pm \sqrt{\frac{2E_b}{T_b}} \sin\left(\frac{\pi t}{2T_b} \right) \quad 0 < t < 2T_b \tag{7.27}$$

During even bit intervals, the in-phase component consists of positive cosine waveform for phase of 0, while negative cosine waveform for phase of π; during odd bit interval, the Q-component consists of positive sine waveform for phase of $\pi/2$, while negative sine waveform for phase of $-\pi/2$. Any of four states can arise: 0, $\pi/2$, $-\pi/2$, π. However, only state 0 or π can occur during any even bit interval and only $\pi/2$ or $-\pi/2$ can occur during any odd bit interval. The transmitted signal is the sum of I- and Q-components and its phase is continuous with time.

    Two important characteristics of MSK are as follows: each data bit is held for two bit period, meaning the symbol period is equal two bit period ($h = 1/2$) and the I-component and Q-component are interleaved. I- and Q-components are delayed by one-bit period with respect to each other. Therefore, only I- or Q-component can change at a time (when one is at zero crossing, the other is at maximum peak).

7.4.4.4 Configuration of Photonic MSK Transmitter Using Two Cascaded E-OPMs

E-OPMs and interferometers operating at high frequency using resonant-type electrodes have been studied and proposed in [33–35]. In addition, high-speed electronic driving circuits evolved with the ASIC technology using 0.1 μm GaAs P-HEMT or InP HEMTs* enable the feasibility in realization of the optical MSK transmitter structure. The baseband equivalent optical MSK signal is represented in Equation 7.25.

The first E-OPM enables the frequency modulation of data logics into USB and LSB of the optical carrier with frequency deviation of f_d. Differential phase precoding is not necessary in this configuration due to the nature of the continuity of the differential phase trellis. By alternating the driving sources $V_d(t)$ to sinusoidal waveforms for simple implementation or combination of sinusoidal and periodic ramp signals that was first proposed by Amoroso in 1976 [23], different schemes of linear and nonlinear phase shaping MSK transmitted sequences can be generated whose spectra are shown in Figure 7.34.

The second E-OPM enforces the phase continuity of the lightwave carrier at every bit transition. The delay control between the E-OPMs is usually implemented by the phase shifter shown in Figure 7.25. The driving voltage of the second E-OPM is precoded to fully compensate the transitional phase jump at the output $E_{01}(t)$ of the first E-OPM. Phase continuity characteristic of the optical MSK signals is determined by the algorithm given in Equation 7.25.

$$\Phi(t,k) = \frac{\pi}{2}\left(\sum_{j=0}^{k-1} a_j - a_k I_k \sum_{j=0}^{k-1} I_j \right) \tag{7.28}$$

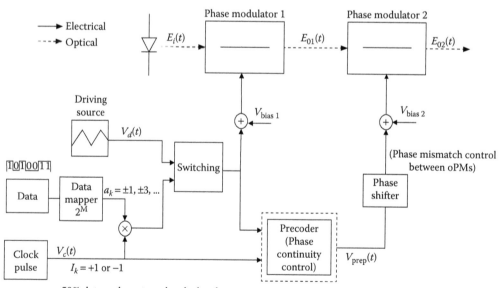

FIGURE 7.25 Block diagram of the optical MSK transmitter employing two cascaded optical phase modulators.

---

* Fujitsu Optical Components Ltd., Japan, www.fujitsu.com

where:

$a_k = \pm1$ are the logic levels

$I_k = \pm1$ is a clock pulse whose duty cycle is equal to the period of the driving signal

$V_d(t) f_d$ is the frequency deviation from the optical carrier frequency

$h = 2f_dT$ is previously defined as the frequency modulation index

In case of optical MSK, $h = 1/2$ or $f_d = 1/(4T)$. The phase evolution of the continuous phase optical MSK signals is explained in Figure 7.25. To mitigate the effects of unstable stages of rising and falling edges of the electronic circuits, the clock pulse $V_c(t)$ is offset with the driving voltages $V_d(t)$.

7.4.4.5 Configuration of Optical MSK Transmitter Using Mach–Zehnder Intensity Modulators: I-Q Approach

The conceptual block diagram of optical MSK transmitter is shown in Figure 7.25. The transmitter consists of two dual-drive electro-optic MZM modulators generating chirpless I- and Q-components of MSK modulated signals, which is considered as a special case of staggered or offset QPSK. The binary logic data is precoded and deinterleaved into even and odd bit streams that are interleaved with each other by one bit duration offset (Figures 7.26 and 7.27).

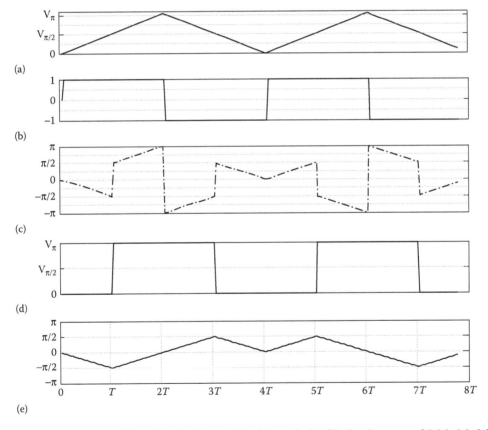

FIGURE 7.26 Evolution of time-domain phase trellis of the optical MSK signal sequence [-1 1 1 -1 1 -1 1 1] as inputs and outputs at different stages of the optical MSK transmitter. The notation is denoted in Figure 7.25 accordingly; (a) $V_d(t)$: periodic triangular driving signal for optical MSK signals with duty cycle of 4 bit period; (b) $V_c(t)$: the clock pulse with duty cycle of $4T$; (c) $E_{01}(t)$: phase output of oPM1; (d) $V_{prep}(t)$: precomputed phase compensation driving voltage of oPM2; and (e) $E_{02}(t)$: phase trellis of a transmitted optical MSK sequence at output of oPM2.

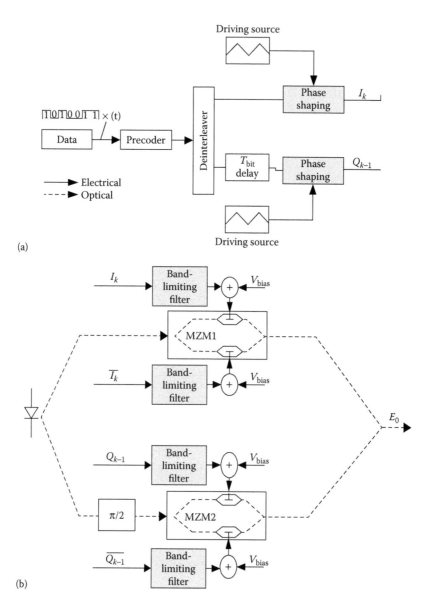

FIGURE 7.27   (a,b) Block diagram configuration of band-limited phase shaped optical MSK.

The upper and lower arms of the dual-drive MZM modulator are biased at $V_\pi/2$ and $-V_\pi/2$ respectively, and driven with data and its complementary $\overline{\text{data}}$. The driving voltage for phase shaping of MSK format can follow a periodic triangular function or simply a sinusoidal signal which can be approximated as a linear plus some partial nonlinear part or MSK-like signal which also attain the linear phase trellis property and possess small ripples introduced in this partial nonlinearity. The magnitude fluctuation level depends on the magnitude of the phase shaping driving source. High spectral efficiency can be achieved with tight filtering of the driving signals before modulating the electro-optic MZMs. Three types of pulse shaping filters are investigated including Gaussian, raised cosine, and squared-root-raised-cosine filters. The optical carrier phase trellis of linear and nonlinear optical MSK signals is shown in Figure 7.28a and b, respectively.

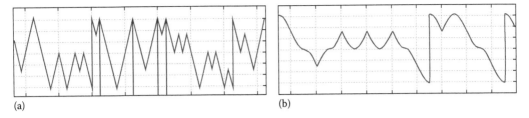

(a)   (b)

FIGURE 7.28   Phase trellis of (a) linear and (b) nonlinear MSK transmitted signals.

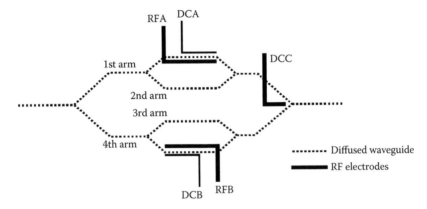

FIGURE 7.29   Schematic diagram (not to scale) of an SSB FSK optical modulator formed by nested MZ modulators.

7.4.5   SINGLE-SIDEBAND (SSB) OPTICAL MODULATORS

7.4.5.1   Operating Principles

An SSB modulator can be formed using a primary interferometer with two secondary MZM structures, the optical Ti-diffused waveguide paths that form a nested primary MZ structure as shown in Figure 7.29 [15]. Each of the two primary arms contains MZ structures. Two RF ports are for RF modulation and three DC ports are for biasing the two secondary MZMs and one primary MZM. The modulator consists of X-cut Y-propagation $LiNbO_3$ crystal, where you can produce an SSB modulation just by driving each MZ. DC voltage is supplied to produce the π phase shift between upper and lower arms. DC bias voltages are also supplied from DC_B to produce the phase shift between third and fourth arms. A DC bias voltage is supplied from DC_C to achieve a $\pi/2$ phase shift between $MZIM_A$ and $MZIM_B$. The RF voltage applied $\Phi_1(t) = \Phi\cos\omega_m t$ and $\Phi_2(t) = \Phi\sin\omega_m t$ are inserted from RF_A and RF_B respectively by using a wideband $\pi/2$ phase shifter. Φ is the modulation level and ω_m is the RF angular frequency.

7.4.6   MULTICARRIER MULTIPLEXING (MCM) OPTICAL MODULATORS

Another modulation format that can offer much higher single channel capacity and flexibility in dispersion and nonlinear impairment mitigation is the employment of multicarrier multiplexing (MCM). When these subcarrier channels are orthogonal, then the term orthogonal frequency division multiplexing (OFDM) is used [19].

Our motivation in the introduction of OFDM is due to its potential as an ultrahigh-capacity channel for the next generation Ethernet, the optical internet. The network interface cards for 1- and 10-Gb/s Ethernet are readily commercial available. Traditionally, the Ethernet data rates have grown in 10-Gb/s increments, so the data rate of 100 Gb/s can be expected as the speed of the next generation of Ethernet. The 100-Gb/s all-electrically time-division-multiplexed transponders are becoming increasingly important because they are viewed as a promising technology that may be able to meet speed requirements of the new generation of Ethernet. Despite the recent progress in high-speed electronics, electrically time-division-multiplexing (ETDM) [20] modulators and photodetectors are still not widely available, so that alternative approaches to achieving a 100-Gb/s transmission using commercially available components and QPSK are very attractive. However, the use of polarization division multiplexing (PDM) and 25 GBd QPSK would offer 100 Gb/s transmission with superiority under coherent reception incorporating digital signal processing (DSP) [21]. This PDM-QPSK technique will be described in later chapters.

OFDM is a combination of multiplexing and modulation. The data signal is first split into independent subsets and then modulated with independent subcarriers. These subchannels are then multiplexed to form OFDM signals. OFDM is thus a special case of FDM but instead like one stream, it is a combination of several small streams into one bundle.

A schematic signal flow diagram of an MCM is shown in Figure 7.30. The basic OFDM transmitter and receiver configurations are given in Figure 7.31a and b, respectively [22]. Data streams (e.g., 1 Gb/s) are mapped into a two-dimensional signal point from a point signal constellation such as quadrature amplitude modulation (QAM). The complex-valued signal points from all subchannels are considered as the values of the discrete Fourier transform of a multicarrier OFDM signal. The serial-to-parallel converter arranges the sequences into equivalent discrete frequency domain. By selecting the number of sub-channels sufficiently large, the OFDM symbol interval can be made much larger than the dispersed pulse width in a single-carrier system, the symbol interval can be made much larger than the width of the dispersed pulse in a single-carrier system. This can lead to smaller ISI. The OFDM symbol, shown in Figure 7.32, is generated under software processing, as follows: input QAM symbols are zero-padded to obtain input samples for inverse fast Fourier transform (IFFT), the samples are inserted to create the guard band, and the OFDM symbol is multiplied by the window function that can be represented by a raised cosine function. The purpose of cyclic extension is to preserve the orthogonality among subcarriers even when the neighboring OFDM symbols partially overlap due to dispersion.

A system arrangement of the OFDM for optical transmission in laboratory demonstration is shown in Figure 7.32. The data sequences are arranged in the sampled domain and then to the frequency domain and then to the time domain representing the OFDM waves that would look like analog waveforms as shown in Figure 7.32d. Each individual channel at the input would carry the same data rate sequence. These sequences can be generated from an arbitrary waveform generator. The multiplexed channels are then combined and converted to time domain using the IFFT module and then converted to the analog version via the two DAC. These orthogonal data sequences are then used to modulate I- and Q-optical waveguide sections of the electro-optical modulator to

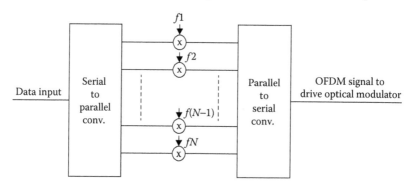

FIGURE 7.30 Multicarrier FDM signal arrangement. The middle section is the RF converter as shown in Figure 7.26.

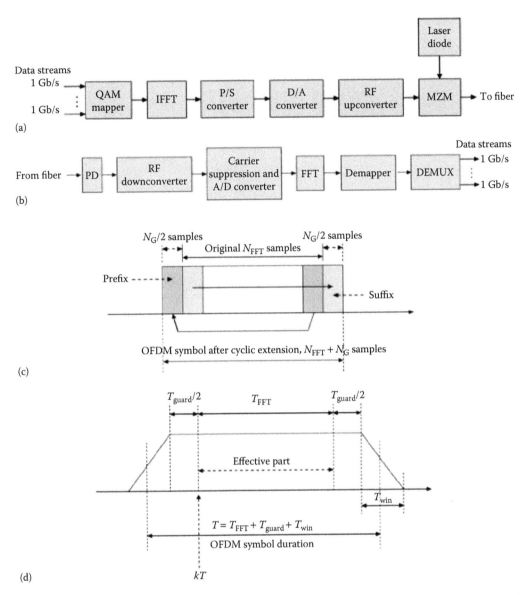

FIGURE 7.31 Schematic diagram of the principles of generation and recovery of OFDM signals: (a) Electronic processing and optical transmitter; (b) optoelectronic and receiver configurations; (c) OFDM symbol cyclic extension and (d) OFDM symbol after windowing.

generate the orthogonal channels in the optical domain. Similar decoding of I- and Q-channels is performed in the electronic domain after the optical transmission and optical-electronic conversion via the photodetector and electronic amplifier.

In comparison with other optical transmitters, the optical OFDM transmitter will require a DSP (digital signal processor), DAC (digital-to-analog converter), and ADC (analog-to-digital converter) for shaping the pulse spectrum and then the IFFT operation so as to generate the OFDM analog signals to modulate the optical modulator. The schematic of the OFDM generator and coherent receivers is shown in Figure 7.33.

In OFDM, the serial data sequence, with a symbol period of T_s and a symbol rate of $1/T_s$, is split up into N parallel substreams (subchannels).

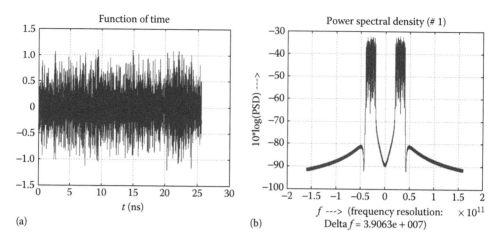

FIGURE 7.32   Generated spectra of OFDM via the use of an optical FFT/IFFT-based OFDM system including representative waveforms. (a) OFDM signals in the time domain. (b) Power spectrum of OFDM signal with 512 subcarriers a shift of 30 GHz for the line rate of 40 Gb/s and QPSK modulation.

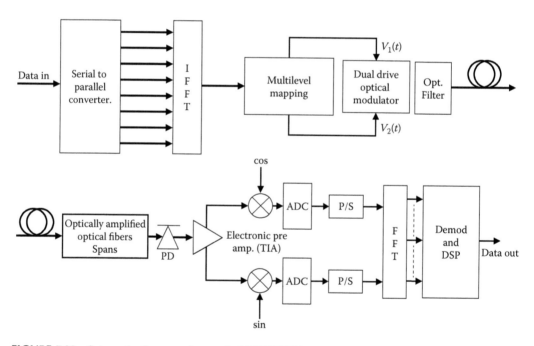

FIGURE 7.33   Schematic diagram of an optical FFT/IFFT-based OFDM system. S/P and P/S = serial-to-parallel conversion and vice versa. PD = photodetector, TIA = transimpedance amplifier.

7.4.7   Spectra of Modulation Formats

Utilizing this double phase modulation configuration, different types of linear and nonlinear CPM phase shaping signals including MSK, weakly-nonlinear MSK, and linear-sinusoidal MSK can be generated. The third scheme was introduced by Amoroso [23] and its side lobes decay with a factor of f-8 compared to f-4 of MSK. The simulated optical spectra of DBPSK and MSK schemes at 40 Gb/s are contrasted in Figure 7.34. Table 7.5 outlines the characteristics and spectra of all different modulation schemes.

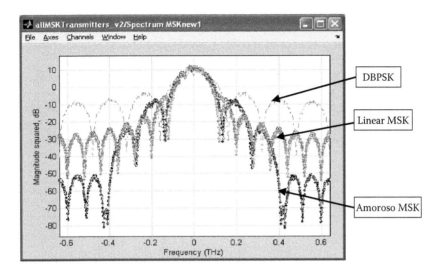

FIGURE 7.34   Spectra of 40 Gbps DBPSK, and linear and nonlinear MSK.

TABLE 7.5
Typical Parameters of Optical Intensity Modulators for Generation of Modulation Formats

Modulation Techniques	Spectra	Formats	Definition/Comments
Amplitude modulation – ASK-NRZ	DSB + carrier	ASK – NRZ	Biased at quadrature point or offset for prechirp
AM – ASK-RZ	DSB + carrier	ASK-RZ	Two MZIMs required—one for RZ pulse sequence and other for data switching.
ASK – RZ-Carrier suppressed	DSB – CSRZ	ASK-RZCS	Carrier suppressed, biased at π phase difference for the two electrodes.
Single sideband	SSB + carrier	SSB NRZ	Signals applied to MZIM are in phase quadrature to suppress one sideband. Alternatively an optical filter can be used.
CSRZ DSB	DSB – carrier	CSRZ-ASK	RZ pulse carver is biased such that a π phase shift between the two arms of the MZM to suppress the carrier and then switch on and off or phase modulation via a data modulator.
DPSK-NRZ DPSK-RZ, CSRZ-DPSK		Differential BPSK RZ or NRZ/RZ-carrier suppressed	
DQPSK		DQPSK – RZ or NRZ	Two bits per symbol
MSK	SSB equivalent	Continuous phase modulation with orthogonality	Two bits per symbol and efficient bandwidth with high side-lobe suppression.
MCM (multicarrier multiplexing—e.g., OFDM)	Multiplexed bandwidth – base rate per subcarrier		
Duobinary	Effective SSB		Electrical low-pass filter required at the driving signal to the MZM
Phase modulation (PM)	Chirped carrier phase		Chirpless MZM should be used to avoid inherent crystal effects, hence carrier chirp.

7.5   SPECTRAL CHARACTERISTICS OF DIGITAL MODULATION FORMATS

Figure 7.35 shows the power spectra of the DPSK modulated optical signals with various pulse shapes including NRZ, RZ33, and CSRZ types.

For the convenience of the comparison, the optical power spectra of the RZ OOK counterparts are also shown in Figure 7.34.

Several key notes observed from Figures 7.35 and 7.36 are outlined as follows. (i) The optical power spectrum of the OOK format has high power spikes at the carrier frequency or at signal modulation frequencies, which contribute significantly to the severe penalties caused of the nonlinear

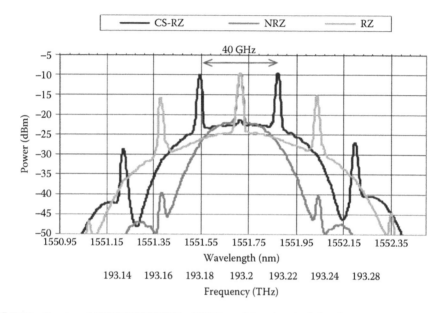

FIGURE 7.35   Spectra of CSRZ/RZ33/NRZ—DPSK modulated optical signals.

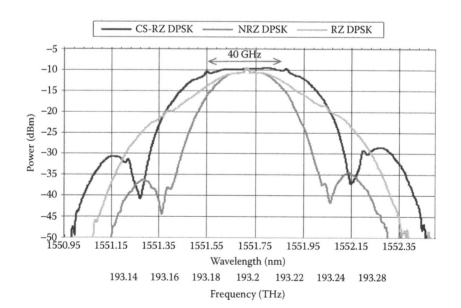

FIGURE 7.36   Spectra of CSRZ/RZ/NRZ—OOK modulated optical signals.

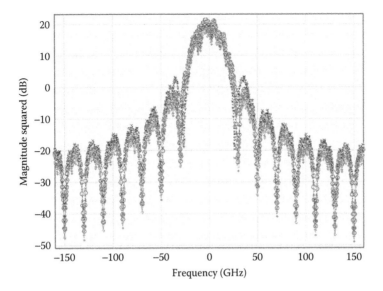

FIGURE 7.37   Optical power spectra of three types of I-Q optical MSK formats: linear (gray), weakly nonlinear (light gray), and strongly nonlinear (black).

effects. Meanwhile, the DPSK optical power spectra do not contain these high power frequency components. (ii) RZ pulses are more sensitive to the fiber dispersion due to their broader spectra. In particular, RZ33 pulse type has the broadest spectrum at the point of −20 dB down from the peak. This property of the RZ pulses thus leads to faster spreading of the pulse when propagating along the fiber. Thus, the peak values of the optical power of these CSRZ or RZ33 pulses decrease much more quickly than the NRZ counterparts. As the result, the peak optical power quickly turns to be lower than the nonlinear threshold of the fiber, which means that the effects of fiber nonlinearity are significantly reduced. (iii) However, NRZ optical pulses have the narrowest spectrum. Thus, they are expected to be most robust to the fiber dispersion. As a result, there is a trade-off between RZ and NRZ pulse types. RZ pulses are much more robust to nonlinearity but less tolerant to the fiber dispersion. The RZ33/CSRZ-DPSK optical pulses are proven to be more robust against impairments especially SPM and PMD compared to the NRZ-DPSK and the CSRZ/RZ33-OOK counterparts.

Optical power spectra of three I-Q optical MSK modulation formats that are linear, weakly nonlinear, and strongly nonlinear types are shown in Figure 7.37. It can be observed that there are no significant distinctions of the spectral characteristics between these three schemes. However, the strongly nonlinear optical MSK format does not highly suppress the side lobe as compared to the linear MSK type. All three formats offer better spectral efficiency compared to the DPSK counterpart as shown in Figure 7.38. This figure compares the power spectra of three modulation formats: 80 Gb/s dual-level MSK, 40 Gb/s MSK, and NRZ-DPSK optical signals. The normalized amplitude levels into the two optical MSK transmitters comply with the ratio of 0.75/0.25.

Several key notes can be observed from Figure 7.38 and are outlined as follows. (i) The power spectrum of the optical dual-level MSK format has identical characteristics to that of the MSK format. The spectral width of the main lobe is narrower than that of the DPSK. The base-width takes a value of approximately ±32 GHz on either side compared to ±40 GHz in the case of the DPSK format. Hence, the tolerance to the fiber dispersion effects is improved. (ii) High suppression of the side lobes with a value of approximately greater than 20 dB in the case of 80 Gb/s dual-level MSK and 40 Gb/s optical MSK power spectra; thus, more robustness to interchannel crosstalk between dense wavelength division multiplexed channels. (iii) The confinement of signal energy in the main lobe of spectrum leads to a better SNR. Thus, the sensitivity to optical filtering can be significantly reduced [2–6]. A summary of the spectra of different modulation formats is given in Table 7.5.

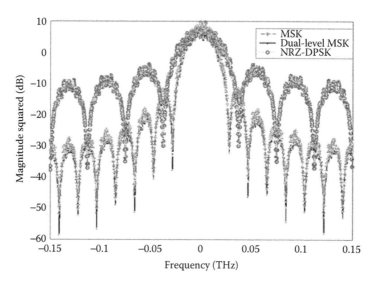

FIGURE 7.38 Spectral properties of three modulation formats: MSK (gray, dash), dual-level MSK (black, solid), and DPSK (light gray, dot).

7.6 I-Q INTEGRATED MODULATORS

7.6.1 IN-PHASE AND QUADRATURE PHASE OPTICAL MODULATORS

We have described in the above sections the modulation of QPSK and M-ary-QAM schemes using single-drive or dual-drive MZIM devices. However, we have also witnessed another alternative technique to generate the states of constellation of QAM, using I-Q modulators. I-Q modulators are devices in which the amplitude of the in-phase and that of the quadrature components are modulated in synchronization as illustrated in Figure 7.39b. These components are $\pi/2$ out of phase; thus, we can achieve the in-phase and quadrature of QAM. In optics, this phase quadrature at optical frequency can be implemented by a low-frequency electrode with an appropriate applied voltage as observed by the $\pi/2$ block in the lower optical path of the structure given in Figure 7.39a. This type of modulation can offer significant advantages when multilevel QAM schemes are employed for example 16-QAM (see Figure 7.39c) or 64-QAM. Integrated optical modulators have been developed in recent years, especially in electro-optic structures such as LiNbO$_3$ for coherent QPSK and even with polarization division multiplexed optical channels. In summary, the I-Q modulator consists of two MZIMs performing ASK modulation incorporating a quadrature phase shift.

Multilevel or multicarrier modulation formats such as QAM and OFDM have been demonstrated as the promising technology to support high capacity and high spectral efficiency in ultrahigh-speed optical transmission systems. Several QAM transmitter schemes have been experimentally demonstrated using commercial modulators [24–26] and integrated optical modules [27,28] with binary or multilevel driving electronics. The integration techniques could offer a stable performance, and effectively reduce the complexity of the electrics in QAM transmitter with binary driving electronics. The integration schemes based on parallel structures usually require hybrid integration between LiNbO$_3$ modulators and silica-based planar lightwave circuits (PLCs). Except the DC electrodes for the bias control of each sub-MZM (child MZM), several additional electrodes are required to adjust the relative phase offsets among embedded sub-MZMs. Shown in Figure 7.39a is a 16-QAM transmitter using a monolithically integrated quad Mach–Zehnder in-phase/quadrature (QMZ-I-Q) modulator. As distinguishable from the parallel integration, four sub-MZMs are integrated and arranged in a single I-Q superstructure, where two of them are cascaded in each of the arms

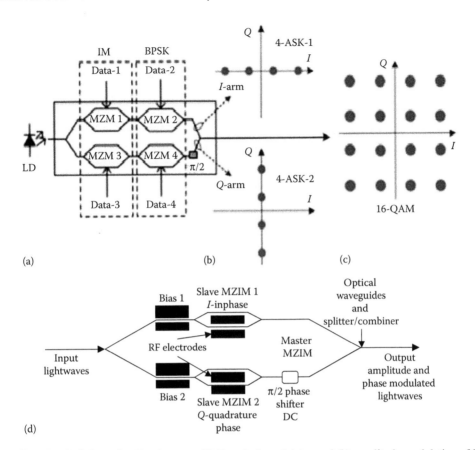

FIGURE 7.39   (a) Schematic of an integrated I-Q optical modulator and (b) amplitude modulation of lightwave path in the in-phase and quadrature components. (c) Constellation of a 16-QAM modulation scheme. (d) Alternate structure of an I-Q modulator using two salve MZIMs and one master MZIM.

(I and Q arms). These two pairs of child MZMs are the combined to form a master or parent MZ interferometer with a $\pi/2$ phase shift incorporated in one arm to generate the quadrature phase shift between them; thus, we have the in-phase arm and the quadrature optical components. In principle, only one electrode is required to obtain orthogonal phase offset in these I-Q superstructures, which makes the bias control much easier to handle, and thus stable performance. A 16-QAM signal can be generated using the monolithically-integrated QMZ-I-Q modulator with binary driving electronics, shown in Figure 7.39c, by modulating the amplitude of the lightwaves guided through both I (in-phase) and Q (quadrature) paths as indicated in Figure 7.39b. Hence, we can see that the QAM modulation can be implemented by modulating the amplitude of these two orthogonal I- and Q-components so that any constellation of Mary-QAM, for example, 16-QAM or 256-QAM, can be generated. Alternatively, the structure of Figure 7.39d gives an arrangement of two slave MZIMs and one master MZIM for the I-Q modulator that is commonly used in Fujitsu type.

   In addition, a number of electrodes would be incorporated so that biases can be applied to ensure the amplitude modulation operating at the right point of the transfer curve characteristics of the output optical field versus the applied voltage of the MZIM in each interferometric branch of the master MZ interferometer. This can be commonly observed and simplified as in the I-Q modulator manufactured by Fujitsu[*] shown in Figure 7.40 (top view only).

---

[*] Fujitsu Optical Components Ltd., Japan, www.fujitsu.com

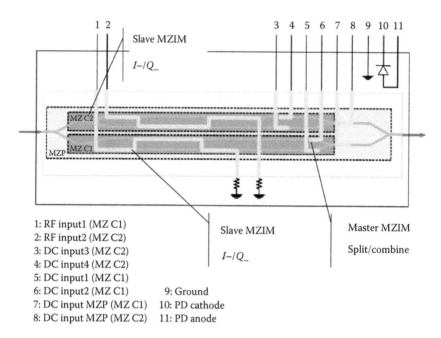

1: RF input1 (MZ C1)
2: RF input2 (MZ C2)
3: DC input3 (MZ C2)
4: DC input4 (MZ C2)
5: DC input1 (MZ C1)
6: DC input2 (MZ C1)
7: DC input MZP (MZ C1)   9: Ground
8: DC input MZP (MZ C2)   10: PD cathode
                            11: PD anode

FIGURE 7.40   Schematic diagram of a Fujitsu PDM-I-Q modulator with assigned electrodes.

The arrangement of the high-speed I-Q Mach–Zehnder modulator using Ti-diffused lithium nio-bate (LiNbO$_3$) waveguide technology in which the DC bias can be adjusted at the separate DC port of the modulator. This type of modulator can be employed for various modulation formats such as NRZ, DPSK, Optical Duobinary, DQPSK, DP-BPSK, DP-QPSK, M-ary QAM, and so on. Built-in PD monitor and coupler function for auto-bias control. 100 Gb/s optical transmission equipment (NRZ, DPSK, Optical Duobinary, DQPSK, DP-BPSK, PDM-QPSK) can be generated by this I-Q modulator in which four wavelength carriers are used with 28–32 Gb/s per channel to form 100 Gb/s including extra error coding bits and payload 25 Gb/s for each channel.

7.6.2 I-Q Modulator and Electronic Digital Multiplexing for Ultrahigh Bit Rates

Recently, we [29–33] have seen reports on the development of electrical multiplexer whose speed can reach 165 Gb/s. This multiplexer will allow the interleaving of high-speed sequence so that we can generate higher bit rate with the symbol rate as shown in Figure 7.41. This type of multiplexing in the electrical domain has been employed for generation of superchannel in [34]. Thus, we can see that the data sequence can be generated from the DACs and then their analog outputs at the output can be conditioned to the right level required by the digital multiplexer by the assistance of a clock generator at the multiplexing instants. Note that the multiplexer operates in digital mode so any predistortion of the sequence for dispersion precompensation will not be possible unless some predistortion can be done at the output of the digital multiplexer.

This time-domain interleaving can be combined with the I-Q optical modulator to generate M-level QAM signal and thus further increasing the bit rate of the channels. Several of these high bit rate channels can be combined with pulse shaping, for example the Nyquist shape, to generate superchannels that will be illustrated in Chapter 7. With a digital multiplexer operating higher than 165 Gb/s, 128 Gb/s bit rate can be generated with 32 GSy/s data sequence. Thus, with polariza-tion multiplexing and 16-QAM, we can generate 8 × 128 Gb/s for one channel or 1.32 Tb/s per channel. If eight of these 1.32 Gb/s form a superchannel, then the bit rate capacity reaches higher than 10 Tb/s. This is all possible with a symbol rate of 32 GBd and within the bandwidth available electronic components as well as photonic devices at the transmitter and receiver.

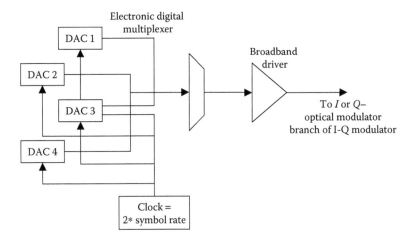

FIGURE 7.41   Time division multiplexing using high-speed sequences from DACs.

7.7   DAC FOR DSP-BASED MODULATION AND TRANSMITTER

Recently, we have witnessed the development of ultrahigh sampling rate DAC and ADC by Fujitsu and NTT of Japan. Figure 7.42a shows an IC layout of the InP-based DAC produced by NTT [35].

7.7.1   DIGITAL-TO-ANALOG CONVERTER

A differential input stage incorporating D-type flip-flops, and summing up circuits can be combined to produce analog signals representing the digital samples as shown in Figure 7.42a, b and c. These DACs allow the generation and programmable sampling and digitalized signals to form analog signals to modulate the I-Q modulator as described in Section 5. These DACs allow shaping of the pulse sequence

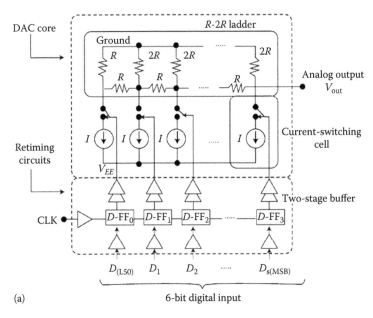

FIGURE 7.42   NTT InP-based DAC with 6-bit (a) schematic. (Extracted with permission from Yamazaki, H. et al., 64QAM modulator with a hybrid configuration of silica PLCs and LiNbO$_3$ phase modulators for 100-Gb/s applications. In *ECOC*, paper 2.2.1, 2009. With permission.)   (*Continued*)

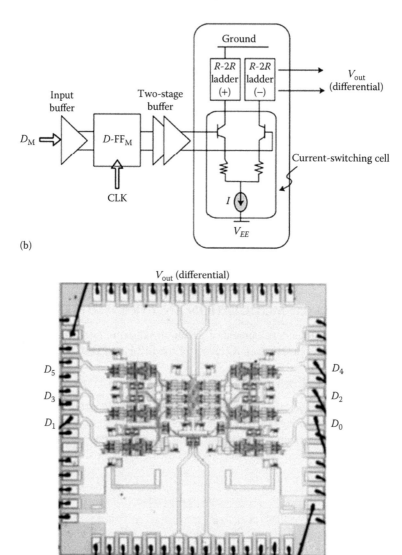

FIGURE 7.42 (Continued)   NTT InP-based DAC with 6-bit (b) differential input stage, and (c) IC layout. (Extracted with permission from Yamazaki, H. et al., 64QAM modulator with a hybrid configuration of silica PLCs and LiNbO$_3$ phase modulators for 100-Gb/s applications. In *ECOC*, paper 2.2.1, 2009. With permission.)

for predistortion to combat dispersion effects when necessary to add another dimension of compensation in combination with that function implemented at the receiver. Test signals used are ramp waveform for assurance of the linearity of the DAC at 27 GS/s as shown in Figure 7.43a and eye diagrams of generated 16-QAM signals after modulation at 22 and 27 GSy/s are shown in Figure 7.43b.

7.7.2 STRUCTURE

DSP-based optical transmitter can incorporate DAC for pulse shaping, pre-equalization, and pattern generation as well as digitally multiplexing for higher symbol rates. A schematic structure of the DAC and functional blocks fabricated by SiGe technology is shown in Figure 7.44a and b respectively. An external sinusoidal signal is fed into the DAC so that multiple clock sources can be generated for sampling at 56–64 GSa/s. Thus, the noises and clock accuracy depend on the stability

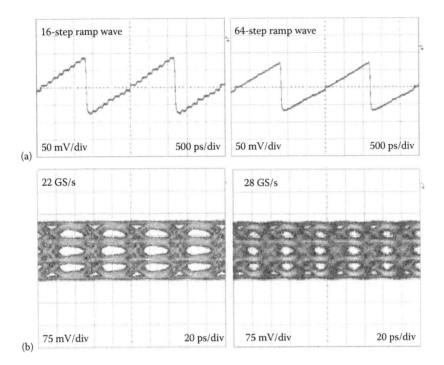

FIGURE 7.43   (a) Test signals and converted analog signals for linearity performance of the DAC; and (b) 16-QAM 4-level generated signals at 22 and 27 GSy/s.

and noise of this synthesizer/signal generator. Four DACs submodules are integrated in one IC with four pairs of eight outputs of $(V_I^+, V_Q^+)(H_I^+, H_Q^+)$ and $(V_I^-, V_Q^-)(H_I^-, H_Q^-)$.

7.7.3   GENERATION OF I- AND Q-COMPONENTS

The electrical outputs from the quad DACs are in mutual complementary pair of positive and negative signs. Thus, it would be able to form two sets of four output ports from the DAC development board. Each output can be independently generated with offline uploading of pattern scripts. The arrangement of the DAC and PDM-I-Q optical modulator is depicted in Figure 7.45. Note that we require two PDM I-Q modulators for generation of odd and even optical channels.

As the Nyquist pulse shaped sequences are required, a number of pressing steps can be implemented by using (i) the transfer characterization of the DAC; and (ii) pre-equalization in the RF domain to achieve equalized spectrum in the optical domain, that is at the output of the PDM I-Q modulator.

The characterization of the DAC is conducted by launching the DAC sinusoidal wave at different frequencies and measuring the waveforms at all eight output ports. As observable in the insets of Figure 7.45 the electrical spectrum of the DAC is quite flat provided that pre-equalization is done in the digital domain launching to the DAC. The spectrum of the DAC output without equalization is shown in Figure 7.46a and b. The amplitude spectrum is not flat due to the transfer function of the DAC, as given in Figure 7.47, which is obtained by driving the DAC with sinusoidal waves of different frequencies. This shows that the DAC acts as a low-pass filter with the amplitude of its passband gradually decreasing when the frequency is increased. This effect can exist when the number of samples is reduced and the frequency is increased as the sampling rate can only be set in the range of 56–64 GSa/s. The equalized RF spectra are depicted in Figure 7.46c and d. The time-domain waveforms corresponding to the RF spectra are shown in Figure 7.46e and f and thence (g) and (h) for

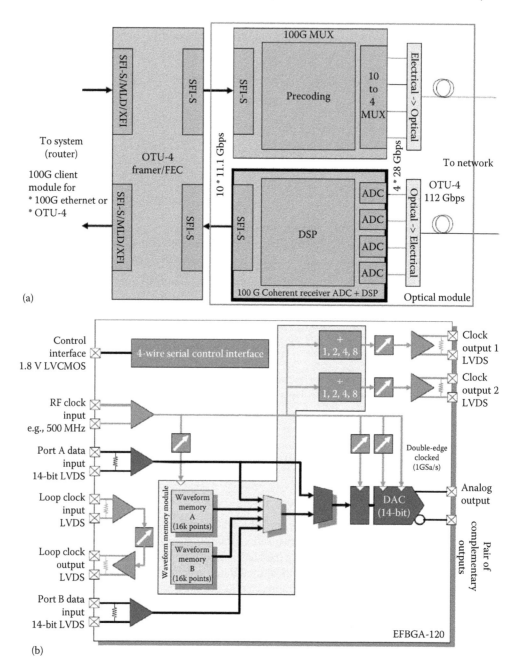

FIGURE 7.44 (a) Structure of the Fujitsu DAC; note that four DACs are structured in one integrated chip. (b) Functional block diagram.

the CoD after the conversion back to electrical domain from the optical modulator via the real-time sampling oscilloscope Tektronix DPO 73304A or DSA 720004B. Furthermore, the noise distribution of the DAC is shown in Figure 7.47b indicating that the sideband spectra of Figure 7.46 come from these noise sources.

It is noted that the driving conditions for the DAC are very sensitive to the supply current and voltage levels with resolution of even down to 1 mV. With this sensitivity, care must be taken when new patterns are fed to DAC for driving the optical modulator. Optimal procedures must be

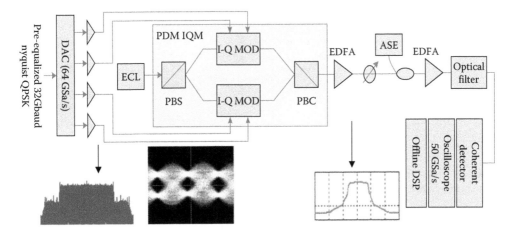

FIGURE 7.45 Experimental setup of a 128 Gb/s Nyquist PDM-QPSK transmitter and B2B performance evaluation.

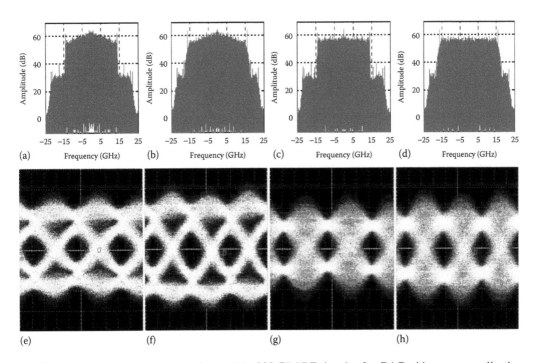

FIGURE 7.46 Spectrum (a) and eye diagram (e) of 28 GBd RF signals after DAC without pre-equalization, (b) and (f) for 32 GBd; spectrum (c) and eye diagram (g) of 28 GBd RF signals after DAC with pre-equalization, (d) and (h) for 32 GBd.

conducted with the evaluation of the constellation diagram and BER derived from such constellation. However, we believe that the new version of the DAC supplied from Fujitsu Semiconductor Pty Ltd of England Europe has overcome somehow these sensitive problems. But we still recommend that care should be taken and inspection of the constellation after the coherent receiver must be done to ensure that there is no error in the B2B connection. Various time-domain signal patterns obtained in the electrical time domain generated by DAC at the output ports can be observed. Obviously, the variations of I- and Q-signal components lead to an increase of the noise, and hence blurry constellations.

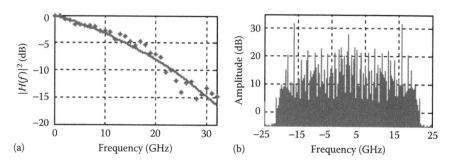

(a)   Frequency (GHz)   (b)   Frequency (GHz)

FIGURE 7.47 (a) Frequency transfer characteristics of the DAC. Note the near-linear variation of the magnitude as a function of the frequency. (b) Noise spectral characteristics of the DAC.

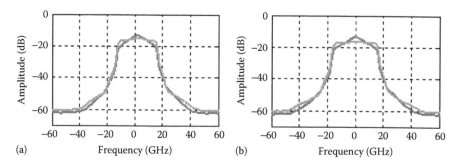

(a)   Frequency (GHz)   (b)   Frequency (GHz)

FIGURE 7.48 Optical spectrum after PDM IQM; light gray line for without pre-equalization, gray line for with pre-equalization. (a) 28 GBd; (b) 32 GBd.

7.8 REMARKS

Currently, photonic transmitters play a principal part in the extension of the modulation speed into several GHz range and make the modulation of the amplitude, the phase, and frequency of the optical carriers and their multiplexing possible. Photonic transmitters using LiNbO$_3$ have been proven in laboratory and installed systems. The principal optical modulator is the intensity one that can be a single-drive or dual-drive MZIM type where by to generate binary or multilevel amplitude or phase modulation and even more effective for discrete or continuous PSK techniques. The effects of the modulation on transmission performance will be given in the subsequent chapters.

Spectral properties of the optical 80 Gb/s dual-level MSK, 40 Gb/s MSK, and 40 Gb/s DPSK with various RZ pulse shapes are compared. The spectral properties of the first two formats are similar. Compared to the optical DPSK, the power spectra of optical MSK and dual-level MSK modulation formats have more attractive characteristics. These include the high spectral efficiency for transmission, higher energy concentration in the main spectral lobe, and more robustness to interchannel crosstalk in dense wavelength-division multiplexing due to greater suppression of the side lobes. In addition, the optical MSK offers the orthogonal property, which may offer a great potential in CoD, in which the phase information is reserved via I- and Q-components of the transmitted optical signals. In addition, the multilevel formats would permit the lowering of the bit rate and hence substantial reduction of the signal effective bandwidth and the possibility of reaching the highest speed limit of the electronic signal processing, the DSP, for equalization and compensation of distortion effects. The demonstration of ETDM receiver at 80 G and higher speed would allow the applications of these modulation formatted scheme that is very effective in ultrahigh-speed transmission.

In recent years, research for new types of optical modulators using silicon waveguides has attracted several groups. In particular, thin layer of graphene deposited [36] on silicon waveguides enables the improvement of the EA effects and enables the modulator structure to be incredibly compact and potentially performing at speeds up to 10 times faster than current technology allows, reaching higher than 100 GHz and even to 500 GHz. This new technology will significantly enhance our capabilities in ultrafast optical communication and computing. This may be the world's smallest optical modulator, and the modulator in data communications is the heart of speed control. Furthermore, these grapheme silicon modulators can be integrated with Si- or SiGe-based microelectronic circuits such as DAC, ADC, and DSP so that the operating speed of the electronic circuits can reach much higher than the 25–32 GSy/s of today technology.

Noises and distortion noises as well as polarization interferences in optical modulators are critical in the generation of higher order modulation schemes for enhancing the spectral efficiency, especially in the 400 Gb/s 16-QAM schemes or Tbps channels employing multisubcarrier with Nyquist pulse shaping. These issues and solutions will be addressed in the subsequent chapters, especially in Chapter 12.

REFERENCES

1. P.K. Tien, Integrated Optics and New Wave Phenomena in Optical Waveguides, *Rev. Mod. Phys.*, 49, 361–420, 1977.
2. R.C. Alferness, Optical Guided-Wave Devices, *Science*, 234(4778), 825–829, Nov 14, 1986.
3. M. Rizzi and B. Castagnolo, Electro-Optic Intensity Modulator for Broadband Optical Communications, *Fiber Integrated Opt.*, 21, 243–251, 2002.
4. Takara H, High-speed Optical Time-Division-Multiplexed Signal Generation, *Opt. Quant. Electron.*, 33(7–10), Jul 10, 2001, 795–810.
   E.L. Wooten, K.M. Kissa, A. Yi-Yan, E.J. Murphy, D.A. Lafaw, P.F. Hallemeier, D. Maack, D.V. Attanasio, D.J. Fritz, G.J. McBrien, and D.E. Bossi, A Review of Lithium Niobate Modulators for Fiber-Optic Communications Systems, *IEEE J. Sel. Topics Quant. Elect.*, 6(1), Jan/Feb 2000, 69–80.
5. K. Noguchi, O. Mitomi, H. Miyazawa, and S. Seki, A Broadband Ti: LiNb03 Optical Modulator with a Ridge Structure, *J. Lightwave Technol.*, 13(6), Jun 1995.
6. J. Noda, Electro-optic Modulation Method and Device using the Low-energy Oblique Transition of a Highly Coupled Super-grid, *IEEE J. Lightwave Technol.*, LT-4, 1986.
7. M. Suzuki, Y. Noda, H. Tanaka, S. Akiba, Y. Kuahiro, and H. Isshiki, Monolithic Integration of InGaAsP/InP Distributed Feedback Laser and Electroabsorption Modulator by Vapor Phase Epitaxy, *IEEE J. Lightwave Tech.*, LT-5(9), Sept 1987, 127.
8. H. Nagata, Y. Li, W.R. Bosenberg, and G.L. Reiff, DC Drift of X-Cut LiNbO3 Modulators, *IEEE Photonics Technol. Lett.*, 16(10), 2233–2335, Oct 2004.
9. H. Nagata, DC Drift Failure Rate Estimation on 10 Gb/s X-Cut Lithium Niobate Modulators, *IEEE Photonics Technol. Lett.*, 12(11), 1477–1479, Nov 2000.
10. R. Krahenbuhl, J.H. Cole, R.P. Moeller, and M.M. Howerton, High-Speed Optical Modulator in $LiNbO_3$ with Cascaded Resonant-Type Electrodes, *J. Lightwave Technol.*, 24(5), 2184–2189, 2006.
11. A. Sano, T. Kobayashi, A. Matsuura, S. Yamamoto, S. Yamanaka, Z. Yoshida, Y. Miyamoto, M. Matsui, M. Mizoguchi, and T. Mizuno, 100 × 120-Gb/s PDM 64-QAM Transmission over 160 km Using Linewidth-Tolerant Pilotless Digital Coherent Detection. In 36*th European Conference and Exhibition on Optical Communication*, Los Angeles, CA, 2010.
12. P. Dong, C. Xie, L. Chen, L.L. Buhl, and Y.-K. Chen, 12-Gb/s Monolithic PDM-QPSK Modulator in Silicon. *Opt. Express*, 20(26), B624–B629, 2012.
13. A. Hirano, Y. Miyamoto, and S. Kuwahara, Performances of CSRZ-DPSK and RZ-DPSK in 43-Gbit/s/ch DWDM G.652 Single-mode-fiber Transmission. In *Proceedings of OFC'03*, 2, pp. 454–456, 2003.
14. A.H. Gnauck, G. Raybon, P.G. Bernasconi, J. Leuthold, C.R. Doerr, and L.W. Stulz, 1-Tb/s (6/spl times/170.6 Gb/s) Transmission Over 2000-km NZDF using OTDM and RZ-DPSK Format, *IEEE Photonics Technol. Lett.*, 15(11), 1618–1620, 2003.
15. Y. Yamada, H. Taga, and K. Goto, Comparison Between VSB, CS-RZ and NRZ Format in a Conventional DSF based Long Haul DWDM System. In *Proceedings of ECOC'02*, 4, 1–2, 2002.
16. A.H. Gnauck, X. Liu, X. Wei, D.M. Gill, and E.C. Burrows, Comparison of Modulation Formats for 42.7-Gb/s Single-Channel Transmission Through 1980 km of SSMF, *IEEE Photonics Technol. Lett.*, 16(3), 909–911, 2004.

17. A. Hirano, Y. Miyamoto, and S. Kuwahara, Performances of CSRZ-DPSK and RZ-DPSK in 43-Gbit/s/ch DWDM G.652 Single-mode-fiber Transmission. In *Proceedings of OFC'03*, 2, 454–456, 2003.

18. K.K. Pang, *Digital Transmission*. Melbourne: Mi-Tec Publishing, 2002, p. 58.

19. I.B. Djordjevic and B. Vasic, 100-Gb/s Transmission Using Orthogonal Frequency-Division Multiplexing, *IEEE Photonics Technol. Lett.*, 18(15), Aug 1, 2006.

20. C. Schubert, R.H. Derksen, M. Möller, R. Ludwig, C.-J. Weiske, J. Lutz, S. Ferber, A. Kirstädter, G. Lehmann, and C. Schmidt-Langhorst, Integrated 100-Gb/s ETDM Receiver, *IEEE J. Lightwave Technol.*, 25(1), Jan 2007.

21. L. Nguyen Binh, *Digitl Processing: Optical Transmission and Coherent Reception Techniques*. CRC Press, Taylor and Francis Grp, New York, 2013.

22. A. Ali, Investigations of OFDM transmission for direct detection optical systems, Dr. Ing. Dissertation, Albrechts Christian Universitaet zu Kiel, 2012.

23. F. Amoroso, Pulse and Spectrum Manipulation in the Minimum Frequency Shift Keying (MSK) Format, *IEEE Trans. Commun.*, 24, 381–384, Mar 1976.

24. P.J. Winzer, A.H. Gnauck, C.R. Doerr, M. Magarini, and L.L. Buhl, Spectrally Efficient Long-haul Optical Networking using 112-Gb/s Polarization-Multiplexed 16-QAM, *J. Lightwave Technol.*, 28, 547–556, 2010.

25. M. Nakazawa, M. Yoshida, K. Kasai and J. Hongou, 20 Msymbol/s, 64 and 128 QAM Coherent Optical Transmission Over 525 km using Heterodyne Detection with Frequency-Stabilized Laser, *Electron. Lett.*, 43, 710–712, 2006.

26. X. Zhou and J. Yu, 200-Gb/s PDM-16QAM generation using a new synthesizing method. In *ECOC*, paper 10.3.5, 2009.

27. T. Sakamoto, A. Chiba, and T. Kawanishi, 50-Gb/s 16-QAM by a quad-parallel Mach–Zehnder modulator. In *ECOC*, paper PDP2.8, 2007.

28. H. Yamazaki, T. Yamada, T. Goh, Y. Sakamaki, and A. Kaneko, 64QAM modulator with a hybrid configuration of silica PLCs and LiNbO3 phase modulators for 100-Gb/s applications. In *ECOC*, paper 2.2.1, 2009.

29. J. Hallin, T. Kjellberg, and T. Swahn, A 165-Gb/s 4:1 Multiplexer in InP DHBT Technology, *IEEE J. Solid-St. Circ.*, 41(10), 2209–2214, Oct 2006.

30. K. Murata, K. Sano, H. Kitabayashi, S. Sugitani, H. Sugahara, and T. Enoki, 100-Gb/s Multiplexing and Demultiplexing IC Operations in InP HEMT Technology, *IEEE J. Solid-St. Circ.*, 39(1), 207–213, Jan 2004.

31. M. Meghelli, 132-Gb/s 4:1 Multiplexer in 0.13-μm SiGe-bipolar Technology. *IEEE J. Solid-St. Circ.*, 39(12), 2403–2407, Dec 2004.

32. Y. Suzuki, Z. Yamazaki, Y. Amamiya, S. Wada, H. Uchida, C. Kurioka, S. Tanaka, and H. Hida, 120-Gb/s Multiplexing and 110-Gb/s Demultiplexing Ics, *IEEE J. Solid-St. Circ.*, 39(12), 2397–2402, Dec 2004.

33. T. Suzuki, Y. Nakasha, T. Takahashi, K. Makiyama, T. Hirose, and M. Takikawa, 144-Gbit/s selector and 100-Gbit/s 4:1 multiplexer using InP HEMTs. In *IEEE MTT-S International Microwave Symposium Digest*, pp. 117–120, Jun 2004.

34. X. Liu, S. Chandrasekhar, P.J. Winzer, T. Lotz, J. Carlson, J. Yang, G. Cheren, and S. Zederbaum, 1.5-Tb/s Guard-Banded Superchannel Transmission over 56 GSymbols/s 16QAM Signals with 5.75-b/s/Hz Net Spectral Efficiency, Paper Th.3.C.5.pdf ECOC Postdeadline Papers. In *ECOC*, the Netherlands, 2012.

35. M. Nagatani and H. Nosaka, High-speed Low-power Digital-to-Analog Converter Using InP Hetero-junction Bipolar Transistor Technology for Next-generation Optical Transmission Systems, *NTT Technical Review*, 9(4), 1–8, 2011.

36. M. Liu, X. Yin, E. Ulin-Avila1, B. Geng, T. Zentgraf, L. Ju, F. Wang, and X. Zhang, A Graphene-based Broadband Optical Modulator, *Nature*, 474, 64, Jun 2, 2011.

8 Differential Phase-Shift Keying Photonic Systems

This chapter studies the phase-shift keying (PSK) modulation for noncoherent transmission and detection optical communications systems. The differential mode of detection is most appropriate for systems in which the phases contained in two consecutive bit periods are assigned with discrete values, for example, 0, π or 0, $\pi/2$, π, $3\pi/2$, and so on. Experimental demonstration of the discrete differential phase-shift keying (DPSK) modulation formats is also described. A MATLAB®–Simulink® model for DPSK modulation format and optical fiber optically amplified multispan transmission system is included in the appendix in which noise models of optical reception are described.

8.1 INTRODUCTION

Owing to the tremendous growing demand of high-capacity transmission over the Internet and between data centers (DCs) of the DC-centric networks, high data rate of 40 Gb/s per wavelength channel has appeared to be an attractive feature in the next generation of the light wave communications system [1–3]. Under the current 10 Gb/s dense wavelength-division multiplexing (DWDM) optical system, overlaying 40 Gb/s on the existing network can be considered to be the most cost-effective method for upgrading purpose. However, there are a number of technical difficulties confronted by communications engineers that involve interoperability requiring a 40 Gb/s line system to have signal optical bandwidth, tolerance to chromatic dispersion (CD), resistance to nonlinear crosstalk, and susceptibility to accumulated noise over multispan of optical amplifier to be similar to a 10 Gb/s system. It is noted here that the current 100 Gb/s optical transport networks employ coherent reception techniques in association with the multiplexing of the polarized modes of the linearly polarized mode and the 2-bit/symbol quadrature phase-shift keying (QPSK), thus the baud rate is only 25 Gbaud or 28 GBd if 12% forward error coding is used. We can still use 40 Gb/s self-coherent detection as the case study without much alterations if we can lift the receiver sensitivity under the mixing of the signals with the local oscillator (LO) laser.

In view of this, advanced modulation formats have been demonstrated as an effective scheme to overcome impairments of a 40 Gb/s system. DPSK modulation format has attracted extensive studies due to its benefit over the conventional on-off keying (OOK) signaling format, including 3-dB lower optical signal-to-noise ratio [1] at a given bit-error rate (BER), more robustness to narrow band optical filtering, and more resilience to some nonlinear effects such as cross-phase modulation and self-phase modulation. Moreover, spectral efficiencies can be improved by using multilevel signaling. On top of that, coherent detection (CoD) is not critical as DPSK detection requires comparison of two consecutive pulses; hence, the source coherence is required only over 1-bit period.

Nevertheless, DPSK format involves rapid phase change causing intensity ripples due to CD that induce pattern-dependent self-phase modulation (SPM) – group velocity dispersion (GVD) effect [2]. Therefore, return-to-zero (RZ) pulse can be employed in conjunction with DPSK to generate more tolerance to the data pattern-dependent SPM-GVD effect. In addition, RZ improves dispersion tolerance and nonlinear effects particularly in a long-haul network at a high data rate. Specific RZ format, like carrier-suppressed RZ (CSRZ), helps to reduce the inherent wide spectral bandwidth.

At 40 Gb/s, generation of RZ pulse is not feasible because it is at 10 Gb/s due to the large bandwidth requirement. Thus, 40 Gb/s RZ signals are produced optically in "pulse carver" by driving the modulator with 20-GHz RF signal. With the remarkable advancement in external modular, especially Mach–Zehnder interferometic (intensity) modulator (MZIM), this is easily achieved by utilizing microwave optical transfer characteristic of MZIM.

Despite the telecom boom and subsequent bust, internet traffic has been growing steadily. This growth requires new investment in telecommunications infrastructure to provide long-haul communications capacity between major population centers in the world. The transmission route between Melbourne to Sydney and vice versa is one of the most intensive demanding upgrading to higher capacity, especially from the current 10 to 40 Gb/s. One of the most cost-effective ways to provide such upgrades is to use the existing fiber infrastructure and upgrade the transmitters and receivers at either end of the long-haul links. That means the transmission rate of 40 Gb/s over the existing 10 G RZ format dense multiwavelength optical communications systems.

Because of the properties of the installed fiber (which is older and also degraded than state-of-the-art fibers used in laboratory "hero" experiments), the transmission methods must be highly tolerant to CD and polarization mode dispersion (PMD). This favors the use of advanced modulation formats (such as variants of PSK) rather than ultrahigh rate time division multiplexed schemes because the effects of CD and second-order PMD are proportional to bit rate squared. However, various optical filters installed throughout the DWM optical transmission systems such as optical multiplexers (mux), demultiplexers (demux), and reconfigurable optical add-drop muxes (ROADMs) would affect the spectral properties of the multiplexed channels, particularly in case of a hybrid transmission of 10 and 40 Gb/s channels.

Recently, advanced modulation formats are considered to play a significant role in enhancing the effectiveness of bandwidth reduction and effective transmission over long distances [1,2]. Although new to optical communications, these advanced modulation formats are very well known in the area of wire and wireless communications systems. Accordingly, the modulation can be achieved by manipulating the amplitude, frequency, or phase of the carriers correspondingly to the coding of the input data sequence. Either coherent or incoherent transmission and detection techniques have become possible over the years for optical systems. However, incoherent detection is preferred to minimize the linewidth obstacles of the lasers at the transmitter and the LOs required for homodyne or heterodyne CoD systems. Furthermore, the phase comparison in differential detection can be easily implemented using delay interferometric photonic component. This is extremely hard if implemented in the electrical domain at very high bit rate. Therefore, differential discrete or continuous phase modulation formats can be the preferred techniques.

This chapter investigates the transmission of 40 Gb/s channels over 10 Gb/s DWDM optical system in which "standard" single optical filters (SSMFs) normally employed for 10 Gb/s systems are used. DPSK is transmitted and compared with those employing RZ and nonreturn-to-zero (NRZ) amplitude shift keying (ASK). We demonstrate that with the passband of optical filters in the order of 0.5 nm the transmitted 40 Gb/s channels are not penalized and vice versa for 10 Gb/s DWDM channels. Transmission BER and receiver sensitivities are reported for different transmission scenarios.

Due to high demand of data transmission capacity, with the advent of wavelength division multiplexing (WDM) technology, data transmission has maneuvered into a revolutionary stage for increasing higher system capacity. Higher bit rates up to 40 Gb/s and growing numbers of WDM channel are increasingly challenging the limit of ASK format.

8.2 OPTICAL DPSK MODULATION AND FORMATS

Figure 8.1 shows the structure of a 40 Gb/s DPSK transmitter in which two external LiNbO$_3$ optical interferometric modulators (MZIM) are used [3]. Operational principles of on MZIM have been presented in Chapter 7. The MZIM used in Figure 8.1 can be either a single- or dual-drive type. As previously described, the optical DPSK transmitter consists of a narrow linewidth laser source to generate a lightwave of wavelength conformed to the International Telecommunication Union (ITU) grid, which is then modulated via two cascade MZIMs, one to generate a periodic (clock like) RZ or CSRZ formats before feeding through the data modulator that would switch on and off for amplitude

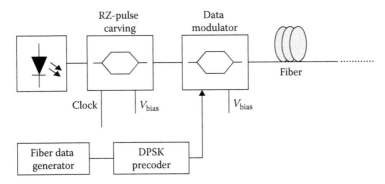

FIGURE 8.1 Schematic diagram of a discrete PSK optical transmitter using cascaded MZI modulators.

modulation of exerting a phase shift to the lightwave carrier for discrete phase modulation. Unlike Binary phase-shift keying or QPSK the phase states change at the transitional instant of the symbols, this phase shift can be continuous at the transition for the case of continuous-phase frequency shift keying or minimum-shift keying (MSK) modulation formats, which are presented in later chapters.

8.2.1 GENERATION OF RZ PULSES

The first MZIM, commonly known as the pulse carver, is used to generate the periodic pulse trains with a required RZ format. The suppression of the lightwave carrier can also be carried out at this stage if necessary, and it is known as CSRZ. CSRZ pulse shape is found to have attractive attributes in long-haul WDM transmissions compared to other RZ types including optical-phase difference of π in adjacent bits, suppression of the optical carrier (OC) component in optical spectrum, and smaller spectral width. Figure 8.2 shows the structure of a 40 Gb/s ASK transmitter in which two external LiNbO$_3$ optical interferometric modulators (MZIM) [4] are used. Operational principles of MZIM are presented in Chapter 7. The MZIM used in Figure 8.2 can be either a single- or dual-drive type. The optical ASK transmitter would consist of a narrow linewidth laser source to generate a lightwave of wavelength conformed to the ITU grid.

Different types of RZ pulses can be generated depending on the driving amplitude of the RF voltage and the biasing schemes of the MZIM. The equations governing the RZ pulses electric field waveforms are [1]

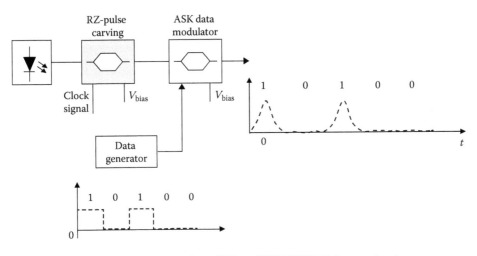

FIGURE 8.2 Cascaded MZIM for generation of RZ- or CSRZ-DPSK lightwave signals.

$$E(t) = \begin{cases} \dfrac{1}{\sqrt{E_b}} \sin\left[\dfrac{\pi}{2}\cos\left(\dfrac{\pi t}{T_b}\right)\right] & \text{67\% duty-ratio RZ pulses} \\[3ex] \dfrac{1}{\sqrt{E_b}} \sin\left[\dfrac{\pi}{2}\left(1+\sin\left(\dfrac{\pi t}{T_b}\right)\right)\right] & \text{33\% duty-ratio RZ pulses} \end{cases} \qquad (8.1)$$

where E_b is the pulse energy per transmitted bit.

The first type 33% duty-ratio RZ pulses are commonly known as "carrier max or conventional" (RZ33) pulses, whereas the 67% duty-cycle RZ pulse is the CSRZ pulse type. The art in generation of these two RZ pulse types stays at the difference of biasing point on the transfer curve of an MZIM. The bias voltage conditions and the pulse shape of these two RZ types, the carrier suppression, and nonsuppression of maximum carrier can be implemented with the biasing points at the minimum and maximum transmission point of the transmittance characteristics of the MZIM, respectively. The peak-to-peak amplitude of the RF driving voltage is $2V\pi$, where $V\pi$ is the required driving voltage to achieve a π phase shift of the lightwave carrier. Another important point is that the RF signal is operating at only half of the transmission bit rate. Hence, pulse carving is actually implementing the frequency doubling. The generation of RZ33 and CSRZ pulse train is shown in Figure 8.3a and b, respectively.

The pulse carver can also utilize a dual-drive MZIM, which is driven by two complementary sinusoidal RF signals. This pulse carver is biased at $-V_{\pi/2}$ and $+V_{\pi/2}$ with the peak-to-peak amplitude of $V_{\pi/2}$. Thus, a π phase shift is created between the state "1" and "0" of the pulse sequence and hence the RZ with alternating phase 0 and π. If the carrier suppression is required, then the two electrodes are applied with voltages V_π and swing voltage amplitude of V_π.

RZ-ASK optical systems are proven to be more robust against impairments especially self-phase modulation and PMD. Although RZ modulation offers improved performance, RZ optical systems usually require more complex transmitters than those in the NRZ ones. Compared to only one stage for modulating data on the NRZ optical signals, two modulation stages are required for generation of RZ optical pulses.

8.2.2 Phasor Representation

Recalling the derivation of a phase-modulated interferometer as given in Chapter 5, we have the output field of the MZIM given by

$$E_o = \frac{E_i}{2}\left[e^{j\varphi_1(t)} + e^{j\varphi_2(t)}\right] = \frac{E_i}{2}\left[e^{j\pi v_1(t)/V_\pi} + e^{j\pi v_2(t)/V_\pi}\right] \qquad (8.2)$$

It can be seen that the modulating process for generation of RZ pulses can be represented by a phasor diagram as shown in Figure 8.4. A phasor is represented by a vector of an optical field amplitude and its phase in a rectangular coordinate system. The rotation angular speed of this vector in a circle represents the variation with respect to time or the angular optical frequency ω and time variant phase ωt. This rotation is extremely fast, so one can see it as stationary on the Cartesian plane. This technique gives a clear understanding of the superposition of the fields at the coupling output of two arms of the MZIM. Here, a dual-drive MZIM is used, that is, the data $[V_1(t)]$ and the complementary data (data: $V_2(t) = -V_1(t)$) are applied into each arm of the MZIM, respectively, and the RF voltages swing in inverse directions. Applying the phasor representation, vector addition, and simple trigonometric calculus, the process of generation RZ33 and CSRZ is explained in detail and verified.

The width of these pulses is commonly measured at the position of full-width at half maximum (FWHM). It is noted that the measured pulses are intensity pulses, whereas we are considering the addition of the fields in the MZIM. Thus, the normalized E_0 field vector has the value of $\pm 1/\sqrt{2}$ at the FWHM intensity pulse positions, and the time interval between these FWHM points gives the FWHM values.

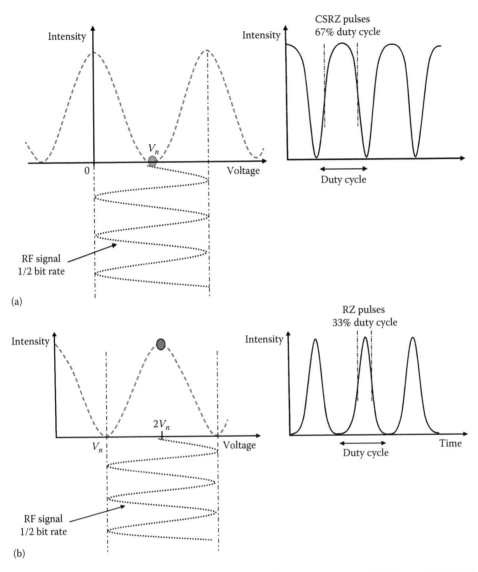

(a)

(b)

FIGURE 8.3 Biasing and driving sinusoidal electrical waveform to generate (a) CSRZ and (b) RZ-33%.

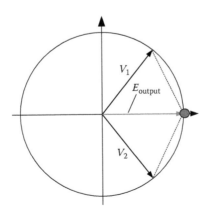

FIGURE 8.4 Phasor representation for generation of output field in dual-drive MZIM.

8.2.3 PHASOR REPRESENTATION OF CSRZ PULSES

The key parameters for representing an optical modulator type MZIM, including DC phase created by the V_{bias}, the amplitude of driving voltage, and its corresponding initialized phasor representation are shown in Figure 8.5a and b.

Swing voltage of driving RF signal on each arm has the amplitude of $V_{\pi/2}$ (i.e., $V_{p-p} = V_\pi$). RF signal operates at half of bit rate ($B_R/2$). At the FWHM position of the optical pulse, the $E_{\text{output}} = \pm 1/\sqrt{2}$ and the component vectors V_1 and V_2 form a phase of $\pi/4$ with vertical axis as shown in Figure 8.6.

Considering the scenario for generation of 40 Gb/s CSRZ optical signals (25 ps pulse width), the modulating frequency is $f_m = 20\,\text{GHz} = B_R/2$. At the FWHM positions of the optical pulse, the phase is given by the following expression

$$\frac{\pi}{2}\sin(2\pi f_m) = \frac{\pi}{4} \Rightarrow \sin 2\pi f_m = \frac{1}{2} \Rightarrow 2\pi f_m = \left(\frac{\pi}{6}, \frac{5\pi}{6}\right) + 2n\pi \tag{8.3}$$

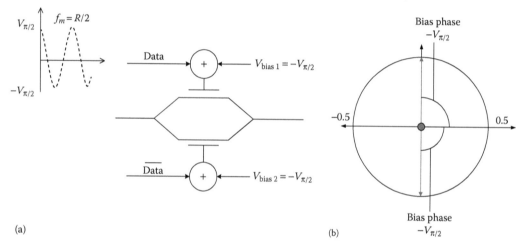

FIGURE 8.5 Biasing voltages of (a) MZIM structure and (b) phasor representation for RZ pulse generation. V_{bias} is $\pm V_{\pi/2}$.

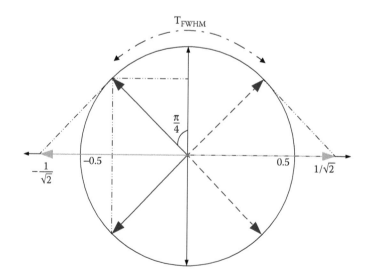

FIGURE 8.6 Phasor representation of evolution of CSRZ pulse generation using dual-drive MZIM.

Thus, the calculation of T_{FWHM} can be carried out and hence, the duty cycle of the RZ optical pulse can be obtained as given in the following expression:

$$T_{FWHM} = \left(\frac{5\pi}{6} - \frac{\pi}{6}\right)\frac{1}{B_R 2\pi} = \frac{1}{3}\pi \times \frac{1}{B_R} \Rightarrow \frac{T_{FWHM}}{T_{BIT}} = \frac{1.66 \times 10^{-11}}{2.5 \times 10^{-11}} = 66.67\% \tag{8.4}$$

The result clearly verifies the generation of CSRZ optical pulses.

8.2.4 PHASOR REPRESENTATION OF RZ33 PULSES

The key parameters including the V_{bias}, the amplitude of driving voltage, and its correspondent initialized phasor representation are shown in Figure 8.7a and b.

At the FWHM position of the optical pulse, the $E_0 = \pm 1/\sqrt{2}$ and the component vectors V_1 and V_2 form a phase of $\pi/4$ with horizontal axis as shown in Figure 8.8.

Considering the scenario for generation of 40 Gb/s CSRZ optical signal (25 ps pulses), the modulating frequency is $f_m = 20$ GHz $= B_R/2$. At the FWHM positions of the optical pulse, the phase is given by the following expressions

$$\frac{\pi}{2}\cos(2\pi f_m t) = \frac{\pi}{4} \Rightarrow t_1 = \frac{1}{6 f_m} \tag{8.5}$$

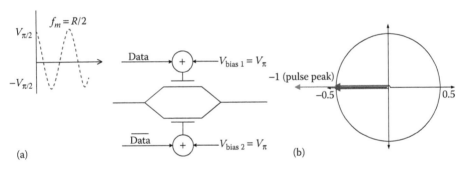

FIGURE 8.7 Generation of 33% RZ lightwave pulse sequence. V_{bias} is V_π for both arms. Swing voltage of driving RF signal on each arm has the amplitude of $V_{\pi/2}$ (i.e., $V_{p-p} = V_\pi$) and RF signal operates at half bit rate ($B_R/2$). V_{bias}, bias voltage at DC to create the initial phase shift on one optical path; V_π, the bias voltage required to create a π phase shift; $V_{\pi/2}$, likewise for $\pi/2$ phase shift at DC; V_{p-p}, peak-to-peak voltage amplitude; $B_R/2$, half of bit rate.

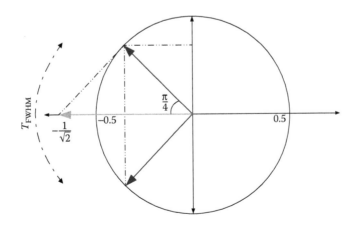

FIGURE 8.8 Phasor representation of RZ33 pulse generation using dual-drive MZIM.

$$\frac{\pi}{2}\cos(2\pi f_m t) = -\frac{\pi}{4} \Rightarrow t_2 = \frac{1}{3f_m}$$ (8.6)

Thus, the calculation of T_{FWHM} can be carried out, and hence the duty cycle of the RZ optical pulse can be obtained as given in the following expression:

$$T_{\text{FWHM}} = \frac{1}{3f_m} - \frac{1}{6f_m} = \frac{1}{6f_m} \therefore \frac{T_{\text{FWHM}}}{T_b} = \frac{1/6f_m}{1/2f_m} = 33\%$$ (8.7)

Therefore, the result clearly verifies the generation of RZ33 optical pulses.

8.2.5   Discrete Phase Modulation—DPSK

8.2.5.1   Principles of DPSK and Theoretical Treatment of DPSK and DQPSK Transmission

For PSK signals, the data information is contained in the discrete phase shifts versus the phase of the unmodulated carrier. The required absolute phase at the receiver must be supplied by the recovered carrier such as the OPLL or relatively the electrical PLL that would be complicated. Thus, it is more effective to process the phase difference of the two consecutive symbols.

Information encoded in the phase of an OC is commonly referred to as optical PSK. In early days, PSK required precise alignment of the transmitter and demodulator center frequencies. Hence, PSK system was not widely deployed. With the introduction of DPSK scheme, CoD is not critical because DPSK detection only requires source coherence over 1-bit period by comparison of two consecutive pulses.

A binary 1 is encoded if the present input bit and the past encoded bit are of opposite logic and a binary 0 is encoded if the logic is similar. This operation is equivalent to XOR logic operation. Hence, an XOR gate is usually employed in differential encoder. NOR can also be used to replace XOR operation in differential encoding as shown in Figure 8.9a. The decoding logic circuit is also shown in this figure.

In optical application, electrical data "1" is represented by a π phase change between the consecutive data bits in the OC, while state "0" is encoded with no phase change between the consecutive data bits. Hence, this encoding scheme gives rise to two points located exactly at π phase difference with respect to each other in signal constellation diagram. For continuous PSK, such as the minimum-shift keying, the phase evolves continuously over a quarter of the section, thus a phase change of $\pi/2$ occurs between one state to the other [5] as indicated by bold circles in Figure 8.9b.

For differential quadrature phase-shift keying (DQPSK), another quadrature component or subsystem would be superimposed on the DPSK (the in-phase) because they are orthogonal to each other. In another words, the phase constellation of Figure 8.9b would have another imaginary ($\pi/2$) axis with the states of $+\pi/2$ and $-\pi/2$.

8.2.5.2   Optical DPSK Transmitter

Figure 8.10 shows the structure of a 40 Gb/s DPSK transmitter in which two external LiNbO$_3$ optical interferometric modulators [3] (MZIM) are used. The MZIM used in Figure 8.10 can be either a single- or dual-drive type. The optical DPSK transmitter would consist of a narrow linewidth laser source to generate a lightwave of wavelength conformed to the ITU grid.

As shown in Figure 8.11, the optical RZ pulses are then fed into the second MZIM through which the RZ pulses are modulated by the precoded NRZ binary data to generate RZ-DPSK optical signals. Electrical data pulses are differentially precoded in the precoder using the XOR coding scheme. Without pulse carver and sinusoidal RF signal, the system becomes NRZ-DPSK transmitter.

In DPSK modulation, the MZIM is biased at the minimum transmission point. The precoded electrical data has the peak-to-peak amplitude equal to $2V_\pi$ and operates at the transmission bit rate. The modulation principles are demonstrated in Figure 8.11.

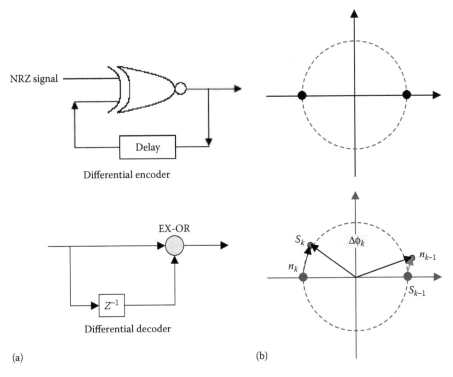

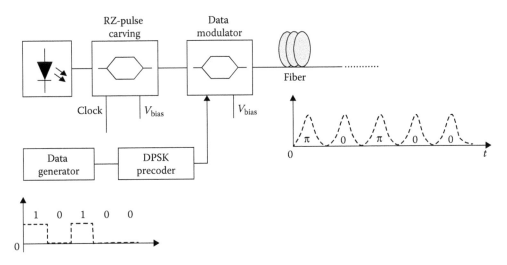

FIGURE 8.9 (a) DPSK precoder and decoder for encoding and decoding of differential data. (b) Signal constellation diagrams.

FIGURE 8.10 Generation of RZ formats DPSK modulation.

Electro-optic (EO) phase modulator (PM) might also be used for generation of DPSK signals, but Mach–Zehnder wave guide structure has several advantages over PM described elsewhere [1]. Use of optical PM produces chirped transmitted optical signals, whereas use of EO intensity modulator—MZIM, especially X-cut versions, produces chirp-free signals. In practice, a small amount of chirp might be useful for transmission [6].

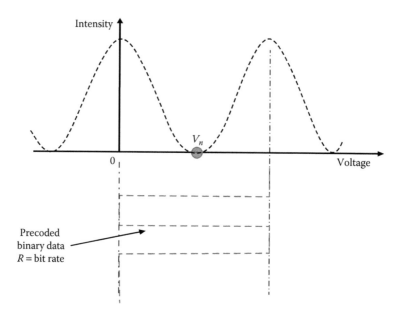

FIGURE 8.11   Voltage bias and amplitude swing levels for NRZ-DPSK modulation formats.

8.2.6   DPSK BALANCED RECEIVER

Consider two consecutive symbols whose signals can be represented by

$$s_{k-1} = A\sin(2\pi f_c t + \phi_{k-1}) \quad \text{for } (k-1)T_b \le t \le kT_b$$

$$s_k = A\sin(2\pi f_c t + \phi_k) \qquad \text{for } kT_b \le t \le (k+1)T_b$$

$$(8.8)$$

where A is an arbitrary amplitude and f_c the carrier frequency. At the receiver side, the phase difference $\Delta\phi_k = \phi_k - \phi_{k-1}$ needs to be determined. If there is no distortion, then the phase difference is that of the original phase difference generated at the transmitter. However, in general, the signals are distorted when reaching the receiver, thus we have

$$x_{k-1}(t) = s_{k-1}(t) + n_{k-1}(t)$$

$$x_k(t) = s_k(t) + n_k(t)$$

$$(8.9)$$

Figure 8.9b shows the position of the data signals at the receiver side, including the additional contribution of the noise vectors of the two consecutive intervals. The phase angle between the two vectors can then be written as

$$\cos\Delta\phi_k = \frac{x_k x_{k-1}}{|x_k||x_{k-1}|} \quad \text{and} \quad \sin\Delta\phi_k = \frac{x_k x^*_{k-1}}{|x_k||x^*_{k-1}|}$$

$$(8.10)$$

$$\text{with} \quad x_k x_{k-1} = \int_0^T x_k(t)x_{k-1}(t)\mathrm{d}t \quad \text{and} \quad x_k x^*_{k-1} = \int_0^T x_k(t)x^*_{k-1}(t)\mathrm{d}t$$

The above equation represents the scalar products between the two received signal vectors x_k and x_{k-1}. For nondistortion case of binary DPSK modulation format, the phase difference would be 0 and π, thus $\Delta\phi_k = \pi$. Thus, the received signals are in push–pull states and bipolar in nature due to the fact that if the phase difference is zero, the received waveform would be maximum and minimum in the negative sense if a π phase difference, which can be represented, following Equation 8.8, as

$$b_k = \text{sgn}(x_k x_{k-1})$$
$$\rightarrow b_k = \text{sgn}(\cos\Delta\phi_k)$$
(8.11)

The receiver for determining the differential phase can be implemented using a balanced receiver operating in the push–pull mode. The delay balanced receiver is shown in Figure 8.12a in which the received optical field is split into two arms, one delayed in the optical domain by a symbol period and another with no delay. The delay time must be matched to the propagation time of the lightwave carrier over the guided region of the optical fiber or integrated waveguide. A thermal tuning section is normally used to adjust to any mismatched time delay section. Two photodetectors are connected in balanced configuration, hence producing a push–pull eye diagram (Equation 8.10) as shown in Figure 8.12b.

Figure 8.12 shows the structure of an MZDI-based DPSK balanced receiver with an insert of a typical eye diagram at the output of the MZDI balanced receiver for 50 ps, that is 20 Gb/s DPSK optical pulses. Note that the diagram shown is of a RZ format, so there is no continuous base line. The received signal at the output of the balanced receiver is given by [7]:

$$P_D(t) = \left|E_D(t) + E_D(t-\tau)\right|^2 - \left|E_D(t) - E_D(t-\tau)\right|^2$$
$$= 4\Re\left\{E_D(t)\right\}E_D^*(t-\tau) = \cos\left(\Delta\phi + \varsigma\right)$$
(8.12)

where $E_D(t)$ and $E_D(t-\tau)$ are the current and the 1-bit delay version of the optical DPSK pulses, respectively. $\Delta\phi$ and ς represent the differential phase and the phase noise caused by the MZ delay interferometer imperfections (MZDIs), respectively. The later issue is not a severe degradation factor in modern MZDI-based DPSK receiver due to the use of a thin-film heater for tuning any waveguide path mismatch. The tuning has high stability with the implementation of the electronic feedback control circuit.

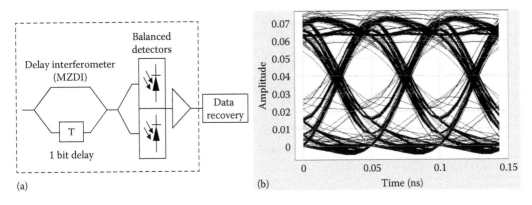

(a)   (b)

FIGURE 8.12 Schematic diagram of the balanced receiver for detection of the phase difference of two consecutive bits (a) structure and (b) a sample of phase detected eye diagram. Bit rate = 20 Gb/s format RZ.

8.3   DPSK TRANSMISSION EXPERIMENT

8.3.1   COMPONENTS AND OPERATIONAL CHARACTERISTICS

The components and operating remarks of the optical transmission system shown in Figure 8.13 for ASK and DPSK modulation and related formats of RZ, NRZ, and CSRZ are given in Table 8.1.

8.3.2   SPECTRA OF MODULATION FORMATS

The center wavelength is set at 1551.72 nm. The experimental spectra of NRZ, RZ, and CS-RZ of ASK and DPSK formats are shown in Figure 8.14. It is noted that for DPSK spectra, the carrier at the center is at the same level as the signal. The RZ spectra are wider the NRZ spectra as expected due to the shorter pulse width. The influence of the spectra of the optical signals on the dispersion and nonlinear impairments will be discussed in later sections. For CSRZ, the carrier is clearly suppressed. Due to the resolution of the spectrum analyzer, the deep suppression is not observed.

8.3.3   DISPERSION TOLERANCE OF OPTICAL DPSK FORMATS

A typical experimental testbed set up is shown in Figure 8.15. There are significant points for consideration in the setup of the testbed: (i) 40 Gb/s Tx is SHF5003 DPSK Transmitter, which can generate both ASK and DPSK data; (ii) in case of ASK, the Discovery Semiconductor "Lab Buddy"

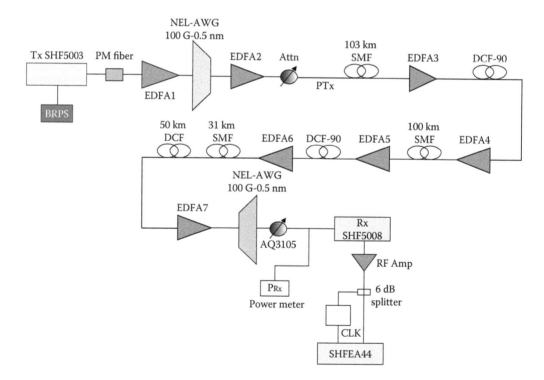

FIGURE 8.13 Experimental setup for evaluation of modulation formats of ASK and DPSK modulation, 230 km three optically amplified spans transmission, 234 km SMF, plus 230 km dispersion-compensating fiber (DCF).

TABLE 8.1
Photonic Components and Operating Characteristics

Photonic Sub-systems	Description	Remarks
Laser source	Anritsu tunable laser set at 1551.72 nm for center wavelength (lamda 5); 1548.51 nm for lamda 1 and 1560 nm for lamda 16	Max output power of 8 dBm
Multiplexer and demultiplexer	NEL-AWG = frequency spacing 100G 3 dB BW of 0.5 nm. Circulating property so spectrum would appear cyclic (note the spectrum)—note of the channel input and output—for example, input at port 5 of Lamda 5 then output at port 8. Input Lamda 1 at port 1 then output at port 8 (Lamda 1). This demux is a NEL array waveguide product. The AWG has a circulant property and thus can be used as cascade filters	
Photonic transmitters (Tx) for different modulation format generation	The Tx can be set at CSRZ-DPSK or RZ-DPSK, no DQPSK format facility is available	A pair of external modulators of MZIM single drive. SHF model 5003
Clock recovery module	Clock recovery using 6 dB splitter	SHF module—SHF 1120A
Optical receiver	Balanced receiver used when DPSK format is used. Otherwise Discovery Semiconductor Lab Buddy DS-10H is employed. A MZD interferometer with thermal tuning of the delay path included	Tuning of MZ phase decoder at the Rx is necessary when wavelength channels are changed
Inline optical amplifiers	EDFA 1 driven at 178 mW pump power (saturation mode). EDFA 2 driven at 197 mW (saturation mode)	All EDFAs are driven in the saturation mode
Optical filters	NEL AWG mux/demmux 0.45 nm bandwidth[a] AWG JDS 8 channel mux/demux 0.35 nm bandwidth JDS—1.2 nm bandwidth. Santec – 0.5 nm bandwidth with slow roll-off	
PRBS	Bit pattern generator SHF-BPG 44	
Error analyzer	SHF model EA-44	
"Directly derived clock"	Direct synchronization of the sync signals to the error analyzer. If no "directly derived clock" is indicated then a clock recovery module is used for synchronization and sampling of the received data sequence	
Signal monitoring	Oscilloscope: Tektronix oscilloscope with 70 GHz optical plug-in and Agilent oscilloscope with 50 GHz optical plug-in. Spectrum Analyzer—Photonetics OSA	
Other components	Optical attenuator is used to vary the optical power input at the receiver for setting the operating condition in the linear or nonlinear region. About 5 dBm is required to set the onset level of nonlinear operation. Optical attenuator is inserted in front of the receiver to ensure no damage of the photodetector.	

[a] The bandwidth here is considered as the 3dB bandwidth.

Optical Receiver (45 GHz with built-in amplifier) is used; (iii) in case of DPSK, SHF 5008 DPSK receiver is used (MZ—1-bit delay). Also, the Electrical Amp is utilized to drive the Error Analyzer; (iv) JDS 1.2-nm filter was utilized after EDFA to decrease the ASE noise level; and (v) power launched into SMF (right after the transmitter Tx) is low, hence nonlinearities do not have any impact. The launched power is recorded as shown in Figure 8.16.

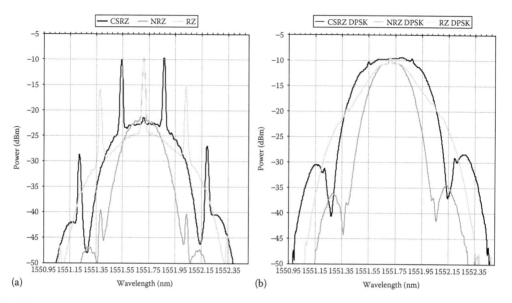

FIGURE 8.14 Optical spectra of transmitted optical signals of formats NRZ, RZ, and CSRZ for (a) ASK and (b) DPSK.

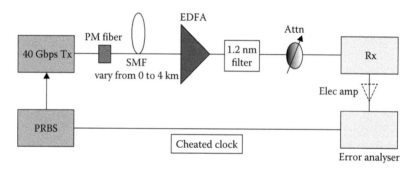

FIGURE 8.15 Experimental testbed for investigation of dispersion tolerance of transmission system.

Figure 8.16 displays the eye diagram under self-coherent detection at different transmission distance from back-to-back to 4-km dispersive SSMF under different modulation formats, NRZ-DPSK, RZ-DPSK, and CSRZ-DPSK. The CSRZ-DPSK format offers the best performance, but with a disadvantage that the detection circuits need to be twice as fast as that in the NRZ format. Current 56-Gbaud speed of optical channels makes the RZ format quite difficult due to unavailability of such ultrahigh electronic or digital circuits (Table 8.2).

Using "directly derived clock," that is, directly from the BPG and hence without the Clock Recovery Unit, the Error Analyzer directly uses clock pulse of the PRBS. Laser source center wavelength is set at 1551.72 nm and the pulse pattern of PRBS is $2^{31} - 1$. Figure 8.16 shows the performance of CSRZ/RZ/NRZ-DPSK formats over lengths of 0–4 km (from top to bottom) of SSMF using the eye diagrams of 40 Gb/s transmission. Hence, the dispersion tolerance to the transmission fiber SSMF with dispersion factor of D = +17 ps/nm.km for DPSK format versus the conventional ASK can be obtained. It can be clearly seen that DPSK system offers approximately 3 dB better in-receiver sensitivity compared to the ASK system. The obtained results confirm the theoretical analysis.

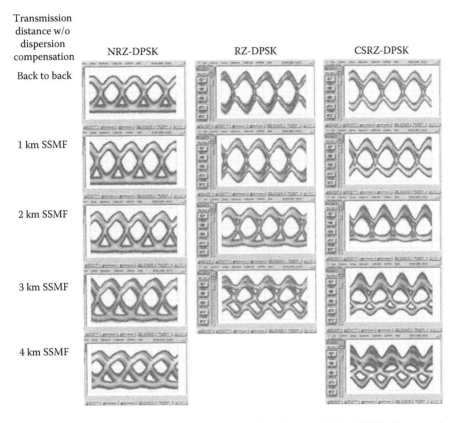

FIGURE 8.16 Eye diagrams of DPSK transmission after 0, 1, 2, 3, and 4 km SSMF for evaluation of the dispersion tolerance.

TABLE 8.2
Maximum Power Pounced into Input of the Optical Fiber Transmission for Different Modulation ASK and DPSK with RXZ, NRZ, and CSRZ Formats

	NRZ-ASK	RZ-ASK	CSRZ-ASK	NRZ-DPSK	RZ-DPSK	CSRZ-DPSK
Launched power (dBm)	−4.2	−7.7	−6.2	−3.1	−6.2	−4.8

ASK modulation format is also employed in this experimental platform under the pulse shaping NRZ and RZ as well as CSRZ. The obtained eye diagrams at different distances are shown in Figure 8.17. The corresponding performances of the ASK and DPSK under different pulse shaping and transmission system with the BER plotted against the receiver sensitivity (in dBm) at the receiver is shown in Figure 8.18. Furthermore, the receiver penalty for different transmission distances (residual dispersive distance) for different modulation formats is plotted in Figure 8.19 under no and with Gaussian filter employed at the output of the receiver.

We observe that no noise floor can be found except for the case of RZ-DPSK after 3-km SSMF transmission. This is due to the limited speed of the receiving circuitry due the RZ pulse shaping. The bandwidth required may reach 70 GHz for 40 Gb/s RZ-DPSK.

Noncompensation Transmission distance	RZ-ASK	NRZ-ASK	CSRZ-ASK
Back to back			
1 km SSMF			
2 km SSMF			
3 km SSMF			
4 km SSMF			

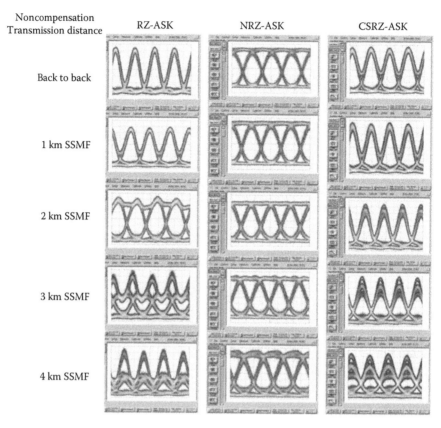

FIGURE 8.17   Eye diagrams of ASK transmission after 0, 1, 2, 3, and 4 km SSMF for evaluation of the dispersion tolerance of the transmission system.

8.3.4   OPTICAL FILTERING EFFECTS

Initially, the impacts of the optical filtering characteristics of the mux are evaluated using a back-to-back set up as shown in Figure 8.20. We set the sampling clock directly from the auxiliary clock output of the $2^{31} - 1$ pattern generator. Although two array waveguide gratings (AWGs—optical multichannel filters) can be used as muxes and demuxes, respectively, we use only one AWG at either the transmitter or the receiver sites. The other filter can be substituted with a multilayer thin-film optical filter. Two optical filters acting as mux and demux at the transmitting and receiving ends, respectively, are then used to evaluate their impacts on 40 Gb/s channels. We did not observe significant degradation of the BER as shown in Figure 8.21.

Figure 8.22 shows the BER versus the receiver sensitivity curves as obtained for ASK and DPSK and DQPSK with RZ, NRZ, or CSRZ formats as shown in Figure 8.21. The sensitivities do not change significantly under the influence of the 0.5-nm optical filter on 40 Gb/s channels operating under different modulation formats. We can now observe the optical spectra of ASK and PSK formats as shown in Figure 8.21a and b. The Gaussian-like or cos^2 profile of the pulses generated at the output of both optical modulators and the parabolic passband properties of the AWG can tolerate a wide signal spectrum. We do not observe any degradation of the BER versus the sensitivity in the cases of wideband optical filters (1.2 nm) and 0.5 nm optical mux filtering. The detected eye diagrams under RZ-DPSK and CSRZ-DPSK transmission are shown in Figure 8.16. The MZDI acting as a phase comparator is thermally tuned so as to obtain a maximum eye opening, thence optimum BER.

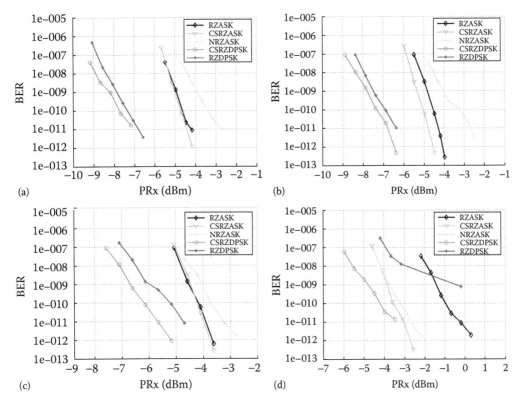

FIGURE 8.18 Performance of dispersion tolerance of advanced modulation formats in case of (a) 0 km, (b) 1 km, (c) 2 km, and (d) 3 km of SSMF.

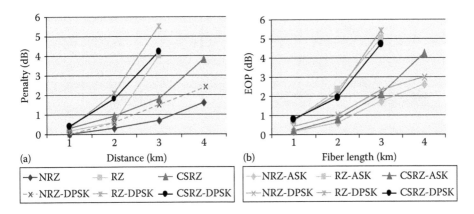

FIGURE 8.19 40Gb/s Modulation formats: (a) Experimental results on power penalties-induced chromatic dispersion limits of ASK and DPSK formats. (b) Simulation results on power penalties-induced chromatic dispersion limits of ASK and DPSK formats.

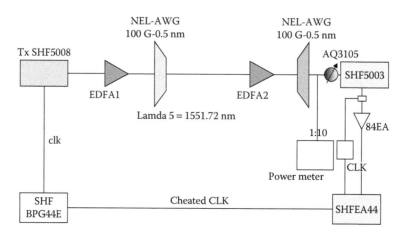

FIGURE 8.20 Back-to-back experimental system for investigation of the optical AWG filtering impacts on modulation formats.

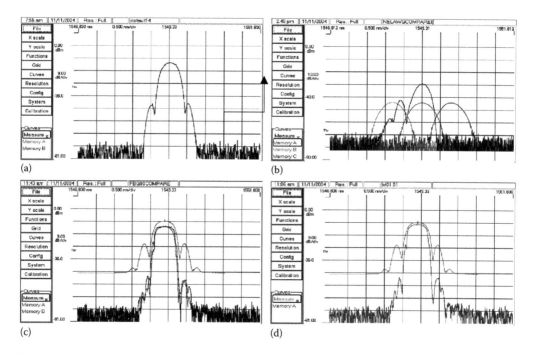

FIGURE 8.21 Optical passbands of AWG filters (a) top left corner: signal spectrum of a channel; (b) right top: optical passbands of the multiplexed output of the AWG—note the parabolic passband characteristics and the "black" curve of the output spectra of a wavelength channel; (c) signal spectrum and its output of the AWG; and (d) same as (c) but different wavelength region.

The characteristics of multiplexer/demultiplexers currently utilized for 10 Gb/s systems are still sufficiently offering good performance for 40 Gb/s transmission systems. The experimental results suggest the feasibility of upgrading the currently deployed system to higher transmission rate −40 Gb/s. Higher bit rate such as 80 or 160 Gb/s need to be further investigated.

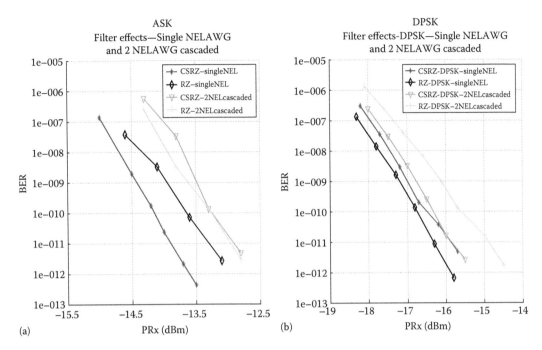

FIGURE 8.22   BER versus input power level of (a) ASK and (b) DPSK modulation with various formats.

8.3.5   PERFORMANCE OF CSRZ-DPSK OVER A DISPERSION-MANAGED OPTICAL TRANSMISSION LINK

A dispersion-managed optical transmission link over 328-km length is set up as shown in Figures 8.13 and 8.24a with three optically amplified spans. The dispersion factor of the transmission fiber is matched with that of a length of dispersion-compensation fiber. The bit pattern generator (BPBG) is used to drive the modulators of the transmitter. RZ and CSRZ can also be generated with an insert of sinusoidal wave generated from a signal synthesizer. The nonlinear effect is explored by varying the optical power launched into the SMF from 0 to 15 dBm as shown in Figure 8.23. The power launched into the DCF is kept unchanged at 0 dBm (nonlinearities are generated in SMF not DCF).

It is noted here that the mopping up of the residual dispersion using a tunable dispersion compensator is used and set at the wavelength 1555.75 nm, which can be shown to be optimum as compared with the BER performance curve obtained at $\lambda_5 = 1551.72$ nm. Hence, laser source is tuned to this wavelength.

8.3.6   MUTUAL IMPACT OF ADJACENT 10G AND 40G DWDM CHANNELS

The 320-km transmission is also conducted to assess the performance of 10G NRZ-ASK and a CSRZ-DPSK 40G channel. The transmission of adjacent and nonadjacent channels is demonstrated and evaluated with a 100-GHz AWG mux and a 1.2-nm tunable filter at the input of the Rx. Insignificant power penalty is observed when adjacent 40G channel is cotransmitted. The setup of the transmission system is shown in Figure 8.24a, with 320-km SSMF and dispersion compensating module with two optical filters and splitters, the 100-GHz AWG, which are used as muxes

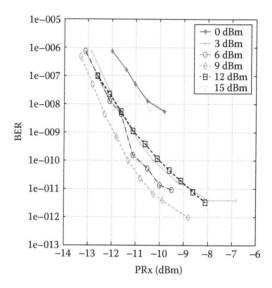

FIGURE 8.23 Transmission performance of CSRZ-DPSK (230 km transmission) under different power launched levels.

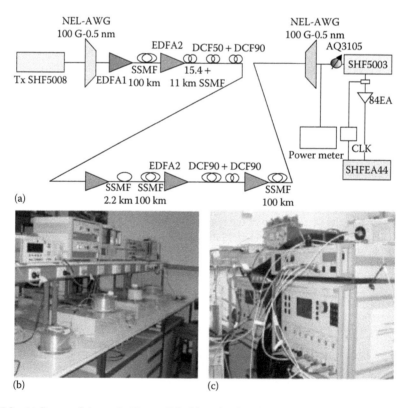

FIGURE 8.24 (a) Set up of the optically amplified long haul optical transmission system—the span length can vary from 50 to 320 km, (b) optically amplified and dispersion compensated fiber transmission line (left)—transmitter and receiver plus, and (c) 40 Gb/s bit pattern generator and error analyzer (right).

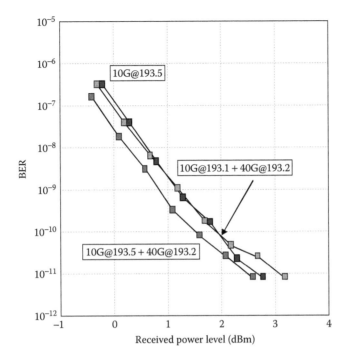

FIGURE 8.25 A 320-km transmission 40-G impact on 10-G channel: BER versus receiver sensitivity (dBm)—effects of 40 G (CSRZ-DPSK) with 10 G (NRZ-ASK) channel simultaneously transmitted for NRZ-ASK and CSRZ-DPSK formats—light gray square dots for 1.2 nm thin film filter and dark gray square dots for 0.5 nm AWG filter (demux with 100 GHz spacing).

and demuxes. An optical filter is inserted at the front end to measure the impact of adjacent CSRZ-DPSK 40G channel on 10G NRZ-ASK performance. No significant impact was noted when 40G channel, adjacent or nonadjacent, is cotransmitted as evaluated in Figure 8.25.

8.4 DQPSK MODULATION FORMAT

When digitizing data for transmission across many hundreds of kilometers, many digital modulation formats have been proposed and investigated. This section investigates the DQPSK modulation format. This modulation scheme, although having been in existence for quite some time, has only been implemented in the electrical domain. Its application to optical systems proved difficult in the past as constant phase shifts in OC were required to be maintained. However, with the improvement in optical technology and alternative transmitter design setup, these difficulties have been eliminated.

8.4.1 DQPSK

One of the main attractive features of the DQPSK modulation format is that it offers both twice as much bandwidth and increased spectral efficiency compared to OOK. As an example, comparisons between 8 × 80 Gb/s DQPSK systems and 8 × 40 Gb/s OOK systems show that DQPSK modulation offers more superior performance, that is, the spectral efficiency. The very nature of the signaling process also allows noncoherent detection at the receiver, thus reducing the overall cost of the system design.

The idea behind DQPSK, as the name suggests, modulation format is to apply a differential form of phase-shift modulation to the OC, which encodes the data. DQPSK is an extension to the

TABLE 8.3

DQPSK Modulation Phase Shifts

Di-bit	Phase Difference $\Delta\phi = \phi_2 - \phi_1$ (°)
00	0
01	90
10	180
11	270

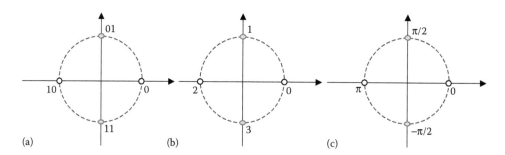

FIGURE 8.26 DQPSK signal constellation and assignment of phase symbols (a) quaternary digits, (b) two-digit binary word, and (c) phase complex plane.

simpler DPSK format. Rather than having two possible symbol phase states (0 or π phase shift) between adjacent symbols, DQPSK is a four-symbol equivalent {0, $\pi/2$, π, or $3\pi/2$}. Depending on the desired di-bit combination to be encoded, the difference in phase, between the two adjacent symbols (OC pulses), is varied systematically. Table 8.3 outlines this behavior.

The parameters ϕ_1 and ϕ_2 as indicated in Table 8.3 are the phases of adjacent symbols. The table can also be represented in the form commonly known as a "constellation diagram" (see Figure 8.26), which graphically explains the signals state in both amplitude and phase.

The DQPSK transmitter design is based on a design previously experimentally proposed [8]. First, a RZ pulse carving MZM, which generates the desired RZ pulse shape, is implemented. Next, an MZIM (generating 0 or π phase shift of OC) is coupled with a PM, which induces a 0 or $\pi/2$ phase shift of the OC. When placed in this configuration, the four phase states of the OC required by the DQPSK modulation format {0, $\pi/2$, π, or $3\pi/2$} can be obtained. Both the MZM and PM are to be driven by random binary generators operating at 40 Gb/s. Figure 8.27 is a schematic of the system transmitters implementing the RZ-DQPSK modulation. We originally started with the NRZ-DQPSK design; however, the RZ format is used more readily in practice and has proven to be more robust to system nonlinearities. The pulse sequence generated after the pulse carver can be observed as shown in Figure 8.28.

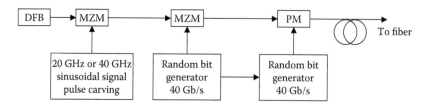

FIGURE 8.27 Schematic of a channel implementing RZ-DQPSK modulation.

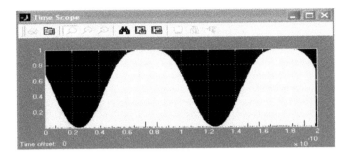

FIGURE 8.28   The OC after RZ pulse carving using an MZM driven by a 10 GHz electrical signal.

For DQPSK, the signal constellation can be considered to be two DPSKs orthogonal to each other. That is, there is a phase shift of $\pi/2$ between the constellation points. Thus, the transmitter and receiver can be implemented with the following phase difference:

$$\Delta\phi_k = \left\{0, \frac{\pi}{2}, \pi, -\frac{\pi}{2}\right\} \quad \text{or} \quad \Delta\phi_k = \left\{\frac{\pi}{4}, \frac{3\pi}{4}, -\frac{\pi}{4}, -\frac{3\pi}{4}\right\} \qquad (8.13)$$

In the latter case, the receiver simply gives the outputs of the binary bits of the imaginary and real parts of the DQPSK signals.

$$b_{k\_I} = \cos\Delta\phi_k = \frac{x_k x_{k-1}}{|x_k||x_{k-1}|} \quad \text{and} \quad b_{k\_Q} = \sin\Delta\phi_k = \frac{x_k x^*_{k-1}}{|x_k||x^*_{k-1}|}$$

$$\text{with} \quad x_k x_{k-1} = \int_0^T x_k(t)x_{k-1}(t)\mathrm{d}t \quad \text{and} \quad x_k x^*_{k-1} = \int_0^T x_k(t)x^*_{k-1}(t)\mathrm{d}t \qquad (8.14)$$

This can be explained as follows. The data information transmitted during the time interval $[(k-1)T_s, kT_s]$, with T_s as the symbol period, is carried by the in-phase and quadrature components of the signal and thus given by the sign of the terms *sin* and *cos* as given in Equation 8.13. An offset of the signal constellation by $\pi/4$ would assist the simplification of the transmitter and receiver due to the fact that

$$\Delta\phi_k = \left\{0, \frac{\pi}{2}, \pi, -\frac{\pi}{2}\right\} \rightarrow \sin\left\{0, \frac{\pi}{2}, \pi, -\frac{\pi}{2}\right\} \rightarrow \{0, \pm 1\} \qquad (8.15)$$

Or effectively, there would be three output levels, while the $\pi/4$ shift would give binary levels.

For a noiseless case, the correlation between the two bits are given as

$$s(t)s(t-T_b) = A\sin\left(2\pi f_c t + \phi_k\right)A\sin\left(2\pi f_c t - 2\pi f_c T_b + \phi_{k-1}\right)$$

$$= \frac{A^2}{2}\{\cos\left(\phi_k - \phi_{k-1} + \alpha\right) - \cos\left(4\pi f_c + \phi_k + \phi_{k-1} - \alpha\right)\} \qquad (8.16)$$

$$\text{detected signals} \qquad\qquad \text{filtered by lowpass filter}$$

The phase difference $\Delta\phi_k = \phi_k - \phi_{k-1} \in \{0, \pi\}$ leading to $\cos(\Delta\phi_k + k\pi) \in \{1, -1\}$ ensures a maximum distance between the two difference symbols.

In the rest of this section, some of the key principles allowing for the successful decoding of the DQPSK-modulated signal are described. We describe an exact receiver model before fiber

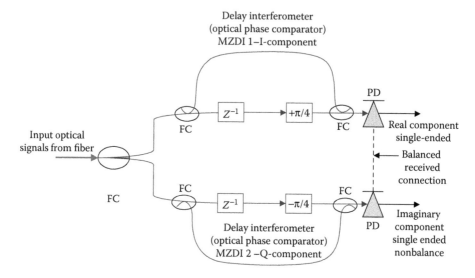

FIGURE 8.29 DQPSK receiver configuration using noncoherent direct detection. FC = fiber coupler, MZDI = Mach–Zehnder Delay Inteferometer.

propagation to allow comparisons between prefiber and postfiber effects of the received eye diagram. The comparison of the phase of the lightwave carrier contained within the two consecutive bits is the optical delay interferometer shown in Figure 8.29. Note that due to the differential nature of the modulation process, the demodulation and detection stage can be considered as "noncoherent" or "direct detection," or effectively, this can be known as a delayed autocorrelation or self-heterodyne detection scheme. The absence of a LO and any other extra hardware required in conventional detectors makes this demodulation technique attractive. The receiver configuration shown in Figure 8.28 used is capable of demodulating the signal transmitted along the 96 km total fiber span. Because the modulation format used in this section typically encodes two bits of data per symbol, it is necessary to extract both bits (termed "real" and "imaginary" bits [8]) from the one received symbol.

We can now outline the purpose of the $\pm\pi/4$ additional phase shift of the OC implemented in the MZDI of the receiver. Because the two bits to be encoded at the Tx are implemented using a MZIM (0 or π phase shift), the second bit is encoded via the PM (0 or $\pi/2$ phase shift). The two devices are said to be in "quadrature" to one another. This implies that there is a $\pi/2$ phase difference between all signaling phase states (see Figure 8.26). The additional $+\pi/4$ and $-\pi/4$ give a total $\pi/2$ phase difference between the upper and lower receiver branches as shown in Figure 8.22. Thus, data recovery can be implemented by comparing the real and imaginary bits received to those transmitted. However, one can consider only the "real component" of the received signal and assess the overall performance of the system via eye diagram analysis (using Q factor method and BER). In practice, the demodulation of the DQPSK signal using the two-branch configuration in Figure 8.29 can be successful demonstrated [9]. A balanced detection structure proven to be more sensitive to optical transmission and detection of noncompensated SSMF is obtained as shown in Figure 8.30, with 10 and 5 km for the Q- and I-components, as shown in Figure 12.47 later in Chapter 6.

8.4.2 Offset DQPSK Modulation Format

In this section, we introduce the offset-DQPSK (O-DQPSK) transmission. ODQPSK has been applied in the transmission over nonlinear satellite channel that offers a smoother transition between the phase states and hence avoids the π phase jumps. In contrast to DQPSK, which

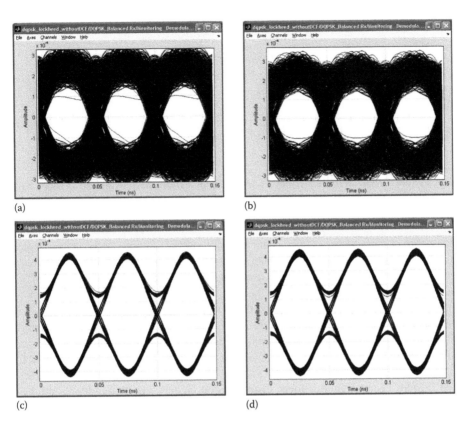

FIGURE 8.30 Eye diagrams achieved for I-component: (a) 10 km SSMF and (c) 5 km SSMF and for Q-components; (b) 10 km SSMF and (d) 5 km SSMF mismatch of dispersion over two 200 km span dispersion compensated and optically amplified spans—RZ-DQPSK 50% RZ pulse shape.

requires detection at the symbol rate, the demodulation of ODQPSK can be achieved on the bit rate, thus allowing the detection of the ODQPSK with only one set of balanced receiver and one set of MZDI. The optical ODQPSK can be generated by two binary RZ-DPSK signals operated at half the bit rate, which can be merged optically by a 3-dB coupler with a 1-bit delay in one path as shown in Figure 8.31.

Similar to the case of DQPSK, a carrier phase difference of $\pi/2$ has to be guaranteed between two parallel signals to ensure the $\pi/2$ phase shift between the states of the constellation points similar to the phase plane shown in Figure 8.32. The receiver for DQPSK modulation format can be implemented using only one MZDI and balanced receiver, but only one MZDI is required and

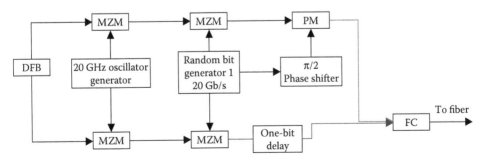

FIGURE 8.31 Schematic diagram of the optical offset DQPSK transmitter.

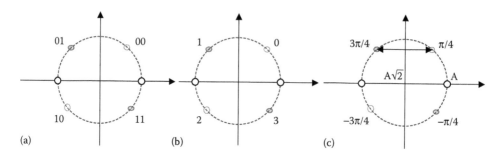

FIGURE 8.32 ODQPSK signal constellation and assignment of phase symbols (a) quaternary digits, (b) two-digit binary word, and (c) phase complex plane with two state (horizontal line) circles are for DPSK constellation points.

a $\pi/2$ phase shifted must be inserted into one path of the MZDI to ensure the demodulation of the $\pi/2$ phase difference between consecutive bits, and hence the in-phase and quadrature components. If there is a cosine function is used instead of a triangular one then the continuity of the phase shift between the constellation points, then the scheme offset DQPSK becomes a MSK modulation scheme. This modulation format is described in Section 7.4.5.4.

The receiver sensitivities of the DQPSK, ODPSK, and DPSK have also been measured for transmissions over 400-km span with optical amplifiers and DCF as shown in Figure 8.32c and d. The bit rate for DPSK is 40 and 20 GB/s for DQPSK and the input power of the fiber is set at 0 dBm to ensure the systems operate under the linear regime. There are no differences between the DQPSK and offset DQPSK because the Euclidean distance between the constellation points are the same.

A 6-dB difference between DQPSK and DPSK is obvious. There is also a difference of the reception sensitivity of 3 dB between receiver employing balanced detector pair and that using a single detector for ODQPSK systems. This is expected as the distance between the "1" and "0" is double for balanced reception to that of a single detector as shown by the detected eye diagrams given in Figure 8.16.

There are two principal differences between the receiver sensitivity of RZ-DPSK and RZ-DQPSK. (i) Using the same total average power (i.e., the same radius on the phasor diagram [Figure 8.32]), the binary level to the quaternary level would require an increase of a factor of $\sqrt{2}$ as we can observe from the signal constellations of the DPSK and DQPSK. Thus, there is a 3-dB increase in the power required for DPSK as compared with DQPSK. (ii) For the MZDI self-heterodyne detection, the detection seems to add an additional power penalty in the splitting and combining of the received optical fields. This is the complexity of the self-homodyne detection as compared to the CoD in which the polarization can be diversified, while in MZDI integrated lightwave circuitry, the polarization of the input lightwave coupled with the fiber would be reduced due to the strong polarization dependence of the rib waveguide of the MZDI.

These original mechanisms of the reduction in the receiver sensitivity can further be explained as follows.

8.4.2.1 Influence of the Minimum Symbol Distance on the Receiver Sensitivity

Assuming the noise pdf distribution at the receiver is Gaussian, the CoD with matched filter would follow the well-known rule of the bit-error probability, which depends on the Euclidean distance between the symbols and the variances at the levels of the symbols. In general, noise power would have a mean of zero and variance σ^2. With a minimum distance d between the symbols, the bit-error probability P_e is given by

$$P_e = \operatorname{erfc}\left(\frac{d/2}{\sigma}\right) \tag{8.17}$$

If the total average power of the signals is A^2, then a binary DPSK has a P_e of

$$P_e = \text{erfc}\left(\frac{A}{\sigma}\right) \tag{8.18}$$

While the DQPSK would have a P_e of

$$P_e = \text{erfc}\left(\frac{A/\sqrt{2}}{\sigma}\right) \tag{8.19}$$

The DQPSK allows the receiving of 4 bits in a symbol period, while DPSK allows only two. Thus, the capacity of the DQPSK is double of that of the DPSK without increasing the noise contribution at the receiver output.

8.4.2.2 Influence of Self-Homodyne Detection on the Receiver Sensitivity

The complexity of CoD is rest on the use of a local laser and the synchronization of the phase and frequency of both the signals and the LO laser. With the advancement in integrated lightwave technology, the MZDI can be implemented without difficulty and the self-heterodyne detection can be assisted by the phase comparison in the digital domain. Indeed the frequency offset between the signal carrier and the LO can be resolved by many DSP algorithms. Thus, the self-heterodyne detection has attracted attention in current DPSK and DQPSK receivers or any receiver that would require the detection of I- and Q-components using balanced receiver. Under a matched filter condition of binary DPSK, the BER is approximately given by

$$P_e = \frac{1}{2}\exp\left(-\frac{A^2}{\sigma^2}\right) \tag{8.20}$$

For QPSK signals using MZDI balanced receiver as a self-homodyne detection, the BER is given by [10]

$$P_e = Q(\vartheta,\mu) - \frac{1}{2}\exp\left(-\frac{\vartheta^2}{\sigma^2}\right)I_0\left(\frac{\vartheta^2\sqrt{2}}{\sigma^2}\right) \tag{8.21}$$

$$\text{with } \mu = A'\sqrt{2}\cos\frac{\pi}{8} \quad \vartheta = A'\sqrt{2}\sin\frac{\pi}{8}$$

where Q represents the Marcum function and I_0 the modified Bessel function of zeroth order, and A' is the amplitude of the signal at the input of the PD. The BER versus the signal-to-noise ratio as given in Equations 8.18–8.21 would be enhanced by CoD, especially when the balanced receiver is employed [11]. However, a 2-dB improvement is obtained for DQPSK for the case of coherent over self-heterodyne detection. These results are considered without considerations of the noise contribution in both the positive and negative electrical signal levels (Figure 8.33).

8.4.3 MATLAB Simulink Model

8.4.3.1 The Simulink Model

MATLAB Simulink models of the transmitter and receivers for DQPSK modulation format are shown in Figures 8.34 and 8.35, respectively, with a $\pi/4$ phase offset of the signal constellation. A RZ-DQPSK optical transmission system is modeled with a pulse carver inserted in front of the data modulator MZIM DQPSK precoder of 2-bit per symbol to provide two output signals (in

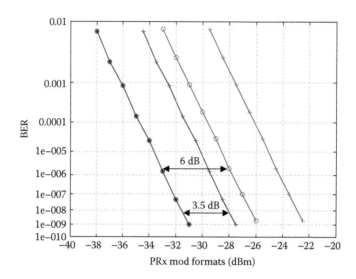

FIGURE 8.33 Optical DPSK and ODPSK transmission over 4 × 100 km SSMF of DWDM 8 × 40 Gb/s transmission. Simulated results of BER versus receiver sensitivity of modulation formats: (dark gray square ○) offset DQPSK _balanced Rx, (dark gray square +) offset DQPSK _single detector, (light gray square *) DQPSK-balanced Rx, (light gray square +) DQPSK-single PD. Black *—offset DQPSK.

electrical domain) to drive the two arms of the optical MZIM. A $\pi/2$ PM is also used to assign the in-phase or quadrature phase components of the DQPSK format.

In the receiver side, differential detection or self-homodyne receiving structure is used with a phase offset of $\pi/4$ to directly obtain the amplitude and phase of the receiving signals as discussed above.

8.4.3.2   Eye Diagrams

The transmission of the DQPSK with a bit rate of 40 Gb/s or 20 G symbols/s is conducted over SSMF length of 10 and 5 km, as shown in Figure 8.36.

The length of SSMF is set to be 10 km and length of DCF 2 km with dispersion factor of DCF is to be five times more negative than that of SSMF (full compensation), as shown in Figure 8.37. The following Simulink model in Figure 8.38 illustrates how to set, in the DCF block in the model, the length of DCF at 2 km and the dispersion factor at five times more negative than that of SSMF (full dispersion compensation). The measured eye diagrams of DQPSK are shown in Figure 8.40 at total bit rate of 108 and 110 Gb/s (or around 50 G symbols/s) when the I- and Q-components are detected separately and when both channels are turned on.

8.5   COMPARISONS OF DIFFERENT FORMATS AND ASK AND DPSK

8.5.1   BER AND RECEIVER SENSITIVITY

8.5.1.1   RZ-ASK and NRZ-ASK

It is obvious that the average power of NRZ and RZ (50% duty) ASK pulse sequences is the same, but the FWHM of the RZ is half of that of NRZ; thus, its peak power is higher. This is proportional to 33% and 67% duty cycle RZ pulses. The higher the peak power of the RZ pulses, the longer the distance between the levels 0 and 1 and hence a possibility of an enhancement of the eye opening and thus BER.

Assuming now that or an equal average power of the two random NRZ and RZ data sequences can be set. Theoretically speaking, a higher peak power might not lead to better receiver sensitivity. However, the matched filtering concept states that [12,13] for any signal corrupted by additive white Gaussian noises in transmission systems with impulse responses of transmitter and receiver $h_T(t)$ and $h_R(t)$, the maximum signal-to-noise ratio at the output of the matched filter is achieved when

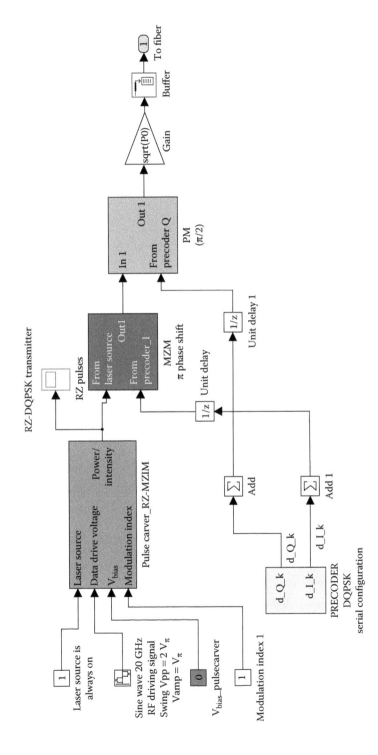

FIGURE 8.34 Transmitter MATLAB Simulink model for DQPSK optical transmission.

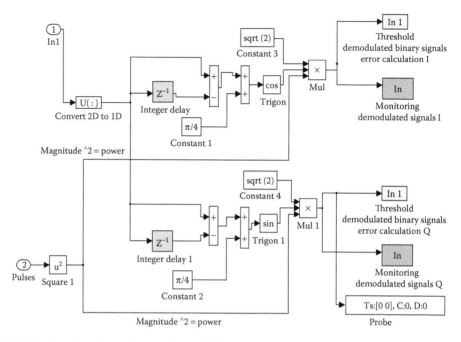

FIGURE 8.35   Receiver Simulink model for DQPSK optical transmission.

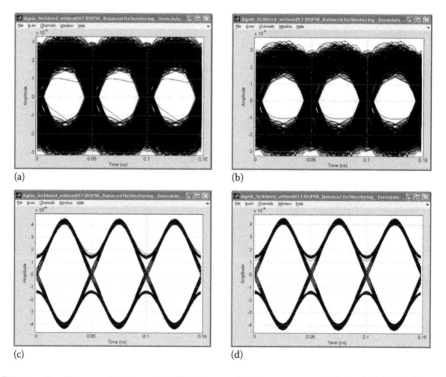

FIGURE 8.36   Eye diagram of balanced DQPSK detected (a) I-component of 10 km SSMF, (b) Q-component of 10 km SSMF, (c) I-component of 5 km SSMF, and (d) Q-component of 5 km SSMF.

FIGURE 8.37 Arrangement of dispersion compensated spans for transmission of DQPSK signals.

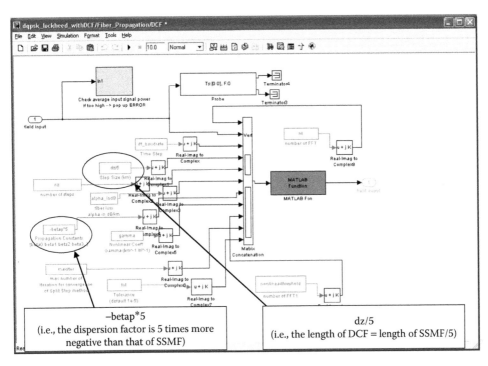

FIGURE 8.38 Simulink model of the fiber sections of dispersion compensating fiber for fully compensated span.

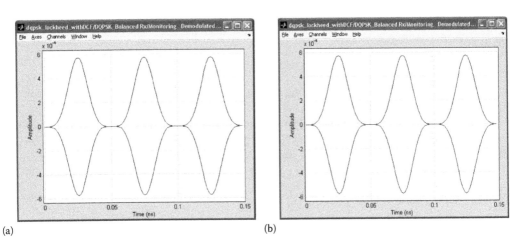

(a)   (b)

FIGURE 8.39 Simulated eye diagram of DQPSK-detected (a) in-phase component message (10 km SSMF and 2 km DCF with DCF dispersion factor about −85 ps/nm.km) and (b) quadrature component of a fully dispersion-compensated SSMF plus DCF (10 km SSMF and 2 km DCF with DCF dispersion factor about −85 ps/nm.km). SSMF, standard single mode fiber; DCF, dispersion compensating fiber.

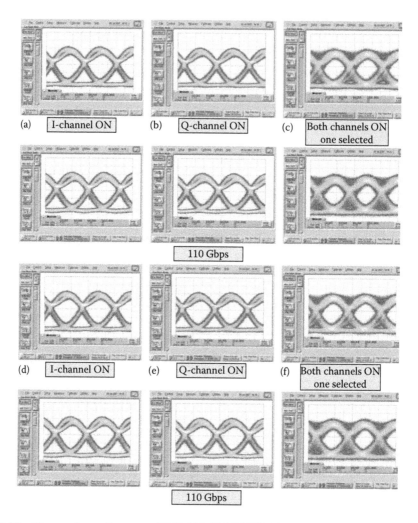

FIGURE 8.40 Measured eye diagrams of 108 Gb/s RZ-DQPSK detected (50 Gsymbols/s). (a) In-phase component, (b) quadrature component, and (c) both channels on and for 110 Gb/s (d) in-phase component, (e) quadrature component, and (f) both channels transmitted, and one component is selected by tuning the MZDI.

$h_R(t) = h_T(t-T)$, with T being the sampling delay time. Alternatively, in the frequency domain, we have $H_R(f)$. $H_T(f)\exp(-j2\pi fT) = 1$.

In the optical domain, such matched filters can be implemented using chirp fiber Bragg grating (FBG), thin-film filter, or microfilters for RZ pulses. Experimentally, it has been observed, as described above, that an improvement of about 2 dB can be achieved for the receiver sensitivity for RZ-ASK as compared with NRZ-ASK in contrast to 3 dB from a theoretical expectation.

8.5.1.2   RZ-DPSK and NRZ-DQPSK

Figures 8.39 and 8.40 show, respectively under simulation and experiment, the difference between the receiver sensitivity of the modulation formats RZ-DPSK and RZ-DQPSK under the detection structures of single and balanced detectors. A 3-dB improvement in the balanced receiver over that of the single detector is as per expectations from the push–pull mechanism of the balanced detection. This is known to be the maximum performance achievable with balanced receiver with narrow band optical filtering at the input of the MZDI [14].

For the detected signals of RZ-DPSK and RZ-DQPSK formats, under balanced and non-balanced (or single detector) there is a difference of 6–7 dB between these schemes. This is due to two possibilities: (i) for the same averaged signal power, the distance between "1" and "0" by a factor of sqrt(√2) for DQPSK as compared to that of DPSK, can be obviously observed from the signal constellations. (ii) but, using MZDI at the input of the balanced detector pair the optical power is split twice at the MZDI so no gain in power and even loss due to unused optical ports and scattering loss in the power split coupler MZDI.

8.5.1.3 RZ-ASK and NRZ-DQPSK

The comparison between ASK and DPSK and DQPSK is important for the case that the differential phase with balanced receiving is used in the upgrading for 10–40 G rates. RZ-DPSK is expected, theoretically and confirmed experimentally as shown in Figure 8.22, to have a 3-dB enhancement over the RZ-ASK and about 6 dB over NRZ-ASK. Thus, the performance of RZ-DQPSK would nearly be the same as that of RZ-ASK but with a symbol rate twice that of the NRZ or RZ-ASK. The Simulink MATLAB model for RZ-DQPSK is given in Section 8.7.

This can be explained as follows: (i) The distance between the "1" and "0" of the ASK is nearly the same as that of DQPSK signals of the same average power as seen from the signal constellation shown in Figure 8.41; (ii) RZ format outperforms the NRZ format by 3 dB (maximum). Thus, RZ-DQPSK would be expected to improve over the NRZ-ASK by the same amount or less. (iii) The use of MZDI as an optical phase comparator at the input of the balanced detector pair would introduce optical power loss due the splitting and unused ports, which can be improved using planar lightwave circuit technology.

In summary, the use of multilevel modulation formats may suffer the optical loss and splitting when MZDI balanced receiving technique is used. These multilevel schemes would offer an effective high information capacity with the symbol rates much lower than that of the binary schemes, for example, 100 Gb/s can be reduced to 25 G symbols/s if a two-level star-QAM (quadrature amplitude modulation) is used. These multilevel modulation schemes are described in Chapter 8.

8.5.2 Dispersion Tolerance

Dispersion tolerance is the measure of the penalty of the eye opening at the receiver after the transmission whose dispersion over the number of spans is completely compensated with some extra residual dispersion. This is very important in practice that would specify the maximum length of uncompensated fiber allowable during the installation of the transmission link. Figure 8.42 shows

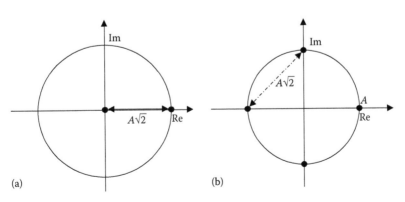

(a) (b)

FIGURE 8.41 Signal constellation of (a) ASK and (b) DQPSK for the same average power.

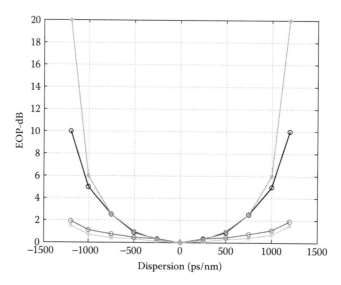

FIGURE 8.42   Measured and simulated dispersion tolerance modulation format of ASK and DQPSK (medium gray asterisks—RZ-DQPSK, black circles—RZ-ASK)—both RZ format of 50% duty cycle and NRZ-ASK (light gray asterisks), NRZ-DPSK (medium gray circles).

the simulated and measured dispersion tolerance of 10 Gb/s RZ-ASK and RZ-DPSK at 20 Gb/s, effectively equivalent in the symbol rate. The tolerance of both modulation formats seems very much the same. This can be explained by observing the spectral distribution of these two modulation formats shown in Figure 8.14. The spectra of RZ-ASK and RZ-DPSK are very much the same except the peak power at the center of the passband of the RZ-ASK, which is understood as the contribution of the amplitude rising from the 1 to 0 or vice versa. The 3-dB bandwidth is thus the same and this contributes to the interference between the sidebands of the modulated lightwave. The RZ-DQPSK at 20 Gb/s is equivalent to that of 10 Gb/s DPSK, and thus we could state that the dispersion tolerance of the RZ-DQPSK would be equivalent to that of the 10 Gb/s DPSK modulation format. A 1-dB penalty of the eye diagram or the eye open penalty (EOP) is at a dispersion of 650 ps/nm for the two modulation formats, equivalent to 650/17 km of SSMF. The EOP of the NRZ-ASK and NRZ-DPSK formats are also plotted against the dispersion to give very much similar pattern as expected from their spectra. Note that for NRZ-DPSK, the data modulator is dual modulator and no chirp is introduced.

8.5.3   Polarization Mode Dispersion Tolerance

Other studies [15–19] have investigated the impacts of PMD on DPSK, DQPSK direct detection formats. In systems with impairments dominated by first-order PMD, smaller duty cycle RZ formats would improve the resilience to the PMD by the ASK and DPSK systems as the ISI is smaller for narrower pulses. RZ-DQPSK is much more resilient to the first-order PMD. DQPSK allows a DGD at least twice higher for the same EOP due to PMD as compared to that suffered by ASK and DPSK.

Systems in which the PMD is compensated, the second-order PMD would be minimized for modulation formats that offer narrower spectra due to the fact that the second order is wavelength dependent and hence narrower spectrum would offer more resilience due to second-order PMD effects.

8.5.4   Robustness toward Nonlinear Effects

8.5.4.1   Robustness toward SPM

The setup for investigation, both experimental and simulation, consists of four spans with 100-km SSMF and dispersion-compensating modules (DCMs) and associated EDF amplifiers at the end of SSMF and at the output of DCM with a booster EDFA and attenuator to adjust the launched input power to the transmission link. Four wavelength channels in the middle of the C band are used. The EOP of a wavelength channel at 1552.95 nm is monitored and plotted against the launched power is shown in Figure 8.43 for RZ-ASK and RZ-DQPSK modulation formats. It is observed that the nonlinear SPM threshold for 1-dB EOP is 3 dBm, which is consistent with a 0.1π phase shift due to the nonlinear phase effect of a fiber with a nonlinear coefficient $n^2 = 2.3 \times 10^{-20}$ m^{-1}. However, the nonlinear threshold for RZ-DQPSK is observed at around 7 dBm. By inspecting the spectra of the ASK and DPSK, we notice that switching the amplitude on and off causes transitional time or equivalently the sampling time of the sequence and the Fourier transformation of the sequence would have spikes of the sampling frequency. This peak power level and hence the average power of the ASK sequence is contained mostly in the carrier peaks. Unlike in DPSK or DQPSK, envelop is constant and the energy is contained mostly in the signal as we could observe no peaks in their spectra. Thus, RZ-DQPSK and RZ-DPSK data sequence would be more tolerant to the SPM effects.

8.5.4.2   Robustness toward XPM

In 40-Gb/s systems, the nonlinear SPM effects dominate the impairments due to fiber nonlinearity. However, in 10-Gb/s DWDM systems, the situation is different, especially for 100-GHz spacing [20]. The XPM is caused by the nonlinear phase effects of a channel in the DWDM system, especially when its intensity is fluctuating, which would cause a phase disturbance in other channels. This PM disturbance would then be transferred to other channels with amplitude modulation. The XPM effects can be considered to contribute to the frequency spectrum via a transfer function [21], as given by

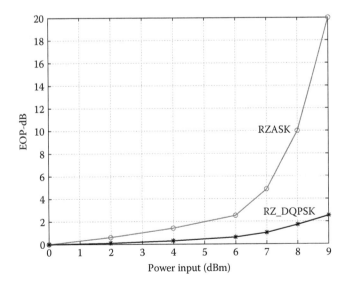

FIGURE 8.43   EOP induced by SPM versus input power at the beginning of the fiber transmission links of 4×100 km fully compensated SSMF spans at a bit rate of 40 Gb/s. RZ-ASK (circle) and RZ-DQPSK (star).

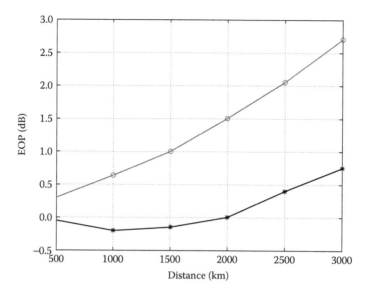

FIGURE 8.44 EOP induced by XPM versus total transmission distance (dispersion compensated and attenuation equalized spans). RZ-ASK (circle) and RZ-DQPSK (star).

$$H_{12} \approx \frac{2\gamma_1}{\alpha - j2\pi f d_{12}} \qquad (8.22)$$

where γ_1 is the nonlinear factor due to the intensity fluctuation of wavelength λ_1 channel, α is the attenuation factor, and $d_{12}(d_{12}z = D(\lambda_1 - \lambda_2)z$ is the group delay difference between the two channels λ_1 and λ_2 over a distance z. Thus, for 100-GHz spacing and over SSMF of $D = 17$ ps/nm/km, the 3-dB passband of this low-pass filter transfer function is at about 540 MHz, which would fall into the signal bands of 10 Gb/s bit rate system.

Figure 8.44 shows the EOP due to XPM of 10 Gb/s 100 GHz spacing DWDM transmission of RZ-ASK and RZ-DPSK over multispan link in which both dispersion and attenuation effects are equalized with DCM and EDFAs. Clearly, the DPSK format is much more tolerant to the XPM effects as compared to the RZ-ASK format. XPM effect could be reduced by using electrical filter of a corner frequency of about 540 MHz with a sharp roll-off factor placed at the output of the electronic preamplifier.

8.5.4.3 Robustness toward FWM

Unlike the impairments due to SPM and XPM mainly caused by the optical power of the channels, FWM effects come from the relative phase difference between the channels. We can summarize the influence of FWM on DPSK and DQPSK signals as follows.

For dispersion-shifted fibers, the phase velocity of different channels around the zero dispersion spectral region is the same and thus phase matching would happen and enhance the generation of the fourth wave, which would fall within the active band of equally spaced DWDM system. Figure 8.45a and b shows simulated results of the EOP of the fourth channel versus the peak and

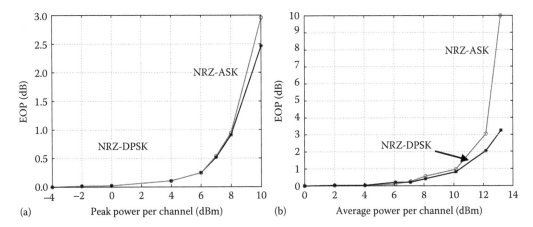

FIGURE 8.45 EOP induced by FWM versus (a) launched power (peak) and (b) average power of three NRZ-ASK and NRZ-DPSK channels after 100 km DSF with zero dispersion wavelength at 1552.93 nm. RZ-ASK (circle) and RZ-DQPSK (star).

average power of three NRZ-ASK and NRZ-DPSK channels with 100-GHz spacing in the region of zero dispersion wavelength of DSF.

8.5.4.4 Robustness toward SRS

The impact of stimulated Raman scattering (SRS) on different modulation formats has been reported [22] including the formats DPSK and DQPSK, which have been shown to significantly improve the resilience to SRS crosstalk. In particular, when the RZ formats are employed, the SRS effects are negligible. The principal reasons are due to the periodic variation of the optical power or field propagating through the fiber; hence, no low-frequency components exist to enhance the SRS effects. The effect is very much like that of the XPM but at a lower frequency. Thus, RZ-DPSK would exhibit no low frequency generated by the SRS. Simulated results of the effects of SRS on the broadening of NRZ pulse ASK modulation over the number of spans of SSMF and nonzero dispersion-shifted fiber (NZ-DSF), completely optically amplified and dispersion compensated, are shown in Figure 8.46. The SRS effects on the pulse broadening of RZ-DPSK is also shown to be significantly undisturbed by the periodic variation of the phase of the carrier, thereby demonstrating strong resilience of the modulation format to SRS. The standard deviation of the pulse amplitude over the average signal level is measured as the percentage.

8.5.4.5 Robustness toward SBS

All DPSK and DQPSK do not exhibit the carrier component at the center of the passband of the spectrum, thus it can be stated that the SBS would play no role in PSK modulation formats.

8.6 REMARKS

This chapter gives the fundamental aspects of the optical transmission of discrete phase modulation formats (DPSK, DQPSK, and ASK) as well as the structures of the transmitters and receivers. Experimental setup for transmission of modulation formatted signals, NRZ, RZ, CSRZ, NRZ-DPSK,

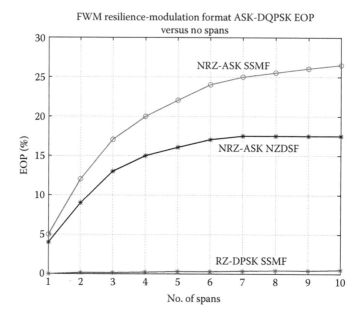

FIGURE 8.46   Percentage broadening of the "1" level (% of pulse period) due to SRS crosstalk for 11 channels 4 nm spacing 10 Gb/s bit rate under SSMF and NZDSF fiber spans. Modulation format NRZ-ASK. Span length = 100 km SSMF fully dispersion compensated.

RZ-DPSK, CS-RZ DPSK, and so on, over the optically amplified dispersion compensating system has been described. The experimental results have been accomplished to investigate the filtering effects on performance of the transmission systems, the dispersion tolerance and receiver sensitivity performance of the RZ-ASK and RZ/CSRZ-DPSK. The filtering properties of the muxes and demuxes do not affect significantly the transmission performance in terms of BER and receiver sensitivity of the 40-Gb/s RZ or CSRZ-DPSK modulation formats.

We have also reported and simulated the nonlinear effects including SRS, SPM, XPM, and SBS on the eye opening at the receiver of the PSK modulation, particularly the DPSK and DQPSK with RZ formats. PSK modulation is much more resilient to the nonlinear effects as compared to ASK.

The negligible mutual effects of adjacent 40-Gb/s DPSK channels on 10-Gb/s ASK-modulated transmission channel enable the feasibility of the upgradeability of the 10-Gb/s current system to 40-Gb/s optical DPSK-modulated transmission. We have also measured the transmission quality of both 40- and 10-Gb/s channels and observed no significant degradation of either channel by the other. Our next step is to launch the 40-Gb/s modulation format over a commercial multiwavelength transport Tranxpress system, as shown in Figure 8.47, and then over an installed multiwavelength 10-Gb/s transmission link between cities, for example, Melbourne and Sydney in Australia.

APPENDIX: MATLAB SIMULINK MODEL FOR DQPSK OPTICAL SYSTEM

A number of models in MATLAB Simulink are given in Figures 8.48 through 8.51. Figure 8.48 shows the block diagram of the whole transmission system, generally an optical transmitter, the optically amplified fiber spans, and the receiver including subsystems for evaluation of BER versus parameters

FIGURE 8.47 A multiwavelength optical transport systems.

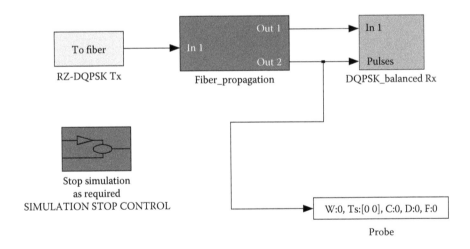

FIGURE 8.48 General schematic diagram of the DQPSK system.

such as receiver sensitivity, eye opening penalty, and so on. Figure 8.49 shows the model of optical transmitters in association with Figure 8.50 as a precoder in electrical domain to generate signals for driving the electrodes of the MZIM. Figure 8.51 gives the model of balanced receiver incorporating MZDI as the phase comparator for DPSK. A sample model of this chapter, including noise models of receivers, can be downloaded from the publisher's Web site www.crcpress.com. Contact the publisher for details of the location. Readers should contact the publisher for details of the location of the Web site to download.

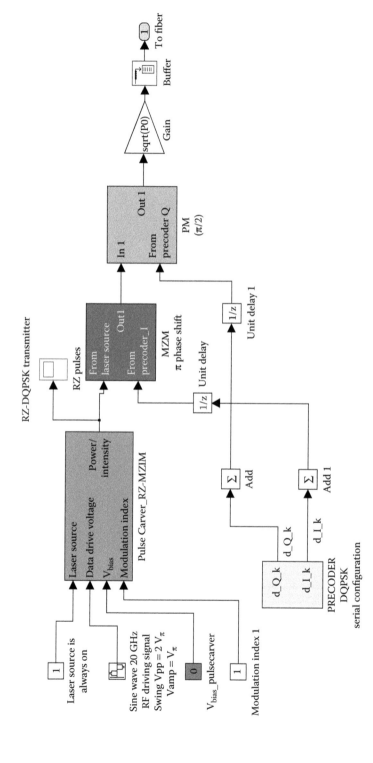

FIGURE 8.49 Simulink model of the optical transmitter for RZ DQPSK.

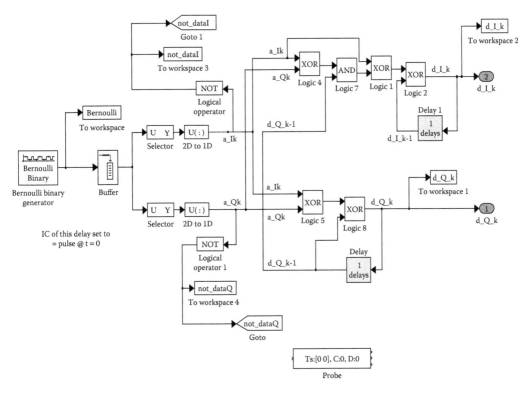

FIGURE 8.50 Simulink model of the electrical precoder for optical transmitter for RZ-DQPSK.

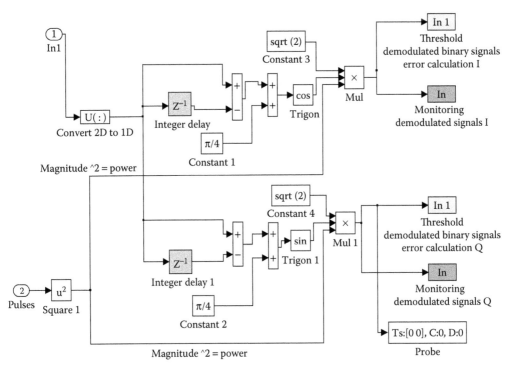

FIGURE 8.51 Simulink model of the MZDI optical balanced receiver for RZ DQPSK.

REFERENCES

1. P. J. Winzer, Optical transmitters, receivers, and noise, in *Wiley Encyclopedia of Telecommunications*, J. G. Proakis, ed. New York: Wiley, 2002, pp. 1824–1840.
2. P. J. Winzer and R.-J. Essiambre, Advanced optical modulation formats, in *Proc. ECOC* , Rimini, Italy, 2003, Invited paper Th2.6.1, pp. 1002–1003.
3. L. N. Binh, Lithium niobate optical modulators, in *Proc. Int. Conf on Material and Technology Symposium M: Material and Devices*, Singapore, July 2005.
4. R. A. Linke and A. H. Gnauck, High-capacity coherent lightwave systems, *J. Lightwave Technol.*, 6(11), 1750–1769, Nov. 1988.
5. J.-X. Cai, D. G. Foursa, C. R. Davidson, Y. Cai, G. Domagala, H. Li, L. Liu et al., A DWDM demonstration of 3.73 Tb/s over 11 000 km using 373 RZ-DPSK channels at 10 Gb/s, *Proc. OFC*, Atlanta, GA, 2003, Postdeadline paper PD22. OFC 2013, Annaheim, CA.
6. J. C. Livas, E. A. Swanson, S. R. Chinn, E. S. Kintzer, R. S. Bondurant, and D. J. DiGiovanni, A onewatt, 10-Gbps high-sensitivity optical communication system, *IEEE Photon. Technol. Lett.*, 7, 579–581, 1995.
7. A. H. Gnauck, X. Liu, X. Wei, D. M. Gill, and E. C. Burrows, Comparison of modulation formats for 42.7-Gb/s single-channel transmission through 1980 km of SSMF, *IEEE Photon. Technol. Lett.*, 16, 909–911, 2004.
8. C. Wree, RZ-DQPSK format with high spectral efficiency and high robustness towards fiber nonlinearities, Technical report, University of Kiel, Germany.
9. C. Wree, Experimental investigation of receiver sensitivity of RZ-DQPSK modulation format using balanced detection, in *Proc. OFC 2003*, Paper ThE5, vol. 2, 2003, p. 456.
10. Y. Okunev. In *Phase and Phase Difference in Digital Communications*. Boston, MA: Artec House, 1997.
11. C. Wree, Differential phase shift keying for long haul fiber-optic transmission based on direct detection, Dr. Ing. Dissertation, 2004, CAU University zu Kiel, Kiel Germany, Figure 4.6 p. 75.
12. S. Benedetto and E. Biglieri. In *Principles of Digital Communications with Wireless Applications*. New York: Kluwer Academics, 1999.
13. K. K. Pang, *Digital Communications*, Lecture notes. Melbourne, Australia: Monash University, 2002.
14. A. H. Gnauck and P. J. Winzer, Optical phase-shift-keyed transmission, *IEEE J. Lightwave Technol.*, 23(1), 115–130, Jan 2005.
15. C Xie, L. Moeller, H. Haustein, and S. Hunsche, Comparison of system tolerance to polarization mode dispersion between different modulation formats, *IEEE Photon. Technol. Lett.*, 15, 1168–1170, 2003.
16. J. Wang and J. M. Kahn, Impact of Chromatic and polarization mode dispersion on DPSK systems using interferometric demodulation and direct detection, *IEEE J. Lightwave Technol.*, 10, 96–100, 2004.
17. H. Kim, C. R. Doerr, R. Pafchek, L. W. Stulz, and P. Bernasconi, Polarization-mode dispersion impairments in direct-detection differential phase-shift-keying systems, *Electron. Lett.*, 38(18), 1047–1048, 2002.
18. R. A. Griffin, R. I. Johnstone, R. G. Walker, S. D. Wadsworth, K. Kerry, A. C. Carter, M. J. Wale, J. Hughes, P. A. Jerram, and N. J. Parsons, 10 Gb/s optical differential quadrature phase shift keying (DQPSK) transmission using GaAs/AlGaAs integration, in *Proc Opt. Fiber Communication OFC*, postdeadline paper FD-6, 2002.
19. R. A. Griffin and A. C. Carter, Optical differential quadrature phase shift keying (oDPSK) for high capacity optical transmission, in *Proc Opt Fiber Communication Conf. OFC*, paper WX-6, 2002.
20. P. Mitra and J. Stark, Nonlinear limits to the information capacity of optical fiber communications, *Nature*, 411, 1027–1070, 2001.
21. C. Wree, Differential phase shift keying for long haul transmission based on direct detection, Dr. Ing. Dissertation, 2002, CAU University zu Kiel, p. 24.
22. S. Schoemann, J. Leibrich, C. Wree, and W. Rosenkranz, Impact of SRS-induced crosstalk for different modulation formats, in *Proc Optical Fiber Conf., OFC*, paper FA5, 2004.

9 Multilevel Amplitude and Phase-Shift Keying Optical Transmission Systems

This chapter presents the modulation formats that combine the amplitude modulation and differential phase modulation schemes, that is, the multilevel amplitude-shift keying (ASK), and phase-shift keying (PSK). Comparisons between multilevel and binary modulations are made. Critical issues of transmission performance for multilevel are identified. A simulation platform based on MATLAB®–Simulink® [1,2] for multi-level modulation format is described. Furthermore, for reaching the 100 Gb/s Ethernet, a number of multilevel modulations such as PSK and orthogonal frequency division multiplexing (OFDM) are proposed and given in the last section of the chapter.

9.1 INTRODUCTION

Under the conventional on-off keying (ASK) modulation format, the transmission bit rate beyond 40 Gb/s per optical channel is very costly, because the electronic signal processing technology may have reached its fundamental speed limit. It is expected that advanced photonic modulation formats such as M-ary ASK and differential PSK would replace ASK in the near future. These advanced formats would offer efficient spectral properties and would thus be able to increase transmission rate without placing stringent requirements on high-speed electronics and to use the same existing photonic communication infrastructure.

Coherent communication system developed in the mid 1980s has extensively exploited different modulation techniques to improve the optical signal-to-noise ratio (OSNR). However, the coherence detection has faced considerable difficulties due to the stability of the source spectrum and the laser linewidth for a mere gain in the receiver sensitivity of 3 dB for heterodyne detection and 6 dB for homodyne detection in order to extend the repeaterless distance 60–80 km of standard single-mode fiber (SSMF).

The invention of the optical amplifiers (OAs) in the early 1990s has overcome the fiber attenuation limit, and thus they offer a significant improvement in optical transmission technology. Owing to this, ultra-long-haul, ultra-high-capacity optical transmission systems have been deployed widely around the world in the last decade. The technology has been matured, with ASK modulation reaching 10 Gb/s per optical channel, total channel count of hundreds, and 100/50 GHz channel spacing.

Based on the proven efficient spectra and transmission technology, especially the controllable total dispersion of the transmission and compensating fibers, it is much more advantageous that these spectral regions be efficiently used. Therefore, the contribution of advanced modulation techniques and formats would offer higher spectral efficiency for photonic transmission.

Furthermore, digital modulation techniques have been well established over the last half century with amplitude, frequency, or phase modulations. These techniques, especially phase modulations, relying principally on the detection schemes—that is, on whether they are coherent or pseudo-coherent differential detections—have been heavily exploited in wireless communication networks. In the photonic domain, for a long time, the technological difficulties associated with manufacturing narrow linewidth lasers have prevented the use of coherent and differential phase modulations. Only over the last several years, owing to the maturity of the laser technology, particularly the successful development of distributed feedback laser, laser linewidth has reached a level that is much smaller than the modulation bandwidth. The coherence of the sources is now sufficient for differential phase modulating and detecting applications, which require the phase of the sources to remain stable over at least two consecutive symbol periods.

Recently, advanced modulation techniques have attracted significant interests from photonic transmission research and system engineering community. Several modulation schemes and formats such as binary differential phase-shift keying (BDPSK), differential quadrature phase-shift keying (DQPSK), duobinary ASK associated with nonreturn-to-zero (NRZ), return-to-zero (RZ), carrier-suppressed return-to-zero (CSRZ) formats have been widely studied. However, what have not been widely explored are optical multilevel modulation schemes. Although multilevel schemes have been intensively exploited in wireless communications, there are minimum works to date that incorporate the incoherent multilevel optical amplitude- and phase-shift keying modulation schemes [3], which offer the following advantages: (a) Lower symbol rate: Hence, for the same available spectral region, a multilevel modulation scheme would offer a transmission capacity higher than binary modulation counterparts. (b) Efficient bandwidth utilization: Photonic transmission of these multilevel signals could be implemented over the existing optical fiber communications infrastructure without significant alteration of the system architecture, thus saving the cost of capital investment and easing the system management. (c) The complexity of the coder and demodulation subsystems falls within the technological capabilities of current microwave and photonic technologies.

The principal objectives of this chapter are as follows: (1) To evaluate different modulation and coding techniques and signal pulse formats for long-haul, ultra-high-capacity transmission, thus determining novel modulation schemes, the multilevel amplitude- and phase-shift keying, and others for the research studies. (2) To develop analytical, simulation, and experimental testbeds to demonstrate the uniqueness and superiority of our novel schemes. (3) To unveil the principal directions for photonic modulation and transmission technologies for the next transmission generation by a comparative study of the modulations formats. (4) To demonstrate the effectiveness and superiority of a novel photonic communication system on its counterparts, based on advanced multilevel optical modulation format and implementation of the system on Simulink platform. The aim is to demonstrate it as a useful platform for desktop computer simulation.

Hence, a conceptual photonic transmission system is proposed, based on a hybrid technique that combines the phase and amplitude modulations, the multilevel amplitude-differential phase-shift keying (MADPSK) format. This technique combines two modulation formats, the well-known M-ary ASK and the M-ary DPSK, to take the advantages of high receiver sensitivity and dispersion tolerance (DT) (DPSK) and the enhancement of total transmission capacity (M-ASK) as compared with the traditional ASK format.

The models of MADPSK transmitter and receiver have been structured for MADPSK signaling. A simulation model based on MATLAB–Simulink platform has been developed for the proof of concept. The system performance is evaluated for back-to-back and long-haul transmission. Analytical and simulation results of the transmission configurations are demonstrated. The followings are presented: (1) Noise mechanisms, for example, quantum shot noises, quantum phase noises, optically amplified noises, noise statistics, nonlinear phase noises (NLPNs); hence, design of an optimum detection and decision-level schemes for MADPSK; (2) Linear, nonlinear, and polarization dispersion impairments and their impacts on MADPSK system performance; (3) Matched filter design for optimum MADPSK signal detection; (4) Offset MADPSK (O-MADPSK) modulation schemes; (5) Multilevel amplitude-minimum shift keying (MAMSK) modulation; (6) MADPSK modulation for applications in subcarrier transmission systems, especially for metropolitan wide-area multiadd/drop networks; and (7) Other issues or additional modulation formats suitable for MADPSK.

This chapter is thus organized as follows: Section 9.1 gives a brief review of a number of advanced photonic modulation formats. Section 9.2 reviews and compares different modulator structures used for generating advanced photonic modulation signals and emphasizes the advantages of dual-drive Mach–Zehnder intensity modulator (MZIM) as modulator for generating MADPSK signal, the main object of the research. In Section 9.4, a novel photonic transmission system with MADPSK modulation format is proposed. Section 9.5 summarizes the preliminary works and results. Section 9.3 identifies a number of critical issues and alternative multilevel signaling for optical systems.

9.1.1 AMPLITUDE AND DIFFERENTIAL PHASE MODULATION

9.1.1.1 Amplitude Shift Keying Modulation

9.1.1.1.1 NRZ-ASK Modulation

ASK has been the dominant modulation technique from the early days of optical communications. The main advantage of this modulation is that ASK signal is not sensitive to the phase noise. The ASK modulation can take two principal formats: The first one is called NRZ-ASK, in which the "1" optical bit occupies the whole bit period, and the second one, RZ-ASK, has the "1" bit presented in only the first half of the bit period.

Figure 9.1 shows the spectrum of a 40 Gb/s NRZ-ASK signal, with the carrier seen at the highest peak and the 3 dB bandwidth reaching the bit rate. The main advantage of NRZ-ASK signal is that its spectrum is generally the most compact compared with that of other formats such as RZ-ASK and CSRZ-ASK. On the other hand, however, NRZ-ASK signal is affected by fiber chromatic dispersion (CD) and is more sensitive to fiber nonlinear effects as compared with its RZ-ASK and CSRZ-ASK counterparts.

9.1.1.1.2 RZ-ASK Modulation

An RZ-ASK signal, shown in Figure 9.2, is similar to an NRZ-ASK signal, except for the "1" bit occupying only the first half of the bit period. This signal can be generated by a transmitter shown in the same figure in which an NRZ-ASK transmitter is followed by a pulse caver driven by a pulse train synchronized with the data source. The pulse train has frequency equal to the data rate. The RZ-ASK pulse width can take the form of 33%, 50%, 66% duty ratio. Because of its narrower pulse width, the spectrum of RZ-ASK signal, shown in Figure 9.3, is larger than that of NRZ-ASK signal, hence leading to less spectrum efficiency. In this spectrum, the carrier is seen as highest peak; the two side peaks are RF-modulating signals positioned 80 GHz apart.

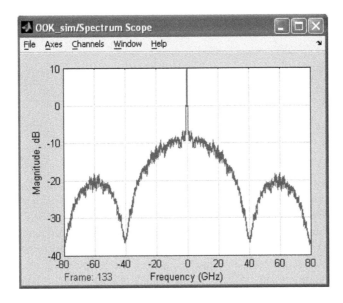

FIGURE 9.1 Spectrum of 40 Gb/s NRZ-ASK signal.

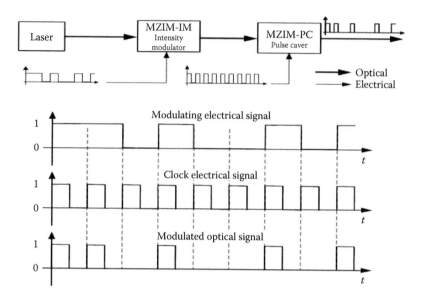

FIGURE 9.2 RZ-ASK transmitter and electrical signalling for generation of optical signal.

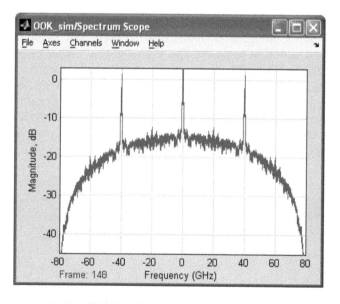

FIGURE 9.3 Spectrum of 40 Gb/s 50% RZ-ASK signal.

9.1.1.1.3 CSRZ-ASK Modulation

The CSRZ-ASK modulation format is similar to standard RZ-ASK modulation format, except that the neighboring optical pulses have π-phase difference. The carrier in neighboring time slots is thus cancelled out and effectively excluded from the signal spectrum. The CSRZ-ASK signal can be generated by a transmitter, with scheme shown in Figure 9.4. In this scheme, the first MZIM modulates the intensity of optical signal coming from a laser source, whereas the second MZIM, driven by a clock signal at the haft data rate, caves the NRZ pulses into RZ pulses. Because the second MZIM is biased at the minimum-intensity point, it provides an RZ pulse train at the data rate with alternating phases 0 and π for neighboring time slots. The CSRZ-ASK signal can be also detected by a direct-detection receiver, as it would not be phase sensitive.

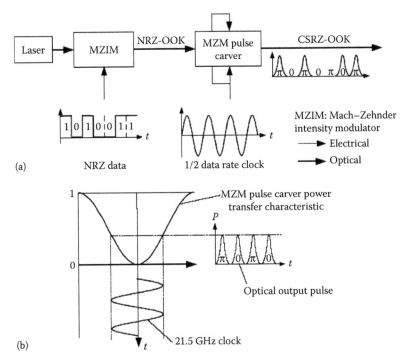

(a)

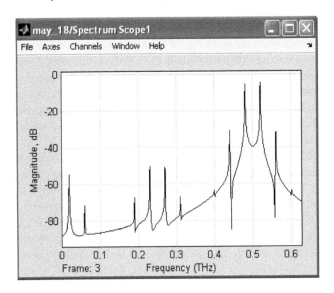

(b)

FIGURE 9.4 (a) Block diagrams of CSRZ-ASK transmitter and (b) generation of optical pulse with alternative phase using biasing control and amplitude. (Adapted from L.N. Binh, *Advanced Digital Optical Communications*, 2nd edn., CRC Press, Boca Raton, FL, 2015.)

The main advantages of CSRZ-ASK include narrower spectrum, higher tolerance to dispersion, and stronger robustness against fiber nonlinear effects as compared with standard RZ-ASK. Because its peak optical power is much lower than that of other formats, it is less affected by both self-phase modulation (SPM) and crossphase modulation (XPM). Figure 9.5 shows the spectrum of 40 Gb/s CSRZ-ASK signal with very low level of carrier power.

FIGURE 9.5 Spectrum of 40 Gb/s CSRZ-ASK. (Extracted from L.N. Binh, *Advanced Digital Optical Communications*, 2nd edn., CRC Press, Boca Raton, FL, 2015.)

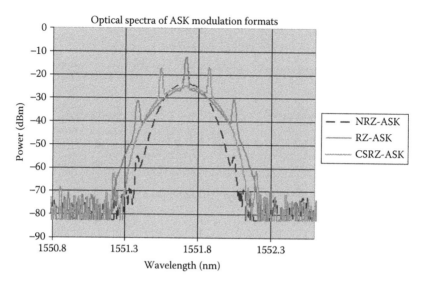

FIGURE 9.6 Spectrum of NRZ-ASK, RZ-ASK, and CSRZ-ASK signals.

ASK is a modulation technique that generates a signal $s(t)$ by multiplying a digital signal $m(t)$ by a carrier f_c:

$$s(t) = Am(t)\cos 2\pi f_c t; \quad \text{for } 0 < t < T, \tag{9.1}$$

where A is amplitude envelop and digital signal $m(t)$ may take one of M levels $\left[b_0, b_1, ..., b_M \right]$. When $M = 2$, $s(t)$ is a binary ASK signal with ASK as a special case. ASK is also implemented in NRZ, RZ, and CSRZ formats, whose spectra are shown in Figure 9.6 in the same graph for the purpose of comparison. Like their ASK analogs, NRZ-ASK has the most compact spectrum, whereas RZ-ASK has the broadest spectrum. In term of energy, CSRZ-ASK has the lowest peak power because the carrier signal has been effectively removed.

9.1.1.2 Differential Phase Modulation

Under ASK/ASK modulation schemes with the associated NRZ, RZ, CSRZ formats, the amplitude of optical carrier varies accordingly. Phase modulation, on the other hand, modulates carrier phase and thus facilitates the use of bipolar signals "±1". This distinguished feature means that phase modulation offers significant improvement in receiver sensitivity as compared with ASK modulation. With the recent advancement in photonic lightwave technology, especially integrated optic delay interferometer (DI), differential phase modulation and demodulation and balanced receiver have become realizable. This section gives a brief overview on the differential modulation techniques and their implementations in photonic domain, especially the MADPSK.

The term NRZ-BPSK, or traditionally NRZ-DPSK, is commonly used for denoting a modulation technique in which optical carrier is always present with a constant power and only its phase is alternated between 0 and π. The modulation rule is as follow: (1) At the transmitter, initially, a reference "0" bit is entered as the present encoded bit. Then, the next data bit is compared with the present encoded bit. If they are different, then the next encoded bit is "1," for which a phase change of π occurs, else the next encoded bit is "0," which causes no (or 0) phase change. (2) At the receiver, the phase of the carrier at the present bit slot is compared with that of the previous one. If the phase difference is π, then the data is decoded as "1"; otherwise, the data is "0" when phase difference is 0.

One of the NRZ-DPSK transmitter structures is shown in Figure 9.7. User data are first encoded by a differential encoder into the driving voltage, which then alternates phase of the carrier signal between 0 and π. In detecting an NRZ-DPSK signal, a delay Mach–Zehnder interferometer (MZI)

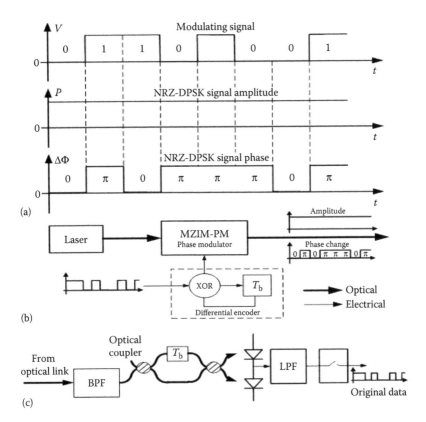

FIGURE 9.7   (a) NRZ-DPSK signal, (b) transmitter, and (c) receiver. (Adapted from L.N. Binh, *Advanced Digital Optical Communications*, 2nd edn., CRC Press, Boca Raton, FL, 2015.)

in combination with balanced optoelectronic receiver can be used. The interferometer acts as the phase comparator with constructively and destructively interfered outputs. As shown in Figure 9.7, the received optical signal is split into two arms of an MZI, one of which has a one-bit optical delay. The MZI compares the phase of each bit with the phase of the previous bit, and the photodetector (PD) converts the phase difference to intensity. When there is no phase shift between two bits, they are added constructively and give maximum rise to the output signal; otherwise, they cancel out when the phase shift is equal to π. If the differential phase shift is Δϕ, then the differential current at the output of the balanced PD can be written as

$$i = A^2 \cos \Delta\phi \tag{9.2}$$

Because the balance receiver uses both constructive and destructive ports of the MZI, the detected signal level can swing from "1" to "−1." Compared with ASK or with the use of unbalanced receiver, where signal amplitude is limited between "1" and "0," DPSK can offer a 3 dB improvement in receiver sensitivity.

Owing to its constant envelop, NRZ-DPSK signal is less sensitive to power modulation-related nonlinear effects, such as SPM and XPM, than its NRZ-ASK counterpart. On the other hand, however, long-haul DPSK systems, including both NRZ and RZ, with OA are affected by NLPN. Amplified spontaneous emission (ASE) noise of OAs is converted into phase noise, leading to the waveform distortion and, consequently, signal degradation. The spectrum of NRZ-DPSK signal is shown in Figure 9.8, together with other DPSK formats. It can be seen that NRZ-DPSK signal has the most compact spectrum compared with other DPSK formats. This can be explained by the fact that the NRZ-DPSK signal amplitude remains constant, regardless of whether bit "1" or bit "0"

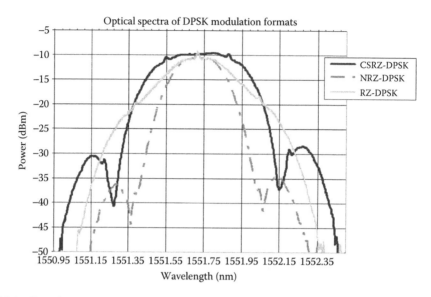

FIGURE 9.8 Experimentally measured spectra of NRZ-DPSK, RZ-DPSK, and CSRZ DPSK signals.

is transmitted, and thus, the energy is distributed more equally when comparing with RZ-DPSK and CSRZ-DPSK signals.

The RZ-DPSK format is similar to NRZ-DPSK format, with the only difference that instead of constant optical power, pulse narrower than bit period appears in each bit slot, as shown in Figure 9.9. However, the RZ-DPSK transmitter resembles an RZ-ASK transmitter with the phase modulator (PM) and replaces the intensity modulator (IM). The RZ-DPSK signal can also be detected by the same receiver used for NRZ-DPSK signal. Owing to its narrow pulse, RZ-DPSK format is expected

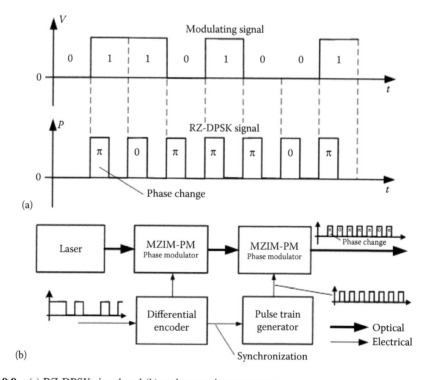

FIGURE 9.9 (a) RZ-DPSK signal and (b) and transmitter structure.

to minimize the effects of intersymbol interference and thus is capable of achieving a longer transmission distance. Narrow pulse, however, spreads spectrum of RZ-DPSK signal wider than that of NRZ-DPSK, making RZ-DPSK systems more susceptible to CD. To reduce the effect of this impairment, CD compensation devices are used.

The RZ-DPSK signal energy does not distribute equally, as in the case of NRZ-DPSK. Most of it concentrates in only a fraction of bit duration, while reducing nearly to zero for the rest of time. This large energy fluctuation makes the signal more susceptible to fiber nonlinearity and signal detection more difficult.

The carrier suppression technique can also be used in conjunction with RZ-DPSK modulation to produce CSRZ-DPSK signal, which has been demonstrated as one of the most attractive modulation formats in high spectral efficiency wavelength division multiplexing (WDM) and dense WDM systems.

It is due to higher energy and spectral efficiency, increased tolerance to fiber nonlinearity induced impairments, and increased CD and polarization mode dispersion (PMD) of CSRZ-DPSK as compared with the RZ-DPSK counterpart. The spectra of CSRZ-DPSK, RZ-DPSK, and NRZ-DPSK are shown together in Figure 9.8 for comparison.

The CSRZ-DPSK signal can be generated by a transmitter whose scheme, shown in Figure 9.10, can consist of the ASK parts similar to that of CSRZ-ASK and an additional phase modulator (PM). The main difference is for CSRZ-DPSK transmitter a PM is replacing the intensity modulator (IM) in the CSRZ-ASK transmitter. The receiver for CSRZ-DPSK has the same structure as that of the NRZ-DPSK reception scheme.

To increase transmission bit rate without suffering bandwidth requirement, one can code more than one bit into a data symbol. The DQPSK modulation is the first step in the realization of this idea.

A signal constellation or signal space (Figure 9.11) is the best way to represent a DQPSK signal, in which the points representing phase-modulated signals are located in two orthogonal axes called I and Q (for in-phase and quadrature components, respectively). Each of the two data bits $[D_0 D_1]$

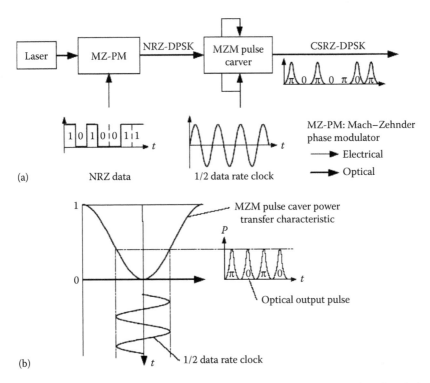

FIGURE 9.10   (a) Block diagrams of CSRZ-DPSK transmitter and (b) and generation of optical pulse with alternative phase by driving the dual-drive MZIM with a 2 V_π voltage swing.

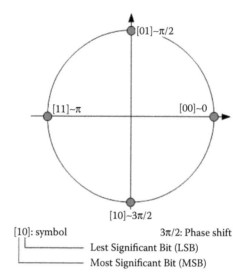

[10]: symbol $3\pi/2$: Phase shift
       └────────── Lest Significant Bit (LSB)
       └────────── Most Significant Bit (MSB)

FIGURE 9.11 DQPSK signal constellation.

are first pre-encoded into a symbol and then the symbol is encoded into phase shift, which may take one of the four values $[0, \pi/2, \pi, 3\pi/2]$, depending on the bit combination it represents. The DQPSK symbol rate is thus equal to only half of the bit rate. Intuitively, one can say that with the same bandwidth available, DQPSK can offers twice transmission capacity compared with ASK and binary DPSK counterparts.

The DQPSK signal can be generated by a transmitter shown in Figure 9.12. This structure consists of two MZIMs connected in parallel. A $+\pi/2$ phase shift is introduced in one of these MZIMs, making optical signals in two paths orthogonal to each other. A precoder encodes user

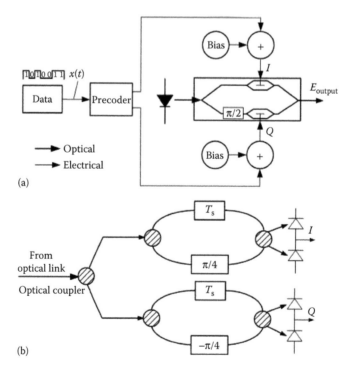

FIGURE 9.12 Parallel structure of DQPSK transmitter, T_s = symbol duration.

TABLE 9.1
DQPSK Signal Bit-Phase Mapping

D_1	D_0	I	Q	Phase Shift
0	0	0	0	0
0	1	0	1	$\pi/2$
1	1	1	1	π
1	0	1	0	$3\pi/2$

data in accordance with the differential rule to generate I- and Q- driving voltages, which then modulate carrier's phase in two optical paths. Modulated carrier components are then combined at the output of the MZI. If the two normalized driving signals are denoted by I and Q, respectively, then the output signal is

$$E_{\text{output}} = I \cos 2\pi f_c t + Q \sin 2\pi f_c t, \qquad (9.3)$$

where f_c is the frequency of optical carrier. The coding and mapping bits $[D_1, D_0]$ into I and Q and signal constellation points follow the rule (Table 9.1).

The DQPSK receiver uses two set of MZDI and balance receivers to detect in-phase (I) and quadrature-phase (Q) components of the received signal; each set is similar to the one used in NRZ-DPSK receiver. There are, however, two main differences: First, the delay introduced in the first branches of interferometers is now replaced by the symbol duration T_s; second, the phases of signal in the second branches are shifted by $+\pi/4$ and $-\pi/4$ for I- and Q-components, respectively. These additional phase shifts are needed to separate two orthogonal-phase components, I and Q.

Figure 9.13a shows the spectrum of 40 Gb/s NRZ-DQPSK signal with the single-sided bandwidth of the main lobe equal 20 GHz, which is only half of the transmitted bit rate. The spectra of RZ-DQPSK signal, Figure 9.13b, is much broader, with strong harmonics beside the main lobe.

Despite the numerous advancements in optical modulation techniques, the number of levels encoded in a signal symbol falls far behind 256 or 1024 achieved in microwave modulation schemes. The phase noises associated with optical sources and OAs have hindered the use of phase-related modulation schemes to current fluctuations in the photodetection, thus resulting in the degradation of the bit-error rate (BER). Differential phase demodulation process based on the phase comparison of two consecutive symbols requires that the phase should remain stable over two symbol periods. Thus, narrow linewidth lasers are critical for phase-modulated systems. It has been shown that to achieve a power penalty less than 1 dB, $\Delta v/B < 1\%$, with Δv and B indicating the laser linewidth and system bit rate, respectively. In optical transmission systems where OAs are used, the ASE noises intermingle with the fiber nonlinear phase effect, thus enhancing the NLPN. The SPM-induced NLPN is dominant phase noise in single optical channel systems, whereas the XPM-induced phase noise is main phase noise in multichannel (WDM) systems.

Significant phase noises caused by optical sources and OA have prevented optical DPSK schemes from having many levels in each symbol. To increase the number of levels in the signal space, and thus, the number of bits per symbol, to more than four, one of the most preferred solution is a combination of DQPSK and ASK.

Recently, Hayase et al. [3] have demonstrated experimentally that a 30 Gb/s 8-states per symbol optical modulation system using a combined ASK and DQPSK modulation scheme, as shown in Figure 9.14, maps three bits into a symbol, thus creating transmission bit rate three times higher than symbol rate. The transmitter consists of two cascaded PMs and an amplitude modulator (AM). The first PM, driven by data bit D_0, creates 0 and π phase shifts, whereas the second PM, driven by D_1, forces two further phase shifts, 0 and $\pi/2$, the quadrature phase to generate four distinct phases of

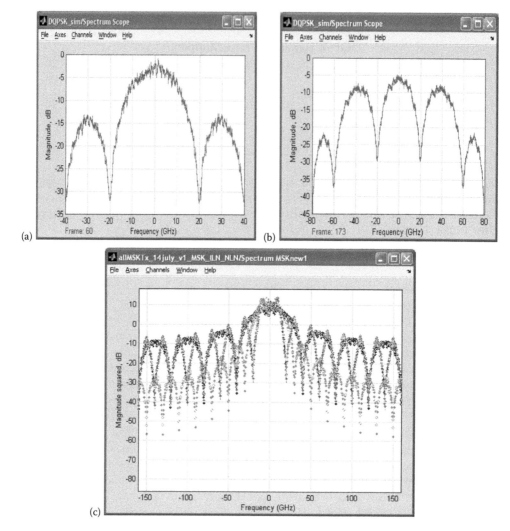

FIGURE 9.13 Optical spectra of 40 Gb/s (a) NRZ-DQPSK, (b) 50% RZ-DQPSK, and (c) DPSK, as compared with MSK (light-gray curve).

DQPSK signal. The AM, driven by D_2 bit, shifts the four phases between two amplitudes to create in total eight signal points.

At the receiver side, optical signals are detected in both amplitude and differential phases. An ASK demodulator detects the D_2 bit. The other is a DQPSK demodulator and detects D_1 and D_0 bits. Sekine et al. [4] reported an experimentally similar scheme, but with four bits $[D_3, D_2, D_1, D_0]$ mapped into a symbol; $[D_1, D_0]$ bits are used to generate a "normalized" DQPSK signal, whereas $[D_3, D_2]$ bits manipulate the amplitude of this DQPSK signal between four concentric circles. Thus, a 16-ary MADPSK signal can be generated. This would offer 40 Gb/s bit rate with the symbol rate of only 10 Gbauds.

9.1.2 COMPARISON OF DIFFERENT OPTICAL MODULATION FORMATS

Different amplitude and phase optical modulation formats are summarized in Table 9.2. In most cases, NRZ-ASK parameters are used as references. From the comparison, it can be concluded that MADPSK takes advantage over other modulation formats in term of spectral efficiency

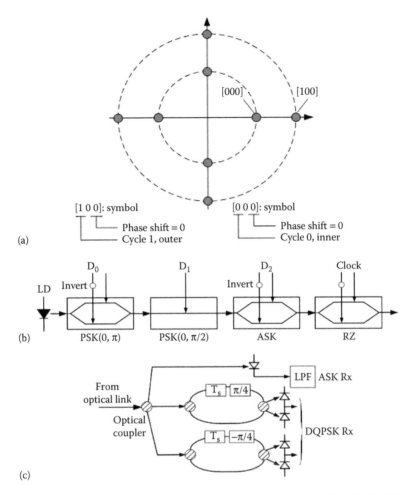

FIGURE 9.14 8-ary APSK modulation experimental configuration (extracted from 0), 10 GHz clock assign synchronization of symbol rate, data modulator, and quadrature phase shift in optical domain using the PM; two balanced receivers for differential phase-shift detection and direct detection for amplitude detection. (a) 8-ary ASK-DPSK signal, (b) transmitter configuration, and (c) receiver configuration. (Adapted from L.N. Binh, *Advanced Digital Optical Communications*, 2nd edn., CRC Press, Boca Raton, FL, 2015.)

and ability to significantly increase transmission bit rate, which are very, if not most, important parameters for an optical transmission system. It is also expected that MADPSK inherits good properties (and, of course, the bad ones—if any) from two basic ASK and DPSK modulation formats. The spectra of different modulation formats such as NRZ-DQPSK, 67% CSRZ-DQPSK, 100 Gb/s CSRZ 16-ADPSK are shown in Figure 9.15, indicating their spectral efficiencies.

9.1.3 MULTILEVEL OPTICAL TRANSMITTER

In this section, several optical transmitter structures used for generating DQPSK signal are described. It is necessary because based on DQPSK modulation format, a novel optical transmission system will be developed. All these structures have MZIM as their base component, which can be single- or dual-electrode structure.

Unlike single-drive MZIM, a dual-drive electrode structure with two traveling wave RF electrodes can modulate the phase of optical signals in both of its branches, resulting in push–pull operation. Interference at the output of dual-drive MZIM will produce phase-modulated signal. However, when the effects of phase modulation in the two branches are exactly equal but opposite

TABLE 9.2

Comparison of Different Optical Modulation Formats

Modulation Format	Spectral Width	Receiver Sensitivity	Resilience to Dispersion	Resilience to SPM	Resilience to XPM	Current Transmission Bit Rate Limits
NRZ-ASK	Narrowest	Lowest	Worst	High	High	40 Gb/s
RZ-ASK	2xRZ-ASK (at 50% duty ratio)	Higher than NRZ-ASK	Higher NRZ-ASK	High	High	40 Gb/s
CSRZ-ASK	Same as RZ-ASK	Higher than NRZ-ASK	Higher than NRZ-ASK	High	High	40 Gb/s
NRZ-DPSK	Same as NRZ-ASK	3 dB better than NRZ-ASK	Higher NRZ-ASK	Worse than ASK	Worse than ASK	40 Gb/s
RZ-DPSK	Same as RZ-ASK	3 dB better than RZ-ASK	Higher NRZ-ASK	Worse than ASK	Worse than ASK	40 Gb/s
CSRZ-DPSK	Same as CSRZ-ASK	3 dB better than CSRZ-ASK	Higher CSRZ-ASK	Higher than ASK	Higher than ASK	40 Gb/s
DQPSK	1/2 DPSK	1.5 dB better than ASK	UR*	Worse than ASK	Worse than ASK	2xDPSK
MADPSK	1/M DPSK					$(\log_2 M)$ xDPSK – expected

in sign, the output signal becomes intensity modulator. In this manner, dual-drive can be used for both phase and intensity modulations. The relationship between input and output signals of a dual-drive MZIM can be described by

$$E_{\text{output}} = \frac{E_{\text{input}}}{2} \left[\exp\left(j\pi \frac{V_1(t)}{V_\pi} \right) + \exp\left(j\pi \frac{V_2(t)}{V_\pi} \right) \right] \qquad (9.4)$$

where:

$V_1(t)$ and $V_2(t)$ are driving voltages applied to modulator

V_π is voltage required to provide a π phase shift of the carrier in each branch of MZIM

It is also noted here that unlike in single-drive MZIMs, chirp effect does not exist in dual-drive MZIMs.

The transmitter structure shown in Figure 9.12 is called parallel type. It is only one of the several structures, namely parallel structure, serial structure, single-PM structure, and dual-drive MZIM structure, that can be used specifically for generating DQPSK signal. These terms are used to indicate the structuring of MZIMs, whether they are connected in tandem, parallel, or just a pure PM with a single-electrical-drive port. Another possible modulation structure for generating optical multilevel amplitude and phase is shown in Figure 9.16. It is an electro-optic transmitter of the serial type: an MZIM-generating in-phase component and a PM-generating quadrature component are connected in tandem. Pre-encoded data generate two signals: one is used for driving the MZIM and the other for driving the PM. Usually, the square shapes of the pre-encoded waveforms are replaced by the raised cosine one before being fed to the modulators. Furthermore, the biasing conditions and the amplitude of the modulators can be used to generate 33% to 67% pulse width RZ formats. It is also noted that the pulse shape would follow a $\cos^2$ profile due to the property of the IM. This transmitter would suffer the chirping effects due to the rise time of the electrical driving signals and hence would contribute to the distortion of the lightwave signals, in particular when switching between the lowest level and the highest level.

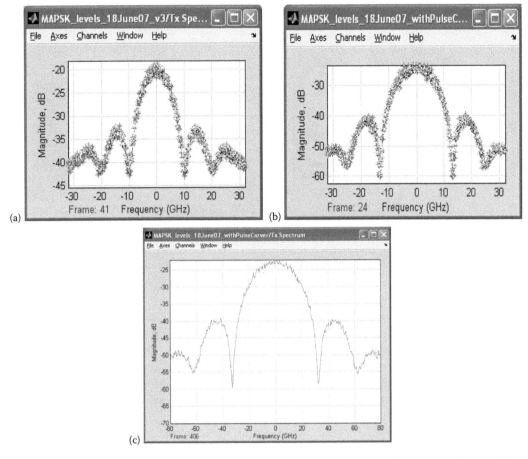

FIGURE 9.15 Optical Spectra of 40 Gb/s: (a) NRZ-DQPSK, (b) 67% CSRZ-DQPSK, and (c) 100 Gb/s CSRZ 16-ADPSK.

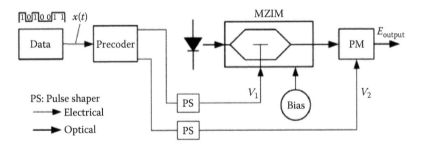

FIGURE 9.16 Cascade PM and MZIM for DQPSK signal generation.

A typical multilevel eye diagram of a multilevel modulation scheme is shown in Figure 9.17. The accuracy and noise levels are critical to ensure the degree of eye opening and the distance of the eye levels. Owing to the threshold of the nonlinear effects in the transmission fibers, the maximum level of a multilevel level must not exceed, so there is a maximum eye opening in multilevel amplitude and phase. Thus, minimum noise contributions due to electronics and optical energy at the reception subsystems as well as at the transmitter must be minimized.

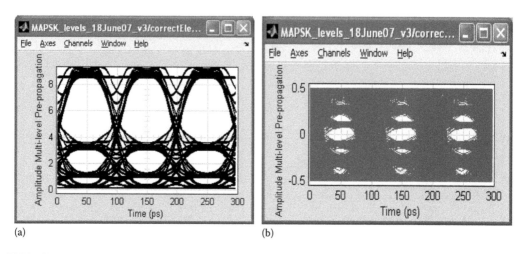

(a)   (b)

FIGURE 9.17 Eye diagram—amplitude detection section of 40 Gb/s after transmission of (a) 5 km SSMF transmission under direct detection and (b) quadrature phase—coherent detection (CoD) without digital signal processing.

The single-PM structure, Figure 9.18, uses only one MZIM as PM. Pre-encoded data are added up to create a single-driving voltage. One of the two pre-encoded data is amplified, and together with the other signal, it represents four positions of DQPSK signal.

The dual-drive MZIM structure, Figure 9.19, uses two driving voltages for modulating optical carrier phase in two branches of an MZIM. Data are first pre-encoded following the differential rule and then used to create driving voltages $V_1(t)$ and $V_2(t)$, corresponding to the signal constellation points.

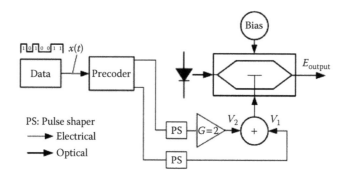

FIGURE 9.18 Single-drive PM structure for MZIM.

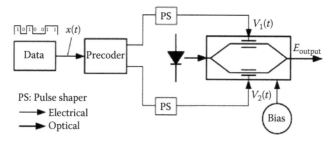

FIGURE 9.19 Dual-drive MZIM structure.

TABLE 9.3

Comparison of DQPSK Transmitter Structures

Parameters for Comparison	Parallel MZIM	Serial MZIM & PM	Single PM	Dual-Drive MZIM
Complexity of circuit design	Complicated in matching of ultrahigh-frequency electrical paths; high insertion loss. Flexible in biasing.	Complicated in matching of ultrahigh-frequency electrical paths; high insertion loss. Flexible in biasing.	Simple in photonic but complicated in realization of ultrahigh-frequency signal connections.	Simplest but require multilevel voltage switching at symbol rate (microwave speed).
Ability to create MADPSK signal	Not possible. A separate ASK modulator required.	Not possible. A separate ASK modulator required	Impossible. A separate ASK modulator required.	Dual-drive MZIM acts as ASK and DPSK simultaneously.

In the four transmitter structures described above, the parallel and serial structures are the most complex and difficult to implement because they have discrete devices connected together. On the other hand, the dual-drive MZIM and single-PM structures are much simpler because they require less discrete devices. Furthermore, as it will be shown in the next section, dual-drive MZIM can be configured to work as both PMs and AMs at the same time, so it can easily generate not only DQPSK but also, generally, MADPSK signals. Thus, a dual-drive MZIM is the principal part of the MAPSK transmission system. Table 9.3 gives a comparison between different transmitter structures; the dual-drive modulator is outstanding as a combined amplitude and phase switching between the states of a multicircular constellation.

9.1.3.1 Single Dual-Drive MZIM Transmitter for MADPSK

The main reason of explaining why the dual-drive MZIM structure has attracted our attention in this chapter is that it can play the role of both AM and PM simultaneously, which is impossible with other transmitter structures. This means that to generate an MADPSK signal, there is no need to employ separate PMs and AMs, as has been implemented in the works of Sekine et al. [4] 0 and Hayase et al. [3]. This section describes a method for generating 16-ary MADPSK signal by using this dual-drive MZIM structure.

The 16-ary MADPSK signal constellation of interest is shown in Figure 9.20. It is actually a combination of a 4-ary ASK and a DQPSK signal, in which four bits $[D_3, D_2, D_1, D_0]$ are mapped into a symbol; among them, two bits $[D_1 D_0]$ are coded into four phases $[0, \pi/2, \pi, 3\pi/2]$ and two bits $[D_3 D_2]$ are coded into four amplitude levels $[I_3, I_2, I_1, I_0]$. With the use of balanced receiver and DI, which is a solely available practical optical phase demodulator today, the MADPSK signal produces clear DQPSK eye patterns centered at zero-voltage decision level, only when constellation points are positioned in a radial pattern.

Recall that the signal at the output of the dual-drive MZIM can be represented as

$$E_o = \frac{E_i}{2} e^{j\phi_1} + \frac{E_i}{2} e^{j\phi_2} \tag{9.5}$$

with $\phi_1 = \pi V_1(t)/V_\pi$, $\phi_2 = \pi V_2(t)/V_\pi$, where E_i and E_o are electrical fields of the input and output optical signals, respectively; $V_1(t), V_2(t)$ are driving voltages applied to paths 1 and 2, respectively, of the modulator, and V_π is the voltage required to provide a π phase shift of the carrier propagating through the MZIM.

Equation 9.7 suggests that with properly chosen input signal E_{input} and driving voltages $V_1(t), V_2(t)$, all signal points of the constellation in Figure 9.20 can be constructed from two phasor signals

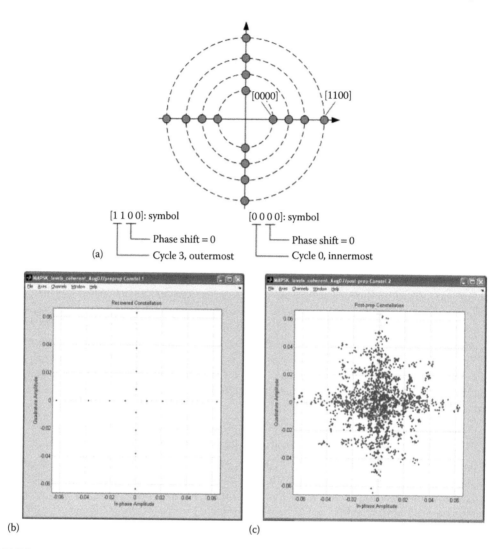

FIGURE 9.20 16-ary MADPSK signal bit-phase mapping: (a) design, (b) Simulink scattering plot before transmission, and (c) after 200 km transmission with 2 km mismatch in dispersion.

$E_i/2\,e^{j\phi_1}$ and $E_i/2\,e^{j\phi_2}$. Indeed, if E_i is chosen to be equal, that is, the electrical field corresponds to signal points in the largest circle of the constellation, then a constellation signal point E_{output} with the phase shift θ_i in the circle n can be found as a sum of two vectors $E_i/2\,e^{j\phi_{ni1}}$ and $E_i/2\,e^{j\phi_{ni2}}$, where $\phi_{ni1} = \theta_i + \arccos\left(E_n/E_{\text{input}}\right)$ and $\phi_{ni2} = \theta_i - \cos^{-1}\left(E_n/E_i\right)$. The subscripts i and n denote the phase position and the order of the circle of interest, respectively. Figure 9.21 illustrates the relationship between E_i, E_0, ϕ_{ni1}, and ϕ_{ni2}. For simplicity, the signal point is chosen with $\theta_i = 0$.

By substituting ϕ_1 and ϕ_2 into Equation 9.7, the driving voltages for this point can be obtained as

$$V_{ni1}(t) = \frac{V_\pi}{\pi}\left[\theta_i + \cos^{-1}\left(E_n/E_i\right)\right], V_{ni2}(t) = \frac{V_\pi}{\pi}\left[\theta_i - \cos^{-1}\left(E_n/E_i\right)\right] \qquad (9.6)$$

Figure 9.22 shows, in general, a schematic block diagram of both the photonic transmitters and balanced receivers for 16ADQPSK transmission scheme. The branches for detection of amplitude, in-phase and quadrature phase components are also given. The reception block diagram can also be employed for MADPSK, but with slightly DSP algorithms.

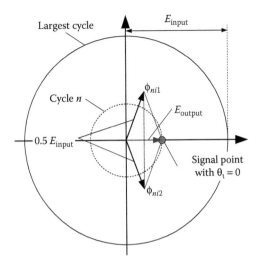

FIGURE 9.21 Relationship between E_i, E_o, ϕ_{ni1}, and ϕ_{ni2} using phasor representation: (a) photonic transmitter and (b) amplitude-phase detection-balanced receiver and decoding.

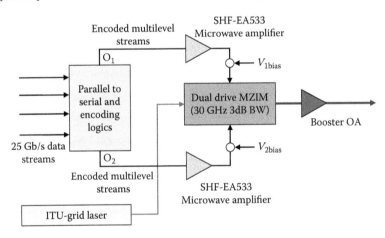

(a)

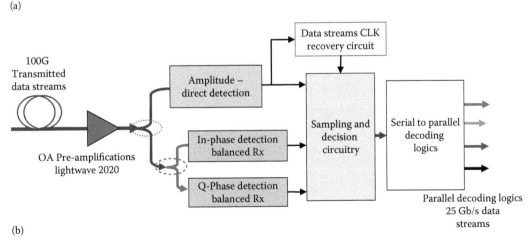

(b)

FIGURE 9.22 Schematic diagram of the photonic transmitters and receivers for the 16ADQPSK transmission scheme: (a) transmitter and (b) receivers with branches for detection of amplitude, in-phase and quadrature phase components.

9.2   MADPSK OPTICAL TRANSMISSION

In general, the structures of the MADPSK can be given as shown in Figure 9.23. A model has been constructed for investigating the performance of systems based on MADPSK modulation format. It consists of signal-coding model, transmitter model, receiver model, and transmission and dispersion-compensation fiber models.

In the subsequent sections, the 16-ary MADPSK signal model, described in Section 9.1.3.1, will be used. To balance the ASK and DQPSK sensitivities, ASK signal levels are preliminary adjusted to the ratio $I_3 / I_2 / I_1 / I_0 = 3/2/1.5/1$, as shown in Figure 9.24. These level ratios can be determined from the SNRs at each separation distance of the eye diagram or q-factor. The noise is assumed to be dominated by the beat noise resulted from the mixing of the signals and the ASE noise.

The transmitter model shown in Figure 9.23 is used to produce the 16-ary MADPSK signal. It consists of a distributed feedback laser source generating continuous wave (CW) light (carrier), which is then modulated in both phase and amplitude by a dual-drive MZIM. Each four bits of user data $[D_3D_2D_1D_0]$ are first grouped into a symbol and then encoded to generate two electrical driving signals, $V_1(t)$ and $V_2(t)$, under which amplitude and phase of the carrier in two optical paths of the dual-drive MZIM will be modulated to produce NRZ 16-ary MADPSK signal. The following RZM-PC caves NRZ pulse train into RZ pulse for minimizing the effects of intersymbol interference.

The receiver model shown in Figure 9.25 consists of two phase demodulators: an amplitude demodulator (AD) and a data multiplexer (MUX). Two phase demodulators are used for extracting $[D_1D_0]$ bits, and they work exactly in the same way as the ones in DQPSK receiver, described in the above section. The AD is used for detecting four amplitude levels of the MADPSK signal. It is a

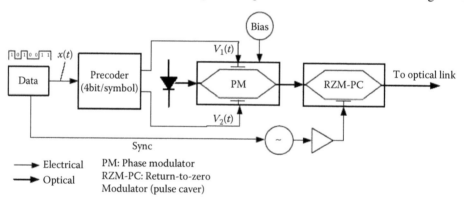

FIGURE 9.23   MADPSK transmitter.

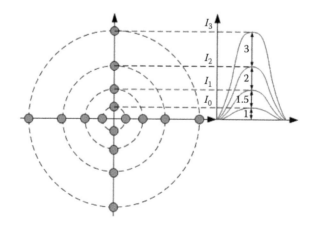

FIGURE 9.24   ASK interlevel spacing.

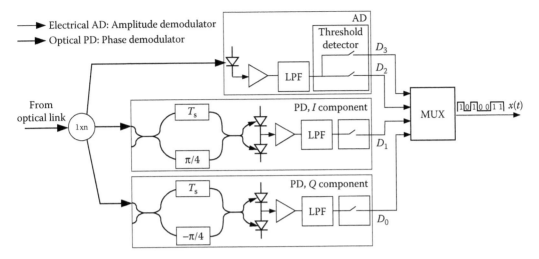

FIGURE 9.25 Amplitude direct detection and photonic phase comparator with balance receiver for MADPSK demodulation.

well-known direct-detection scheme consisting of a photodiode, followed by an electronic receiver. The amplitude-modulated signal is then threshold detected in association with a clock recovery circuit to recover two bits $[D_3 D_2]$. Two bits $[D_3 D_2]$ are interleaved with two bits $[D_1 D_0]$ by the MUX to reconstruct the original binary data stream.

9.2.1 PERFORMANCE EVALUATION

Under performance evaluation, the following main parameters are investigated: (1) System BER versus SNR: a solution for system BER will be found analytically, and BER will be computed against different SNR values and bit rates. System BER versus SNR will also be obtained by system simulation and cross-checked with analytical results. Graphs of BER versus SNR will be plotted; (2) System BER versus receiver sensitivity: BER versus receiver sensitivity will be obtained analytically and by simulation, and the results will be cross-checked. Graphs of BER versus receiver sensitivity will be plotted; (3) DT: transmission over fibers of types ITU-G652, ITU-G.655, and LEAF with corresponding dispersion factors will be considered. Graphs of the power penalty due to the dispersion as compared with back-to-back transmission will be plotted against the dispersion factor in ps/nm; and (4) Tolerance to other system impairments: like with DT, power penalty due to other impairments such as laser, and OA phase noise, receiver phase error will be investigated. Corresponding graphs will be plotted.

Performance evaluation is conducted under the effects of the following conditions or contexts: (1) Different pulse shapes: raised cosine, rectangular, and Gaussian; (2) Modulation formats: NRZ, RZ, and CSRZ; (3) ASE noise of OAs; (4) Transmitter impairments: laser noise; (5) Receiver impairments: phase error of DI-based phase demodulators; (6) Change of ASK interlevel spacing; (7) Optical and electrical filtering; and (8) Multichannel environment: system performance in combination with dense WDM technology would be reported in future.

9.2.2 IMPLEMENTATION OF MADPSK TRANSMISSION MODELS

9.2.2.1 System Modeling

The following simulation models have been built on the MATLAB–Simulink platform for proving the working principles and for investigating the performance of systems using optical MADPSK modulation. A transmitter is simulated to generate 16-ary MADPSK signal, and a receiver is used to

TABLE 9.4

Phase and Driving Voltages for 16-ary MADPSK Constellation

Positions	θ_i	V_{i1}	V_{i2}
		Circle 3	
1100	0°	$0.0V_\pi$	$0.0V_\pi$
1101	90°	$0.5V_\pi$	$0.5V_\pi$
1111	180°	$1.0V_\pi$	$1.0V_\pi$
1110	270°	$1.5V_\pi$	$1.5V_\pi$
		Circle 2	
1000	0°	$0.2952V_\pi$	$-0.2952V_\pi$
1001	90°	$0.7949V_\pi$	$0.2046V_\pi$
1011	180°	$1.2947V_\pi$	$0.7043V_\pi$
1010	270°	$1.7944V_\pi$	$1.2041V_\pi$
		Circle 1	
0100	0°	$0.3919V_\pi$	$-0.3919V_\pi$
0101	90°	$0.8917V_\pi$	$0.1078V_\pi$
0111	180°	$1.3914V_\pi$	$0.6076V_\pi$
0110	270°	$1.8912V_\pi$	$1.1073V_\pi$
		Circle 0	
0000	0°	$0.4575V_\pi$	$-0.4575V_\pi$
0001	90°	$0.9573V_\pi$	$0.0422V_\pi$
0011	180°	$1.4570V_\pi$	$0.5420V_\pi$
0010	270°	$1.9568V_\pi$	$1.0417V_\pi$

reconstruct the original binary signal. These models run over a simulated single-mode optical fiber. Laser chirp, OA phase noise, nonlinearities, CD, PMD, and other impairments will be involved in later stages to evaluate different performance characteristics of the modulation format: system BER, receiver sensitivity, and power penalties due to different impairments.

The phases and the driving voltages for creating signal points of the 16-ary MADPSK constellation is computed and tabulated in Table 9.4.

9.2.3 TRANSMITTER MODEL

MATLAB–Simulink model of the system is shown in Figure 9.26. The transmitter model using the dual-drive MZIM structure is shown in Figure 9.27. The purposes of blocks are as follows: (1) "User data and ADPSK pre-coder" block generates a pseudo-random data sequence to simulate user data stream and encodes each group of 4 data bits into a symbol. (2) "Voltage driver 1" and "Voltage driver 2" blocks map pre-encoded data into driving voltages for modulating amplitude and phase of the carrier in the dual-drive MZIM. (3) Two "complex phase shift" blocks simulate two optical paths of the dual-drive MZIM. (4) "Sum block" simulates the combiner at the output of MZIM. (5) "Gaussian noise generator" block simulates noise source. (6) "Amplifier" block simulates OA.

9.2.4 RECEIVER MODEL

The receiver structure is shown in Figure 9.28. The functions of the blocks are as follows: (1) Each DI is simulated by a set of two "Magnitude-Angle" blocks, a "Delay block," and a "Sum" block. The "Delay block" stores the phase of the previous symbol, the "Magnitude-Angle" blocks extract

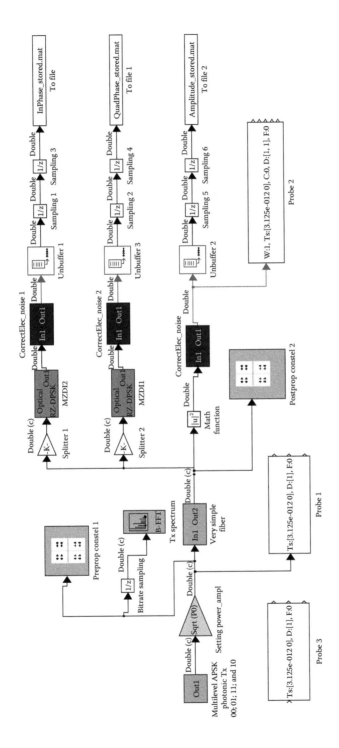

FIGURE 9.26 MATLAB-simulated system model.

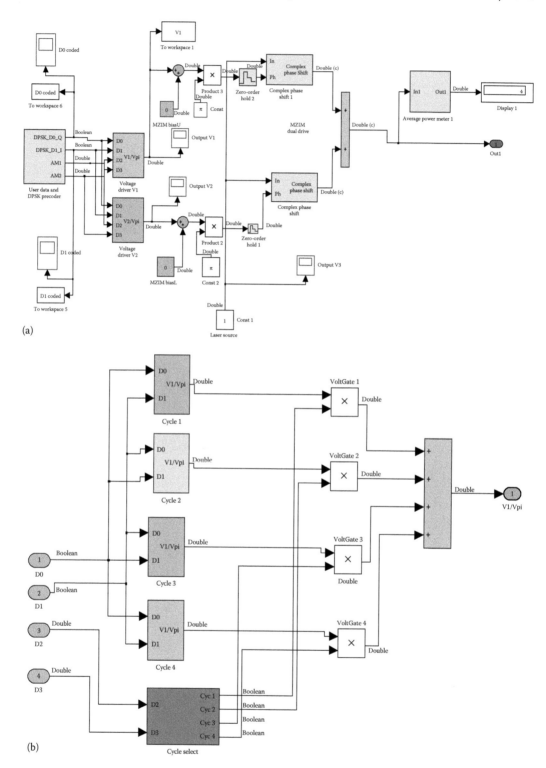

FIGURE 9.27 MATLAB-simulated MADPSK: (a) transmitter and (b) logic precoder.

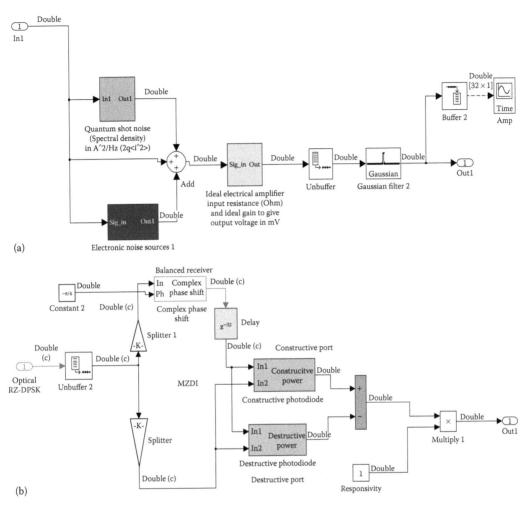

FIGURE 9.28 MATLAB-simulated MADPSK receiver: (a) amplitude direct detection and (b) balanced receiver detection—in-phase and quadrature.

the phase and amplitude of present and previous symbols, which will be used in the followed different phase demodulation and detection operations. (2) "Constant $\pi/4$" and "Constant $-\pi/4$" and next two "Sum" blocks simulate extra phase delay in each branch of DI. (3) Two "Cos" blocks and two "Product" blocks simulate two balanced receivers. (4) "Amplitude Detector D2 and D3" block simulates ASK detector for D_2 and D_3 bits. (5) Three "Analog Filter Design" blocks simulate electrical low-pass (LP) filters. (6) "Phase Detector D0_I" and "Phase Detector D1_Q" blocks simulate the threshold detectors for D_0 and D_1 bits (I- and Q-components of a DQPSK signal), respectively.

9.2.5 TRANSMISSION FIBER AND DISPERSION-COMPENSATION FIBER MODEL

The propagation of optical signal in a fiber medium that is dispersive and nonlinear is best described by the nonlinear Schrödinger equation (NLSE) [2], as described in Chapter 2. Other parameters are explained below. Transmission fiber model, as shown in Figure 9.29, is used to simulate the propagation of optical signal. This fiber model simulates the impairments that have impacts on the system performance.

All characteristic parameters of the fiber medium together with optical input signal are taken by the "Matrix Concatenation" block and then processed by a MATLAB function that solves the NLSE

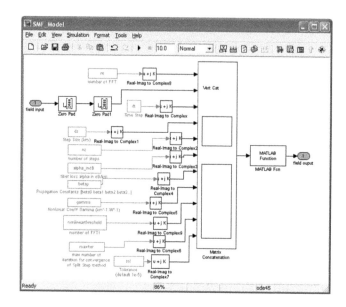

FIGURE 9.29 Input parameters for single-mode fiber SSMF model.

by using the split-step Fourier method [5]. The dispersion-compensation fiber model has the same structure of the propagation fiber model, except that the signs of the propagation constant β_2 in the two models are opposite.

9.2.6 TRANSMISSION PERFORMANCE

9.2.6.1 Signal Spectrum, Signal Constellation, and Eye Diagram

The spectrum of 40 Gb/s 16-ary MADPSK signal obtained by running transmitter model is given in Figure 9.30. As it has been seen clearly in the graph, the single-sided bandwidth of the main lobe equals 10 GHz. Numerically, it amounts to only one-fourth of the transmission bit rate, and from that, it can be concluded that MADPSK is a high-bandwidth-efficient modulation format.

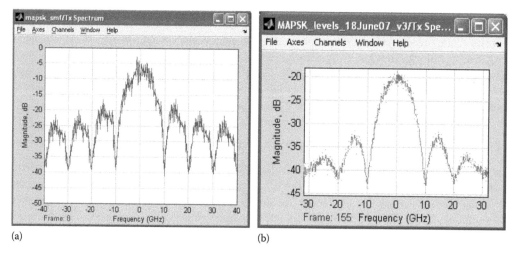

FIGURE 9.30 40 Gb/s MADPSK spectrum with sampling rates of (a) 10 GSamples/s and (b) 20 GSamples/s.

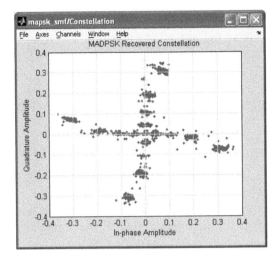

FIGURE 9.31 40 Gb/s MADPSK constellation recovered at the receiver.

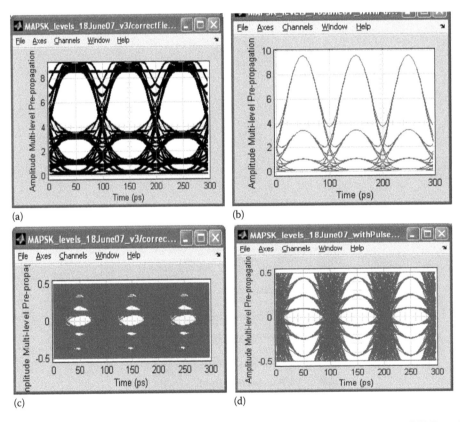

FIGURE 9.32 40 Gb/s MADPSK eye diagram at OSNR = 20 dB: (a) NRZ amplitude, (b) CSRZ amplitude, (c) NRZ in phase, and (d) CSRZ in phase.

Figure 9.31 shows the signal constellation recovered at the receiving end. Noise and nonlinear property of fiber cause amplitude and phase fluctuations and scatter signal points around some mean value. The MADPSK eye diagram is shown in Figure 9.32 for the I component. (The Q component should have the similar diagram.) This eye diagram clearly shows four amplitude levels associated with two phase shifts, 0 and π.

9.2.6.2   BER Evaluation

The MADPSK system can be considered as consisting of two subsystems, ASK and DQPSK, and its error probability can be evaluated as a join error probability of the two:

$$P_{ADPSK} = \left[\frac{1}{2} P_{ASK} + \frac{1}{2} P_{DPSK} - \frac{1}{2} P_{ASK} \cdot \frac{1}{2} P_{DPSK} \right] = \frac{1}{2} \left[P_{ASK} + P_{DPSK} - P_{ASK} \cdot P_{DPSK} \right] \qquad (9.7)$$

where P_{ASK} and P_{DPSK} are error probabilities of ASK and DQPSK subsystems, respectively.

9.2.6.2.1   ASK Sub-system Error Probability

Figure 9.33 shows four ASK signal levels, b_0, b_1, b_2, b_3, three decision levels, d_1, d_2, d_3, and standard deviation of noise, $\sigma_0, \sigma_1, \sigma_2, \sigma_3$, at different signal levels. The error probability of the ASK subsystem can be evaluated by

$$P_{ASK} = \frac{2}{M+1} \sum_{1}^{M} Q\left(\frac{b_i - d_i}{\sigma_i} \right) = \frac{2}{3+1} \left[\begin{array}{c} Q\left(\dfrac{b_1 - d_1}{\sigma_1} \right) \\[2ex] + Q\left(\dfrac{b_2 - d_2}{\sigma_2} \right) \\[2ex] + Q\left(\dfrac{b_3 - d_3}{\sigma_3} \right) \end{array} \right] \qquad (9.8)$$

For example, in our system: (1) $b_1 = 8.08e-2$, $b_2 = 1.45e-1$, $b_3 = 2.42e-1$, (2) $d_1 = 5.11e-2$, $d_2 = 1.08e-1$, $d_3 = 1.88e-1$, and (3) $\sigma_1 = 5.00e-3$, $\sigma_2 = 6.70e-3$, $\sigma_2 = 8.65e-3$ at an OSNR = 20 dB.

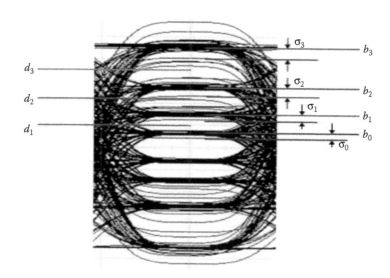

FIGURE 9.33   MADPSK eye diagram: signal levels, decision levels, and standard deviation of noise.

The error probability of the ASK subsystem thus equals:

$$P_{ASK} = \frac{1}{2}\left[\begin{array}{l} Q\left(\frac{(8.08e-2)-(5.11e-2)}{5.00e-3}\right)+Q\left(\frac{(1.45e-1)-(1.08e-1)}{6.70e-3}\right) \\ +Q\left(\frac{(2.42e-1)-(1.88e-1)}{8.65e-3}\right) \end{array}\right] \rightarrow$$

$$P_{ASK} = \frac{1}{2}\left[Q(5.94)+Q(5.52)+Q(6.24)\right]$$

$$= \frac{1}{2}\left[(1.47e-9)+(1.73e-8)+Q(2.26e-10)\right]$$

$$= 9.49e-9$$

The error probability of ASK subsystem over a range of OSNR from 6 to 24 dB is evaluated and shown in Figure 9.34.

9.2.6.2.2 DQPSK Subsystem Error Probability Evaluation

In term of DPSK modulation, the system can be broken up into four independent DQPSK subsystems, corresponding to circle 0, circle 1, circle 2, and circle 3 of the signal constellation. First, the error probability of each subsystem is evaluated and then all the error probabilities are averaged to obtain the error probability of the DQPSK subsystem.

Each DQPSK subsystem in turn can be thought of as made from two 2-ary DPSK subsystems. The error probability of each 2-ary DPSK subsystem is evaluated and then all error probabilities are averaged to get the error probability of DQPSK subsystem:

$$P_{DQPSK} = 1-(1-P_{DPSK\_I})(1-P_{DPSK\_Q}) = P_{DPSK\_I} + P_{DPSK\_Q} - P_{DPSK\_I} \cdot P_{DPSK\_Q} \tag{9.9}$$

where P_{DPSK_I} and P_{DPSK_Q} are the error probabilities of in-phase (I) and quadrature-phase (Q) components, respectively, of each DPSK subsystem (circle). Because I is coded by bit D_0, Q is coded by

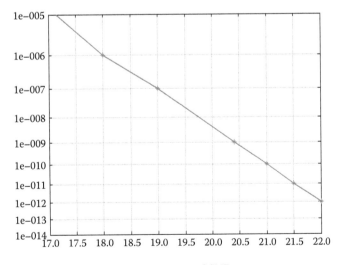

FIGURE 9.34 Error probability of ASK subsystem versus OSNR.

bit D_1, I and Q are detected in the same way, and D_0 and D_1 are supposed to be equally probable, then Equation 9.20 becomes:

$$P_{DQPSK} = 2P_{DPSK\_I} - P^2_{DPSK\_I} = 2P_{DPSK\_Q} - P^2_{DPSK\_Q} \quad (9.10)$$

P_{DPSK_I} is evaluated based on the δ-factor [5] as

$$P_{DPSK\_I} = \frac{1}{2}\left(\frac{\delta}{\sqrt{2}}\right) \approx \frac{\exp\left(-\delta^2/2\right)}{\delta\sqrt{2\pi}} \quad (9.11)$$

where $Q = i_H - i_L/\sigma_H + \sigma_L$, and i_H, i_L and σ_H, σ_L are mean value and standard deviation of signal currents at high and low levels at the input of the receiver, respectively. For example, the transmission parameters can be set as follows: $i_H = 3.23e-02$, $i_L = (-3.23e-02)$, and $\sigma_H = \sigma_L = 3.16e-3$ at OSNR = 20 dB. The δ-factor for a single DQPSK subsystem of circle 0 thus equals and the corresponding error probability is $P_{DPSK\_I\_CYCLE\,0} = 1/2\,\text{erfc}\left(10/\sqrt{2}\right) \approx 7.7e-24$.

The error probability of circle 0 (inner most circle) is $P_{DQPSK\_CYCLE\,0} = 2*(7.7e-24) - (7.7e-24)^2 = 1.54*10e-23$. Thus, the error probability of all four circles is

$$P_{DQPSK} = \frac{1}{4}\left[P_{DQPSK\_CYCLE\,0} + P_{DQPSK\_CYCLE\,1} + P_{DQPSK\_CYCLE\,2} + P_{DQPSK\_CYCLE\,3}\right] \quad (9.12)$$

P_{DQPSK} over a range of OSNR from 6 to 24 dB is evaluated and shown in Figure 9.35.

9.2.6.2.3 MADPSK System BER Evaluation

The MADPSK system error probability is evaluated based on Equation 9.17. Figure 9.36 shows the graphs of error probabilities for the ASK subsystem, the DQPSK subsystem, and the MADPSK system in the same coordinates for comparison purpose. As it can be observed from Figure 9.36, at OSNR = 24 dB, the MADPSK thus requires higher signal average power as compared with QPSK or DPSK schemes. Thus due to the nonlinear threshold of the fiber there is a limit of the maximum power level which this multilevel modulation scheme can transmit, hence lowering the transmission distance. It is also clear that for the same value of OSNR, especially when it is high, DQPSK subsystem outperforms ASK counterpart and the overall performance of MADPSK system is dominated by the performance of the ASK subsystem. Thus, the spaces between ASK levels could be adjusted for a better balance between BER ASK and BER DQPSK to achieve a better

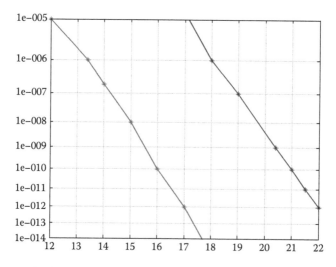

FIGURE 9.35   Error probability of DQPSK subsystem versus OSNR.

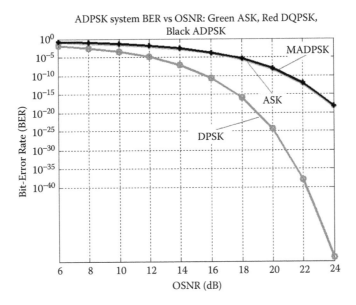

ADPSK system BER vs OSNR: Green ASK, Red DQPSK, Black ADPSK

FIGURE 9.36 Error probability of MADPSK (black) system versus OSNR, logarithm scale. Error probability of ASK (light gray) and MADPSK are nearly coinciding dark gray circle curve indicates BER versus OSNR for DPSK.

overall MADPSK BER performance. This probably is caused mainly by the intersymbol interference during the transition of different levels.

Figure 9.36 shows the simulation results of 16ADPSK at 100 Gb/s transmission (extreme left graph) in comparison with other modulation formats such as duobinary 50 and duobinary 67 and experimental results of CSRZ-DPSK. The bit rates of these other transmission results are at 40 Gb/s. It is observed that for 16MADPSK, the receiver sensitivity is close to the −28 dBm performance standard used in 10 Gb/s NRZ transmission and performs better at 100 Gb/s than the other modulations operating at the lower rate of 40 Gb/s. However, this superior performance at 100 Gb/s is still at penalty of approximately 3 dB compared with 10 Gb/s transmission systems. Fortunately, this penalty can easily be compensated for using a low-noise optical preamplifier at the receiver end. For example, a 15 dB gain in optical preamplifier with a 3 dB noise figure (NF) would adequately resolve the issue. The comparison of the BER with the receiver sensitivity of 16ADPSK and duobinary formats and ASK is shown in Figure 9.37. It indicates a 2–3 dB improvement of the MADPSK.

The detection of the lowest level may have been affected by the noise level of the optical preamplifier when only the amplitude information is used. This can be improved significantly if both phase detection and amplitude detection are used, as we can observe from Figure 9.32c and d.

9.2.6.2.4 Chromatic Dispersion Tolerance

The residual CD of the optical link is characterized by the DL product, defined as product of the dispersion coefficient D and the total fiber length L. Figure 9.38 shows the signal phase evolution under the effect of CD. It can be seen that with a predetermined $DL = 50$ ps/nm, all signal points are rotated around the [0,0] origin by the same angle of approximately 0.125 rad. This confirms the parabolic phase shift due to the CD. This phenomenon is called linear phase distortion in contrary to the nonlinear phase distortion caused by the fiber nonlinearity.

Figure 9.39 shows the BER penalty versus different values of DL product. It can be seen very clearly that the BER performance of NRZ format is severely affected by fiber dispersion. When DL increases from 0 ps/nm (fully CD compensated) to 35 ps/nm, its BER performance is improved by 1.5 dB; however, it is sharply degraded by 28 dB penalty at $DL = 50$ ps/nm and should be worse for higher value of dispersion. This leads to the conclusion that it is undesirable to use NRZ format in

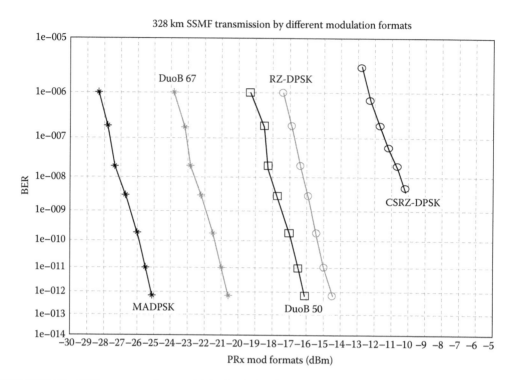

FIGURE 9.37 BER versus receiver sensitivity for MADPSK format and other duobinary, ASK (simulation), CSZ, and CSRZ-DPSK (experimental). Legend: lightest * is the MADPSK.

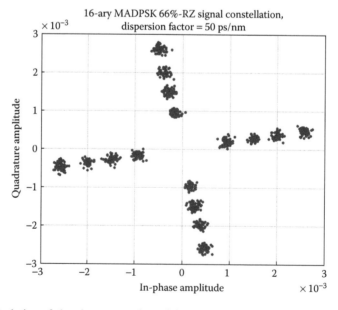

FIGURE 9.38 Evolution of the phase scattering of the MADPSK signal constellation under chromatic dispersion effects.

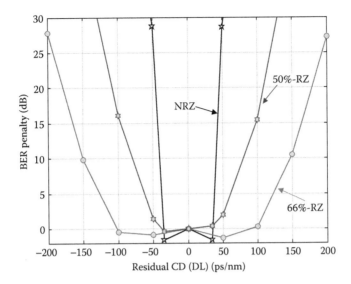

FIGURE 9.39   Error probability of MADPSK system versus the dispersion-length DL product.

MADPSK systems, because the optical link residual dispersion usually cannot be compensated to a small amount and ineffective dispersion management and control plan could lead to a very high BER.

The 66% RZ format, on the other hand, can tolerate much higher degree of CD. Its BER performance is even slightly improved at $DL \approx 50$ ps/nm, and the BER penalty is just less than 1 dB at $DL = 100$ ps/nm. This is equivalent to the transmission over 6 km of uncompensated SSMF fiber without significantly giving up the BER performance.

The MAPSK offers lower symbol rate and hence higher channel capacity, which would allow the upgrading of higher rate merging in a low-bit-rate optical fiber transmission system without modifying the photonic infrastructure of the optical networks.

9.2.7   CRITICAL ISSUES

This section outlines the critical issues to evaluate the performance of MADPSK systems.

9.2.7.1   Noise Mechanism and Noise Effect on MADPSK

Although receiver noises in multilevel amplitude modulation were investigated intensively in the 1980s, little has been reported for multilevel phase and differential phase modulations. One of the principal goals in the system design, especially for long-haul transmission systems, is to achieve high receiver sensitivity. At a given optical power, the error probability depends on the noise power and hence the receiver sensitivity.

Quantum shot noise is the fundamental noise mechanism in photodiode, which leads to the fluctuation in the detected electrical current even when the incident optical signal has a constant or varied power. Thus, it is signal dependent. Furthermore, the beating of the currents of the signal and the optical phase noise would generate an amplitude-dependent noise at different level signals of the MADPSK. It is caused by random generation of electrons, contributing to the photoelectric current, which is a random variable. All photodiodes generate some current even in the absence of optical signal because of the stray light and/or thermal generation of electron-hole pairs, the dark current.

In MADPSK, the amplitude of the signal of the outermost circle of the constellation would be affected by the quantum shot noises, which are strongly signal amplitude dependent, especially when there is an optical preamplifier. On the other hand, it is desirable that the innermost constellation would have the largest magnitude to maximize the optical signal energy for long-haul transmission. Therefore, an optimum receiving scheme must be developed both analytically and by modeling and eventually by experimental demonstration.

However, the amplitude of the outermost constellations is limited by the nonlinear SPM effects that would be further explained in the next few sections. Thus, the lower and upper limits of the amplitude of the MADPSK would be extensively investigated in the next phases of the research.

The electronic equivalent noise, as seen from the input of the electronic preamplifier following the PD, can be measured and taken into account to the total noise process caused by thermal noises of the input impedance, the biasing current shot noise, and the noises at the output of the electronic preamplifier. These noises are combined with the signal-dependent quantum shot noises to gauge their contributions to the MADPSK receiver. Thus, we may consider new structures of electronic amplifiers or matched filter at the input of the receiver to achieve optimum MADPSK receiver structure.

For long-haul transmission systems, ASE of OA is probably the most important noise mechanism. In OAs, even in the absence of input optical signal, spontaneous emission always occurs stochastically when electron-hole pairs recombine and release energy in the form of light. This spontaneous emission is noise, and it is amplified by the OAs together with the useful optical signal and accumulated along optical transmission link.

Noises reduces the SNR and hence system BER and receiver sensitivity. Noise models also affect the design of optimum detection schemes such as decision thresholds. To the best of my knowledge, a thorough investigation of the noise mechanism and its impacts on multilevel signaling has never been reported, except some preliminary results for 10 Gb/s 4-ary ASK schemes. Thus, all noise sources and mechanism by which they affect the system performance must be thoroughly investigated. These noises are used to estimate the optimum decision level of the detection of the amplitude of the multilevel eye diagram.

9.2.7.2 Transmission Fiber Impairments

For optical signals, the transmission medium is an optical fiber with associated OAs and dispersion compensation devices or is a leased wavelength running on top of a dense WDM system. Impairments are always part of the transmission medium; among them, CD, PMD, and nonlinearity are critical.

When an optical pulse propagates along a fiber, its spectral components disperse owing to the differential group delay and the output pulse broadens. Chromatic dispersion is proportional to the fiber length and the laser linewidth, especially the spectrum of the lightwave modulate signals. Chromatic dispersion may cause optical pulses to overlap each other, thus leading to intersymbol interference and increase system BER, especially for ASK systems. The DPSK systems, on the other hand, are more CD tolerant. For the MADPSK systems, the phase constellation, as shown in Figure 9.40, is rotating when the MADPSK is under the linear CD effect. It is also well known and developed in our model that this CD can be compensated by dispersion-compensating fiber modules. However, the mismatching of the dispersion slopes of the transmission and compensating fibers is very critical for multichannel multilevel modulation schemes.

Optical pulse is also broadened by PMD, which is actually the time mismatching between two orthogonal polarizations of the optical pulse when they traverse along a fiber. In the ideal optical fiber having truly homogeneous glass and truly coaxial geometry of the core, the two optical polarizations would propagate with the same velocity. However, it is not the case for a real fiber, so the two polarizations have different speeds and will reach the fiber end at different times.

Similar to the CD effects, PMD can cause pulse overlapping and thus increase system BER. However, unlike CD, which is practically constant over time and can be compensated in a large scale, PMD is a stochastic process and cannot be managed easily. It has been well known that PMD has the Maxwellian probability density function with a mean value $\langle \text{PMD} \rangle = K_{\text{PMD}} \cdot \sqrt{L}$, where K_{PMD} is defined as PMD coefficient whose measured values vary from fiber to fiber in the range $[0.01 - 1 ps/\sqrt{km}]$ and L is fiber length. Under the MADPSK, the signal space of the constellations

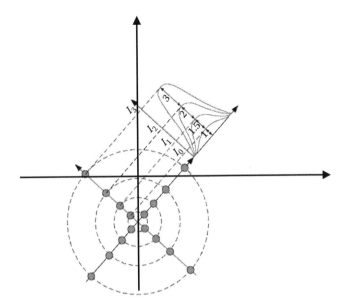

FIGURE 9.40   ASK inter-level spacing and offset modulation and detection line.

would be affected either in the magnitude or in phases by PMD but is expected to be dominated by phase distortion. It is well known that the PMD first and second effects are critical for ASK modulation. For DPSK, it is expected that the principal axes of the polarization modes propagating through the fiber would be minimally affected. Thus, under hybrid amplitude-phase modulation scheme, there are several issues to be resolved. Under the MADPSK scheme, the delay of the polarization modes would generate the phase difference or phase distortion on I- and Q-components, resulting in enhancement of the distortion effects of the intersymbol interference. The amplitude distortion would then be increased but considered to be secondary effect.

9.2.7.3   Nonlinear Effects on MADPSK

Nonlinear effects occur due to the nonlinear response of the fiber glass to the applied optical power. Fiber nonlinearity can be classified into stimulated scattering and Kerr effect. Among several stimulated scattering effects, stimulated Raman scattering, caused by interaction between light and acoustical vibration modes in the fiber glass, is the most critical. Under this mechanism, optical signal is reflected back to the transmitter, and in WDM systems, its power is also transferred from shorter to longer wavelengths, thus attenuating signal and causing crosstalk. Kerr effect is the roots of intensity-dependent phase shift of the optical field. It is shown up in three forms: SPM, XPM, and four-wave mixing (FWM), provided the phase matching is satisfied.

SPM is usually the dominant effect in a single-channel DPSK system. The changes in instantaneous power of optical pulses together with the ASE from associated OAs lead to intensity-dependent changes, the Kerr effect, in the guided-medium refractive index, resulting in the effective index of the guided mode. These changes are converted to the phase shifts or phase noise of the lightwave carriers. At the receiver, the phase noise is transferred back to intensity noise, which degrades BER. As mentioned above, the contribution of noises into different levels of the MADPSK scheme is very critical to determine optimum decision thresholds. This is further complicated by these additional nonlinear effects, especially the NLPN, usually contributed by the SPM, owing to the outer most constellation. These NLPN effects from the outermost constellation to other inner-circle signal spaces have never been investigated.

XPM modulation becomes the most critical nonlinearity in WDM systems, where the phase shifts (noises) in one channel come from refractive index fluctuations caused by power changes in other channels. XPM modulation becomes more pronounced when neighboring channels have equal bit rate. Four-wave mixing is basically a crosstalk phenomenon in WDM systems. When three wavelengths with frequencies, ω_1, ω_2, and ω_3 propagate in a nonlinear fiber medium at which the dispersion is zero, they combine and create a degenerate fourth wavelength, which would fall on the location of an active-wavelength channel. If these parametric wavelengths fall into other channels, they cause crosstalk and degrade the performance of the system. Although FWM is expected to reduce the receiver sensitivity, in the proposed system, to minimize the effects of fiber nonlinearity, the maximum power of optical signal should not be set higher than a certain threshold. This maximum power dictates the amplitude of signal points in the outermost circle (circle 3), and hence other circles, of the signal space. Thus, optimization of the signal amplitude levels for MADPSK is very critical.

9.2.8 OFFSET DETECTION

The 16-ary MADPSK signal model described in Section 9.1.3.1 can be modified. To balance the ASK and DQPSK sensitivities, ASK signal levels are preliminary adjusted to the ratio $I_3 / I_2 / I_1 / I_0 = 3 / 2 / 1.5 / 1$ and rotated by $\pi/4$, as shown in Figure 9.41. These level ratios can be determined from the SNR at each separation distance of the eye diagram or q-factor. The noise is assumed to be dominated by the noise

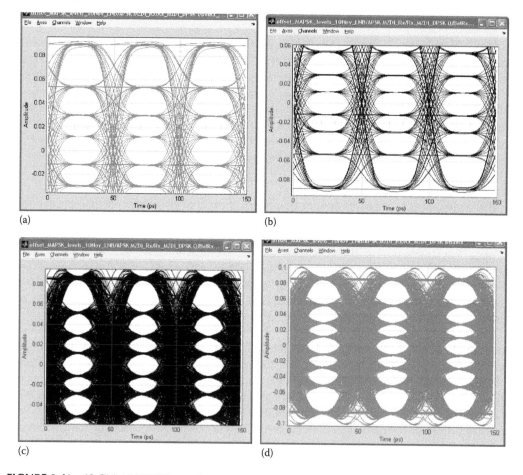

FIGURE 9.41 40 Gb/s MADPSK eye diagrams of the I (a) and Q (b) components: (A) 0 km—back to back (B) 2 km SSMF mismatch over three 100 km SSMF transmission spans (dispersion compensated).

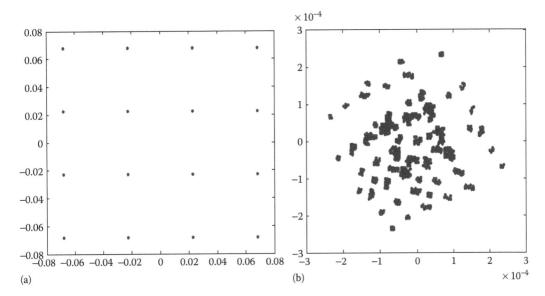

FIGURE 9.42   Constellation of 16-square QAM after two optically amplified spans and 2 km SSMF dispersion mismatch: (a) prepropagation and (b) postpropagation.

resulted from the mixing between the signal and that of ASE noise. The eye opening is expected to improve significantly, as shown in Figure 9.42.

9.3   STAR 16-QAM OPTICAL TRANSMISSION

This section gives a briefing on the simulation of the transmission performances of optical transmission systems over 10 spans of dispersion-compensating and optically amplified fiber transmission systems. The modulation format is focused on the Star 16-quadrature amplitude modulation (QAM) with two-level and eight-phase state constellation. Optical transmitters and coherent receivers are the main transmission terminal equipments—other constellations of the 16-QAM are also given very briefly. Simulation results have shown that it is possible to transmit and detect the data symbols for 43 Gb/s, with the possibility of scaling to 107 Gb/s without much difficulty. The OSNR with 0.1 nm optical filters is achieved with 18 dB and 23 dB for back-to-back and long-haul transmission cases with a DT of 300 ps/nm.

9.3.1   REMARKS

To increase the channel capacity and bandwidth efficiency in optical transmission, the multilevel modulation formats such as QAM formats are of interest. In the digital transmission with multilevel (M-levels) modulation, m bits are collected and mapped onto a complex symbol from an alphabet with $M = 2m$ possibilities at the transmitter side.

The symbol duration is $T_s = mT_B$, with T_B as the bit duration, and the symbol rate is $f_s = f_B/m$, with $f_B = 1/T_B$ as the bit rate. This shows that for a given bit rate that the symbol rate decreases if the modulation level increases. This means that higher bandwidth efficiency can be achieved by a higher-order modulation format. For 16-QAM format, $m = 4$ bits are collected and mapped to one symbol from an alphabet with $M = 16$ possibilities. In comparison with the case of binary modulation format, there only $m = 1$ bit is mapped to one symbol from an alphabet with $M = 2$ possibilities. With 16-QAM format and a data source with a bit rate of $f_B = 40$ Gb/s, a symbol rate of only $f_s = 10$ Gbaud/s is necessary. From commercial point of view, it means that a 40 Gb/s data rate can be transmitted with 10 Gb/s transmission devices. In the case of binary transmission, the transmitter needs a symbol rate of $f_s = 40$ Gbaud/s. It means that 16-QAM transmission requires four times slower transmission devices than the binary transmission does. It is noted here that a 10.7 GSy is used as the symbol rate so as to compare the simulation results with the well-known 10.7 Gb/s modulation schemes such as DPSK and CSRZ-DPSK. For 10.7 Gb/s bit rate, the transmission performance, that is, the sensitivity, and OSNR can be scaled accordingly without any difficulty.

This section gives a general approach regarding the design and simulation of Star 16-QAM with two amplitude levels and eight phase states forming two star circles. We term this Star 16-QAM as 2A-8P Star 16-QAM, two amplitude levels and eight phase states. The transmission format is discussed with theoretical estimates and simulation results to determine the transmission performance. The optimum Euclidean distance is defined for the design of star 16-QAM. Then, in the second section, the two detection schemes, namely the direct detection and the CoD, for Star-QAM constellations are discussed.

9.3.2 Design of 16-QAM Signal Constellation

There are many possibilities to design 16-QAM signal constellation. Three most popular constellations can be introduced. For 16-QAM modulation schemes, constellations are (1) Star 16-QAM, (2) Square 16-QAM, and (3) Shifted-square 16-QAM. The first two of these constellations are implemented. However, only the Star 16-QAM with two amplitudes and eight phases per amplitude level is employed in this section.

9.3.3 Signal Constellation

The signal constellation for Star 16-QAM with Gray coding is shown in Figure 9.43. The binary presentation of the symbols in the figure is shown in mapping Table A.1 of Table 9.5. As can be seen from this figure, the symbols are evenly distributed on two rings and the phase difference between the neighboring symbols on the same ring is equal ($\pi/4$). In order to detect a received symbol, its phase and amplitude must be determined. In other words, between two amplitude levels of the rings and among eight phase possibilities, there are a number of ways to form this constellation.

The ring ratio (RR) for this constellation is defined as: $RR = b/a$, where a and b are the ring radii, as shown in Figure 9.1. The RR can be set to different values to optimize the transmission performance.

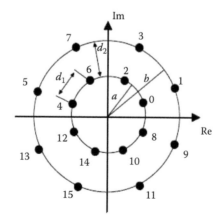

FIGURE 9.43 Theoretical arrangement of the modulation constellation for Star 16-QAM.

TABLE 9.5

Symbol Mapping and Coding for Star 16-QAM

A.1: Symbol-to-bit presentation

$0 \rightarrow 0000$	$4 \rightarrow 0100$	$8 \rightarrow 1000$	$12 \rightarrow 1100$
$1 \rightarrow 0001$	$5 \rightarrow 0101$	$9 \rightarrow 1001$	$13 \rightarrow 1101$
$2 \rightarrow 0010$	$6 \rightarrow 0110$	$10 \rightarrow 1010$	$14 \rightarrow 1110$
$3 \rightarrow 0011$	$7 \rightarrow 0111$	$11 \rightarrow 1011$	$15 \rightarrow 1111$

A.2: Gray coding for Star 16-QAM

$0 \rightarrow 1$	$4 \rightarrow 7$	$8 \rightarrow 15$	$12 \rightarrow 9$
$1 \rightarrow 0$	$5 \rightarrow 6$	$9 \rightarrow 14$	$13 \rightarrow 8$
$2 \rightarrow 3$	$6 \rightarrow 5$	$10 \rightarrow 13$	$14 \rightarrow 11$
$3 \rightarrow 2$	$7 \rightarrow 4$	$11 \rightarrow 12$	$15 \rightarrow 10$

A.3: Mapping for Star 16-QAM

$0 \rightarrow 1$	$4 \rightarrow 9$	$8 \rightarrow 13$	$12 \rightarrow 5$
$1 \rightarrow 0$	$5 \rightarrow 8$	$9 \rightarrow 12$	$13 \rightarrow 4$
$2 \rightarrow 3$	$6 \rightarrow 11$	$10 \rightarrow 15$	$14 \rightarrow 7$
$3 \rightarrow 2$	$7 \rightarrow 10$	$11 \rightarrow 14$	$15 \rightarrow 6$

A.4: Gray coding for square 16-QAM

$0 \rightarrow 12$	$4 \rightarrow 11$	$8 \rightarrow 13$	$12 \rightarrow 4$
$1 \rightarrow 10$	$5 \rightarrow 2$	$9 \rightarrow 5$	$13 \rightarrow 3$
$2 \rightarrow 15$	$6 \rightarrow 8$	$10 \rightarrow 14$	$14 \rightarrow 7$
$3 \rightarrow 9$	$7 \rightarrow 1$	$11 \rightarrow 6$	$15 \rightarrow 0$

A.5: Mapping for square 16-QAM

$0 \rightarrow 10$	$4 \rightarrow 11$	$8 \rightarrow 4$	$12 \rightarrow 15$
$1 \rightarrow 6$	$5 \rightarrow 1$	$9 \rightarrow 14$	$13 \rightarrow 3$
$2 \rightarrow 5$	$6 \rightarrow 2$	$10 \rightarrow 13$	$14 \rightarrow 0$
$3 \rightarrow 9$	$7 \rightarrow 8$	$11 \rightarrow 7$	$15 \rightarrow 12$

9.3.4 OPTIMUM RING RATIO FOR STAR CONSTELLATION

From Figure 9.43, it can be seen that there are many possibility to choose the RR for Star 16-QAM constellation. Here, the theoretical best RR is defined to minimize the error probability in an arbitrary white Gaussian noise (AWGN) channel by maximizing the minimum distance d_{min} between the neighboring symbols. The results for AWGN channel can be used approximately for optical transmission. For Star 16-QAM, the minimum distance d_{min} is maximized, when:

$$d_1 = d_2 = b - a = d_{min} \tag{9.13}$$

With some geometrical calculations, it can be obtained that:

$$d_{min} = 2\,a \cdot \sin(22.5°) \tag{9.14}$$

which leads to the optimal RR of:

$$RR_{opt} = b/a = (d_{min} + a)/a = (2 \cdot a \cdot \sin 22.5 + a)/a \approx 1.77 \tag{9.15}$$

The average power of Star 16-QAM constellation can be determined as

$$P_0 = (8a^2 + 8b^2)/16 = (a^2 + b^2)/2 \tag{9.16}$$

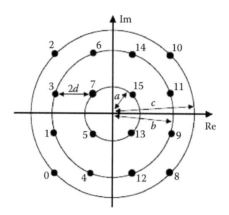

FIGURE 9.44   Square 16-QAM signal constellation.

Thus, we have the relationship between the average optical power and the minimum distance between the two rings of the two amplitude levels as

$$d_{min} \approx 0.53 \left(P_0 \right)^{1/2} \tag{9.17}$$

The obtained $RR_{opt} = 1.77$ does not depends on P_0 and is constant for each P_0 value. For an average power of 5 dBm (3.16 mW), $d_{min} = 2.98 \cdot 10^{-2} \sqrt{W}$, $a = 3.89 \cdot 10^{-2} \sqrt{W}$, and $b = 6.87 \cdot 10^{-2} \sqrt{W}$ are obtained.

9.3.4.1   Square 16-QAM

The signal constellation of the square 16-QAM with Gray coding is shown in Figure 9.44. The square 16-QAM can be generated by a combination of two rings and a QPSK in the outermost ring that can also be a ring. Thus, the circular phasor generation scheme can be used to form signals amplitude and phases to be applied to the optical modulators. The binary presentation of the symbols in the figure is shown in mapping Table A.1 of Table 9.5. In the constellation of the square 16-QAM, the 16 symbols have equal distance with direct neighbors and in total 12 different phases, that is, three phases per quarter, distributed on three rings. The phase differences between neighboring symbols on the inner and outer rings are equal ($\pi/2$), but the phase differences between neighboring symbols on the middle ring are different ($37°$ or $53°$). If the distance between direct neighbors in the square 16-QAM be rotated as $2d$, the average symbol power (P_0) of the constellation is

$$P_0 = 10 \cdot d_2 \tag{9.18}$$

For an average power of 5 dBm (3.16 mW), it can be computed that $d = 1.77 \cdot 10^{-2} \sqrt{W}$, and from it, $a = 2.5 \cdot 10^{-2} \sqrt{W}$, $b = 5.6 \cdot 10^{-2} \sqrt{W}$, and $c = 7.5 \cdot 10^{-2} \sqrt{W}$. In comparison with Star 16-QAM, here the distances between the middle ring and the outer ring are much smaller. It means, to achieve the same BER, the square 16-QAM needs a higher average power than Star 16-QAM. The decision method for the square 16-QAM is more complicated than that for Star 16-QAM. First, the decision between three amplitude possibilities of each ring should be made, and then, depending on the ring level, the decision is made between four or eight phase possibilities.

9.3.4.2   Offset-Square 16-QAM

To optimize the phase detection of the middle ring, it is envisaged that the phase differences between neighboring symbols on the middle ring in the square 16-QAM should be equal. Thus, the shifted-square 16-QAM is introduced by shifting (rotation) of symbols on the middle ring to obtain equal phase differences between all neighboring symbols, as shown in Figure 9.7. After shifting the symbols on the middle ring, the distances between all direct neighbors are not necessarily equal. In comparison with square 16-QAM, this constellation may offer more robust detection against phase distortions, according to our amplitude and phase detection method introduced in later section.

9.3.5 Detection Methods

In the case of differential encoding for the 16-QAM format, there are two different detection methods to demodulate and recover the data in the receiver: (1) DD and (2) CoD.

In this section, DD means detection with MZDI or (2 × 4) 90° hybrid, and CoD is similar except that a local oscillator (LO) which a very narrow linewidth laser is used to mix the signal and its lightwaves to generate the intermediate frequency (IF) or base band signals, with preservation of the modulated phase states. Both these two receiving methods have different implementations, which can be introduced as follows

9.3.5.1 Direct Detection

In contrast to CoD, differential decoding is done for DD in the optical domain. Indeed, this is equivalent to a self-homodyne CoD. This has the disadvantage that the transmitted absolute phase is lost after differential decoding. However, the relative phase (the phase of differential decoded signal) remains in electrical domain, and it makes the electrical equalization still possible. The equalization with relative phases is more difficult, and the results are worse than those of the equalization with absolute phases. The advantage of DD, compared with CoD, is that the synchronization of local laser with that of the signal light wave is omitted. There are two methods to implement the DD: One is with MZDI and the other is with a (2 × 4) 90° hybrid coupler.

9.3.5.2 Coherent Detection

In a coherent receiver, an LO is used to mix its signal with the incoming signal light wave for demodulation. As a result, it makes it possible to preserve the phase in electrical domain. This makes the electrical equalizing very effective in coherent detectors. For coherent detectors, the differential decoding is done in electrical domain. Dependent on the intermediate frequency (f_{IF}), defined as $f_{IF} = f_s - f_{LO}$, three different coherent methods can be distinguished: (1) Homodyne receiver, (2) heterodyne receiver, and (3) intradyne receiver. Only homodyne receiver is included in this section and the other two are only briefly mentioned.

9.3.5.2.1 Homodyne Receiver

A receiver is called homodyne when the carrier frequency (f_s) and the LO frequency (f_{LO}) are the same. It means:

$$f_{IF} = f_s - f_{LO} = 0 \tag{9.19}$$

In practice, because of the laser linewidth, carrier synchronization must be implemented to set the center frequency and the phase of LO to the same values as those in the incoming signal. For homodyne receivers, carrier synchronization can be implemented in the optical domain via an optical phase locked loop. Carrier synchronization failure causes degradation in the receiver's performance, but in this document, this effect is not considered and a perfect synchronization in the receiver (a perfect single spectrum line) is assumed. Alternatively, as mentioned later, heterodyne receiver using only one 90° hybrid coupler with associated electronic demodulation circuitry can be used to simplify the receiver configuration for CoD. Polarization control is another critical difficulty in all coherent receivers, which is also not included in this chapter. The implementation of homodyne receiver for Star 16-QAM is described in several text books.

9.3.5.2.2   Heterodyne Receiver

For this kind of receiver, it applies:

$$f_{IF} = f_s - f_{LO} \neq 0 > B_{opt} \qquad (9.20)$$

B_{opt} is the optical bandwidth of the transmitted signal. The IF will be mixed in electrical domain with a synchronous or asynchronous method in the LP domain. In the case of the synchronous demodulation, the phase synchronization can be done in the electrical domain. The implementation complexity of heterodyne receivers in optical domain is less than that of homodyne receivers.

9.3.5.2.3   Intradyne Receiver

Intradyne receiver requires:

$$f_{IF} = f_s - f_{LO} \neq 0 < B_{opt} \qquad (9.21)$$

The phase synchronization in intradyne receiver can be done in the digital domain. This makes intradyne receiver less complex in optical domain than homodyne receiver. Intradyne receiver, compared with heterodyne receiver, has the advantage that the processing bandwidth of intradyne receiver is smaller. The disadvantage of intradyne receiver is the higher requirement of intradyne receiver on laser linewidth than that of heterodyne receiver.

9.3.6   STAR 16-QAM FORMAT

In this section, the transmission performance of Star 16-QAM is evaluated by simulations and compared. The implementation of the transmitter for Star 16-QAM is introduced in the first section. Both direct and coherent homodyne receivers for Star 16-QAM are described in the second section. In the case of homodyne receiver (without laser phase noise and ideal frequency locking), two different receiver models, one without phase estimation and the other with phase estimation, are realized. The simulation results of all three receivers are shown in the third section and their comparison is shown in the fourth section.

9.3.6.1   Transmitter Design

There are many possibilities to implement the transmitter for Star 16-QAM described in Section 9.2.3. For the simulations in this work, the parallel transmitter shown in Figure 9.45 is implemented. The bit stream enters the differential encoder module after serial-to-parallel converting.

The differential encoder provides the following processes: (1) The four bits which have parallelly arrived at the module are mapped by the well-known Gray coding into symbols according to the mapping given in Table A.2 of Table 9.5; (2) The precoded symbols are differentially encoded (differential coding); and (3) The differentially encoded symbols are mapped again to other symbols to drive the Mach–Zehnder modulators (MZMs) according to the mapping in Table A.3 of Table 9.5.

Each symbol at the output of differential encoder module is represented by four bits. The bits are sent to pulse formers. The first two bits drive the first two MZMs, with lightwaves generated from the CW laser. If the input bit is equal to "1," then the output of MZM is "−1," and in other case, the output of MZM is a "1" (after sampling). After combining the output signals of these two MZMs and considering the 90° phase delay in one arm, we obtain the QPSK signal shown in Figure 9.45.

The third bit from the differential encoder output drives a PM to obtain the 8-PSK signal constellation from the QPSK signal. If this bit is equal to "1," then the QPSK symbol will rotate by $\pi/4$. The 8-PSK signal constellation is shown in Figure 9.46. To achieve the two-level Star 16-QAM signal constellation, another MZM is used to generate the second amplitude. If the fourth bit of differential encoder output is "1," then this output symbol is set on the outer ring of the constellation; otherwise, it is set on the inner ring. This MZM sets the RR of constellation. The signal constellation after MZM3 is shown in Figure 9.47.

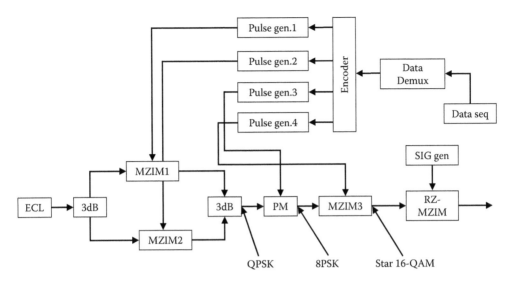

FIGURE 9.45 Schematic diagram of the optical transmitter for Star 16-QAM. Legend: Sig gen = signal generator; pulse gen = pulse generator; ECL = external cavity laser; MZIM = optimum MZ intensity modulators and PM = optimum phase modulator.

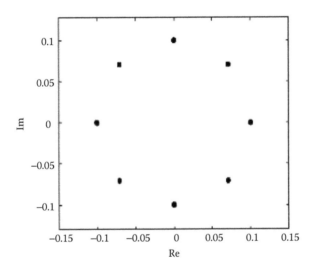

FIGURE 9.46 Constellation of the first amplitude level generated from the optical transmitter for Star 16-QAM.

The signal constellation in Figure 9.47 can be constructed from the whole constellation in Figure 9.43 with a rotation of $\pi/6$. The advantage of this rotation is that on the real and imaginary axis of the constellation, only eight different amplitude levels, instead of nine levels, exist. Therefore, another PM can be used between MZM3 and MZM-RZ to rotate the constellation by $\pi/6$. This additional PM is not shown in Figure 9.45. To increase the receiver sensitivity and reduce the signal chirp, an RZ pulse curving with a duty cycle of 50% should be implemented at the end of the transmitter, with an MZM driven by a sinus signal generator. In our simulations, the MZMs in Figure 9.45 work in push–pull operation and the PMs are MZMs working as PM. The received constellation of this Star-QAM, after transmission through 10 spans of 100 km and optically amplified, is shown in Figure 9.48, using coherent reception and digital signal processing (DSP) to recover the original constellation.

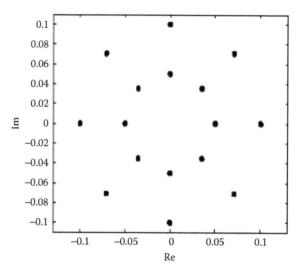

FIGURE 9.47 Constellation of the first and second amplitude levels generated from the optical transmitter for Star 16-QAM.

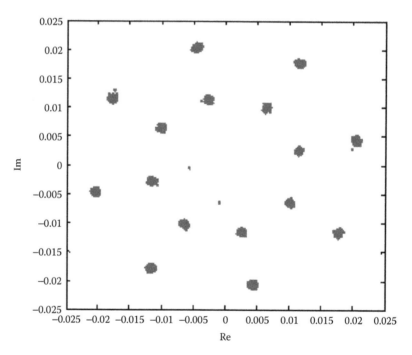

FIGURE 9.48 Constellation of the first and second amplitude levels at the receiver of Star 16-QAM after 10-spans of dispersion compensated SSMF links.

9.3.6.2 Receiver for 16-Star QAM

In this section, the implementations of DD receiver and coherent receiver for Star 16-QAM are explained. For CoD, there are many possibilities in digital domain of receiver to recover the data. The two methods implemented in this study detect the symbol before realizing the differential decoding. The difference is that one detects the symbol directly using the method described in Section 9.4.2.1, whereas the other employs a phase estimation algorithm, described in Section 9.4.2.2, before the symbol detection in order to cancel the phase distortions (phase synchronization between LO and the received signal). Another possibility, which is not implemented in this work, is doing at first the differential decoding of the incoming signal and then symbol detection.

9.3.6.3 Coherent Detection Receiver without Phase Estimation

The structure of CoD receiver is shown in Figure 9.49a. After transmission over fiber, the signal is amplified by an erbium doped fiber amplifier (EDFA). The input power of EDFA can be changed via an attenuator to set the OSNR. The output signal of EDFA is sent to a band pass (BP) filter in order to reduce the noise bandwidth.

An attenuator is used to set the OSNR, as required. The output signal of EDFA is sent to a BP filter in order to reduce the noise bandwidth. The signal from LO and the output of BP filter are sent to a (2×4) $\pi/2$-hybrid and after it to two balanced detectors. The (2×4) 90°-hybrid and the balanced detectors demodulate and separate the received signal into in-phase (I) and quadrature (Q) components. The structure of (2×4) $\pi/2$-hybrid coupler, the balanced detectors, and their mathematical description can be found in many published works for coherent optical communication technology. This coherent detector can be simplified further if heterodyne detection is used, as shown in Figure 9.49b, and is commercially available by Discovery Semiconductor [6]. However, the electrical signals should be demodulated into I- and Q-components, as shown above, rather than the balanced receiver of Discovery Semiconductor.

Furthermore, an amplitude direct detector in the electronic digital processor can process the magnitude of the complex vector formed by the real I- and imaginary Q-components to obtain the amplitude and phase of the received signals. Thus their corresponding positions on the constellation can be determined and thence the decoding of the data symbols

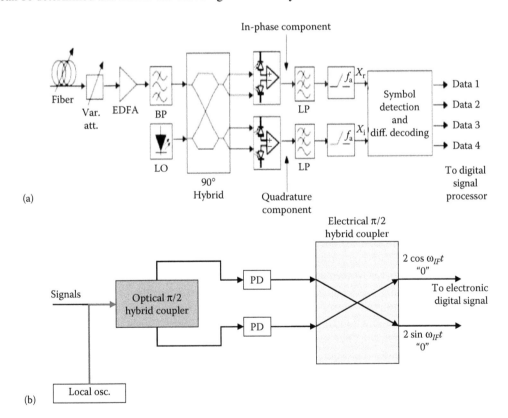

FIGURE 9.49 Coherent receiver for star 16-QAM: (a) homodyne and heterodyne I- and Q-detection model and (b) heterodyne model with optical and electrical $\pi/2$ hybrid couplers-electrical detection of I- and Q-components.

After LP filtering and sampling of I- and Q-components, the samples are sent to the symbol detection and differential decoding module. The sampling is done in the center of the eye diagram. In the symbol detection and differential decoding module, we first recover the symbols from the incoming samples and then make the differential decoding of symbols. In order to recover the symbols, the I- and Q-components are added together to a complex signal. Now, according to the original signal constellation, a decision must be made as to which symbol our complex sample must be mapped. This decision has two parts. At first, an amplitude decision is made to determine to which ring the sample belongs. After this, a phase decision takes place to determine to which of the eight possible symbols on a ring our sample belongs.

For amplitude decision, a known bit sequence for receiver (training sequence) is used and an amplitude threshold a_{th} is defined according to:

$$a_{th} = \{\max 1 \le k \le n \mid s_1(k)\mid + \min 1 \le k \le m \mid s_2(k)\mid\}/2 \qquad (9.22)$$

where $s_1(k)$ is the kth complex received samples of symbols on the inner ring and n is the whole number of symbols on the inner ring. In addition, $s_2(k)$ is the kth complex received sample of symbols on the outer ring and m is the whole number of symbols on the outer ring. If the amplitude of one sample is larger than the threshold, the symbol is decided to be on the upper ring; otherwise, it is decided to be on the inner ring. For the phase decision, the complex plane is divided into eight equiphase intervals. According to the interval in which the phase of sample falls, an index from 0 to 7 is assigned to this sample. The next steps are differential decoding and mapping. Amplitude differential decoding and phase differential decoding are done separately. From their results, the symbol detection and, after it, symbol-to-bit mapping are done in the inverse to that of the encoding in the transmitter.

Alternatively, the detection can be conducted with a heterodyne receiver that uses only one single $\pi/2$ hybrid optical coupler and then detected by the two photodiodes and then coupled through a $\pi/2$ electrical hybrid coupler to detect the I- and Q-components for the phase and amplitude reconstructions of the received signals, as shown in Figure 9.49b. The phase estimation can then be estimated by processing I- and Q-signals in the electronic domain, as described in the next section.

Owing to the two levels of Star 16-QAM, there must be an amplitude detection subsystem, which can be implemented using a single PD, followed by an electronic preamplifier, as shown in Figure 9.51. An electronic processor would be able to determine the position of the received signals on the constellation, and hence, decoding can be implemented without any problem. The transmission performance presented in this article would not be affected. Only the technological implementation would affect, and hence, the electronic noise or optical noise contribution to the receiver can be taken into account.

9.3.6.4   Coherent Detection Receiver with Phase Estimation

The method and structure of this receiver are almost the same as for the previous receiver shown in Figure 9.49. The difference here is that a phase estimation is done before the phase decision in the symbol detection and differential decoding module. The dispersion of the single-mode optical fiber is purely a phase effect and thus causes phase rotation and results as phase decision error. The effect of dispersion on Star 16-QAM format is shown in Figure 9.50.

The left plot (a) in the figure shows the sampled input signal into the fiber, and the plot on the right side shows the sampled output of the fiber with a CD of 300 ps/nm. The fiber is considered as linear in this simulation. Comparison of point A in both figures shows that this point is spread and rotated due to dispersion. The spreading causes both phase and amplitude distortion, whereas rotation causes only phase distortion. To solve the problem of phase rotation, the following phase estimation method is implemented. Generally, this phase estimation method is for phase synchronization between LO with linewidth and signal to replace the optical phase lock loop.

The incoming signal after sampling can be described as

$$c(k) = Ae^{j(\Phi'\text{tot}(k)\pm\text{mod}(k))} \qquad (9.23)$$

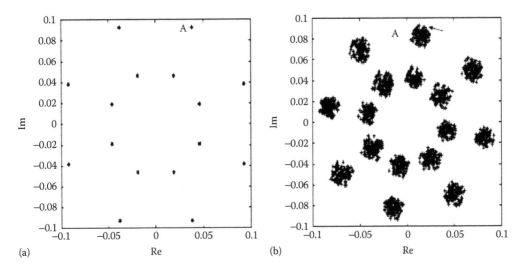

FIGURE 9.50 Signal constellation before (a) and after (b) propagation through the optical fiber link-phase rotation.

Where Φ'_{tot} is the phase distortion due to dispersion and noise and ϕ_{mod} is the signal phase that must be recovered. Now, ϕ_{tot} must be eliminated from $c(k)$. Φ_{mod} values are $\pi/8$, $3\pi/8$, $5\pi/8$, $\pi/8$, $9\pi/8$, $11\pi/8$, $13\pi/8$, and $15\pi/8$. If this phase values are multiplied by 8, then:

$$c_8(k) = c^8(k) = A^8 e^{j(8'\text{tot}(k)+8\_\text{mod}(k))} = A^8 e^{j(8'\text{tot}(k)+\_)} = -A^8 e^{j(8'\text{tot}(k))} \tag{9.24}$$

and from this:

$$\Phi'_{tot}(k) = 1/8\ arg(-c_8(k)) \tag{9.25}$$

$\phi'_{tot}(k)$ is the estimated phase for $\phi'_{tot}(k)$. In the simulations in this work, $arg(c_8(k))$ is filtered (the filter makes average between 20 neighbor symbols) to avoid the phase jumps from symbol to symbol. The filter order of 20 is not optimized for each CD.

Now, the signal phase $\phi_{mod}(k)$ can be estimated as

$$\phi_{mod}(k) = arg(c(k)) - \phi'_{tot}(k) = \Phi_{mod}(k) + \phi'_{tot}(k) - \phi'_{tot}(k). \tag{9.26}$$

After this phase estimation, the signal decision takes place with the same method as for amplitude decision case.

9.3.6.5 Direct Detection Receiver

The block diagram of the DD receiver is shown in Figure 9.51. After the optical filter, the signal is split into two branches via a 3 dB coupler. We name these two branches as the intensity branch and the phase branch. In the phase branch, the phase differential demodulation is done in the optical domain. The signal and the delayed signal at T_s (symbol duration) are sent into the (2×4) $90°$-hybrid and after that again into balanced detectors. At the output of the balanced detectors, the in-phase and quadrature components of the demodulated and differential decoded and received signal can be derived. After electrical filtering, the signal is sampled and then sent into the symbol detection module. In amplitude branch, the amplitude is determined and differentially decoded. After the photo diode, the signal is LP filtered and sampled and then fed into the amplitude detection and differential decoding module; a well-known optical coherent structure can be used to accomplish the amplitude decision and differential decoding. At the end, the in-phase and quadrature components

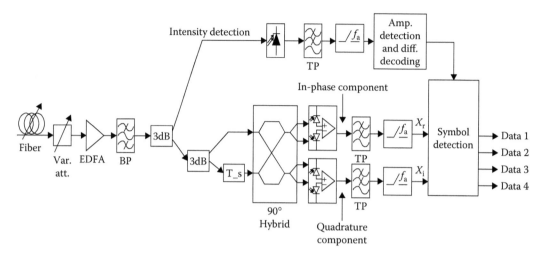

FIGURE 9.51   Direct detection receiver for star 16-QAM.

and the amplitude branch are sent to the symbol detection module for further processes as symbol detection and symbol-to-bit mapping.

9.3.6.6   Coherent Receiver without Phase Estimation

In this section, the required OSNRs at 10^{-4} BER (using Monte Carlo simulation) are determined for the coherent and incoherent DD receivers. For each detection method, the optimum RR is obtained to minimize the OSNR at BER = 10^{-4}. (The BER is determined via Monte Carlo simulations.) After that, the DT at 2 dB OSNR penalty at BER = 10^{-4} is determined. Only in this work, the OSNR penalty is 2 dB; however, it can also have other values. The OSNR penalty is defined as the OSNR difference in dB between the OSNR of back-to-back case and the OSNR of other CD values. DT is the CD interval that can be achieved with a certain OSNR penalty. In practice, DT describes how much dispersion (residual dispersion) a system can tolerate with an OSNR penalty smaller than 2 dB. The simulations in this work are done via the simulation tool. The simulations are done for the linear and nonlinear channels. The simulation parameters are given in Table 9.6. The average input power of nonlinear fiber in this work is always 5 dBm.

9.3.6.6.1   Linear Channel

In Figure 9.52b, the optimum RR (RR_{opt}) can be seen for each CD. The optimum RR is the RR that minimizes the OSNR for the given CD. The optimum RR changes here nearly linear with CD and can be expressed as

$$RR_{opt} = -0.002\,|\,CD\,|+1.92 \quad \text{for } 50 \text{ ps/nm} \leq |\,CD\,| \leq 300 \text{ ps/nm} \tag{9.27}$$

TABLE 9.6

Simulation Parameters for 16-STAR QAM

$f_s = 10.7$ GHz	$\lambda_c = 1550$ nm
$P_{laser} = 7$ dBm	$\alpha_{att} = 0.21$ dB/km
$\gamma = 0.00137$ 1/W/m	$F_n = 5$ dB
$G_{EDFA} = 30$ dB	fil-opt. = Gaussian 1. order
$B_{opt} = 44$ GHz	$\lambda_{LO} = 1550$ nm
$P_{LO} = 0$ dBm	fil-el. = Butterworth 3. degree
$B_e = 11$ GHz	$R_{photodiode} = 1$ A/W

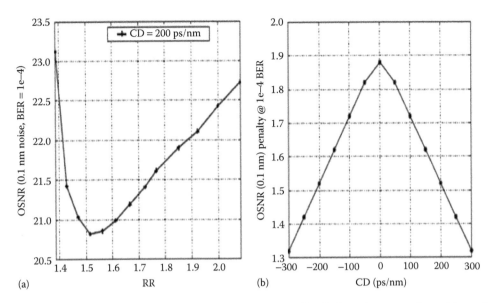

FIGURE 9.52 OSNR versus RR (a) and optimum RR versus CD (b) for coherent receiver without phase estimation.

The optimum RR increases with CD because increasing of CD means increasing of phase rotation due to dispersion. This causes more phase detection errors and thus a higher OSNR requirement. Increase in RR reduces the phase error probability but increases the amplitude error probability. The optimum RR is the best trade-off between phase errors and amplitude errors for each CD.

In the back-to-back case, RR_{opt} is around 1.87. The theoretical value obtained for RR_{opt} is 1.77. The difference is because in introduced coherent receiver, at first, the symbol detection is done and then the differential decoding is done. In the case where differential decoding is done before symbol detection, the optimum RR is around 1.77, as excepted. To determine the RR_{opt} for each CD, RR is changed for each CD. The RR value that yields the smallest OSNR is RR_{opt}. The RR step (which determines the RR accuracy) in simulations is 0.05. The characteristic for other CDs is similar to Figure 9.52a. To compare the OSNRs, the reference in this work is the OSNR from back-to-back case. In case of residual dispersion, it is of interest that how the system performance changes with the change of RR. The simulation results of three different RR can be seen in Figure 9.53. As shown in Figure 9.53a, the OSNR performance for the back-to-back case is decreased if (in simulated interval) the RR from 1.87 decreases. A degradation of 6.5 dB is determined if the RR from 1.87 decreases to 1.32. From other side, the DT at 2 dB OSNR penalty increases (Figure 9.9b). The DT for RR = 1.87 is 220 ps/nm and for RR = 1.32 is 460 ps/nm. To understand the reason for this behavior, Figure 9.52a should be considered again. For each CD, if the RR decreases from the RR_{opt} value, the OSNR increases rapidly. This means that RR = 1.32 for back-to-back case has increased the OSNR. However, at CD = 300 ps/nm, the signal has the minimum OSNR for this RR that leads to an increase of the OSNR for B2B (back-to-back) case. Furthermore, under the case without phase estimation with CD = 300 ps/nm the OSNR decreases. This effect causes a larger DT at a certain OSNR penalty. In practical system, according to higher requirement for OSNR or DT, the RR can be chosen.

In Figure 9.53a, it can be seen that the required OSNR for |CD| = 300 ps/nm makes a jump compared with other CDs. With |CD| = 350 ps/nm, it is not possible to reach a BER of 10^{-4}. The reason is that |CD| = 350 ps/nm is the limit of the system. For this CD, it is not possible to transmit error free even without noise, owing to phase rotations caused by dispersion (phase detection error). The signal constellation of received signal after sampling with CD = 350 ps/nm can be seen in

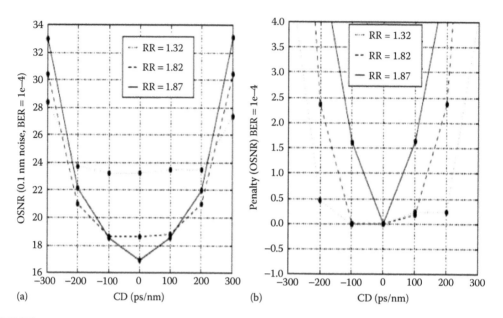

FIGURE 9.53   OSNR versus CD (a) and OSNR penalty versus CD (b) for coherent receiver.

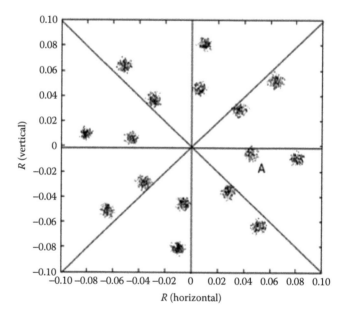

FIGURE 9.54   Received signal constellation with CD = 350 ps/nm without noise for coherent receiver without phase estimation in linear channel.

Figure 9.54; for example, some of the received symbols of A are over the phase threshold line and they generate the detection errors. Typical eye diagram at the output of the coherent receiver at the limit of the distortion is shown in Figure 9.55. Phase estimation can be implemented in the digital signal processor. It is noted that the signal constellation is rotated uniformly, owing to the property of the SMF that is purely phase distortion, and thus, in the processing of the constellation, it is best if the reference frame of the phasor diagram is rotated to align with the constellation, thus simplifying the phase estimation process at $\pi/8$ and its multiple values.

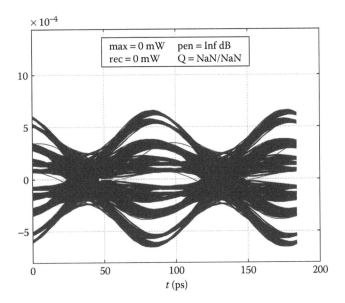

FIGURE 9.55 Typical eye diagram at the output of the coherent receiver under significant distortion limit of the 2A-8P Star 16-QAM.

9.3.6.6.2 Nonlinear Effects

The optimum RR in case of nonlinear channel for different CDs is shown in Figure 9.56. As mentioned for linear channel, here, the RR_{opt} decreases with increase of CD also. The difference in Figure 9.52 is that the diagram is not symmetric. The reason of this is the interaction between dispersion and SPM. For a positive CD, the dispersion and SPM have a constructive interaction, which results in a better OSNR performance. For a negative CD, the dispersion and SPM have destructive interaction, and it results in a much worse OSNR performance. This effect can be appreciated when comparing Figure 9.52b with Figure 9.57. The curve slope for negative CD is higher in nonlinear channel. It means the phase distortion is higher in the nonlinear channel. For the positive CD, the

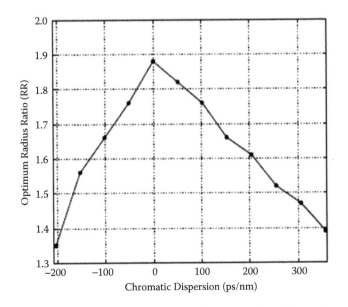

FIGURE 9.56 RR_{opt} versus CD in nonlinear channel for coherent receiver without phase estimation.

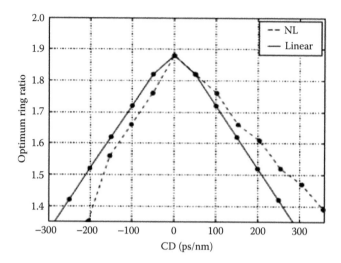

FIGURE 9.57 RR_{opt} comparison in linear and nonlinear channel for coherent receiver without phase estimation.

slope in the nonlinear channel is lower, and it means the phase distortion is less. The simulation results for different RR are shown in Figure 9.58. Here, it can be seen again that the required OSNR for the back-to-back case increases with decrease of RR. Similarly, for the same reason as for the linear case, the DT at 2 dB OSNR penalty increases as well (Figure 9.59b). Figure 9.59 shows the eye diagram of the 16-QAM detected at the output of a balanced receiver at a baud rate of 10 GB/s.

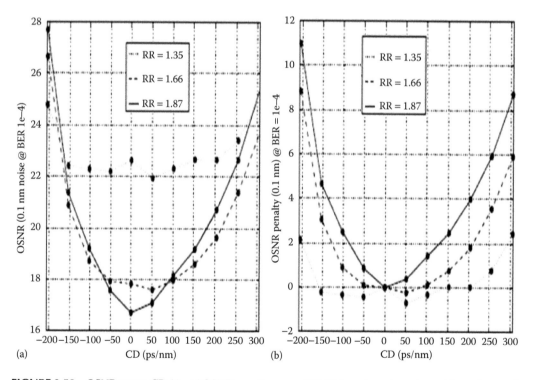

FIGURE 9.58 OSNR versus CD (a) and OSNR penalty versus CD (b) for coherent receiver without phase estimation in nonlinear channel. Note that no equalization, only phase estimation processing of I- and Q-components in electrical domain.

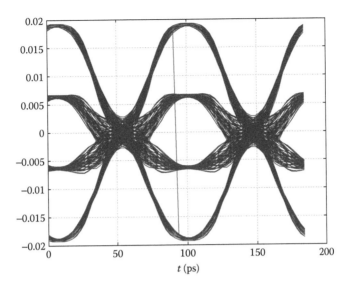

FIGURE 9.59 Eye diagram of the 16-QAM detected at the output of a balanced receiver. Bit rate of 40 Gb/s and baud rate of 10 GB/s.

9.3.6.7 Remarks

The design of a Star 16-QAM modulation scheme is proposed for CoD for ultrahigh-speed and ultrahigh-capacity optical fiber communications schemes. Two amplitude levels and eight phases (2A-8P 16_QAM) are considered to offer simple transmitter and receiver configurations and, at the same time, the best receiver sensitivity at the receiver. An optical SNR of about 22 db is required for the transmission of Star 16-QAM over optically amplified transmission dispersion-compensated link. DT of 300 ps/nm is possible with 1 dB penalty of the eye opening at 40 Gb/s bit rate or 10 GSy/s with an OSNR of 18 dB. An OSNR could be about 22 dB for 107 Gb/s bit rate and a symbol rate of 26.75 G symbols/sec. The transmission link consists of several spans of total 1000 km dispersion-compensated optically amplified transmission link. Optical gains of the in-line OAs are set to compensate for the attenuation of the transmission and compensating fibers with an NF of 3 dB.

The optical transmitter and receivers incorporating commercially available coherent receiver are structured and are sufficient for engineering of the optical transmission terminal equipment for the bit rate of 107 Gb/s and a symbol rate of 26.3 GSy/s. Furthermore, electronic equalization of the receiver PSK signals can be done using blind equalization, which would improve the DT much further. For a symbol rate of 10.7 Gb/s, this DT for 1 dB penalty would reach 300 km of SSMF. This electronic equalization can be implemented without any difficulty at the 10.7 GSy/s. For 107 Gb/s bit rate, similar improvement of the DT can be at 26.5 GSy/s, provided that the electronic sampler can offer more than 50 GSa/s sampling rate.

9.4 OTHER MULTILEVEL AND MULTISUBCARRIER MODULATION FORMATS FOR 100 GB/S ETHERNET TRANSMISSION

Numerous technologies have been introduced in recent years to cope with the ever-growing demand for transmission capacity in optical communications. Although the optical SMF offers enormous bandwidth in the order of magnitude of 10 THz, efficient exploitation of bandwidth had started to become an issue a couple of years ago. Moreover, limited speed of electronics and electro-optic devices such as modulators and photo receivers are considered bottleneck for further increase of data rate based on binary modulation.

For all these reasons, optical modulation formats offering a high ratio of bits per symbol are an essential technology for next generation's high-speed data transmission. This way, data throughput can be increased while required bandwidth in the optical domain and for electronics is kept on a lower level.

Based on the demand for transmission technologies offering high ratio of bits per symbol, two promising candidates to achieve a data rate of 100 Gb/s per optical carrier, namely optical OFDM and 16-ary multilevel modulation, are discussed,

In this section, two promising candidates, namely optical OFDM and 16-ary multilevel modulation are introduced. Performance is analyzed by means of numerical simulation and experiment.

9.4.1 MULTILEVEL MODULATION

Optical modulation formats incorporating four or eight bits per symbol were investigated in numerous contributions in the last couple of years (e.g., DQPSK [7] and 8-DPSK [8]). However, to carry out the step from 10 to 40 Gb/s data rate by using devices designed for 10 Gsymb/s, a number of 16 bits are required per symbol. The main challenge is to find the optimal combination of ASK and DPSK formats. Several approaches, which are reviewed in [3,6], are possible.

The simplest structure can be an extension of a 30 Gb/s 8-DPSK by an additional PM, resulting in 40 Gb/s 16-DPSK; that is, in the complex plane, 16 symbols are placed onto a unit circle, as shown in Figure 9.60a. Depending on the current bit at the data input of the additional PM, the 8-DPSK symbol is shifted by $\pi/8$ in case of a "1," whereas in case of a "0," the incoming phase of the symbol is preserved. Although it seems simple, experimental implementations have shown that the phase stability of the modulators and corresponding demodulators is very critical. Thus, experimental set-up must be stabilized. Moreover, 16-DPSK is suboptimum regarding exploitation of the full area that the complex plane offers; that is, the ratio of symbol distance and signal power is low, resulting in poor receiver sensitivity.

Similar behavior can be found for the other extreme case of 16-ASK, as shown in Figure 9.60b. Here, the 16 symbols are placed onto the positive real axis. The symbol distance is extremely narrow, thus resulting in poor sensitivity.

Much improved performance can be achieved by combining the amplitude- and phase-shift keying modulations of ASK and DPSK, that is, the M-ary ADPSK, as described above. There are a number of combinations of the M-ary ADPSK. One approach can be the extension of an 8-DPSK by two

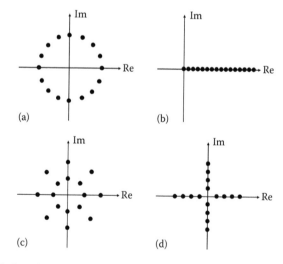

FIGURE 9.60 Constellation of symbols in the complex plane for (a) 16-DPSK, (b) 16-ASK, (c) Star 16-QAM, and (d) 16ADPSK.

rings of ASK levels. Thus, the 16 symbols appear as two rings in the complex plane, with 8 symbols per ring, as shown in Figure 9.60c. Alternatively, this topology is known as Star-16 QAM. A second structure is given by combining the DQPSK with four-level ASK (or equivalently M-ary ADPSK), resulting in four rings with four symbols each, as analyzed in above sections and shown in Figure 9.60d. Both structures utilize effectively the complex plane. However, both structures require that the sensitivities or the magnitudes (diameters) of the rings have to be optimized to compromise the sensitivity performance of the DPSK and the ASK geometrical distribution. Especially, the inner ring has to be of sufficient size to enable distinction of the different phases of the symbols on this ring. They are limited by the nonlinear effects of the transmission fiber and the noises contributed by the receiver and the in-line OAs. Hence, the distance between the constellations is limited by these two limits.

A strategy to mitigate this trade-off was introduced in [7] by using a special pulse shaping called inverted RZ. For binary ASK in conjunction with inverse RZ, a "0" is encoded as temporary breakdown of the optical power, whereas for a "1," the optical power remains at high level. Using this pulse shaping, for the M-ary ADPSK format, the four levels of the QASK part are transmitted by means of four different values for temporary decay of optical power. The DQPSK part, however, is transmitted by modulating the phase of the signal in the time slot between the symbols, which implies that in the transmitter, the phase of the signal can be detected while the signal has maximum power.

Measurement results for this modulation format are depicted in Figure 9.61, where the BER is plotted as function of the OSNR measured within a 0.1 nm optical filter bandwidth. The main outcome is the fact that the DQPSK-part is insignificantly disturbed by an additional QASK part. Moreover, even after transmission over 75 km of SSMF, the DPSK part shows a very low penalty. In contrary, the QASK component inherently shows low performance due to low symbol Euclidean distance. Improvement might be achievable by optimizing the duty cycle of inverse RZ and the RR. A simulated eye diagram of the 16 square QAM with the constellation before and after the transmission over two optical spans with 2 km SSMF mismatch is shown in Figure 9.42.

9.4.2 OPTICAL ORTHOGONAL FREQUENCY DIVISION MULTIPLEXING

The OFDM is a transmission technology that is primarily known from wireless communications and wired transmission over copper cables. It is a special case of the widely known frequency division multiplexing (FDM) technique, for which digital or analog data are modulated onto a

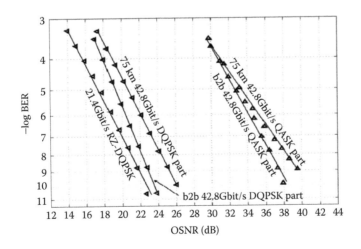

FIGURE 9.61 Measurement of BER results for 16-ary inverse RZ ADPSK modulation format transmission.

certain number of carriers and transmitted in parallel over the same transmission medium. The main motivation for using FDM is the fact that because of parallel data transmission in frequency domain, each channel occupies only a small frequency band. Signal distortions originating from frequency-selective transmission channels, the fiber CD, can be minimized. The special property of OFDM is characterized by its very high spectral efficiency. While for conventional FDM, the spectral efficiency is limited by the selectivity of the bandpass filters required for demodulation, OFDM is designed such that the different carriers are pairwise orthogonal. This way, for the sampling point, the intercarrier interference is suppressed, although the channels are allowed to overlap spectrally.

Orthogonality is achieved by placing the different RF carriers onto a fixed-frequency grid and assuming rectangular pulse shaping. It can be shown that in this special case, the OFDM signal can be described as the output of a discrete inverse Fourier transform, with the parallel complex data symbols as input. This property has been one of the main driving aspects for OFDM in the past, since modulation and demodulation of a high number of carriers can be realized by simple DSP, instead of using many LOs in transmitter and receiver. Recently, OFDM has become an attractive topic for digital optical communications. It is just another example of the current tendency in optical communications to consider technologies that are originally known from classical digital communications. Using OFDM appears to be very attractive, since the low bandwidth occupied by a single OFDM channel increases the robustness toward fiber dispersion drastically, allowing the transmission of high data rates of 40 Gb/s and more over hundreds of kilometers without the need for dispersion compensation. In the same way as for modulation formats such as DPSK or DQPSK, which were introduced in recent years, for OFDM, the challenge for optical system engineers is to adapt a classical technology to the special properties of the optical channel and the requirements of optical transmitters and receivers.

Thus, two approaches have been reported recently. An intuitive approach introduced by Lorente et al. [9] makes use of the fact that the WDM technique itself already realizes data transmission over a certain number of different carriers. By means of special pulse shaping and carrier wavelength selection, the orthogonality between the different wavelength channels can be achieved, resulting in the so-called orthogonal WDM technique. However, in this way, the option of simple modulation and demodulation by means of discrete Fourier transform cannot be utilized, as this kind of DSP is not available in the optical domain.

As an alternative and popular method, an electrical OFDM signal is generated by means of electrical signal processing, followed by modulation onto a single optical carrier. This approach is known as optical OFDM (oOFDM). Here, the modulation is a two-step process: first, the electrical OFDM signal is already a broadband bandpass signal, which is then modulated onto the optical carrier. Second, to increase data throughput, oOFDM can be combined with WDM, resulting in multi-Tb/s transmission system, as shown in Figure 9.62. Nevertheless, oOFDM itself offers different options for implementation. An important issue is optical demodulation, which can be realized by means of either DD or CoD by using an LO. The DD is preferable owing to its simplicity. However, for DD, the optical intensity has to be modulated. Owing to the fact that the electrical OFDM signal is quasi-analog, with zero mean and high peak-to-average ratio, the majority of the optical power has to be wasted for the optical carrier (i.e., an additional direct current value of the complex baseband signal), resulting in low receiver sensitivity. For CoD, in addition, the bandwidth efficiency is twice as high as for DD, since for pure intensity modulation, a double-sideband signal is inherently generated. For CoD, a complex optical I-Q modulator composed of two real modulators in parallel, followed by superposition with $\pi/2$ phase shift, allows for transmission of twice as much data within the same bandwidth. For intensity modulation, the bandwidth efficiency may be increased by suppressing one of the redundant sidebands, resulting in optical OFDM with single-sideband transmission. First, the serial data at the input to the system, can be converted to parallel streams. These parallel data sequences are then mapped to QAM constellation in the frequency domain and then by inverse fast Fourier

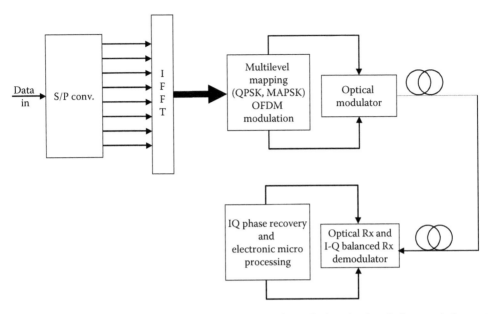

FIGURE 9.62 Schematic diagram of optical OFDM transmitter of a long-haul optical transmission system using multilevel modulation formats.

transform (IFFT), they are converted back to the time domain. The time domain signals are in the I- and Q-components, which are then fed into I and Q optical modulators. This optical modulation can be DQPSK or any other multilevel modulation subsystem. At the end of the optical fiber transmission, I- and Q-components are detected either by direct detection or by CoD. For CoD, a 2×4 $90°$ hybrid coupler is used to mix the polarizations of LO and that of the received signals. The outputs of the couplers are then fed into the balanced optical receivers. The mixing of the local laser source and of the signals preserves the phases of the signals, which are then processed by a high-speed electronic processor. For DD, the I- and Q-components are detected differentially; the amplitude and phase detection are then compared and processed similarly as for the coherent case.

In order to show the robustness of oOFDM toward fiber dispersion and also fiber nonlinearity, numerical simulations are carried out for a data stream of 42.7 Gb/s data rate. The number of OFDM channels can be varied between $N_{min} = 256$ and $N_{max} = 2048$. A guard interval of 12 ns can be inserted, a strategy belonging inherently to OFDM technology; it ensures orthogonality of the different channels in case of a transmission channel with memory. For the optical modulation, intensity modulation using a single MZM in conjunction with single-sideband filtering and DD was implemented. The nonlinear optical transmission channel consisted of eight 80-km non-dispersion-compensating fiber (DCF) spans of SSMF. As a criterion for performance, required OSNR for a BER of 10^{-3} (Monte Carlo) is measured. Using FEC, after decoding, this is transferred into a BER below 10^{-9}, depending on the specific code.

Figure 9.63 shows the required OSNR as function of fiber launch power for different values of N. The most important result is that transmission is possible over 640 km over SSMF without any dispersion compensation. It can be explained by the fact that even for the lowest value of $N_{min} = 256$, each subchannel occupies a bandwidth of approximately 42.7 GHz/256 = 177 MHz, resulting in high robustness toward fiber dispersion.

The principal difficulties of optical OFDM are that the pure delay due to the variation of the refractive index of the fiber with respect to the optical frequency leads to bunching of the subchannels and hence to the increase of the optical power; thus, unexpected SPM may occur in a random manner.

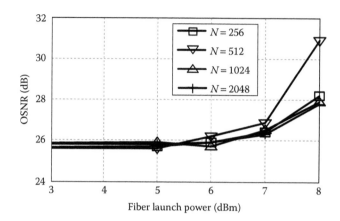

FIGURE 9.63 Simulation result for 42.7 Gb/s oOFDM transmission over 640 km of SSMF; OSNR required for BER = 10^{-3} (Monte Carlo simulation) as function of fiber launch power. (From Wree, C. et al., RZ-DQPSK format with high spectral efficiency and high robustness towards fiber nonlinearities. In *Proceedings of ECOC*, Copenhagen, Denmark, September 2002, paper 9.6.6. With permission.)

9.4.3   100 Gb/s 8-DPSK_2-ASK 16-Star QAM

9.4.3.1   Introduction

Multilevel modulation scheme enables the transmission baud rate to be reduced, thus obtaining the spectral efficiency. Another significant advantage of this modulation scheme is to reduce the requirement of high-speed processing electronics. This is of particular interest for high-speed optical transmission systems.

This part of the report investigates a multilevel modulation scheme, which has eight phases and two amplitude levels. This scheme, which is named in short as 8-DPSK_2-ASK, effectively utilizes 4 bits per one symbol for transmission, in which the first 3 bits are for coding phase information, whereas the coding of the amplitude levels is implemented with the 4th bit. As a result, the transmission baud rate is equivalently a quarter of the bit rate from bit pattern generator.

This section is organized as follows: Section 9.4.3.2 presents detailed description of the optical transmitter for generating 8-DPSK_2-ASK signals. In Section 9.4.3.3, the detailed configuration of the receiver is provided. The configurations of the optical transmitter and receiver are referred from those reported by Djordjevic and Vasic [10]. Section 9.4.3.4 provides study on DT and transmission performance of the 8-DPSK_2-ASK scheme. Finally, a short summary of the report is provided.

9.4.3.2   Configuration of 8-DPSK_2-ASK Optical Transmitter

There have been several different configurations of an optical transmitter for generating multiphase/level optical signals with the use of amplitude or PMs arranged in either serial or parallel configurations. However, the optical transmitters reported in [11–13] require a precoder with high complexity. On the contrary, the configuration reported Djordjevic and Vasic [10] utilizes the Gray mapping technique to differentially encode the phase information, and this significantly reduces the complexity of the optical transmitter. In addition, as elaborated in more detail in Section 9.3, this precoding technique enables the detection scheme using the I-Q demodulation techniques as equivalently in coherent transmission systems.

The optical transmitter of the 8-DPSK_2-ASK scheme employs the I-Q modulation technique, with two MZIMs in parallel and a $\pi/2$ optical PM, as shown in Figure 9.64. At each kth instance, the absolute phase of transmitted lightwaves θ_k is expressed as: $\theta_k = \theta_{k-1} + \Delta\theta_k$, where θ_{k-1} is the phase at $(k-1)$th instance and $\Delta\theta_k$ is the differentially coded phase information. The encoding of this $\Delta\theta_k$ for generating 8-DPSK_2-ASK modulated optical signals (4 bits per one transmitted symbol) follows

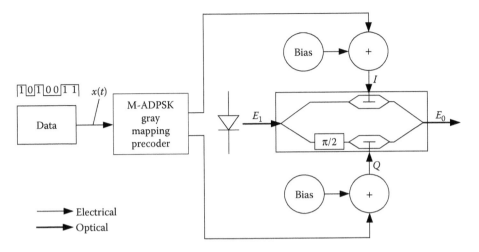

FIGURE 9.64 Optical transmitter configuration of the 8-DPSK_2-ASK modulation scheme.

the well-known Gray mapping rules. This Gray mapping phasor diagram is shown in Figure 9.65. The phasor is normalized with the maximum energy on each branch, that is, $E_{1/2}$.

The amplitude levels are optimized in order that the Euclidean distances d_1, d_2, and d_3 are equal, that is, $d_1 = d_2 = d_3$. After derivation, we obtain: $r_1 = 0.5664$. The I and Q field vectors corresponding to Gray mapping rules from the M-ADPSK precoder (Figure 9.64) are provided in Table 9.7.

The above-described transmitter configuration can be replaced with a dual-drive MZIM. The explanation and derivation for generating 8-DPSK_2-ASK optical signals are also based on the phasor diagram of Figure 9.65. In this case, the output field vector is the summation of two component field vectors, each of which is not only determined by the amplitude but also by initially biased phases.

TABLE 9.7
I and Q Field Vectors in 8-DPSK_2-ASK Modulation Scheme Using Two MZIMs in Parallel

Binary Sequence	$(\Delta\theta_k,$ Amplitude$)$	I_k	Q_k
1000	$(0, 1)$	1	0
1001	$(\pi/4, 1)$	$\sqrt{2}/2$	$\sqrt{2}/2$
1011	$(\pi/2, 1)$	0	1
1010	$(3\pi/4, 1)$	$-\sqrt{2}/2$	$\sqrt{2}/2$
1110	$(\pi, 1)$	-1	0
1111	$(-3\pi/4, 1)$	$-\sqrt{2}/2$	$-\sqrt{2}/2$
1101	$(-\pi/2, 1)$	0	-1
1100	$(-\pi/4, 1)$	$\sqrt{2}/2$	$-\sqrt{2}/2$
0000	$(0, 0.5664)$	$1 * 0.5664$	0
0001	$(\pi/4, 0.5664)$	$\sqrt{2}/2 * 0.5664$	$\sqrt{2}/2 * 0.5664$
0011	$(\pi/2, 0.5664)$	0	$1 * 0.5664$
0010	$(3\pi/4, 0.5664)$	$-\sqrt{2}/2 * 0.5664$	$\sqrt{2}/2 * 0.5664$
0110	$(\pi, 0.5664)$	$-1 * 0.5664$	0
0111	$(-3\pi/4, 0.5664)$	$-\sqrt{2}/2 * 0.5664$	$-\sqrt{2}/2 * 0.5664$
0101	$(-\pi/2, 0.5664)$	0	$-1 * 0.5664$
0100	$(-\pi/4, 0.5664)$	$\sqrt{2}/2 * 0.5664$	$-\sqrt{2}/2 * 0.5664$

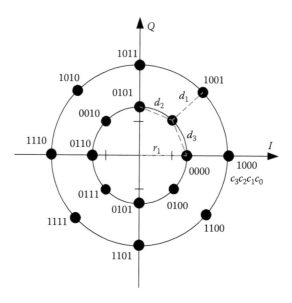

FIGURE 9.65   Gray mapping for optimal 8-DPSK_2-ASK modulation scheme.

9.4.3.3   Configuration of 8-DPSK_2-ASK Detection Scheme

The detection of 8-DPSK_2-ASK optical signals is implemented with the use of two MZDI balanced receivers (Figure 9.66).

Several key notes in this detection structure are stated as follows: (1) The MZDI introduces a delay corresponding to the baud rate. (2) One arm of MZDI has a $\pi/4$ optical phase shifter, whereas the other arm has an optical phase shift of $-\pi/4$. (3) The outputs from two balanced receivers are superimposed positively and negatively, which lead to I- and Q-detected signals, respectively. The I- and Q-detected components are expressed as $I = \mathrm{Re}\{E_k E_{k-1}^*\}$ and $Q = \mathrm{Im}\{E_k E_{k-1}^*\}$. (4) I- and Q-detected

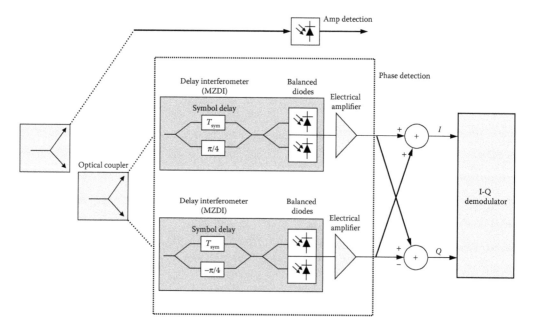

FIGURE 9.66   Detection of configuration for the 8-DPSK_2-ASK modulation scheme.

components are demodulated using the popular I-Q demodulator in the electrical domain. These detected signals are then sampled and represented as shown in the signal constellation.

9.4.3.4   Transmission Performance of 100 Gb/s 8-DPSK_2-ASK Scheme

Performance characteristics of the 8-DPSK_2-ASK scheme operating at 100 Gb/s bit rate are studied in terms of receiver sensitivity, DT, and the feasibility for long-haul transmission. Bit-error rates are the pre-forward error correct (pre-FEC) BERs, and the pre-FEC limit is conventionally referenced at 2e-3. In addition, the BERs are evaluated by the Monte Carlo method.

9.4.3.5   Power Spectrum

The power spectrum of 8-DPSK_2-ASK optical signals is shown in Figure 9.67. It can be observed that the spectral width of the main lobe is about 25 GHz, as the symbol baud rate of this modulation scheme is equal to a quarter of the bit rate from the bit pattern generator. The harmonics are not highly suppressed, thus requiring bandwidth of the optical filter to be necessarily large in order not to severely distort signals.

9.4.3.6   Receiver Sensitivity and DT

The receiver sensitivity is studied by connecting the optical transmitter of the 8-DPSK_2-ASK scheme directly to the receiver to make a back-to-back set-up (Figure 9.68). On the other hand, the DT is studied by varying the length of SSMF from 0 to 5 km (|D| = 17 ps/(nm.km)). Received powers are varied by using an optical attenuator. The optical Gaussian filter has bandwidth-time product (BT) = 3 (B is approximately 75 GHz). Modeling of receiver noise sources comprises shot noise, equivalent noise current density of 20 pA/$\sqrt{Hz}$ at the input of the transimpedance electrical amplifier,

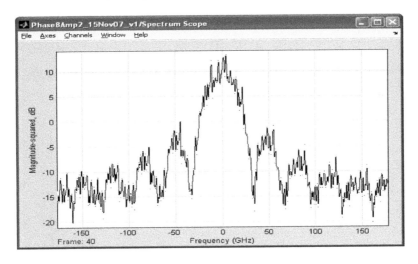

FIGURE 9.67   Power spectrum of 8-DPSK_2-ASK signals.

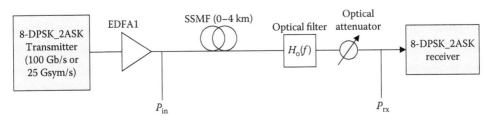

FIGURE 9.68   Set-up for the study of receiver sensitivity (back to back) and dispersion tolerance (0–4 km SSMF) for the 8-DPSK_2-ASK modulation scheme.

and dark current of 10 nA for each of the two photodiodes in balanced structure. A 5th-order Bessel electrical filter with a bandwidth of BT = 0.8 is used.

The numerical BER curves of the receiver sensitivity for cases of 0–5 km SSMF are shown in Figure 9.69. The receiver sensitivity of the 8-DPSK_2-ASK scheme is approximately–8.5 dBm at BER = 1e–4. The receiver sensitivity at BER = 1e–9 can be obtained by extrapolating the BER curve of 0 km case. The power penalty versus residual dispersion results are then obtained and plotted in Figure 9.70. It is realized that the 2 dB penalty occurs for the residual dispersion of approximately 60 ps/nm or equivalently to 3.5 km SSMF.

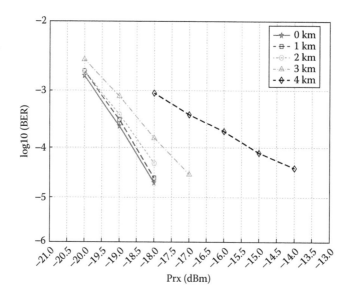

FIGURE 9.69 Plot of BER versus receiver sensitivity (back to back) and dispersion tolerance (0–4 km SSMF) for the 8-DPSK_2-ASK modulation scheme.

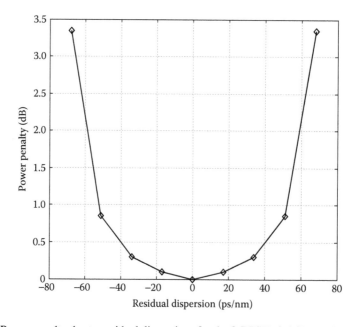

FIGURE 9.70 Power penalty due to residual dispersions for the 8-DPSK_2-ASK modulation scheme.

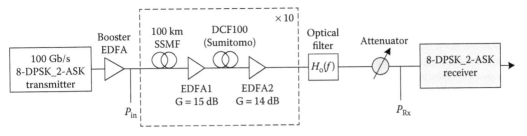

FIGURE 9.71   Transmission set-up of 1100 km optically amplified and fully compensated fiber link.

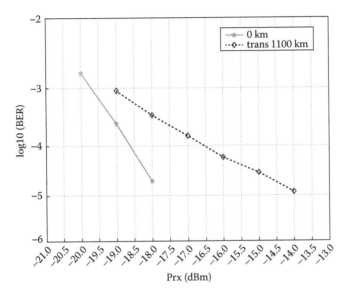

FIGURE 9.72   Plot of BER versus receiver sensitivity for 8-DPSK_2-ASK modulation format transmission.

9.4.3.7   Long-Haul Transmission

The long-haul transmission performance of this modulation format is conducted over 10 optically amplified and fully compensated spans, and each span is composed of 100 km SSMF and 10 km DCF100 (Sumitomo fiber). As a result, the length of the transmission fiber link is 1100 km. This long-haul range is selected to reflect the distance between Melbourne and Sydney. The wavelength of interest is 1550 nm, and the dispersion at the end of the transmission link is fully compensated. The simulation set-up is shown in Figure 9.69. In addition, the fiber attenuation due to SSMF and DCF is also fully compensated by using two EDFAs, with optical gains as depicted in Figure 9.71. These EDFAs have an NF set at 5 dB.

Numerical transmission BERs are plotted against received powers in Figure 9.72 and are compared with the back-to-back BER curve. It can be observed that the BER curve of 1100 km follows a linear trend and feasibly reaches 1e–9 if extrapolated, as shown in Figure 9.72. It should be noted that this transmission performance can be significantly improved with the use of high-performance FEC scheme.

9.5   CONCLUDING REMARKS

The ever-increasing bandwidth hunger in telecommunication networks based mainly on optical fiber communication technology indicates that low-bandwidth-efficient modulation formats such as ASK would no longer satisfy the transmission capacity demands, and new advanced optical modulation schemes should replace ASK in the near future. Advanced optical modulation schemes, especially the multilevel amplitude and phase schemes presented above, are able to: (1) provide long-reach, error-free, and high transmission capacity, (2) have high bandwidth efficiency (no. bits/Hz parameter), (3) push the bit rate well above that could be offered by electronic technology, for example, 100 Gb/s with the detection at the symbol rate, (4) tolerate dispersion and nonlinearity, and (5) maximally utilize the existing optical network infrastructure.

Current developments of photonic technology have enabled the use of differential phase modulation and demodulation in optical domain. At present, BDPSK and DQPSK formats have received a great attention, owing to their improvement in the receiver sensitivity as compared with ASK. Furthermore, the RZ and CSRZ formats would assist the combat with nonlinear effects. However, as long as transmission capacity and bandwidth efficiency are concerned, MADPSK modulation formats would be perform better in trading off for its complexity in the receiver structures.

Alternatively, there are other possible multilevel modulation schemes that would offer further improvement of optical transmission performance [14].

9.5.1 Offset MADPSK Modulation

The binary DPSK and quaternary DPSK (DQPSK) are just special cases of a more general class of differential phase modulation formats that map data bits into phase difference between neighboring symbols. This phase difference $\Delta\phi_i$ can be described as

$$\Delta\phi_i = \theta + \frac{2\pi(i-1)}{M}, \quad i = 1, 2, ... M - 1 \tag{9.28}$$

where θ is the initial phase and M is total number of phase levels. This class of formats is called offset DPSK and is denoted as θ-M-DPSK. Specifically, with $\theta = 0$, $M = 2$ or $M = 4$ θ-M-DPSK become 0-2-DPSK or 0-4-DPSK, which are the conventional binary DPSK and DQPSK mentioned above.

Offset DPSK has been used to transmit over satellite nonlinear channel because its phase transition is smooth and can avoid 180° phase jump. As fiber medium also exhibits nonlinearity, this modulation format attracts our attention as a candidate, together with ASK, for creating a new multilevel modulation format, possibly termed offset MADPSK.

9.5.2 Multilevel Amplitude-Minimum Shift Keying Modulation

Minimum-shift keying (MSK) is a form of OQPSK with sinusoidal pulse weighting. In MSK, data bits are first coded into bipolar signals ± 1, which are then separated into $V_I(t)$ and $V_Q(t)$ streams consisting of even and odd bits, respectively. In the next stage, $V_I(t)$ and $V_Q(t)$ are used to modulate a carrier f_c to create a MSK signal $s(t)$, which can be presented as

$$s(t) = V_I(t)\cos\left(\frac{\pi t}{2T}\right)\cos\left(2\rho\pi f_c t\right) + V_Q(t)\cos\left(\frac{\pi t}{2T}\right)\sin\left(2\pi f_c t\right) \tag{9.29}$$

The MSK format is a well-known modulation format in wireless communications with efficient spectral characteristics owing to high compactness of its main lobe, compared with DPSK, and high suppression of the side lobes. These characteristics indicate that MSK signal is highly dispersion tolerant. A combination of MSK and ASK into MAMSK modulation would even improve the transmission performance without increasing the complexity of the detection scheme.

Multilevel techniques for 100 Gb/s are given; in particular, the OFDM with multicarrier and multilevel amplitude modulations with orthogonality between adjacent channels are proven to be cost-effective and appropriate for current electronic technologies. Two interesting approaches to achieve data transmission of 40 Gb/s and beyond (e.g., 100 Gb/s Ethernet) based on low symbol rate are discussed. On the one hand, oOFDM can combine a large number of parallel data streams into one broadband data stream with high spectral efficiency. Simulation results are shown for different values of the number of parallel data streams in nonlinear environment. On the other hand, 16-ary modulation formats enable 40 Gb/s transmission with 10 GSym/s (i.e., 100 Gb/s with 25 GSym/s). For a special case, namely inverse RZ 16ADQPSK, measurement results are demonstrated.

9.5.3 STAR QAM COHERENT DETECTION

These last two sections of the chapter describe two optical transmission schemes using both coherent and incoherent transmission and detection techniques, the 2A-8P Star QAM and 8DPSK 2ASK 16-Star QAM, for 100 Gb/s or 25 G Symbols/sec.

First, the design of a Star 16-QAM modulation scheme is proposed for CoD for ultrahigh-speed, ultra-high-capacity optical fiber communications schemes. Two amplitude levels and eight phases (2A-8P 16_QAM) are considered to offer significant simple transmitter and receiver configurations and, at the same time, the best receiver sensitivity at the receiver. An optical SNR of about 22 db is required for the transmission of Star 16-QAM over optically amplified transmission dispersion-compensated link. DT of 300 ps/nm is possible with 1 dB penalty of the eye opening at 40 Gb/s bit rate or 10 GSy/s, with an OSNR of 18 dB. An OSNR could be about 22 dB for 107 Gb/s bit rate and a symbol rate of 26.75 G symbols/sec. The transmission link consists of several spans of total 1000 km dispersion-compensated optically amplified transmission link. Optical gains of the in-line OAs are set to compensate for the attenuation of the transmission and compensating fibers with an NF of 3 dB. The optical transmitter and receivers incorporating commercially available coherent receiver are structured and are sufficient for engineering of the optical transmission terminal equipment for the bit rate of 107 Gb/s and a symbol rate of 26.3 Gbauds/sec.

Furthermore, electronic equalization of the receiver PSK signals can be done using blind equalization, which would improve the DT much further. For a symbol rate of 10.7 Gb/s, this DT for 1 dB penalty would reach 300 km of SSMF. This electronic equalization can be implemented without any difficulty at the 10.7 Gbauds/sec. For 107 Gb/s bit rate, similar improvement of the DT can be at 26.5 Gbauds/sec, provided that the electronic sampler can offer more than 50 Gig sampling rate.

Second, the transmitter and receiver configurations for generating 8-DPSK_2-ASK optical signals and DD, as well as the transmission performances, are described. In addition, performance characteristics of this modulation format at 100 Gb/s (equivalently 25 Gbauds/s) has also been investigated in terms of the receiver sensitivity, DT, and the long-haul transmission performance. The simulation results shows that 8-DPSK_2-ASK is a promising modulation for very-high-speed (100 Gb/s) and long-haul optical communications.

REFERENCES

1. L.N. Binh, *Monash Optical Communication System Simulator. Part I: Ultra-long Ultra-High Speed Optical Fiber Communication Systems*, Manual for Technical Development, ECSE Monash University, Melbourne, Australia, 2004.
2. L.N. Binh, *Monash Optical Communication System Simulator. Part II: Ultra-long Ultra-High Speed Optical Fiber Communication Systems*, Manual for Technical Development, ECSE Monash University, Melbourne, Australia, 2004.
3. S. Hayase et al., Proposal of 8-state per symbol (binary ASK and QPSK) 30-Gb/ optical modulation demodulation scheme, *Proceedings of the European Conference on Optical Communication*, paper Th2.6.4, pp. 1008–1009, Rimini, Italy, September 2003.
4. K. Sekine et al., 40 Gb/s 16-ary (4 bit/symbol) optical modulation/demodulation scheme, *IEEE Electron. Lett.*, 41(7), March 2005.
5. L.N. Binh, *Advanced Digital Optical Communications*, 2nd edn., CRC Press, Boca Raton, FL, 2015.
6. C. Wree, et al., *Coherent Receivers for Phase-Shift Keying Transmission*, Discovery Semiconductors.
7. L.N. Binh, *Digital Signal Processing: Optical Transmission and Coherent Reception Techniques*, Chapter 3, Section 3.3.3, CRC Press, Boca Raton, FL, 2014.

8. L.N. Binh, *Digital Signal Processing: Optical Transmission and Coherent Reception Techniques*, Chapter 3, Section 3.3.5, CRC Press, Boca Raton, FL, 2014.

9. R. Llorente, J.H. Lee, R. Clavero, M. Ibsen, and J. Martí, Orthogonal wavelength-division-multiplexing technique feasibility evaluation, *J. Lightwave Technol.*, 23, 1145–1151, March 2005.

10. I.B. Djordjevic and B. Vasic, Multilevel coding in M-ary DPSK/differential QAM high-speed optical transmission with direct detection, *IEEE J. Lightwave Technol.*, 24, 420–428, 2006.

11. M. Serbay, C. Wree, and W. Rosenkranz, Experimental investigation of RZ-8DPSK at 3x 10.7Gb/s. In *LEOS'05*, 2005.

12. M. Seimetz, M. Noelle, and E. Patzak, Optical systems with high-order DPSK and star QAM modulation based on interferometric direct detection, *IEEE J. Lightwave Technol.*, 25, 1515–1530, 2007.

13. H. Yoon, D. Lee, and N. Park, Performance comparison of optical 8-ary differential phase-shift keying systems with different electrical decision schemes, *Opt. Express*, 13, 371–376, 2005.

14. S. Walklin and J. Conradi, Multilevel signaling for increasing the reach of 10Gb/s lightwave systems, *IEEE J. Lightwave Technol.*, 17(11), 2235–2248, November 1999.

15. C. Wree, J. Leibrich, and W. Rosenkranz, RZ-DQPSK format with high spectral efficiency and high robustness towards fiber nonlinearities, *Proceedings of ECOC*, Copenhagen, Denmark, September 2002, paper 9.6.6.

10 Self-Coherent Reception of Continuous Phase Modulated Signals

This chapter describes the modulation scheme that uses the modulation of the phase lightwave carrier; the change of the phase during a one-bit state would be continuous, that is, two frequencies, hence the term continuous-phase frequency-shift keying (CPFSK). When the two frequency signal components are orthogonal, then the CPFSK is considered the minimum-shift keying (MSK). This scheme along with the generation of modulated signals, propagation, and detection schemes is presented in this chapter.

10.1 INTRODUCTION

The generation of MSK requires a linear variation of the phase, resulting in a constant frequency of the optical carrier. However, the generation of the optical phase may be preferred by driving an optical modulator using sinusoidal signal for practical implementation. Thus, a nonlinear variation of the carrier phase, and hence, some distortion effects, is produced. In this chapter, we investigate the use of linear and nonlinear phase-shaping filters and their impacts on MSK modulated optical signals transmission over optically amplified long-haul communications system. The evolution of the phasor of the in-phase and quadrature components is illustrated for lightwave modulated signal transmission. The distinct features of three different MSK modulation formats, linear MSK, weakly nonlinear MSK, and strongly nonlinear MSK, and their transmission are simulated. Transmission performance obtained indicates the resilience of the MSK signals in transmission over multioptically amplified multispans.

In recent years, have witnessed intensive interests in the employment of advanced modulation formats to explore their advantages and performance in high-density and long-haul transmission systems. Return-to-zero (RZ) and nonreturn-to-zero (NRZ) without or with carrier suppression (CSRZ) formats are associated with shift keying (SK) modulation schemes such as amplitude-shift keying (ASK), phase-shift keying (PSK), differential PSK (DPSK), and differential quadrature PSK (DQPSK) [1,2]. Differential detection offers the best technological implementation owing to the nonrequirement of coherent sources and avoid the polarization control of the mixing of the signals and a local oscillator at multi-Tera-Hertz frequency range. Continuous phase modulation (CPM) is another form of PSK in which the phase of the optical carrier evolves continuously from one phase state to the other. For MSK, the phase change is limited to $\pi/2$. Although MSK is a well-known modulation format in radio frequency digital communications, it has been attracting interests only in optical system research in the last few years [3,4,5]. The phase continuous evolution of the MSK has many interesting features: the main lobe of the power spectrum is wider than that of quadrature PSK (QPSK) and DPSK, and the side lobes of the MSK signal spectrum are much lower, allowing the ease of optical filtering, and hence, less distortion due to dispersion effects. In addition, higher signal energy is concentrated in the main lobe of MSK spectrum than in its side lobes, leading to better signal-to-noise ratio (SNR) at the receiver.

Advanced modulation formats for 40 Gb/s and higher-bit-rate long-haul optical transmission systems have recently attracted intensive research, including various amplitude and discrete differential phase modulation and pulse shape formats (ASK-NRZ/RZ/CSRZ, DPSK, and DQPSK-NRZ/

RZ/CSRZ). However, there are only a few reports on optical CPM schemes using external electro-optical modulators [4]. Compared with discrete phase modulation, CPM signals have very interesting and attractive characteristics, including high spectral efficiency, higher energy concentration in the signal bands, and more robustness to interchannel crosstalk in dense wavelength-division multiplexing (DWDM) due to greater suppression of the side lobes, which has recently risen as a critical issue in dense wavelength division multiplexed optical systems [5,6].

In bandwidth-limited digital communication, including wireless and satellite digital transmission, MSK has been proven as a very efficient modulation format owing to its prominent spectral efficiency, high sideband suppression, constant envelop, and high energy concentration in the main spectral lobe. In modern high-capacity and high-performance optical systems, we have witnessed the acceleration of transmission bit rate approaching 40 Gb/s and even to 160 Gb/s. At these ultrahigh bit transmission speeds, the essence of efficient modulation formats for long-haul transmission is critical. There are several technical published works on advanced modulation techniques for optical transmission, which mostly focus on discrete phase modulation, including binary DPSK and DQPSK [3], with various pulse carving formats, including NRZ and CSRZ [2,4]. However, there is only limited number of works on MSK modulation format [7,8]. The features of MSK, compared with other modulation formats, have prompted researchers to investigate its suitability for high-capacity long-haul transmission. The side lobes of MSK power spectrum are greatly suppressed, giving it good dispersion tolerance and avoiding interchannel crosstalks. Thus, this modulation is of interests for further investigations.

Two different proposals of transmitter configurations for generation of linear and nonlinear phase optical MSK modulation are reported in another paper in this conference [9]. The brief operation descriptions of these two transmitter configurations are presented in Section 10.2, whereas the direct detection techniques for both linear and nonlinear optical MSK signals are stated in Section 10.3. Our simulation models for the modulation and system transmission are based on MATLAB®–Simulink® platform [4], with detailed discussions in Section 10.4. In this section, we have also proven that optical MSK signal is capable of propagating over an optically amplified and fully dispersion-compensated system. The performance of not only the conventional MSK format but also the weakly and strongly nonlinear optical MSK sequences is evaluated.

The MSK format exhibits dual alternating frequencies and offers orthogonal property of the two consecutive bit periods. More interestingly, MSK can be considered either a special case of CPFSK or a staggered/offset QPSK in which I and Q components are interleaved with each other [7]. These characteristics are greatly advantageous to optically amplified long-haul transmissions because such a compact spectrum potentially gives a good dispersion tolerance, making MSK a suitable candidate for DWDM systems.

This chapter thus investigates a number of novel structures of photonic transmitters for generation of optical MSK signals. Theoretical background of MSK modulation formats discussed in two different approaches is presented in Section 10.2. Section 10.3 proposes two configurations of optical MSK transmitters that employ (1) two-cascaded electro-optic phase modulators (E-OPMs) and (2) parallel dual-drive Mach–Zehnder modulators (MZMs). Different types of linear and nonlinear phase-shaped optical MSK sequences can be generated. The precoder for I-Q optical MSK structure is also derived in this section. In Section 10.4, we present a simple noncoherent configuration for detection of the MSK and nonlinear MSK modulated sequences. The optical detection of MSK and nonlinear MSK signals employs a $\pi/2$ phase shift in one arm of the Mach–Zehnder interferometric delay (MZIM) balanced receiver. New measurement techniques for the bit-error rates (BERs) under the probability density function (PDF) of the sampled received signals can be computed by superposition of a number of weighted Gaussian PDFs. The following performance results and observations are obtained in Section 10.3: (1) spectral characteristics of linear and nonlinear MSK modulated signals and (2) improvement on dispersion tolerance of MSK and nonlinear MSK over ASK and DPSK counterparts.

Recently, advanced modulation formats have attracted intensive research for long-haul optical transmission systems, including various amplitude and discrete differential phase modulations and pulse shape formats (ASK-NRZ/RZ/CSRZ, DPSK and, DQPSK-NRZ/RZ/CSRZ). For the case of phase modulation, the phases of the optical carrier are discretely coded with "0" and "π" (DPSK) or "0," "π," and "π/2, −π/2" (DQPSK). Although the differential phase modulation techniques offer better spectral properties, higher energy concentration in the signal bands, and more robustness to combat the nonlinearity impairments, as compared with the amplitude modulation formats, phase continuity would offer even better spectral efficiency and at least 20 dB better in the suppression of the side lobes. MSK exhibits a dual alternating frequency between the two consecutive bit periods, which is considered to offer the best scheme, as this offers orthogonal property of the two-frequency modulation of the carrier lightwaves embedded within the consecutive bits. Furthermore, it offers the most simplicity in the implementation of the modulation in the photonic domain.

This chapter thus investigates a number of novel structures of photonic transmitters and differential noncoherent balanced receivers for generation and detection of MSK optical signals, as discussed in the following paragraphs.

Cascaded E-OPMs MSK transmitter: This structure employs two cascaded OPMs. The first OPM plays the role of modulating the binary data logic into two carrier frequencies, deviating from the optical carrier of the laser source by a quarter of the bit rate. The second OPM enforces the phase continuity of the lightwave carrier at every bit transition. The driving voltage of this second OPM is precoded in such as a way that the phase discrepancy due to frequency modulation of the first OPM will be compensated, hence preserving the phase continuity characteristic of MSK signal. Utilizing this double-phase modulation configuration, different types of linear and nonlinear CPM phase shaping signals, including MSK, weakly-MSK, and linear-sinusoidal MSK, can be generated. The optical spectra of the modulation scheme obtained confirm the bandwidth efficiency of this novel optical MSK transmitter.

Parallel dual-drive MZMs optical MSK transmitter: In this second configuration, MSK signals with small ripple of approximately 5% signal amplitude level can be generated. The optical spectrum is demonstrated equivalent to the configuration of Equation 10.1. This configuration can be implemented using commercially available dual-drive intensity interferometric electro-optic modulators.

We also present a simple noncoherent configuration for detection of the MSK and nonlinear MSK modulated sequences. The optical detection of MSK and nonlinear MSK signals employs a π/2 OPM, followed by the MZIM balanced receiver.

The following performance results and observations are obtained: (a) establishing a modeling platform based on MATLAB–Simulink platform for this self-coherent transmission system; (b) confirming spectral characteristics of linear and nonlinear MSK modulated signals; (c) proving better dispersion tolerance of MSK and nonlinear MSK over ASK and DPSK; and (d) presenting novel techniques in the measurement of the BERs with the PDF of the sampled received signals by super-imposing a weighted Gaussian PDFs. This technique has been proven effective, convenient, and accurate for estimating the non-Gaussian PDF for the case of CPM and its transmission over long-haul optically amplified systems. The MSK format has been demonstrated via modeling as a promising candidate in the selection of advanced modulation formats for DWDM long-haul optical transmission system.

10.2 GENERATION OF OPTICAL MSK MODULATED SIGNALS

The two optical MSK transmitter configurations can be referred in more detail in Ref. [6]. The descriptions of these configurations are briefly addressed in Sections 10.2.1 and 10.2.2.

10.2.1 OPTICAL MSK TRANSMITTER USING TWO-CASCADED ELECTRO-OPTIC PHASE MODULATORS

Electro-optic phase modulators and interferometers operating at high frequency using resonant-type electrodes have been studied and proposed in Ref. [7,8]. In addition, the high-speed electronic driving circuits are evolved with that of the ASIC technology employing 18 μm GaAs hot-electron-mobility transistors (HEMT) or InP HEMTs enable the feasibility in realization of the proposed optical MSK transmitter structure. The base-band equivalent optical MSK signal is represented in Equation 10.1.

The first E-OPM enables the frequency modulation of data logics into upperside bands (USBs) and lower side bands (LSBs) of the optical carrier with frequency deviation of f_d. Differential phase precoding is not necessary in this configuration because of the nature of the continuity of the differential phase trellis. By alternating the driving sources $V_d(t)$ to sinusoidal waveforms for simple implementation or combination of sinusoidal and periodic ramp signals, which was first proposed by Amoroso in 1976 [10], different schemes of linear and nonlinear phase shaping MSK-transmitted sequences can be generated.

The second E-OPM enforces the phase continuity of the light wave carrier at every bit transition. The delay control between the E-OPMs is usually implemented by the phase shifter, as shown in Figure 10.1. The driving voltage of the second E-OPM is precoded to fully compensate the transitional phase jump at the output $E_{01}(t)$ of the first E-OPM. Phase continuity characteristic of the optical MSK signals is determined by the algorithm in Equation 10.8.

$$\Phi(t,k) = \frac{\pi}{2}\left(\sum_{j=0}^{k-1} a_j - a_k I_k \sum_{j=0}^{k-1} I_j\right) \tag{10.1}$$

where $a_k = \pm 1$ are the logic levels; $I_k = \pm 1$ is a clock pulse whose duty cycle is equal to the period of the driving signal $V_d(t)$, f_d is the frequency deviation from the optical carrier frequency, and $h = 2f_d T$ is previously defined as the frequency modulation index. In case of optical MSK, $h = 1/2$ or $f_d = 1/(4T)$. The phase evolutions of the CP optical MSK signals are shown with detailed discussion in Figure 10.2. In order to mitigate the effects of unstable stages of rising and falling edges of the electronic circuits, the clock pulse $V_c(t)$ is offset with the driving voltages $V_d(t)$.

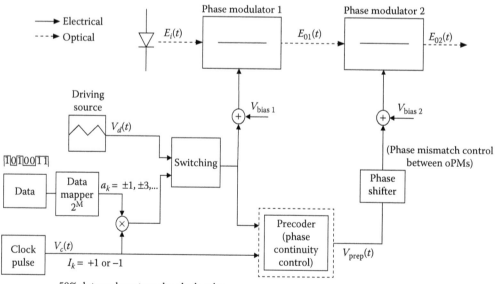

FIGURE 10.1 Block diagram of optical MSK transmitter employing two-cascaded optical phase modulators.

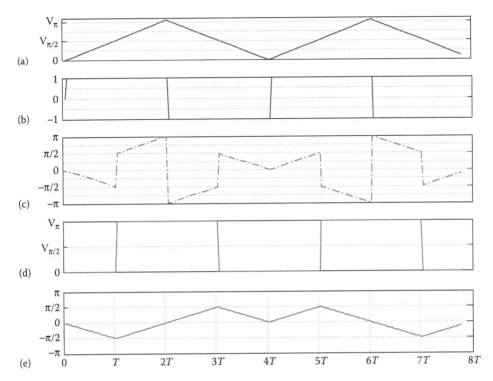

FIGURE 10.2 Evolution of time-domain phase trellis of optical MSK signal sequence [–1 1 1–1 1–1 1 1] as inputs and outputs at different stages of the optical MSK transmitter. The notation is denoted in Figure 10.1 accordingly: (a) $V_d(t)$: periodic triangular driving signal for optical MSK signals with duty cycle of 4 bit period, (b) $V_c(t)$: the clock pulse with duty cycle of 4T, (c) $E_{01}(t)$: phase output of oPM1, (d) $V_{\text{prep}}(t)$: precomputed phase compensation driving voltage of oPM2, and (e) $E_{02}(t)$: phase trellis of a transmitted optical MSK sequences at output of oPM2.

The generated optical signal envelope can be written as

$$\tilde{s}(t) = A \exp\left\{ j \left[a_k I_k f_d \frac{2\pi t}{T} + \Phi(t,k) \right] \right\}, \quad kT \leq t \leq (k+1)T \quad \text{with} \quad a_k = \pm 1, \pm 3, \cdots \quad (10.2)$$

10.2.2 Detection of M-ary CPFSK Modulated Optical Signal

The detection of linear and nonlinear optical M-ary CPFSK utilizes the well-known structure Mach–Zehnder delay interferometer (MZDI)-balanced receiver. The addition of $\pi/2$ phase on one arm of MZDI is also introduced. The time delay being a fraction of bit period enables the phase trellis detection of optical M-ary CPFSK. The detected phase trellis using the proposed technique is shown in Figure 10.3. An optimized ratio of switching frequencies results in the maximum eye open. The detection of optical M-ary CPFSK with delay of $T_b/2$ at $t = (k+1)T_b/2$ is expressed in Equation 10.3.

$$\sin(\Delta\Phi) = \sin\left(a_{k+1} \frac{2\pi f_d (k+1)(T_b/2)}{T_b} - a_k \frac{2\pi f_d k (T_b/2)}{T_b} \right) \quad (10.3)$$

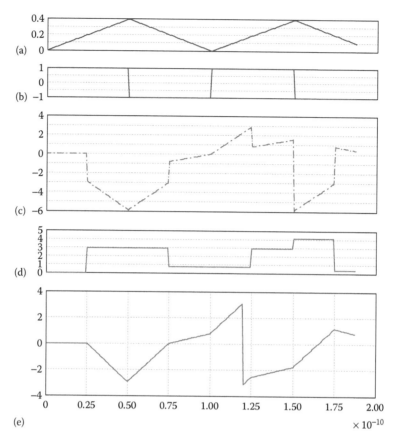

FIGURE 10.3 Demonstration of phase compensation for enforcement of phase continuity at bit transitions: (a) shows the periodic triangular driving signal whose peak voltage is $V\pi/8$ and duty cycle is 4 bit period, (b) shows the clock pulse corresponding to the driving signal, (c) shows frequency switching with $h = \pm1/8, \pm7/8$, (d) shows the computed phase compensation, and (e) shows the phase trellis of an optical quaternary CPFSK.

On the same slope of phase in the phase trellis, the differential phase, and hence the modulated frequency levels, can be mapped to detected amplitude levels via Equation 10.4 (Figure 10.4).

$$\sin(\Delta\Phi) = \sin(a_{k+1}\pi f_d) \tag{10.4}$$

10.2.3 OPTICAL MSK TRANSMITTER USING PARALLEL I-Q MACH–ZEHNDER INTENSITY MODULATORS

The conceptual block diagram of optical MSK transmitter is shown in Figure 10.5a. The transmitter consists of two dual-drive electro-optic MZM modulators generating chirpless I- and Q-components of MSK modulated signals; this is considered a special case of staggered or offset QPSK. The binary logic data are precoded and deinterleaved into even and odd bit streams, which are interleaved with each other by one-bit duration offset. Figure 10.5b shows the block diagram configuration of band-limited phase-shaped optical MSK. Two arms of the dual-drive MZM modulator are biased at $V\pi/2$ and $-V\pi/2$ and are driven with *data* and *data complement*. A phase shaping driving source can be a periodic triangular voltage source in case of linear MSK generation or simply a sinusoidal source for generating a nonlinear MSK-like signal, which also obtains linear phase trellis property but with small ripples introduced in the magnitude. The magnitude fluctuation level depends on the magnitude of the phase shaping driving source. High spectral efficiency can be achieved with tight

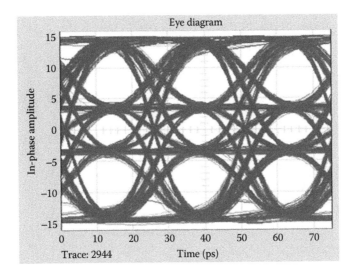

FIGURE 10.4   Eye phase trellis detection of optical M-ary CPFSK modulated signal.

filtering of the driving signals before modulating the electro-optic MZMs. Three types of pulse shaping filters are investigated, including Gaussian, raised cosine, and squared-root-raised-cosine filters. The optical carrier phase trellises of linear and nonlinear optical MSK signals are shown in Figure 10.6.

The generation of linear and nonlinear optical MSK sequences can be briefly discussed as follows:

10.2.3.1   Linear MSK

The pulse shaping waveform for linear MSK is triangular waveform with duty cycle of $4T_b$. The triangular waveforms for quadrature path are delayed by one-bit period with respect to the in-phase path; hence, they are interleaved with each other. The optical MSK signal is the superposition of both even and odd waveforms from the MZIMs. The amplitude of the signal is perfectly constant, clearly displaying the constant amplitude characteristic of CPM. The phase trellis is perfectly linear and the phase transition is continuous, as shown in Figure 10.6a. The signal constellation is a perfect circle.

10.2.3.2   Weakly Nonlinear MSK

The pulse shaping waveform for weakly nonlinear MSK is sinusoidal waveform with amplitude of $1/4V\pi$, and its symbol period is equal to 2-bit period (symbol rate of 20 Gbits/s). The in-phase pulse shaper is a cosine waveform, whereas the quadrature pulse shaper is a sine waveform. There is ripple of approximately 5%. Owing to the nonlinear feature of the sinusoidal function of the pulse shaper, the variation of the phase with respect to time leads to the nonlinearity of the phase trellis. Therefore, the rate of frequency change is not constant. This causes mismatch of MZIM when the modulated waveforms are added up, resulting in the ripple as shown in Figure 10.6b.

10.2.3.3   Strongly Nonlinear MSK

The pulse shaping waveform for strongly nonlinear MSK is sinusoidal waveform, same as for weakly nonlinear MSK. However, the amplitude of pulse shaper is $V\pi/2$. The waveforms are interleaved with each other. The optical MSK signal has ripple of approximately 26%. This ripple is also caused by the mismatch of MZIM, as the modulating waveform is strongly nonlinear. The effect of nonlinearity is obvious in the phase trellis in Figure 10.4b.

The signal state constellations and eye diagrams of optical MSK sequences are shown in Figure 10.7a–c for both of linear and nonlinear schemes. The magnitude ripples. In case of nonlinear configuration, MSK signals with small ripple of approximately 5% signal amplitude level can be

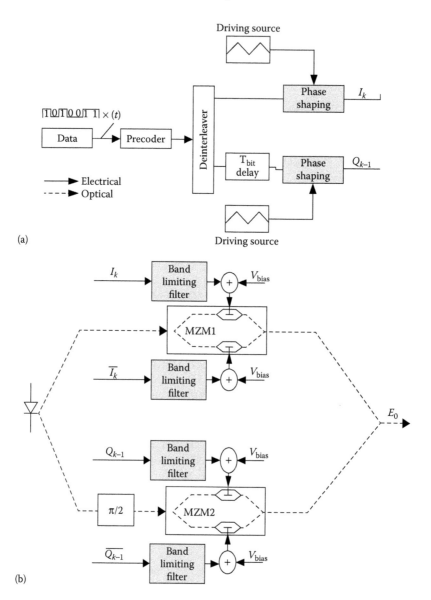

(a)

(b)

FIGURE 10.5   (a,b) Block diagram configuration of band-limited phase shaped optical MSK.

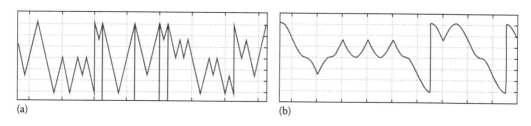

(a)   (b)

FIGURE 10.6   Phase trellis of (a) linear and (b) nonlinear of MSK-modulated and transmitted signals.

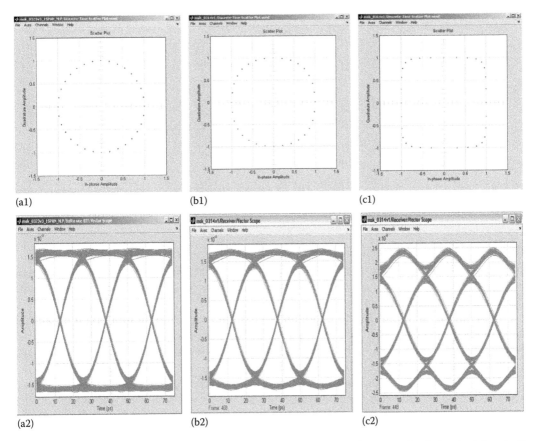

FIGURE 10.7 Constellation diagrams and eye diagrams of optical MSK-transmitted signals: (a1 and a2) linear, (b1 and b2) weakly nonlinear, and (c1 and c2) strongly nonlinear transmission.

generated. This configuration can be implemented using commercially available dual-drive intensity interferometric electro-optic modulators.

The conceptual block diagram of optical MSK transmitter is shown in Figures 10.1 and 10.5. The transmitter consists of two dual-drive electro-optical MZMs modulators generating chirpless I- and Q-components of MSK modulated signals; this is considered as a special case of staggered or offset QPSK. The binary logic data is precoded and deinterleaved into even and odd bit streams, which are interleaved with each other by one-bit duration offset.

Two arms of the dual-drive MZM modulator are biased at $V\pi/2$ and $-V\pi/2$ and are driven with *data* and *inverted data*. A phase shaping driving source can be a periodic triangular voltage source in case of linear MSK generation or simply a sinusoidal source for generating a nonlinear MSK-like signal, which also obtains linear phase trellis property but with small ripples introduced in the magnitude. The magnitude fluctuation level depends on the magnitude of the phase shaping driving source. High spectral efficiency can be achieved with tight filtering of the driving signals before modulating the electro-optic MZMs. Three types of pulse shaping filters are investigated, including Gaussian, raised cosine, and squared-root-raised-cosine filters. Narrow spectral width and high suppression of the side lobes can be achieved.

The logic gates in the precoder are constructed based on the state diagram, as this approach eases the implementation of the precoder. As seen from the state diagram, the current state of the signal is dependent on the previous state, since the state of the signal advances corresponding to the binary data from the previous state. Therefore, memory is needed to store the previous state. The state diagram in Figure 10.8a is developed into a logic state diagram in

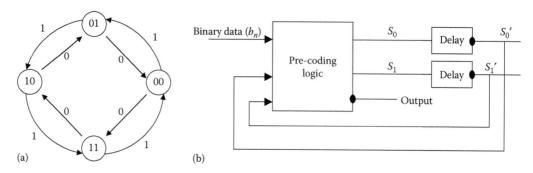

FIGURE 10.8 (a) State diagram for MSK. The arrows indicate continuous increment or decrement of the phase of the carrier. (b) Combinational logic, the basis of the logic for constructing the truth table of the precoder.

Figure 10.8b to enable the construction of truth table. $S_0S_1 = 00$ or $S_0'S_1' = 00$ corresponds to state 0, $S_0S_1 = 01$ or $S_0'S_1' = 01$ corresponds to state $\pi/2$, $S_0S_1 = 10$ or $S_0'S_1' = 10$ corresponds to state π, whereas $S_0S_1 = 11$ or $S_0'S_1' = 11$ corresponds to state $-\pi/2$, with S_0S_1 as current state and $S_0'S_1'$ as previous state.

Two delay units in Figure 10.8b function as memory by delaying the current state and feedback into the precoding logic block, as the previous state. The precoding logic block, which consists of logic gates, would compute the current state and output based on the feedback state (previous state) and binary data from the Bernoulli binary generator.

The truth table is constructed based on the logic state diagram and combinational logic diagram above. For positive half-cycle cosine wave and positive half-cycle sine wave, the output is 1; for negative half-cycle cosine wave and negative half-cycle sine wave, the output is 0. A Karnaugh map can then be constructed to derive the logic gates within the precoding logic block, based on the truth table. The following three precoding logic equations are derived as

$$S_0 = \overline{b_n}\,\overline{S_0'}S_1' + b_n\overline{S_0'}S_1' + \overline{b_n}S_0'S_1' + b_nS_0'S_1' \tag{10.5}$$

$$S_1 = \overline{S_1'} = \overline{b_n}\,\overline{S_1'} + b_n\overline{S_1'} \tag{10.6}$$

$$\text{Output} = \overline{S_0} \tag{10.7}$$

The final logic gates' construction for the precoder is as shown in Table 10.1.

TABLE 10.1
Truth Table Based on MSK State Diagram

$b_nS_0'S_1'$	S_0S_1	Output
100	01	1
001	00	1
010	01	1
101	10	0
110	11	0
111	00	1
000	11	0
011	10	0

10.2.4   OPTICAL MSK RECEIVERS

The optical detection of MSK and nonlinear MSK signals employs a $\pi/2$ OPM, followed by an optical phase comparator, an MZDI, and then a balanced receiver (BalRx). This detection schemes for the linear optical MSK and nonlinear MSK signals can be similar with that of the well-known differential reception structure in which a delay in one arm of the interferometer of the MZDI is the introduced to detect the $\pm\pi/2$ phase difference of two adjacent bits of MSK signal sequence.

A new technique for evaluation of the BERs is implemented. The PDFs of noise-corrupted received signals after decision sampling are computed with superposition of a number of weighted Gaussian PDFs. The technique implements the Expect Maximization theorem and has shown its effectiveness in determining arbitrary distributions.

10.3   OPTICAL BINARY-AMPLITUDE MSK FORMAT

10.3.1   GENERATION

The optical MSK transmitters can be integrated in the proposed generation of optical MAMSK signals. However, in this section, we propose a new simple-in-implementation optical MSK transmitter configuration that employs high-speed cascaded E-OPMs, as shown in Figure 10.2. Electro-optic phase modulators and interferometers that operate at high frequency and use resonant-type electrodes have been studied. In addition, high-speed electronic driving circuits evolved with the ASIC technology using 0.1 μm GaAs P-HEMT or InP HEMTs enable the feasibility in realization of the proposed optical MSK transmitter structure. The base-band equivalent optical MSK signal is represented in Equation 10.6.

$$\tilde{s}(t) = A\exp\left\{j\left[a_k I_k 2\pi f_d t + \Phi(t,k)\right]\right\}, \quad kT \le t \le (k+1)T$$

$$= A\exp\left\{j\left[a_k I_k \frac{\pi t}{2T} + \Phi(t,k)\right]\right\}$$

$$(10.8)$$

where $a_k = \pm1$ are the logic levels, $I_k = \pm1$ is a clock pulse whose duty cycle is equal to the period of the driving signal $V_d(t), f_d$ is the frequency deviation from the optical carrier frequency, and $h = 2f_d T$ is defined in Equations 10.2 and 10.3 as the frequency modulation index. In case of optical MSK, $h = 1/2$ or $f_d = 1/(4T)$.

The first E-OPM enables the frequency modulation of data logics into USBs and LSBs of the optical carrier with frequency deviation of f_d. Differential phase precoding is not necessary in this configuration because of the nature of the continuity of the differential phase trellis. By alternating the driving sources $V_d(t)$ to sinusoidal waveforms, that is, or combination of sinusoidal and periodic ramp signals, different schemes of linear and nonlinear phase shaping MSK transmitted sequences can be generated. The second E-OPM enforces the phase continuity of the light wave carrier at every bit transition. The delay control between the E-OPMs is usually implemented by the phase shifter, as shown in Figure 10.2. The driving voltage of the second E-OPM is precoded to fully compensate the transitional phase jump at the output $E_{01}(t)$ of the first E-OPM. Phase continuity characteristic of the optical MSK signals is determined by the algorithm in Equation 10.2. In order to mitigate the effects of unstable stages of rising and falling edges of the electronic circuits, the clock pulse $V_c(t)$ is offset with the driving voltages $V_d(t)$.

$$\Phi(t,k) = \frac{\pi}{2}\left(\sum_{j=0}^{k-1} a_j - a_k I_k \sum_{j=0}^{k-1} I_j\right) \qquad (10.9)$$

Binary-amplitude MSK (BAMSK) modulation format is proposed for optical communications, as shown in Figure 10.9. We report numerical results of 80 Gb/s 2-bit-per-symbol BAMSK optical

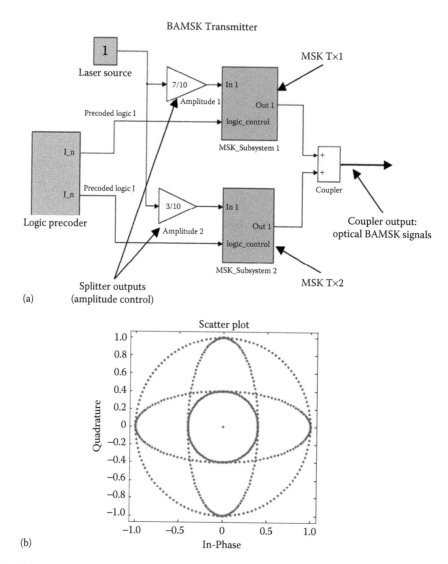

FIGURE 10.9 (a) Block diagram of optical BAMSK transmitter and (b) signal trajectories of optical BAMSK-transmitted signals of multiamplitude MSK.

system on spectral characteristics and residual dispersion tolerance to different types of fibers. A BER of 1e^{-23} is obtained for 80 Gb/s optical BAMSK transmission over 900 km Vascade® fiber multispan optically amplified transmission line. This proves that MSK format is a feasible modulation and transmission technique for long-haul optical system.

10.3.2 OPTICAL MSK

MSK, which is well known in radio frequency and wireless communications, has recently been adapted into optical communications. A few optical MSK transmitter configurations have recently been reported. For optically amplified communications systems, if multilevel concepts can be incorporated in those reported schemes, the symbol rate would be reduced, and hence, the bandwidth efficiency can be achieved. This is the principal motivation for the proposed modulation scheme.

The BAMSK is a special case of M-ary CPM format, which enables binary-level (pulse amplitude modulation- or quadrature amplitude modulation-like) transmission scheme; however, the bandwidth efficiency due to transitional phase continuity properties between two consecutive symbols (CPM-like signals) are preserved. The generation of M-ary CPM sequences can be expressed in Equation 10.1.

$$s(t) = A_n \cos(\omega_c t + \phi_n(t,a)) + \sum_{m=1}^{N-1} B_m \cos(\omega_c t + \phi_m(t,b_m)) \tag{10.10}$$

where

$$\phi_N(t,a) = \pi h a_n q(t - nT) + \pi h \sum_{k=-\infty}^{n-1} a_k \quad nT \leq t \leq (n+1)T \tag{10.11}$$

$$\phi_m(t,b_m) = \pi a_n \left(h + \frac{b_{mn}+1}{2} \right) q(t - nT) + \pi \sum_{k=-\infty}^{n-1} a_k \left(h + \frac{b_{mk}+1}{2} \right) \quad mT \leq t \leq (m+1)T \tag{10.12}$$

In a generalized M-ary CPM transmitter, values of a_n and b_{mn} are statistically independent and are taken from the set of $\{\pm 1, \pm 3, \ldots\}$. A_n and B_m are the signal state amplitude levels, which are either in-phase or π-phase shifts, with the largest level component at the end of nth symbol interval; $q(t)$ is the pulse shaping function; and h is the frequency modulation index. In case of BAMSK, Equations 10.2 and 10.3, which show the constraints of ϕ_m to maintain the phase continuity characteristic of CPM sequences, are simplified to Equations 10.4 and 10.5, respectively, where $h = 1/2$ and the phase shaping function $q(t - nT)$ is a periodic ramp signal with duty cycle of $4T$.

$$\phi_n(t,a) = \pi h a_n \frac{t - nT}{T} + \pi h \sum_{k=-\infty}^{n-1} a_k \quad nT \leq t \leq (n+1)T \tag{10.13}$$

$$\phi_m(t,b_m) = \pi I_n \left(h + \frac{b_{mn}+1}{2} \right) \frac{t - nT}{T} + \pi \sum_{k=-\infty}^{n-1} b_k \left(h + \frac{b_{mk}+1}{2} \right) \quad mT \leq t \leq (m+1)T \tag{10.14}$$

The optical MSK transmitters described in Section 10.2 can be utilized in the proposed generation scheme of optical BAMSK signals. Figure 10.10a shows the block structure of the optical BAMSK transmitter, in which two optical MSK transmitters are integrated in parallel configuration. The amplitude levels are determined from Equations 10.1, 10.4, and 10.5 by the splitting ratio at the output of a high-precision-power splitter. The logic sequences $\{\pm 1, \ldots\}$ of an and b_n are precoded from the binary logic $\{0,1\}$ of d_n as $a_n = d_n - 1$ and $b_n = a_n(1 - d_n - 1/h)$ [9]. The signal-space trajectories of BAMSK signals are shown in Figure 10.10b.

A simple noncoherent configuration for detection of the optical BAMSK sequences consists of phase and amplitude detections. In case of BAMSK, that is, $n = 2$, the system effectively implements 2 bits per symbol with two amplitude levels. Phase detection is enabled with employment of the well-known integrated optic phase comparator MZDI-balanced receiver with one-bit time delay on one arm of MZDI [10]. An additional $\pi/2$ phase shift is introduced. Figure 10.10a and b shows the eye diagrams of the amplitudes and phases of the optical BAMSK signals, respectively. In phase detection, the decision threshold, which is plotted in broken-line style, is at zero level, whereas amplitude levels are determined by different thresholds. New technique in calculation of BER for dispersive and noise-corrupted received signals, which exploits the Expected Maximization theorem, is implemented with superposition of a number of weighted Gaussian probability distribution functions.

A simple noncoherent configuration for detection of a linear and nonlinear optical M-ary MSK sequences consists of phase and amplitude detections, which are very well-known in the discrete PSK schemes such as DPSK or DQPSK. Phase detection is enabled with employment of the well-known integrated optic phase comparator MZDI-balanced receiver with one-bit time delay on one

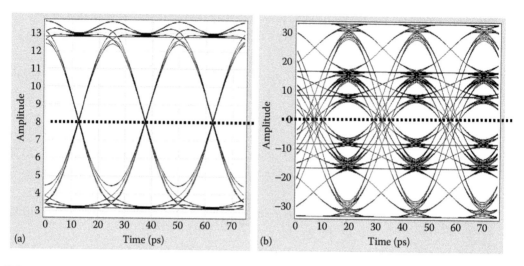

FIGURE 10.10 Eye diagrams of amplitude (a) and phase (b) detection of optical BAMSK-received signals with normalized amplitude ratio of 0.285/0.715. The decision threshold is shown in broken-line style.

arm of MZDI. An additional $\pi/2$ phase shift is introduced to detect the differential $\pi/2$ phase shift difference of two adjacent optical MSK pulses. In case of $N = 2$ and $N = 3$, the system effectively implements 2-bits-per-symbol scheme and 3-bits-per-symbol scheme with 2 and 4 amplitude levels, respectively. In phase detection, the decision threshold, which is plotted in broken-line style, is at zero level because only in-phase and π differential phase are of interests.

10.3.3 Numerical Results and Discussions

The state diagram for MSK is shown in Figure 10.11. The continuous wave carrier source is modulated by the two cascaded MZIMs, which are driven by a voltage level conditioning (broadband microwave amplifiers) fed by the output level of the precoder. The arrows indicate continuous increment or decrement of the phase of the carrier. These information-bearing lightwave signals are then propagated along the fiber spans and detected via a photonic phase comparator, the MZDI, and then detected via a balanced receiver. The obtained eye diagram is then statistically analyzed. More efficient detection scheme using frequency discrimination will be presented in Chapter 8.

10.3.3.1 Transmission Performance of Linear and Nonlinear Optical MSK Systems

The block diagram of simulation set-up is shown in Figure 10.11. The dispersion tolerances of linear, weakly nonlinear, and strongly nonlinear optical MSK signals are numerically investigated, and the results are shown in Figure 10.12. Among the three types, linear MSK is most tolerant to residual

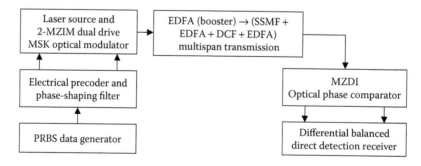

FIGURE 10.11 Schematic diagram of an optically amplified optical transmission system.

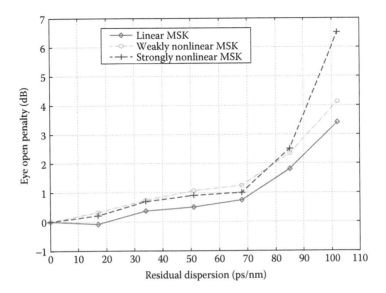

FIGURE 10.12 Dispersion tolerance of 40 Gb/s linear MSK, weakly nonlinear MSK, and strongly nonlinear MSK optical signals.

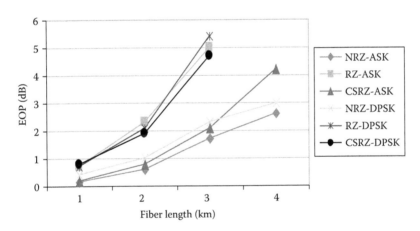

FIGURE 10.13 Simulation of eye open penalty versus transmission distance 1–4 km of SSMF.

dispersion with 1 dB eye-open penalty at 72 ps/nm/km. Strongly nonlinear MSK suffers a severe penalty when residual dispersion exceeds 85 ps/nm/km or equivalently of 5 km SSMF. Figure 10.13 shows the performance of three types of optical MSK modulated signals versus optical SNR (OSNR) in transmission over 540 km Vascade fibers of optically amplified and fully compensated multispan links (6 spans × 90 km/span). The receiver sensitivity of the differential phase comparison balanced receiver is −24.6 dBm. In Vascade fibers, the dispersion factor of the dispersion compensating fiber is negatively opposite to that of the transmission fiber, a standard single-mode fiber (SMF), of +17.5 ps/nm/km at 1550 nm wavelength. In addition, the dispersion slopes of these fibers are also matched. Optical amplifiers of erbium doped fiber amplifier (EDFA) types are placed as follows: one at the end of the transmission fiber and one after the dispersion-compensating fiber (DCF), so that it would boost the optical power to the right level, which is equal to the level of the launched power. The EDFA optical gains of 19 and 5 dB amplification simulated emission noise figures are used. The noise margin reduces severely after the propagation over 6 × 90 km spans. Effects of positive and negative dispersion mismatches and mid-link nonlinearity on phase evolution are shown in Figure 10.12.

The tolerance of these MSK modulation using transmission models to nonlinear effects is also studied, and simulation results are shown in Figures 10.12 and 10.13. The input power into the fiber span is kept increasing, whereas the length of transmission link is constant at 180 km. At BER = 1e^{-9}, linear MSK could tolerate an increase of the input launched power up to 10.5 dBm; weakly nonlinear MSK could tolerate up to 10.2 dBm compared with 9.2 dBm in case of strongly nonlinear MSK. The nonlinear phase shift is proportional to the input power; therefore, increasing the input power would increase nonlinear phase shift as well. This nonlinear phase shift is observed through the asymmetries in the eye diagram and through the scatter plot, which show that the phases of the in-phase and quadrature components have shifted from the x and y axes. The maximum phase shift that could be tolerated is approximately 15° of arc. Although increasing input power increases the noise margin of the eye diagram, it is paid off by the large distortion at sampling time, causing the SNR to decrease. Typical eye diagrams for full compensation and after transmission over 540 km Vascade fibers of optically amplified and fully compensated multispan links (6 spans × 90 km/span) are shown in Figure 10.14.

We note that if the sampling is conducted at the center of the bit period, then the error is minimum, as the ripples of the eyes are fallen on this position. This is the principal reason why the MSK signals can suffer minimum pulse spreading due to residual and nonlinear phase dispersions. Linear MSK, weakly nonlinear MSK, and strongly nonlinear MSK phase shaping functions are investigated. It has been proven that optical MSK is a very efficient modulation that offers excellent performance. With OSNR of about 17 dB, BER is obtained to be 1e−9 and reaches 1e−17 for an optical SNR of 19 dB under linear MSK modulation. The modulation formats of linear and nonlinear phase shaping MSK are also highly resilient to nonlinear effects. Nonlinear distortion appears when the total average power reaches about 9 dBm, that is, about 3–4 dB above that of NRZ-ASK format over an SSMF fiber of 50 μm diameter.

The nonlinear phase shaping filters offer better implementation structures in the electronic domain for driving the dual-drive MZIMs than the linear types but suffer some power penalty;

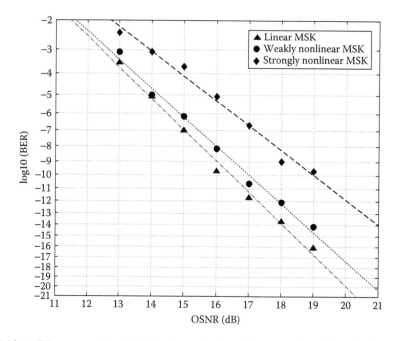

FIGURE 10.14 BER versus optical SNR for transmission of 3 types of modulated optical MSK signals over 540 km Vascade fibers of optically amplified and fully compensated multispan links (6 spans × 90 km/span).

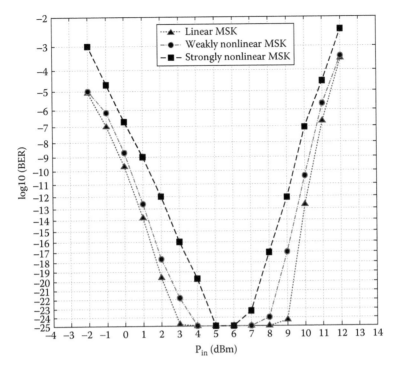

FIGURE 10.15 BER versus input power, showing robustness to nonlinearity of linear, weakly nonlinear, and strongly nonlinear optical MSK signals, with transmission over 180 km Vascade fibers of two optically amplified and fully compensated span links.

however, they are still better than the candidates of other amplitude modulation formats or phase or differential phase modulation formats.

Weakly nonlinear MSK offers much lower power penalty and ease of implementation for precoders and phase shaping filters; thus, it would be the preferred MSK format for long-haul transmission over optically amplified multispan systems. At a BER of $1e^{-12}$, linear MSK is 0.3 and 1.2 dB more resilient than weakly nonlinear MSK and strongly nonlinear MSK, respectively, to nonlinear phase effects (Figure 10.15).

Compared with the DPSK counterpart in 40 Gb/s transmission, various types of filtered MSK modulated signals are more tolerant to residual dispersion. The eye-open penalty of 3 dB is obtained at 4 km of SSMF in case of NRZ–DPSK/ASK or 68 ps/nm accordingly, whereas linear MSK can tolerate up to 98 ps/nm or 6 km accordingly.

It is noted that the pulse shaping using raised-cosine filter offers better dispersion tolerance and lower penalty for RZ-DPSK and CSRZ-DPSK after transmitting 3 km. Approximately 2 dB improvement is observed. Thus, pulse shaping compromises the deficits of RZ pulses because of the broader spectrum compared with NRZ pulse shapes. We observe no significant improvement on NRZ pulses for both DPSK and ASK signals.

10.3.3.2 Transmission Performance of Binary Amplitude Optical MSK Systems

Figure 10.16a numerically compares the power spectra of 80 Gb/s optical BAMSK and 40 Gb/s optical MSK and DPSK signals. The normalized amplitude levels of two optical MSK transmitters take the ratio of 0.715/0.285. Generally, the power spectrum of optical BAMSK format has identical characteristics to that of the MSK format, including narrow spectral width and highly suppressed side lobes, and outperforms the DPSK counterpart.

Figure 10.17b shows numerical results of residual dispersion tolerance in both amplitude and phase of 80 Gb/s optical BAMSK systems with normalized amplitude ratio of 0.285/0.715. Standard

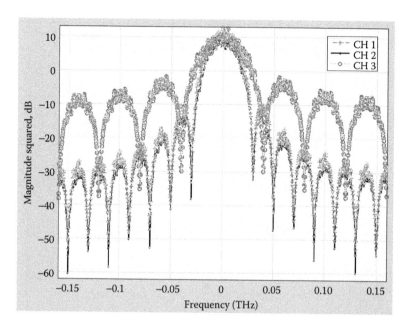

FIGURE 10.16 Comparison of spectra of 80 Gb/s optical BAMSK, 40 Gb/s optical MSK, and 40 Gb/s optical binary DPSK signals.

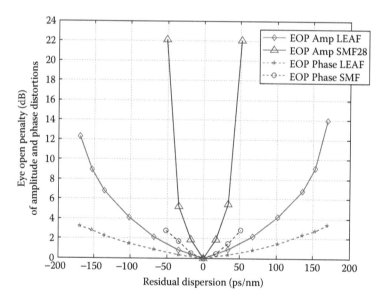

FIGURE 10.17 Numerical results on residual dispersion tolerance of 80 Gb/s optical BAMSK systems (effectively 40 Gb/s symbol rate), with normalized amplitude ratio of 0.285/0.715 in both amplitude and phase detections.

SMFs and Corning large effective area fiber (LEAF) fibers with dispersion factor of ±17 ps/nm/km and ±4.5 ps/nm/km, respectively, are used. As expected, the severe penalty due to fiber dispersion derives from the distortion of the waveform amplitudes whose values dramatically jump over 20 dB penalty compared with 3 dB in case of phase distortion at 50 ps/nm SMF residual dispersion. LEAF fiber enables the system tolerance to residual dispersion of approximately 150 ps/nm for 3 dB penalty. It should be kept in mind that the optical 80 Gb/s system under test has effective transmission rate of approximately 40 Gb/s owing to 2-bit-per-symbol BAMSK scheme.

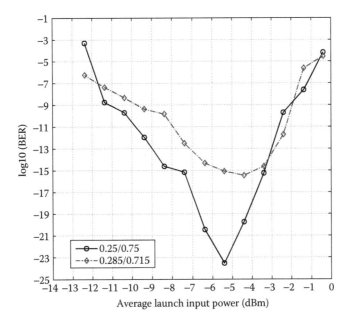

FIGURE 10.18 Transmission performance of 80 Gb/s optical BAMSK over 900 km Vascade fiber systems in two cases of normalized amplitude ratios: 0.25/0.75 (round markers and solid line) and 0.285/0.715 (diamond markers and dashed line), respectively.

Figure 10.16 numerically report the transmission feasibility of 80 Gb/s optical BAMSK signals over 900 km Vascade fibers with possible BER values less than 1e−12 for both normalized input amplitude ratios of 0.285/0.715 and 0.25/0.75, respectively. In Figure 10.18, the performance curve with diamond markers and dashed line represents the first ratio, whereas round markers and solid line curves are used for the latter ratio The single-channel transmission system consists of 80 Gb/s pseudo-random generator with 128-bit sequence, 10 spans of 90 km Vascade fibers (60 km of +17 ps/nm/km and 30 km of −34 ps/nm/km and fully compensated for both chromatic dispersion and dispersion slope), and an optical filter with bandwidth of 80 GHz. Electronic noise of the receiver is modeled with equivalent noise current density of electrical amplifier of 20 pA/(Hz)$^{1/2}$ and a dark current of 2 * 10 nA (2 photodiodes in balanced receiver structure). This configuration yields the back-to-back receiver sensitivity at BER = 1e^{-9}, thus the required OSNR would be approximately −23 dBm. The eye diagrams are obtained after 5th-order Bessel electrical filters with bandwidth of 36 GHz. Launched peak input power is varied from −10 dBm to +3 dBm, whose corresponding average powers range from −12.5 dBm to −0.5 dBm. Phase noise is dominant in total BER in low-average-launched input powers (<−3 dBm). The results raise the necessity of optimizing the amplitude levels for the optimum BER for optical BAMSK signal transmission.

We have proposed, with detailed operation principles, a new configuration of optical MSK transmitter using two cascaded E-OPMs, which reduces the complexity of photonic components. These transmitters can be integrated in parallel configuration for the first-time proposed generation of optical M-ary MSK signals. The number of signal states can increase with the number of transmitters. Spectral properties of optical M-ary MSK are similar to those of optical MSK and better than those of the DPSK counterparts. The main source of penalty of fiber residual dispersion is caused by the waveform amplitude distortions, which can be overcome with advanced dispersion-equalization techniques. Numerical results of transmission performance over 900 km Vascade fibers have been presented in two different cases of normalized amplitude ratios of the BAMSK modulation format. The BER of 1e^{-23} enables the long-haul transmission feasibility of the proposed format. The simulation testbed is successfully developed based on the MATLAB–Simulink platform.

10.4 REMARKS

We have proposed two new schemes of optical MSK generation and detection. These two optical transmitter configurations can generate linear and various types of nonlinear optical MSK modulation formats. The precoder for I-Q optical MSK structure is shown. The direct detection of optical lightwave is utilized, with implementation of the well-known differential noncoherent balanced receivers. Simulated spectral characteristics and dispersion tolerance to 40 Gb/s transmission are presented and compared with those of ASK and DPSK counterparts.

We have proposed the BAMSK modulation format for optical communications. We have reported the configuration for generation and noncoherent detection of the optical BAMSK sequences. The number of signal states can be easily increased with additional optical MSK transmitters. Spectral properties of the 80 Gb/s optical BAMSK are similar to those of the 40 Gb/s optical MSK and better than those of the 40 Gb/s optical DPSK counterparts. The main source of penalty in dispersion tolerance can be caused by the distorted waveform amplitude. This distortion can be overcome by advanced dispersion equalization techniques relevant digital signal processing algorithms in the receiver. Numerical results of transmission performance for 80 Gb/s optical BAMSK system over 900 km Vascade fibers have been reported. The BER of $1e^{-23}$ enables the long-haul transmission feasibility of the proposed format. The simulation testbed is developed on the MATLAB–Simulink platform.

It has been proven that MSK transmitter models for these modulation formats can be implemented using parallel structure of dual-drive MZIM data modulators. The differences in the implementations of these transmitter models are the pulse shaping waveforms, in which linear MSK follows a triangular periodic waveform; weakly nonlinear MSK and strongly nonlinear MSK have sinusoidal waveforms but differ in amplitudes. The models are simulated under different conditions to investigate the effects of fiber characteristics such as fiber loss, nonlinear effects, and dispersion on the performance of the models. The rotation of the scatter plots confirms the behavior of MSK modulated signals due to linear chromatic dispersion and nonlinear phase shaping effects. The simulated optically amplified distance is 540 km fully compensated SSMF. The linear and nonlinear optical MSK modulation formats can thus be offered as an alternative advanced modulation format for long-haul optically amplified transmission.

The visualization of the evolutions of the MSK signal phasor under self-phase modulation is not fully described in this article but will be reported in a future article. Electronic compensation technique can be implemented by design of the predistorted MSK signals at the input of the shaping filters. The optical MSK precompensating transmission system will be reported in the near future.

REFERENCES

1. T. Hoshida et al., Optimal 40 Gb/s modulation formats for spectrally efficient long haul DWDM systems, *IEEE J. Lightwave Technol.*, 20, 2002, 1989–1996.
2. Y. Zhu et al., 1.6 bit/s/Hz orthogonally polarized CSRZ-DQPSK transmission of 8x40 Gbit/s over 320 km NDSF, *OFC* Tu-F1, 2004.
3. T. Sakamoto, T. Kawanishi, and M. Izutsu, Initial phase control method for high-speed external modulation in optical minimum-shift keying format, *ECOC*, 4, 2005, 853–854.
4. M. Ohm and J. Spiedel, Optical minimum shift keying with direct detection, *Proc. SPIE on Optical Transmission, Switching and Systems*, 5281, 2004, 150–161.
5. T. L. Huynh, L. N. Binh, and K. K. Pang, Optical MSK long-haul transmission systems, *SPIE Proc. of APOC*, 6353, 2006, 6353–6386.
6. T. L. Huynh, L. N. Binh, and K. K. Pang, Linear and weakly nonlinear optical continuous phase modulation formats for high performance DWDM long-haul transmission. In *Proceedings of ECOC*, Cannes, France 2006.
7. T. Kawanishi, S. Shinada, T. Sakamoto, S. Oikawa, K. Yoshiara, and M. Izutsu, Reciprocating optical modulator with resonant modulating electrode, *Electron. Lett.*, 41(5), 2005, 271–272.

8. R. Krahenbuhl, J. H. Cole, R. P. Moeller, and M. M. Howerton, High-speed optical modulator in LiNbO3 with cascaded resonant-type electrodes, *J. Lightwave Technol.*, 24(5), 2006, 2184–2189.
9. I. P. Kaminow and T. Li, *Optical Fiber Communications, Volume IVA*: Elsevier Science, New York, Chapter 16, 2002.
10. C. Wree, J. Leibrich, J. Eick, W. Rosenkranz, and D. Mohr, Experimental investigation of receiver sensitivity of RZ-DQPSK modulation using balanced detection, *OFC*, 2, 2003, 456–457.

11 Partial Responses and Single-Sideband Optical Systems

11.1 INTRODUCTION

Optical fiber communication system has continuously evolved over the years. The increasing demand for a higher transmission capacity has driven the development of communication system at ultrahigh capacity ultrahigh bit rates. The fact that 40 Gb/s optical fiber communication system has an extended reach and improved capacity, it has become an attractive alternative to the 10 Gb/s optical fiber communication system. System performance is further enhanced by employing various advanced modulation formats, such as duobinary (DB), return-to-zero differential phase-shift keying (RZ-DPSK), and nonreturn-to-zero differential phase-shift keying (NRZ-DPSK). Research and investigations have been carried out to determine the most appropriate and efficient formats that meet the current, as well as future, demand.

Duobinary modulation can be implemented in either electrical domain or optical domain. Both offer an equivalent effective bandwidth reduction of about 50% as compared with that of the NRZ format. In the electrical domain approach, low-pass band of an electrical filter is used to filter the NRZ signal, thus generating 3-level electrical signals, which are then used to modulate an optical modulator. On the other hand, the DB format can be generated in the photonic domain, offering a three-level coding on the phase of the optical carrier. The "−1" and "+1" are coded using the phase of the lightwave carrier, that is, either "0" or "π." This coding can overcome the dispersion due to its single-sideband (SSB) property as well as the π-phase interference cancellation of any dispersive signal envelopes between adjacent pulses. Furthermore, the optical detection scheme is simpler by direct detection.

Single sideband can also be implemented using the vestigial sideband (VSB) modulation technique, in which an optical filter (OF) is inserted after the optical modulator to filter half of the band of the spectrum. Alternatively, the modulators can be conditioned with two Hilbert transform signals and hence the suppression of half of the band. However, the VSB OF roll-off band must be at the middle of the signal band, otherwise signal distortion occurs. In optical systems, the VSB is not preferred. However, one can employ the Hilbert transformer at the electrical input to obtain the π/2 phase differences feeding into the optical modulators to suppress half of the optical signal band.

The first part of this chapter presents a comprehensive modeling platform for DB modulation (DBM) format of optical fiber transmission system. The modeling of the system is developed on the MATLAB® Simulink® 7.0 or higher. Simulink has been chosen owing to the wide-range availability of subsystem blocks, such as the communication blockset and signal-processing blockset, which ease the implementation process.

Further in this chapter, we demonstrate the transmission of 40 Gb/s alternating phase 0 and π DBM format with 33% and 50% pulse widths, a tolerance of 50 ps/nm, and at least 2 dB improvement in receiver sensitivity, as compared with that under the carrier-suppressed DPSK (CS-DPSK).

The second part of this chapter presents the transmission of optical multiplexed channels of 40 Gb/s using the VSB modulation format over a long-reach optical fiber transmission system. Thus, it is essential that an OF is designed to follow the optical modulator to filter half of the signal band. The effects on the Q factor of fibers dispersion, the passband, and the roll-off frequency of the OFs and the channel spacing are described. The performance of the optical transmission using low and nonzero dispersion fibers or/and dispersion compensation is described. It has been demonstrated that bit-error rate (BER) of 10^{-12} or better can be achieved across all channels, and minimum degradation of the channels can be obtained under this modulation format. OFs are designed with asymmetric roll-off bands. Simulations of the transmission system are also given and compared for channel spacing of 20, 30, and 40 GHz. It is shown that the passband of 28 GHz and 20 dB cut-off band performs best for 40 GHz channel spacing.

11.2 PARTIAL RESPONSES: DB MODULATION FORMATS

11.2.1 REMARKS

The demand for high-capacity, long-haul telecommunication system has increased over the recent years. To achieve high throughput of signals with minimum errors, different advanced modulation formats, such as amplitude-shift keying (ASK), PSK (coherent and differential in-coherent), and frequency-shift keying, have been proposed and comparisons are made to determine which modulation format would offer the best transmission performance. In countering performance degradation, modulation formats aim to narrow down the optical spectrum to enable close channel spacing in the network. They increase symbol duration, so that more uncompensated dispersion accumulates before intersymbol interference (ISI) becomes significant. Furthermore, this format is more resilient to fiber nonlinearities and optical signal distortion.

DB modulation and continuous phase modulation DB are shown to offer high spectral efficiency [1–4]. DB modulation minimizes the ISI impairments in a controlled way instead of eliminating it. It is possible to achieve a signaling rate equal to the Nyquist rate of 2W symbols/s in a channel of bandwidth W Hz. Optical DB technique has received much attention owing to its high dispersion tolerance and high frequency-utilization efficiency by means of spectral narrowing [5–8]. The DBM format is similar to the NRZ format, with the inclusion of dual-phase coding. The phase characteristics of DBM signals compensate for the group velocity dispersions (GVDs) by their reduced spectral components. ISI is reducible, owing to its bit patterns, such as 101, which are transmitted with the "1," carrying opposite optical carrier phase. Therefore, if pulses spread out into the zero time slot, owing to the dispersion of the optical fiber, they tend to cancel each other out. The recovering of signals at the receiver is also relatively simple by conventional direct detection reception. There are two possible types of DBM schemes: constant phase and alternating phase in blocks of logics "1s" [1,9,10].

This section presents the models for photonic transmission, with optical channels operating under DBM format. This includes the development and implementation of the photonic transmitter, the optical fiber propagation, and the opto-electronic receiver. DB modulation encodes two-level electrical signal to three-level electrical signal before modulating the lightwave carrier. The transmitter of the Simulink model will consist of a DB encoder and a dual-drive Mach–Zehnder intensity modulator (MZIM). A baseband modulation is first implemented in the DB encoder, which encodes the binary signal into three levels signals of "1," "0," and "−1." Mach–Zehnder intensity modulator is an electro-optic modulator that converts the electrical signal to optical signal.

The DB or phase-shape binary modulation formats can be generated by modulating a dual-drive MZIM. Recent works have shown that the driving voltages for the modulator can be reduced to generate variable-pulse-width DB optical signals. However, the pulse width of the DB-DPSK has not been thoroughly investigated under the alternating phase of the "1" coded bits. This means that the "0" "π" "0" "π" phases of consecutive "1s" in contrary of conventional DB formats. We also present modeling performance of alternating-phase DBM with a full-width half-mark (FWHM) ratio with respect to the bit period of 100%, 50%, and 33% as contrasted with experimental transmission results of CS-DPSK over 50 km of standard single-mode fiber (SSMF) and dispersion compensation. For the DB case, the transmission without dispersion compensation over the same SSMF length offers better performance for 50% FWHM DBM and slightly worse for 33%. The transmission performance, the BER, and receiver sensitivity of these DBM formats are compared with those of the CS-DPSK experimental transmission.

It is then followed by the description of each component in the 40 Gb/s DBM photonic transmission systems. Section 11.4 is the implementation of Simulink model of the communication system. Lastly, Section 11.5 gives simulated results. Finally, comparisons with theoretical analyses and other modulation formats are given. Although current optical communication systems for access network are based on 25 GBaud, so as to achieve 100 Gb/s by multiplexing four wavelength channels, the techniques presented here at 40 Gb/s are still valid.

11.2.2 DBM FORMATTER

Modulation format aims to modulate one or more field properties to suit system needs. There are four types of field properties, namely intensity, phase, polarization, and frequency. Symbols are constelled in one or more dimensions, in order to carry more information and to travel a further distant. Data modulation format is the information-carrying property of the optical field [11–13].

The DBM schemes can be described as correlative-level coding scheme or partial-response signaling scheme. Correlative-level coding scheme means that by adding ISI to transmitted signal in a controlled manner, a signaling rate equal to the Nyquist rate of 2W symbols/s in a channel of bandwidth W Hz can be achieved. "Duo" in the word DB indicates the doubling of transmission capacity of a conventional binary system. The DBM format is, in fact, NRZ modulation with an inclusion of phase coding. The one bits in the data input are phase modulated. For instance, for a bit pattern of 101, these data are transmitted with the ones carrying opposite phase, 0 and π. If the pulses of the one bits spread out to the zero time slot in between, they will cancel each other. This effect increases the dispersion tolerance and allows the signal to be transmitted over a longer distance.

The DB coding converts a two-level binary signal of 0s and 1s into a three-level signal of "−1," "0," and "+1." This is done by first applying the binary sequence to a pulse-amplitude modulator to produce two-level short pulses of amplitude of −1 and +1, with −1 corresponding to 0 and +1 corresponding to 1. This sequence is then applied to DB encoder to produce a three-level output of "−2," "0," and "2."

As shown in Figure 11.1 input sequence, $\{a_k\}$ of uncorrelated two-level pulses is transformed into $\{c_k\}$, which is a sequence of correlated three-level pulses. The correlation between adjacent pulses is equivalent to introducing ISI into transmitted signal in an artificial manner. The DB encoder is simply a filter involving a single delay element and summer, as shown in Figure 11.2. However, once errors are made, they tend to propagate through the output. This is because a decision made on the current input a_k depends on the decision made on the previous input a_{k-1}. Therefore, precoding is needed to avoid this error propagation phenomenon. Binary sequence, $\{b_k\}$, is converted into another binary sequence, $\{d_k\}$, by modulo-two addition, exclusive OR (XOR) of b_k and d_{k-1}, as show in Figure 11.3.

$$c_k = a_k + a_{k-1} \tag{11.1}$$

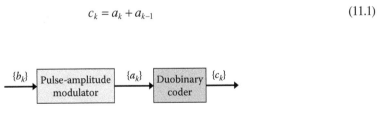

FIGURE 11.1   Generic diagram of coder to transfer pulse amplitude format to DBM.

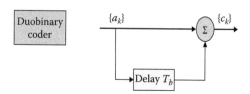

FIGURE 11.2 DB encoder—the block at the left is represented by the signal flow diagram shown on the right.

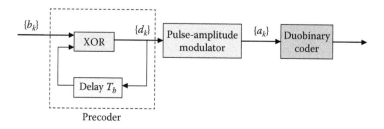

FIGURE 11.3 DBM scheme with precoder.

$$d_k = b_k \oplus d_{k-1} \tag{11.2}$$

The three-level DB output, $\{c_k\}$, is then modulated into a two-level optical signal by an external modulator. The most commonly used external modulator is the MZIM. The optical DB signal has two intensity levels, "on" and "off." The "on" state signal can have one of the two optical phases, 0 and π. The two "on" states correspond to the logic states "1" and "−1" of the DB-encoded signal, $\{a_k\}$, and the "off" state corresponds to the logic state "0." Figure 11.4a shows an example of the original binary signal, the DB-encoded signal, and optical DB signal, and Figure 11.4b shows a summary of coding rule.

The 40 Gb/s DB optical fiber transmission systems (Figure 11.5) show the typical DWDM optical fiber communication system. Signals are modulated at the transmitters and are multiplexed together at the wavelength multiplexer before transmitting them into the fiber. The fiber link is divided into a number of spans. Each span consists of a length of transmission fiber, the SSMF, and a dispersion-compensating fiber (DCF), whose total dispersion is equal and opposite sign of that of the SSMF. Optical amplifiers (OAs), for example, the Erbium-doped fiber amplifier (EDFA) for 1550 nm region, are used to compensate for the fiber transmission loss and that of the DCF. The transmission attenuation of the DCF is reasonably high, about 11–12 dB for a compensating length of 20 km long to compensate 100 km SSMF, hence an extra optical amplifier must be used in cascade to equalize this loss. Thus, another EDFA must be used after the DCF to boost the signal power to an approximate-equal level launched at the input of the optical transmission line. At the end of the fiber, the signals are demultiplexed and detected at the receivers.

It is worth noting here that current long-haul optical transmission systems employ coherent reception and associated digital signal processing (DSP) to overcome several hurdles which were faced by analog electronic processing commonly employed in the first generation coherent systems initiated in 1980s. The modulation formats QPSK are now commonly used in 100G deployed systems over distance from 2000 to 3500 km optical transmission lines consisting of several optically amplified spans of length 80 or 100 km incorporating only one OA and no DCF. THe DSP-based optical receivers compensate the long dispersion distortion effects by processing the received signals in the digital domain. For the DBM transmission system described in this chapter, we consider its use in access with a possibility of a DCF that can be used. Digital signal processing (DSP) can also be used to compensate or equalize the distortion effects on DBM signals. Furthermore, forward error coding (FEC) can be employed to significantly improved the recovery of the originally transmitted symbols, so that a BER of 2.0e−2 or 1.5e−3 can be acceptable for error-free by using hard FEC and soft FEC, respectively.

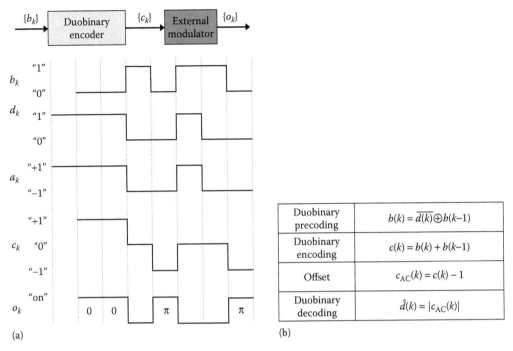

Duobinary precoding	$b(k) = \overline{d(k)} \oplus b(k-1)$
Duobinary encoding	$c(k) = b(k) + b(k-1)$
Offset	$c_{AC}(k) = c(k) - 1$
Duobinary decoding	$\hat{d}(k) = |c_{AC}(k)|$

(a)   (b)

FIGURE 11.4 (a) Example of original binary signal (b_k), precoded signal (d_k), DB-encoded signal (c_k), and optical DB signal (o_k). (b) Summary of coding rule.

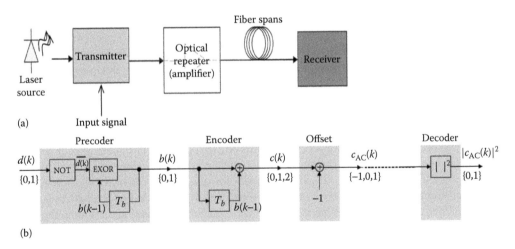

(b)

FIGURE 11.5 Main modules of a DBM optical communication system: (a) generic transmission system and (b) coder and decoder.

The DBM format has become popular compared with other modulation formats, because it extends the transmission distance as limited by fiber loss, without regenerative repeaters. It extends the dispersion limit without additional optical components, such as DCF. Chromatic dispersion has become a main effect that limits the transmission distance. The optical three-level transmission can overcome this limitation, since narrowband signal has higher tolerance to chromatic dispersion compared with broadband signal. Furthermore, DB optical fiber communication system can suppress stimulated Brillouin scattering.

11.2.3 Electro-Optic DB Transmitter

In general, a DBM transmitter is shown in Figure 11.9, consisting of a monochromatic laser source, a coder, and a photonic modulator. Binary data is encoded by DB encoder. This resulting three-level electrical signal is converted into two-level optical signal by using the folding characteristic of an optical MZIM. Depending on the nature of the signal, the resulting modulated light may be turned on and off or may vary linearly in intensity between two levels. The output of the DB transmitter is the modulated lightwaves switched on and off at transitional instances of the input electrical signal.

There are two types of DB transmitters. The conventional DB transmitter, as previously mentioned, includes a dual-drive MZIM driven by three-level electrical signals under push–pull operation. The fact that MZIM is normally driven by two-level signal, the effect of driving it with a three-level signal has its uncertainties. It is proposed that three-level signals may experience significant distortion in electrical amplifiers operating in saturation, leading to penalties for long word lengths. It may also cause the degradation of receiver sensitivity. For these reasons, the second type of DB transmitter has been proposed. This type of transmitter has the MZIM driven by only two-level electrical signals. The optical DB signal generated is the same as DB transmitter type one, that is, constant phase in blocks of 1s.

11.2.4 The DB Encoder

The DB encoder encodes the binary signal, which is a sequence of 0s to a three-level electrical signal. The DB signal is a fundamental correlative coding in partial response signaling. A DB encoder consists of a precoder and a DB coder. A precoder is used before DB coding to allow for easier recovery of binary data at the receiver and to avoid error propagation. The precoder is a simple binary digital circuit that consists of an XOR and a one-bit delay feedback. The DB coder is a filter consisting of a single delay element and a summer.

Binary data input is precoded, with initialization of the one-bit delay to 0. The output of the DB precoder is then modulated by a pulse-amplitude modulator to produce a two-level electrical signal with amplitude of -1 and 1. The DB signal is produced by adding data delayed by one-bit period to the present data. This DB signal is a three-level electrical signal of -2, 0, and 2. Finally, it is converted to a level of -1, 0, and 1. The three-level signal is mapped into optical domain by modulating both amplitude and phase. The "+1" and "−1" levels have the same optical intensity but opposite optical phase.

11.2.5 The External Modulator

The transmittance characteristics of an electro-optic MZIM is shown in Figure 11.6; it shows the relative optical power between the output and input optical waves as a function of the applied electrical voltage (at DC—direct current). The input lightwave is split into two branches and guided travelling through them. The phases of the lightwaves are influenced via the electro-optic effect generated by an applied DC voltage. Thus the interference of these guided lightwaves at the combining section results into depleted or constructed states, thence "0" or "1" intensity modulation. The maximum intensity at the output occurs when there is no or 2π phase difference between the voltages applied to the two branches. However, a minimum is obtained when the phase difference between them is π. The voltage at which the phase difference is π is called the V_π. Therefore, the amplitude of the signals can be designed in such a way that they can swing from the maximum- to minimum-intensity light beam at the MZIM output whose phase states can be either 0 or π.

It is essential here to revisit the operation of the MZIM for DB operation. In an MZIM, the input optical carrier is split into two paths via a Y junction. This Y junction splits the input signal field into $E_i/\sqrt{2}$ each. The resultant signal is

$$E_o = \frac{E_i}{2}\left[1 + \exp\left(j\pi \frac{V(t)}{V_\pi}\right)\right] \qquad (11.3)$$

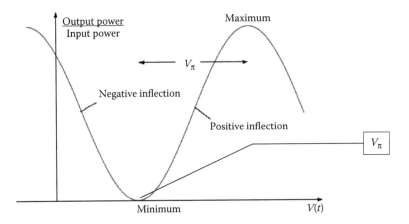

FIGURE 11.6 Transmittance transfer characteristics of the MZIM: output/input power ratio versus applied voltage in single-drive or voltage difference applied to the two electrodes of a dual-drive MZIM.

where:

V_π is the voltage to provide a π-phase shift of each phase modulator

$V(t)$ is the driving voltage

The input and output relationship of this MZIM is as shown in Figure 11.6. It is accompanied by a phase modulation of $\exp(j\varphi(t))$ with $\varphi(t) = \pi V(t)$. For $V(t)$ from 0 to V_π, E_o and E_i have the same phase, and as for $V(t)$ from V_π to $2V_\pi$, E_o and E_i have different phase.

Mach–Zehnder intensity modulator can be single drive or dual drive. Single-drive X-cut LiNbO$_3$ MZM has no phase modulation along with the amplitude modulation. It follows the transfer characteristics of Figure 11.6. Dual-drive X-cut LiNbO$_3$ MZIM (shown in Figure 11.7), on the other hand, has two paths phase modulated with opposite phase shifts in a push–pull operation. The V_π in Figure 11.9 is reduced by half in this case. For a dual-drive Y-cut LiNbO$_3$ MZIM, two paths are driven by complementary signal, with V_1 equals to $-V_2$. The output electric of a dual-drive MZIM is

$$E_o = \frac{E_i}{2}\left[\exp\left(j\pi\frac{V_1}{V_\pi}\right) + \exp\left(j\pi\frac{V_2}{V_\pi}\right)\right] \tag{11.4}$$

The DB optical signal is generated by driving a dual-drive MZIM with push–pull operation, as shown in Figure 11.7. One arm is driven by the DB signal, and the second arm is driven by the inverted DB signal. Figure 11.8 shows the operation of the MZIM. The output electric field $E_o(t)$ can be expressed as

$$E_o(t) = E_i \cos\frac{\Delta\phi(t)}{2} \cdot \exp\left(-j\cdot\frac{\phi_0}{2}\right) \tag{11.5}$$

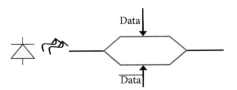

FIGURE 11.7 Dual-drive MZIM.

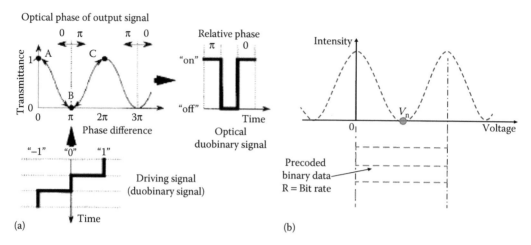

FIGURE 11.8 Driving operation of dual-drive MZM. (a) DB signaling (b) DPSK. Relative phase indicates the optical phase difference at different states of "1." (From Binh, L.N., *Optical Fiber Communication Systems with MATLAB and Simulink Model*, CRC Press, 2nd edn., Boca Raton, FL, 2014. With permission.)

where:

E_i is the input electric field

$\Delta\phi(t)$ is the phase difference between the lightwaves propagating in two optical waveguides

ϕ_0 is a constant when the MZIM is driven in a push–pull operation

At point B of Figure 11.8, the phase of the output optical signal is inverted. The optical DB signal is dependent on the biasing point of the driving signal, which is the electrical DB signal. By biasing at point B in Figure 11.12, "−1" and "+1" levels of the electrical DB signal will correspond to the "on" state of the optical signal, whereas the "0" level will correspond to the "off" state. To achieve the suppression of the lightwave carrier a π-phase difference between the two arms must be created under the DC bias and superimposed by moduulating signals by using high frequency T-bias device.

11.2.6 MATLAB Simulink Structure of DB Transmitters and Precoder

The transmitter model, generally, consists of the DB coder and the MZIM. The DB coder encodes the incoming binary sequence of 0s and 1s to DB electrical signal (Figure 11.9a). This signal is then used to drive the arms of the dual-drive MZIM. One arm is driven by the DB signal, and the other arm is driven by the inverted DB signal. The Bernoulli binary generator generates a random sequence of binary electrical signal. It is set to generate the data at a rate of 40 Gb/s. This signal is encoded by the DB encoder, which consists of a DB precoder and a DB coder. The first output of the encoder is shifted up by 1 to produced levels of "0," "1," and "2." This electrical DB signal is sent to the phase shift block, as shown in Figure 11.9b, to represent these levels with a certain phase. This, in fact, represents the biasing point on the transmittance curve. For dual-drive MZIM, the driving signal is biased at $V_{\pi/2}$. The second output of the DB encoder is the inversion of output 1. This output is used to modulate the second arm of MZIM. The output 2 signal is shifted down by −1, to bias at the point $-V_{\pi/2}$ of the transmittance curve. Mach–Zehnder intensity modulator is an amplitude modulator, accompanied by a shift of phase. This modulation is also called AM-PSK. This lightwave, which is the sine wave produced by the sine wave function, is modulated by the DB signal through the complex phase shift block. The input sine wave is shifted by the amount specified at the *Ph* input.

The DB coder consists of a precoder and a coder, as shown in Figure 11.10a. The precoder is a differential coder, with an XOR gate and a one-bit delay feedback path. The addition of −0.5 and division by 2 function as the amplitude modulator shift levels of the signal from "0" and "1" to "−1"

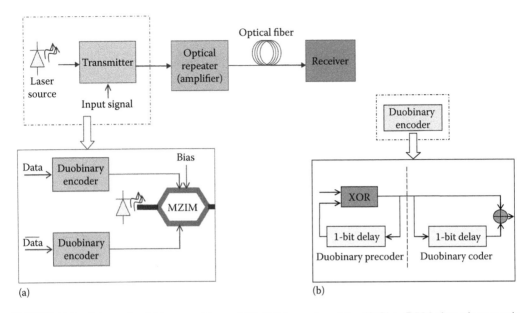

FIGURE 11.9 Schematic of (a) transmitter and (b) DBM encoder of the 40 Gbps DBM photonic transmission system.

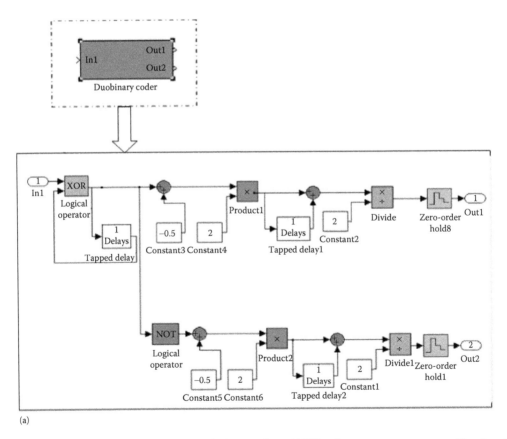

FIGURE 11.10 Simulink models of DB optical transmitter: (a) DB coder. *(Continued)*

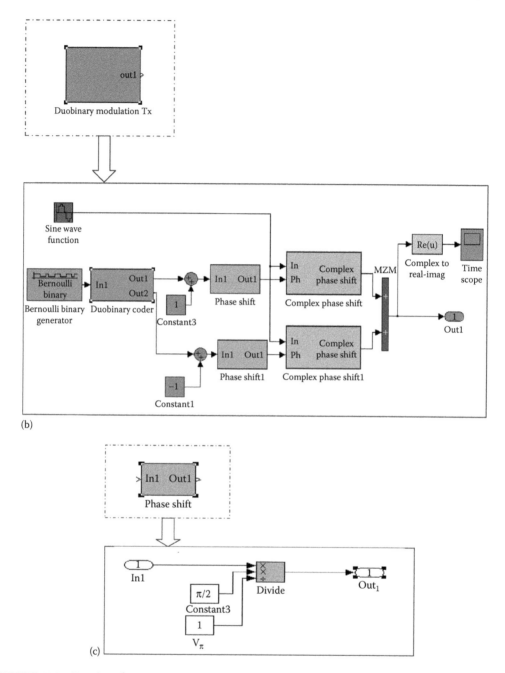

FIGURE 11.10 (Continued) Simulink models of DB optical transmitter: (b) optical MZIM with two phase-shift paths and (c) its phase-shift block.

and "+1." The signal is then added to its one-bit delay to produce a three-level DB signal of level "−2," "0," and "+2," followed by a conversion to a level of "−1," "0," and "+1." The summation of the signal with its one-bit delay is the DB coder. For the second output, the output of the differential coder is inverted, before going through the same operation as Out_1. Zero-order hold is placed before the output of the DB encoder functions to discretize the signals to have a fast-to-slow transition of signals. It holds and samples the signal before transmitting it out. If the signals are transmitted out without the zero-order hold, the transition to "0" level will be overseen. The signal will have only two levels, that is, "−1" and "+1."

To complete the Simulink model of the DB transmitter, Figure 11.10b shows the model of a dual-drive MZIM with two phase paths (see Figure 11.10c for phase shifting in Simulink). Single-drive MZIM can be structured when the phase shift of one path is set to zero or no phase shifting at all. The outputs (inverted and noninverted) of the DB encoder (block "duobinary coder") of Figure 11.10b are connected to the "phase shifter blocks" to modulate the phase of the MZIM paths.

11.2.7 ALTERNATIVE PHASE DB TRANSMITTER

Two types of DB transmitter model are proposed. The conventional DB transmitter, as mentioned previously, uses a dual-drive MZIM driven by three-level DB electrical signals. Mach–Zehnder intensity modulator, shown in Figure 11.11, is usually driven by a two-level electrical signal. In some cases, it occurred that this three-level driving signal might experience significant distortion in electrical amplifiers operating at saturation. This may lead to penalties for long word lengths. It may also occur that there will be a degradation of receiver sensitivity. Owing to these uncertainties, an alternative DB transmitter, as shown in Figure 11.12, is proposed. This second type of DB transmitter has the MZIM driven by two-level electrical signals. It consists of a differential encoder, a one-bit-period electrical time delay, and an MZIM. One arm of the dual-drive MZIM is driven with the signal from the electrical signal generator, whereas the other arm is driven by its inverted version but, delayed by one-bit period. Both DB transmitters produce the same result, which is constant phase in blocks of 1s as shown in Figure 11.13.

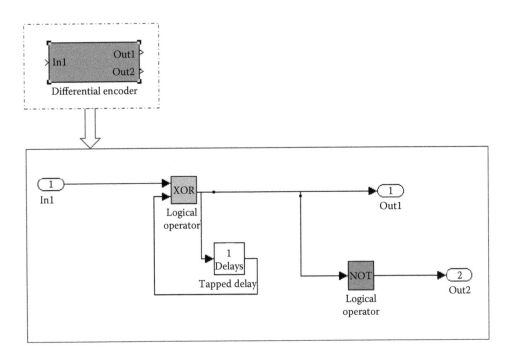

FIGURE 11.11   Simulink model of MZIM.

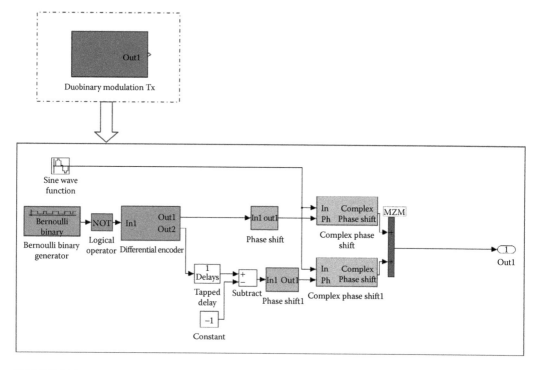

FIGURE 11.12   DB transmitter (type 2).

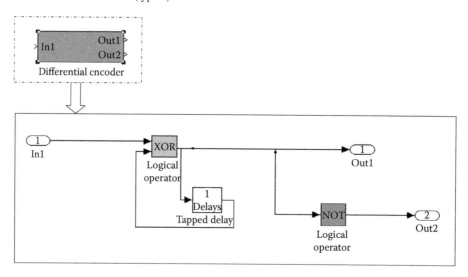

FIGURE 11.13   The differential encoder of type 2 DB transmitter.

11.2.8   FIBER PROPAGATION

As described in Chapter 3, the fiber propagation model models the linear and nonlinear dispersion effects that exist over the entire length of the optical fiber can be represented by the nonlinear Schroedinger equation (NLSE), including the self-phase modulation (SPM), other nonlinear effects can also be integrated, given by

$$\frac{\partial A}{\partial z} = +\beta_1 \frac{\partial A}{\partial t} + \frac{j}{2}\beta_2 \frac{\partial^2 A}{\partial t^2} - \frac{1}{6}\beta_3 \frac{\partial^3 A}{\partial t^3} = j\gamma\, |A|^2\, A \qquad (11.6)$$

where:

β_1 the lightwave group delay

β_2 and β_3 are the first- and second-order GVDs

γ is the nonlinear coefficient

A is the pulse envelope

The NLSE can be solved by the SSF method that integrates two main steps of split-step model: the nonlinear effects that acts alone and the linear effects.

$$\frac{\partial A}{\partial t} = (L + N)A \qquad (11.7)$$

where L and N represents linear and nonlinear operators, respectively, which can be extracted from the NLSE.

The fiber propagation block, as shown in Figure 11.14, consists of Gaussian filter, a gain factor, and the SMF model. The SMF model is shown in Figure 11.15. This model is based on the split step Fourier (SSF) method. It splits the fiber into a number of small sections, dz. All parameters

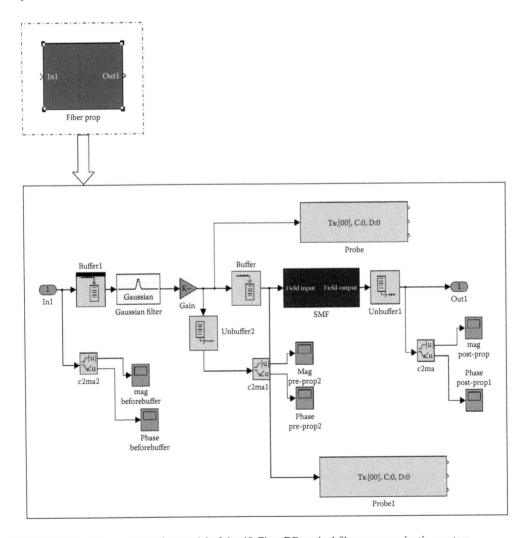

FIGURE 11.14   Fiber propagation model of the 40 Gbps DB optical fiber communication system.

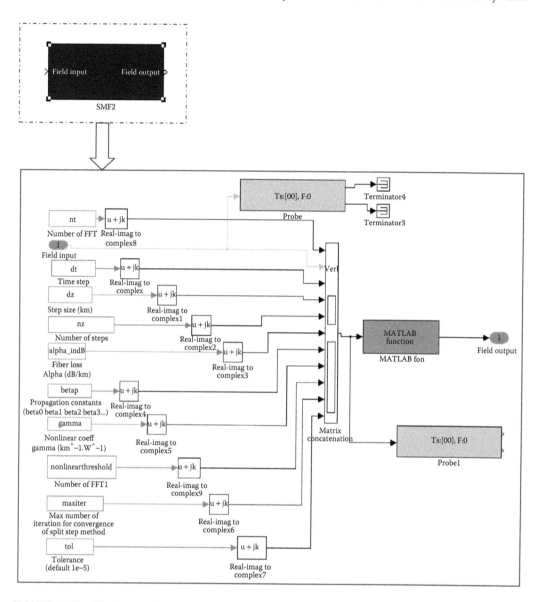

FIGURE 11.15 Single-mode fiber models using the SSF method on NLSE.

needed for the SSF operations are concatenated into a matrix, before passing them to the MATLAB Function. The MATLAB function block solves the NLSE by using the split-step Fourier method. Linear operation is implemented in all steps. When the peak power is greater than the nonlinear (SPM) threshold, the nonlinear operator is activated. In this fiber propagation model, buffer is attached at various points. Buffer is used to redistribute the input samples to a new frame size—in this case, a larger frame than the input frame size. Buffering to a larger frame size yields an output with a slower frame rate than the input. The "Unbuffer" block unbuffers the frame-based input into a sample-based output. The buffer used in this model determines the number of bits sent into the fiber, which is the SMF block.

Probes are connected at two different points of the fiber propagation model. These probes determine the sampling time at these points. Sampling time of 25 ps indicates that the signal is down sample to baseband. At baseband, the complex envelope of the signal will be extracted, and this extracted signal will be transmitting through the fiber, which is represented by SMF in Figure 11.14 [14]. However, the phase contents of the signal are maintained and transmitted through the fiber. These phase components of the DB signal are important, because they increase the dispersion tolerance and thus allow the signal to travel a longer distance.

11.3 DB SELF-COHERENT DETECTION RECEIVER

The receiver model of 40 Gb/s DB optical fiber communication system consists of a Gaussian filter and scopes to observe the performance of the system. The receiver is a conventional direct-detection type. Therefore, a demodulator or decoder is not needed. Power spectrum and eye diagram can be observed directly from this point. The Gaussian filter functions as a baseband filter. The absolute value of the incoming signal is taken, because the DB receiver detects the intensity of the signal. The probe is used to determine the sampling time at that point. "Demodsignals" block shown collects all the data at this point and save them in the workspace. These data are used to plot the histogram, which is used to determine the Q factor and BER of the system.

Demodulation is needed at the receiver, depending on the modulation format used. For instance, DPSK receiver consists of a Mach–Zehnder delay interferometer (MZDI), which demodulates the incoming signal before detecting it by using photodetector. The MZDI lets two adjacent bits interfere with each other at its output port. This interference creates the presence, or absence, of power at the output port, depending on whether the interference is constructive or destructive with each other. The preceding bit in the DPSK signal acts as a phase reference for demodulating the current bit. The delay interferometer (DI) output ports are detected by the balanced detectors. The optical DB signals can be demodulated into a binary signal with a conventional direct-detection-type optical receiver, as shown in Figure 11.16. Decoder is not required in this case. The received signal is directly detected by a photodiode operating as a square-law detector. Optical DB signal consists of two states, "on" and "off." The photodiode works by detecting the incoming intensity of the signal. The recovery of the original electrical signal can be done by simply inverting the signal detected at the photodiode. This inversion is done within the circuit as decision circuit (Figure 11.16a). The signal is observed at this point of the system. Most commonly used parameters to test and observe the performance of a system are the power spectrum, eye diagram, Q factor, BER, and received optical power. The Q factor is the quality factor of the system, under the Gaussian noise distribution. It is determined by the mean voltage level and the standard deviation of the noise.

$$\delta = \frac{\mu_1 - \mu_0}{\sigma_1 + \sigma_0} \tag{11.8}$$

where:
  μ_1 is the mean voltage level of the "1" received
  μ_0 is the mean voltage level of "0" received
  σ_1 is the standard deviation of noise of the "1" received
  σ_0 is the standard deviation of the noise of the "0" received

$$\text{BER} = \frac{1}{2}\text{erfc}\left(\frac{\delta}{\sqrt{2}}\right) \tag{11.9}$$

The 40 Gb/s DB optical fiber communication system can be implemented on Simulink platform. Simulink has been chosen as the computer software for this development of the model because it consists of a variety of communication blocks that can assist in simplifying the process of

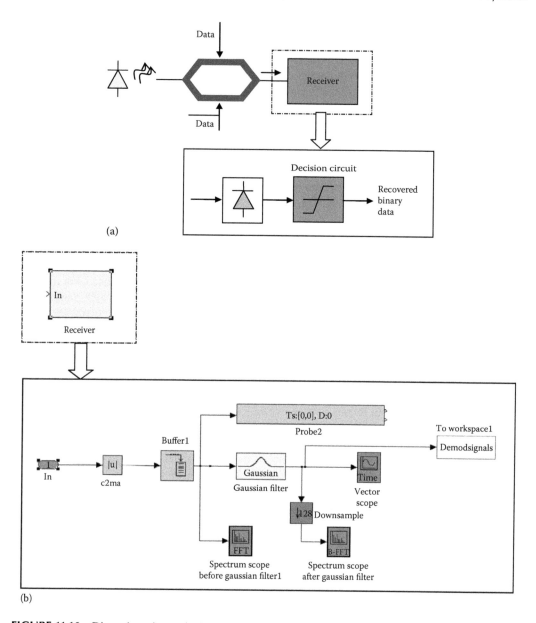

FIGURE 11.16   Direct detection optical receiver (a) and its Simulink model (b).

implementing and improving of this model. Figure 11.17a shows the overall system of the 40 Gb/s DB optical fiber communication system. The main modules of this communication system are the DB transmitter, named as the DBM Tx, the fiber propagation, and the receiver. The DB transmitter, in general, consists of the DB encoder and the MZIM. The fiber propagation block introduces the linear and nonlinear dispersion effects that exist over the entire length of optical fiber to the signal. The receiver is of direct-detection type. Therefore, the output at the receiver can be directly observed using scope, in the form of eye diagrams and power spectrum. Figure 11.17b–d displays the electrical signals at the random signal generator, the Simulink block of Bernoulli, and the two encoded data sequences to feed to the MZIM block. These signals are obtained by inserting the sampling oscilloscope blocks at the output of the encoder block.

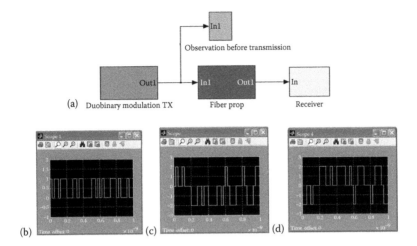

FIGURE 11.17 DB optical fiber communication system Simulink model: (a) generic block diagram of the DBM optical transmission system; and DB encoder outputs (b) Bernoulli binary sequence (c) and (d) DB electrical sequences for driving the electrodes of the MZIM.

11.4 SYSTEM TRANSMISSION AND PERFORMANCE

The overall performance of this 40 Gb/s DB optical fiber communication system can be observed at the receiver of the system. The eye diagram and power spectrum are the parameters that are used to observe the performance of the system. Multiple debugging and testing processes have to be carried out to prove that the system is functioning as expected. Simulation using Simulink has reduced vastly the time and difficulties involved in these processes.

The first step of testing involves the testing of the transmitter. It is important to ensure that DB optical signal is generated at the output of the transmitter. The DB encoder is critical in this case. Its output is checked using the time scope to ensure that the three-level electrical signal is produced. Power spectrum and eye diagram are connected at the output of the transmitter. These obtained results are compared with those obtained experimentally and theoretically to verify that the DB transmitter is generating the correct signals. Fiber propagation model is then connected to the output of the transmitter. Fiber propagation model will introduce the linear and nonlinear effects of fiber, depending on the distance the signal travels. The DB receiver, which is of conventional direct-detection type, is connected after the fiber propagation. The signal is observed directly at the receiver. By observing the eye diagram at this point, the distance for which the signal can propagate without severe distortion can be estimated. The testing is started with a fiber length of 1 km. The distance is increased until the point where significant distortion to the eye diagram can be observed and the "eye" of the eye diagram has closed.

11.4.1 THE DB ENCODER

The DB encoder is the first implemented model in this 40 Gb/s DB optical fiber communication system for generation of three-level DB electrical signals. Scopes probe at various points of the DB encoder. All time scopes are set to the same range to allow for comparisons of the bits within the same range. The temporal sequences at the outputs of the encoder for driving the MZIM are shown in Figure 11.17c and d. Scope 1 shows the data generated by the Bernoulli Binary Generator. Scopes 2–4 show the output at each arm. A bit "0" is encoded as "+2" or "−2," whereas the bit "1" is encoded as "0". This agrees with the DB coding scheme. They also show three-level electrical signal. Scope 4 is the inverted version of scope, as expected, owing to the NOT gate applied to the second arm.

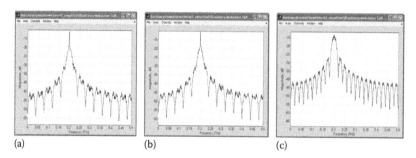

(a)   (b)   (c)

FIGURE 11.18 Power spectrum obtained from each arm of the MZIM (a) and (b), and the output of DB transmitter (c).

11.4.2 THE TRANSMITTER

The transmitter includes the DB encoder and MZIM. After the output of the DB encoder is verified and checked to produce the correct signal, the levels of the signal are represented with a phase. This phase is used to shift the laser source, produced by sine wave function. The testing of the implemented DB transmitter includes observing the eye diagram and power spectrum at the output. Time scopes are attached at various points of the transmitter to observe the signal at these points. Spectrum scopes are connected at the output of the transmitter, which is after the summation of both arms of MZIM, and to each arm of the MZIM to observe the power spectrum at these arms. Some adjustments need to be made to the transmitter model in order to observe the power spectrum. The sine wave function is set to produce a wave of 200 GHz rather than 193 THz. This is to enable the observation of the spectrum centered at a lower frequency and better spectral resolution. The zero-order hold, whose function is to hold and sample the incoming signal, is set to 1e−12, so that the x-axis of the spectrum scope is in the THz range. The obtained results, as shown in Figure 11.18, reflect the results expected from experiments and theories. These are used to verify the power spectrum obtained from Simulink model. Spectrum (c) of Figure 11.18 corresponds to spectrum scope in Figure 11.14. The obtained spectrum reflects on that in Figure 11.18a, with the shape approximately the same. The bandwidth of the obtained spectrum corresponds to the data rate. The spectrums obtained are centered at 200 GHz, which is the carrier frequency of the model. Spectrum (c) is carrier suppressed, whereas the other two spectrums are not. This is as expected from DBM format. Owing to the π-phase difference in the two arms of MZIM, the output is expected to have its carrier suppression.

Further verification of the DB transmitter Simulink model is carried out by monitoring the eye diagram at the output of the transmitter. The observation point before the transmission block in the overall system, shown in Figure 11.16, is used to observe the eye diagram before transmission into the fiber. The eye diagram obtained, as shown in Figure 11.21, has an "open eye" with amplitude of 0.6. This proves that the eye diagram is correct, since the data rate is 40 Gbps, one bit period occurs every 25 ps (Figures 11.19 and 11.20).

11.4.3 TRANSMISSION PERFORMANCE

The overall performance is observed at the receiver of the 40 Gb/s DB optical fiber communication system. The eye diagram and power spectrum are observed by the spectrum scope and vector scope attached inside the receiver. Eye diagrams show the distortion and attenuation of signals at various lengths of the fiber. When the "eye" of the eye diagram closes, the signal is severely distorted and dispersed, and thus, recovering of the signal becomes impossible. The eye diagram can also be used to calculate the received optical power. Q factor and BER can be obtained by plotting the histogram

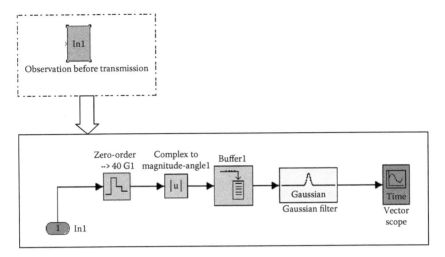

In1
Observation before transmission

Zero-order   Complex to   Buffer1
--> 40 G1   magnitude-angle1

|u|

Gaussian
Gaussian filter

Time

Vector
scope

1   In1

FIGURE 11.19 Signal and eye diagram monitored before transmission block of the 40 Gbps DB optical fiber communication system.

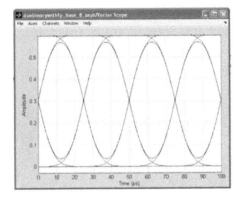

FIGURE 11.20 Eye diagram obtained before propagation block.

of the received signal, followed by some calculations to obtain the mean voltage level and the standard deviation of noise.

The eye diagrams of various lengths of the fiber are obtained. It can be observed from Figures 11.21 through 11.23 that as the length of the fiber increases, the dispersion and noise also increase. The "eye" of the eye diagram closes eventually. The DB signals are expected to be able to travel up to 200 km without the "eye" closing completely. This has been proved by the model. It can be observed that at the length of 250 km, as shown in Figure 11.23b, the "eye" of the eye diagram has yet to be fully closed. Dispersion effects can be observed, and they progressively increase with the distance of the fiber.

Data obtained from the *demodsignals* block, shown in Figure 11.16, is used to plot the histogram, which determines the Q factor and BER. Q factor is the quality factor of the system. From the Q factor, the BER can be calculated. It is expected that for a BER of 10^{-9}, the Q factor is approximately 6. The histograms for 1, 5, and 10 km are shown in Figure 11.24. The two points on the histogram are compared with the histogram at the receiver to obtain the mean values, μ_0 and μ_1, and the standard deviation of noise, σ_0, and the BER can be calculated from the Q factor found. Figure 11.25 shows the plot of BER versus distance. It can be observed that as the distant of propagation of signal

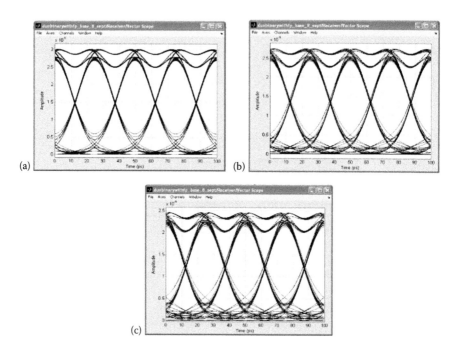

FIGURE 11.21   DB eye diagram for 1 km (a), 5 km (b), and 10 km (c) of SSMF transmission.

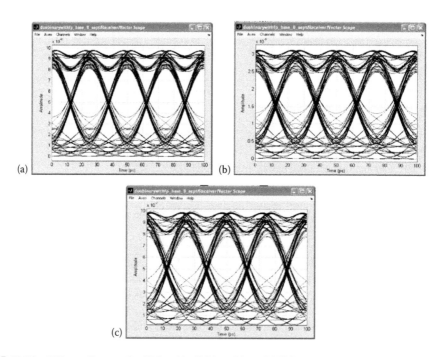

FIGURE 11.22   DB eye diagram for 50 km (a), 100 km (b), and 150 km (c).

increase, the BER increases. This is as expected, because of the linear and nonlinear effects of the fiber-introduced noise and errors to the transmitting signal. Experimental results of the transmission of CSRZ-ASK and RZ-ASK formats are also included for system over 328 km SSMF and DCM modules, completely dispersion compensated with five EDFA modules integrated. Figure 11.26 is the plot of BER versus receiver sensitivity, which is in the range of −23 to −20 dBm.

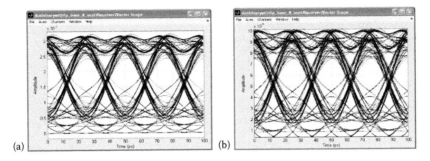

(a)   (b)

FIGURE 11.23 DB eye diagram for 100 km (a) and 150 km (b).

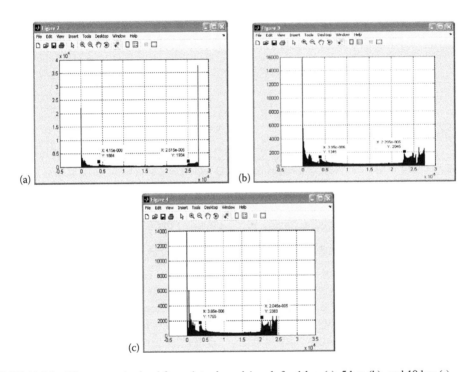

(a)   (b)

(c)

FIGURE 11.24 Histogram obtained from data *demodsignals* for 1 km (a), 5 km (b), and 10 km (c).

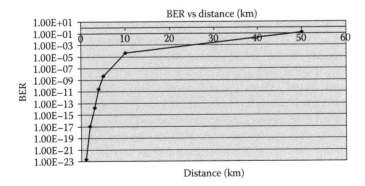

FIGURE 11.25 Plot of BER versus distance (km) of SSMF.

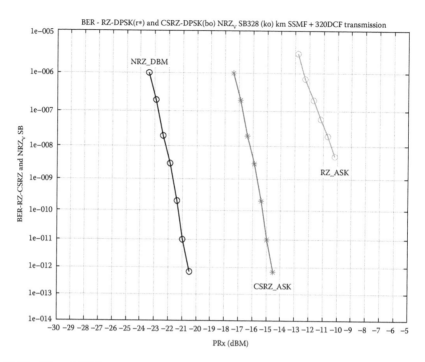

FIGURE 11.26 Plot of BER versus receiver sensitivity (dBm) of simulated NRZ_DBM, (black O) and experimental CSRZ_ASK, RZ_ASK (*). Experimental (O and *) and simulation (continuous) transmission over 328 km optically amplified multi-span system.

11.4.4 ALTERNATING PHASE AND VARIABLE PULSE WIDTH DB: EXPERIMENTAL SET-UP AND TRANSMISSION PERFORMANCE

11.4.4.1 Transmission Set-Up

A transmission system is arranged as shown in Figure 11.27a, consisting of a 50 km of SSMF and associate compensating module; a DPSK transmitter (SHF 5300), optical booster, and preamplifiers; wavelength multiplexer and demultiplexer; an MZDI; an optical balanced receiver SHF-5008; clock recovery module; and error analyzer SHF-EA 44A. A typical eye diagram obtained for 40 Gb/s CSRZ-DPSK modulation format recovered at after transmission is shown in Figure 11.27b. This testbed is also used to investigate the transmission of DWDM channels of various modulation formats. Therefore, the wavelength mux and demux are included here. The BER versus the receiver sensitivity for the CSRZ-DPSK is shown in Figure 11.28. The curves for DBM with alternating phase of the "1" intensity-coded pulses are also obtained by simulation, which are given in the next section.

The DB optical transmission can be experimentally determined via simulation, whose platform can be implemented on MATLAB Simulink for several modulation formats, especially the phase-modulated optical transmission. A typical system arrangement of optical transmitter using a dual-drive MZIM for generation of DPSK and alternating-phase DBM for simulation is shown in Figure 11.29. The fiber propagation model follows the well-known SSF method, with provision for switching between linear and nonlinear power level propagations, so as to minimize the computing time.

The FWHM of the DBM format can be generated by setting the amplitude of the swing voltage levels applied to the dual electrodes of the MZIM. The biasing condition of the MZIM can be varied between the minimum and maximum transmissions and the quadrature points of the transfer characteristics of the modulator to obtain carrier suppression and alternating-phase properties of the modulated lightwaves. The RZ formats and suppression of the carrier can also be generated by using another dual-drive MZIM biased at minimum transmission and half-bit-rate frequency

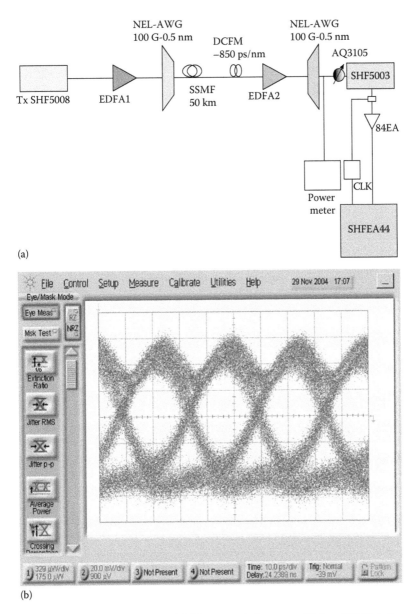

(a)

(b)

FIGURE 11.27 System testbed for CSRZ-DPSK transmission over 50 km SSMF balanced receiver: (a) schematic diagram of the testbed and (b) dispersion compensated received eye diagram of CSRZ-DPSK.

synthesizer. Our Simulink models have been extensively tested, and the system performance agrees well between experiments and modeling. Figure 11.28 shows the agreement between the BER and the receiver sensitivity for CSRZ-DPSK modulation format obtained both in experiment and by simulation. Noises of OAs have also been taken into account in both cases. We select CSRZ-DPSK format to compare with alternating-phase DBM, because it has been proven in practice to offer superior performance as compared with RZ-DPSK and NRZ-DPSK. The MZIM is modulated and biased such that the width of the DBM pulses can be altered with ease.

Simulation results are obtained for alternating-phase DB with an FWHM of 50% and the bit period of 33% for 40 Gb/s, as shown by the light gray circles and star curves in Figure 11.28. An optical Gaussian filter type is also used at the output of the transmitter before transmission. We observe almost 2 dB better receiver sensitivity of the 50% FWHM alternating-phase DB formats

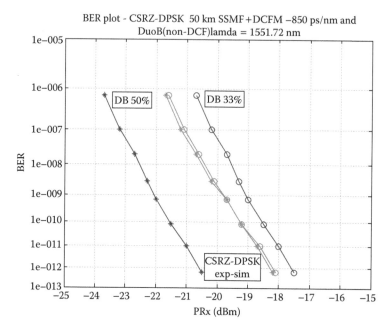

FIGURE 11.28 BER versus receiver sensitivity for 50 km transmission: (a) light gray curve—CSRZ-DPSK transmission with complete compensation; and (b) gray curves—50% and 33% FWHM alternating-phase DBM.

as compared with the CSRZ-DPSK. However, the 33% FWHM DB case offers 1 dB less sensitivity. This indicates the effectiveness of the DBM. It is noted again that the 50 km SSMF without compensation is used for 40 Gb/s DB transmission. A typical eye diagram obtained at the output of the balanced receiver for 33% FWHM DB is shown in Figure 11.29b. The simulation is conducted with 256 random bit pattern and several frames sufficient for measurement of the Q factor without resorting to the Monte Carlo technique. We also assume a Gaussian distribution of the ISI and phase noises. This assumption may have suffered 0.5–1 dB penalty as compared with the chi-square distribution for phase error in DPSK transmission.

The DBM with alternating phase and control of the FWHM of the pulse sequence offer 2 dB improvement as compared with that of the CSRZ-DPSK. The 50% FWHM with a Gaussian profile allows the possibility of lower bandwidth demand on the optical modulators. Still, this format offers better performance owing to the reduction of the signal bandwidth.

11.4.4.2 Testbed for Variable Pulse Width Alternating Phase DB Modulation Optical Transmission

In this section, we investigate the transmission performance of alternating-phase DBM of an FWHM ratio with respect to the bit period of 100%, 50%, and 33%, and compare with experimental transmission of CS-DPSK over 50 km of SSMF and dispersion compensation. For the DB case, the transmission without dispersion compensation over the same SSMF length offers better performance for 50% FWHM DBM and slightly worse for 33%. The transmission performance, the BER, and the receiver sensitivity of these DBM formats are compared with those of the CS-DPSK experimental transmission.

These transmission performances are compared with noncompensating 50 km SSMF transmission of DBM formats with pulse width of 33% and 50%. The latter format offers at least 2 dB improvement in term of the BER and receiver sensitivity over that of CSRZ-DPSK. The 50% and 33% FWHM DBM schemes offer simpler driving circuitry for the optical modulators, owing to their lower swing voltage levels applied to the electrodes of the dual-drive MZIM. This is very important when the bit rate is in the multi-GHz region. Furthermore, the 50% and 33% pulse widths and Gaussian profile will further reduce the demand on the bandwidth of optical modulators, that is, a

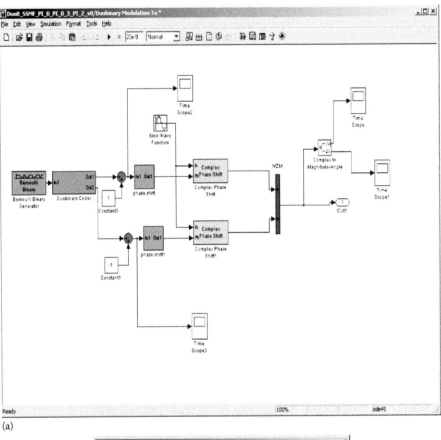

(a)

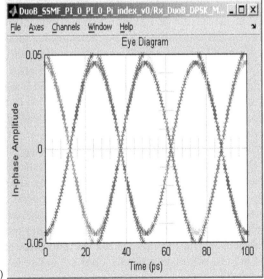

(b)

FIGURE 11.29 (a) Optical transmitter Simulink model for generation of alternating-phase DBM format and (b) nondispersion compensating balanced received eye diagram for DB 33% FWHM.

DuoBinary optically amplified transmission

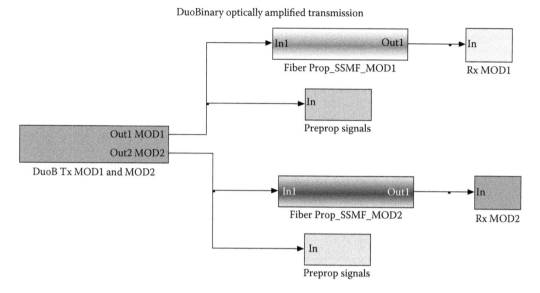

FIGURE 11.30 Generic simulation model for DB variable pulse width transmission. The transmitter is shown in the far left, consisting of two subsystem transmitters—one is the NRZ DB and the other is a different variable pulse width DB format for comparison. Each span consists of a transmission fiber (SSMF) in association with a dispersion compensating fiber and two in-line amplifiers (see Figure 11.34). Cascaded spans are identical with 100 km SSMF and 100 km DCF (negative dispersion factor and matched slope) and 20 dB gain plus 5 dB NF EDFAs.

30 GHz or lower transmittance bandwidth can be used. A Simulink model has also been developed for simulation of the transmission performance of the DBM formats. Simple driving condition can be achieved, and its effectiveness can be demonstrated.

We describe a generic simulation platform on MATLAB Simulink for several modulation formats, especially the phase-modulated and partial response optical transmission, as shown in Figure 11.30. It consists of two DB transmitters, a fiber transmission model, and a direct detection opto-electronic receiver. A typical system arrangement of optical transmitter using a dual-drive MZIM for generation of DPSK and alternating-phase DBM for simulation is shown in Figures 11.31 and 11.32. The fiber propagation model follows the well-known SSF method, with provision for switching between linear and nonlinear power level propagations, so as to minimize the computing time. In the linear regime, a transfer function block is used instead of the SSF method and the NLSE; this saves considerable computing time.

The pulse FWHM of the DBM format can be generated by setting the amplitude of the swing voltage levels applied to the dual electrodes of the MZIM. The biasing condition of the MZIM can be varied between the minimum and maximum transmissions and the quadrature points of the transfer characteristics of the modulator to obtain carrier suppression and alternating-phase properties of the modulated lightwaves. The RZ formats and suppression of the carrier can also be generated by using another dual-drive MZIM biased at minimum transmission points of both sides of the voltage-optical intensity transfer curve of MZIM. This is the pulse carver that would be required to generate variable-pulse-width optical "clock pulses" before feeding into the data modulator (Figures 11.33 through 1.35).

We note also that a half-bit-rate frequency synthesizer is used as driving source applied to the two electrodes for generation of RZ periodic pulse sequence. The optical spectrum of a lightwave 50% RZDB-modulated signals, as shown in Figure 11.36, confirms the estimation of the modulation technique.

The Simulink models have been extensively tested, and the system performance agrees well between experiments and modeling. Figure 11.40 shows the agreement between the BER and the

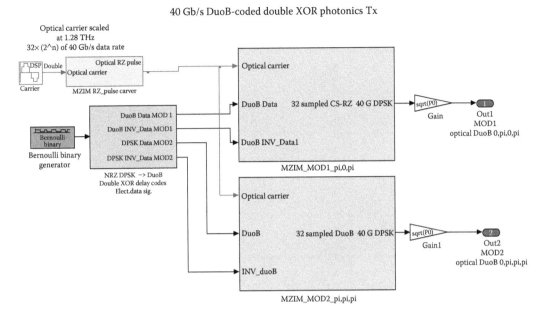

FIGURE 11.31 MATLAB Simulink model showing the pulse carver or optical "clock sources" and the data dual-drive MZIM for generation of variable pulse width RZ DB transmitter.

receiver sensitivity for CSRZ-DPSK modulation format obtained both in experiment and by simulation. Noises of OAs have also been taken into account in both cases. We select CSRZ-DPSK format to compare with alternating-phase DBM, because it has been proven in practice, offering superior performance as compared with RZ-DPSK and NRZ-DPSK. The MZIM is modulated and biased such that the width of the DB-modulated pulses can be altered.

Simulation results are obtained for alternating-phase DB with an FWHM of 50% and the bit period of 33% for 40 Gb/s, as shown by the blue curves indicated in Figure 11.40. An optical Gaussian filter type is also used at the output of the transmitter inserted before transmission system. We observe almost 2 dB better improvement in the receiver sensitivity of the 50% FWHM alternating-phase DB formats as compared with that of the CSRZ-DPSK [7,157]. However, for the 33%FWHM DB case, 1 dB less sensitivity is observed. Single photodetector can also be used to detect and recover the DB-modulated lightwave signals. The simulation is conducted with 256 random bit pattern and several frames sufficient for measurement of the Q factor without resorting to the Monte Carlo technique. We also assume a Gaussian distribution of the ISI and phase noises. This assumption may have suffered 0.5–1 dB penalty as compared with the chi-square distribution for phase error in DPSK transmission [5].

11.4.4.2.1 CSRZ-DPSK Experimental Transmission Platform and Transmission Performance

A transmission system is arranged, as shown in Figure 11.27a, consisting of a 50 km of SSMF and associate compensating module; a DPSK transmitter (SHF 5300), optical booster, and preamplifiers; wavelength multiplexer and demultiplexer; an MZDI; an optical balanced receiver SHF-5008; clock recovery module; and error analyzer SHF-EA 44A. A typical eye diagram obtained for 40 Gb/s CSRZ-DPSK modulation format recovered at after transmission is shown in Figure 11.27b. This testbed is also used to investigate the transmission of DWDM channels of various modulation formats. Therefore, the wavelength mux and demux are included here. The BER versus the receiver sensitivity for the CSRZ-DPSK is shown in Figure 11.40. The curves for DBM with alternating phase of the "1" intensity-coded pulses are also obtained by simulation.

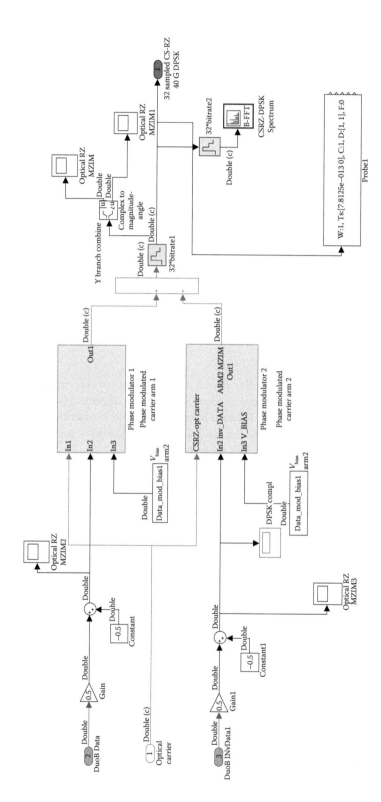

FIGURE 11.32 Electrical filter inserted before modulating the data optical modulator. Electrical filter can be Gaussian or raise cosine types, or Bessel filter of fifth order can be inserted to reduce the required transmitting signal bandwidth.

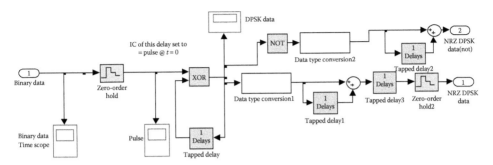

FIGURE 11.33 Simulink simulation model of the data encoder for DB format using differential encoding scheme of DPSK. The differential signals can be delayed or nondelayed and then added with their inverted versions to generate three-level electrical signals to two-level signals for driving the data modulator electrodes.

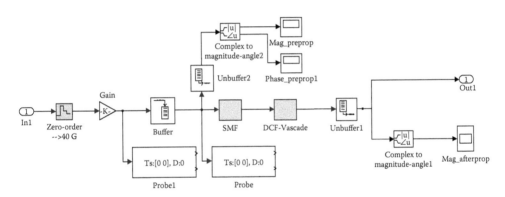

FIGURE 11.34 Simulink simulation model for the propagation of DB-modulated carrier over dispersion compensated transmission link.

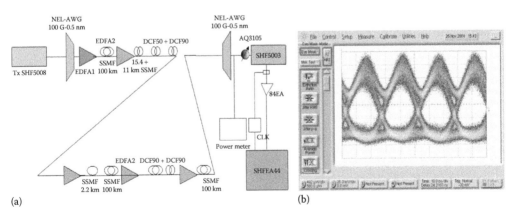

FIGURE 11.35 Experimental and simulated system testbed for CSRZ-, NRZ-, and RZ-DPSK and DB transmissions over 328 km SSMF + 320 km DCF and associated EDFAs using balanced receiver: (a) schematic diagram of the testbed and (b) 34 ps/nm residual dispersion with dispersion compensated received eye diagram of CSRZ-DPSK.

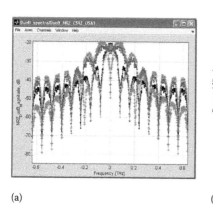
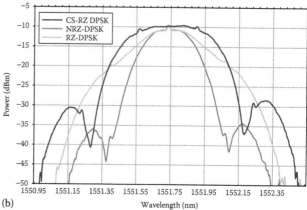

(a)   (b)

FIGURE 11.36   Optical spectrum, as monitored at the output of (a) the 40 Gb/s DB transmitter with formats NRZ (dotted-black), CSRZ 67% pulse width (light gray +), 50% RZ (gray dot), and RZ33% (very light gray *). All normalized with respect to the optical carrier, which is set at 192.52 THz. (b) Spectra of CSRZ-, NRZ-, and RZ-DPSK modulation formats obtained by experiment.

DB simulation set-up is structured close to the experimental set-up for CSRZ-DPSK transmission over 320 km SSMF optical fiber transmission multispan links. Although the modulators employed in the SHF 5008 optical transmitter are of single drive, the simulated modulators are of dual-driven type; signaling and coding blocks are described in Section 11.32. The receiver model is simple, as direct detection receiver circuitry is used, including a wideband photodetector and a microwave amplifier. A Bessel fifth-order or Gaussian (80% or DB signal bandwidth) electrical filter can also be incorporated to minimize the noise contribution.

The optical spectra of DPSK signals under different pulse width and pulse carver operations are shown in Figure 11.36b, whereas Figure 11.36a shows the simulation spectra of their counterparts under DB-coded scheme. The spectra are centered at the optical carrier frequency. The light gray curve (+) shows clearly the suppression of the carrier, whereas the other DB schemes show minimum carrier suppression but substantial signal power level that indicates the specific characteristics of the diphase modulation, especially when the data modulator is biased such that the phase difference between the two arms of the MZIM is π.

Figure 11.37 shows the simulated eye diagrams at the receiver (after the photodetector, including electronic amplifier noises) of the DB-coded sequence of NRZ (a) 33% RZ (b) 50% RZ and (c) 50% DB NRZ pulse shapes. Gaussian electrical filtering has been applied to the NRZ DB and none has been applied to the other RZ formats of the scheme. No electrical filtering is used at the transmitter. The bit rate is 40 Gb/s.

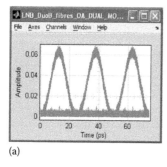
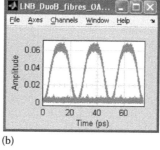
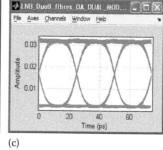

(a)   (b)   (c)

FIGURE 11.37   Received eyes after (3 spans) 320 km SSMF fiber filly compensated, including two in-line optical amplifiers (20 dB gain and 5 dB noise figure) transmission. (a) 33% RZ (b) 50% RZ and (c) 50% DB NRZ pulse shapes.

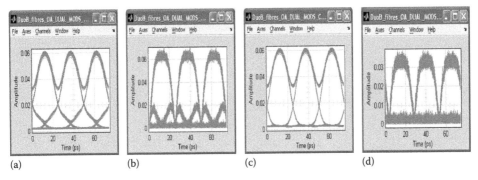

FIGURE 11.38 Received eyes after (~3 spans) 320 km SSMF fiber fully compensated, including two in-line optical amplifiers (20 dB gain and 5 dB noise figure) 40 Gb/s transmission. (a) CSRZ 67% DB with the integration of Gaussian electrical filter with 30% bit rate (b) same as (c) but without filtering at the receiver (d) CSRZ 67% DB with raise cosine filtering at transmitter and no filter at the receiver.

The effects of electrical filtering at the transmitter on the received eye diagrams are shown in Figure 11.38a–d. A 30% bit rate (BT = 0.3) electrical Gaussian filter bandwidth would still sustain the eye opening after 320 km transmission but substantial closure of the temporal eyes (Figure 11.38a and b); that is, the decision sampling time must be very accurate to achieve error-free compensation even under fully compensation transmission. A 50% BT would be, as observed under simulation, much more tolerable to the sampling error. Figure 11.38c and d shows that such filtering effects would not exist if a raise cosine electrical filter with a 0.5 roll-off factor for 40 Gb/s CSRZ 67% DB format transmits over 320 km three-span optically amplified fully compensated transmission.

Full compensated eye diagrams after transmission over the 320 km optically amplified multispan SSMF transmission link are shown in Figure 11.40a, and the dispersion tolerance versus the BER is shown in Figure 11.40b. It is observed that the dispersion tolerance is similar for various pulse-width DB signals. The 33% and 50% RZ-modified DB formats shown indicate error-free transmission, with a residual distance of about 6 km DB format, whereas experimentally, we could only obtain error-free transmission only after 3 km SSMF residual dispersion for CSRZ 67% DPSK format.

Experimentally and under simulation, the launched power is adjusted and the BER versus the receiver sensitivity is obtained in Figure 11.40. Included in this figure is also the performance curve of the CSRZ-DPSK transmission. The BER curves indicate that 50% RZ and 67% NRZ DB formats outperform the CSRZ-DPSK by at least 2 dB. Although these DB performance curves are from simulation, we expect the comparison with those of experimental CSRZ-DPSK would not much be different, as we have taken into account the electronic noise and the signal-dependent shot noises in the models of the photodetector. In both experimental and modeling cases, it is reasonable to assume that the PMD effects are negligible. We expect that the 33% RZ DB performance would fall between these two formats. We have not sufficient time to conduct detailed transmission results for this format, and these results will be reported in the near future.

It is noted that the model represents exactly the photonic behavior of lightwave-modulated modulation formats under the square-law direct detection, incorporating the equivalent electronic noises of 40 GHz 3 dB bandwidth PD-TIA (photodetector-transimpedance amplifier) electronic receiver. Both signal-dependent shot noise and equivalent electronic noise current are modeled. The bandwidth of the receiver is adjusted according to the effective bandwidth of the modulation scheme. This allows us to justify the comparison of practical implementation of the direct detection and balanced receivers. The pulse sequence at the receiver is integrated, and an average power is obtained. Then, its equivalent quantum shot noise is calculated over the electronic amplifier bandwidth. The equivalent electronic noise current at the input of the electronic amplifier is then added to these signal-dependent noises and a signal-to-noise ratio is obtained. Thus, the received optical power can be derived and plotted with respect to the BER obtained from processing the eye diagram obtained after each transmission.

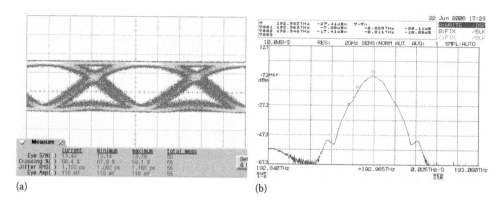

FIGURE 11.39 DB signals: (a) eye diagram; (b) measured optical spectrum. (Courtesy of SHF AG, Berlin Germany.)

The simulation duration is set long enough, so that the number of pulses transmitted and received is sufficiently high. Although we understand that the noise statistic of the phase-modulated optical transmission is asymmetrical and may follow a Maxwellian distribution, it is assumed, in this chapter, that the noise statistic follows a Gaussian distribution. This assumption may suffer only about 0.5–1.0 dB difference from the true distribution.

The DBM with alternating phase and control of the FWHM of the pulse sequence offer 5 dB improvement as compared with that of the CSRZ-DPSK obtained experimentally. For the simulated DB 50% pulse format, there is a 1.2 dB improvement on the receiver power. Equivalent noise current at the receiver of 2.5 pA/Hz$^{1/2}$ is used, which is compatible to that of commercial 50 GHz DPSK optical receiver. The bandwidth of the receiver is adjusted to narrower for 67% DB case, and hence, there are less noises for both the quantum signal-dependent shot noise and total equivalent electronic noise power. The 40 Gb/s DB signals in the time domain and their spectrum are shown in Figure 11.39, using filter of 17 GHz in the transmitter. The 3 dB bandwidth shows clearly the single sideband and embedded carrier. The 50% FWHM with a Gaussian profile allows the possibility of lower bandwidth demand on the optical modulators [2]. Still, this format offers better performance, owing to the reduction of the signal bandwidth. All performance curves shown in Figure 11.40 are obtained when the modulation and the carrier power are operating under linear region, with the nonlinear threshold set at a power that would create a total phase change of 0.1π. A 4.2 dB improvement of 67% as compared with 50% DB is due to the reduction of the electronic and quantum shot noise at the receiver.

11.4.5 DBM AND FEED FORWARD EQUALIZATION USING DIGITAL SIGNAL PROCESSING

Recently, it has been reported that, by experiment, it is demonstrated that by 100-Gb/s time and wavelength division multiplexing (WDM) in passive optical networks with 4λ 100-GHz-spaced 25-Gb/s optical DB channels in the C-band and by using 16-GHz bandwidth Avalanche Photodetector (APD) with postequalization, a loss budget of 37 dB after 20 km SSMF transmission with 1 : 512 power split can be achieved [16]. The platform set-up is shown in Figure 11.41, in which the DBM is a dual-phase modulation format, as described in Section 11.2.6, but with a DSP-based direction detection incorporating an APD with 9 tap feed forward DSP. With an Rx sensitivity of −16.5 dBm and for 25 Gb/s and a BER rate of 1e−7, the distance can be extended to 20 km SSMF, possibly by the equalization for five channels. No noise floor could be observed or reported; therefore, an FEC must be used to achieve error-free transmission over this distance. This shows the tremendous advantage of the DSP with some increase of the latency.

Comparing the performance at BER at 1e−9 in Figures 11.40 and 11.42 using different pulse shaping formats DPSK CSRZ, RZ 50% and 33% and duobinary, we note that BER against receiver sensitivity of the duobinary modulation scheme over the DCF-SSMF transmission system, can be compatible or even better sensitivity.

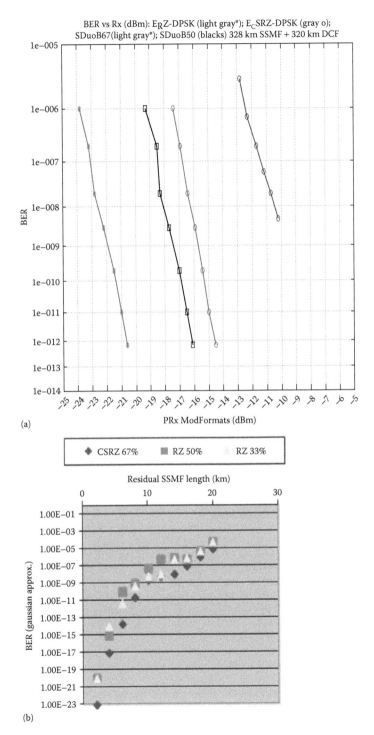

(a)

(b)

FIGURE 11.40 (a) BER versus receiver power for 320 km SSMF and 320 km (effective negative 320 km SSMF) DCF transmission: (a) light gray (*o*) curve—*experimental* CSRZ-DPSK transmission with complete compensation. (b) dark gray (*o*) experimental RZ-DPSK. Light gray (*) curves simulated DB 67% and *square black* simulated DB 50% FWHM alternating-phase modulation. DB 67% shows improvement over DB 50% by 4.2 dB. (c) Simulated dispersion tolerance for 40 Gb/s DB 67% and 50% and 33% FWHM—BER versus residual equivalent SSMF length. Launched power 0 dBm at the output of transmitter into the fiber transmission line.

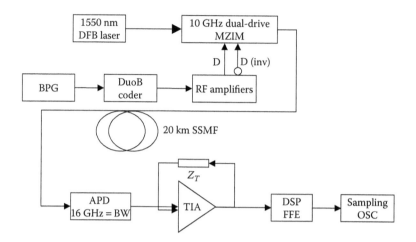

FIGURE 11.41 Set-up schematic of the duobinary by dual-phase modulation using dual-drive MZIM and DSP-based optical receiver with 9-tap FFE equalizer. (After Ye, Z., Li, S., Cheng, N., and Liu, X. Demonstration of High-Performance Cost-Effective 100-Gb/s TWDM-PON Using 4x 25-Gb/s Optical Duobinary Channels with 16-GHz APD and Receiver-Side Post-Equalization. In *ECOC*, paper Mo.3.4.4., Valencia, Spain, 2015.)

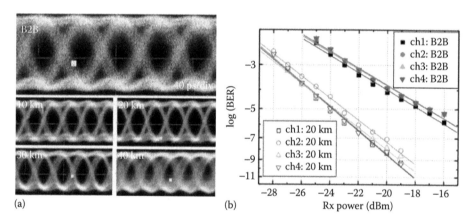

FIGURE 11.42 Experimentally measured (a) eye diagrams and (b) BER performance of the four 25-Gb/s downstream channels at different transmission distance B2B and 10–40 km. B2B = back-to-back Ch1 = channel 1 at 1550.93 nm spacing = 100 GHz or 0.8 nm. (After Ye, Z., Li, S., Cheng, N., and Liu, X. Demonstration of High-Performance Cost-Effective 100-Gb/s TWDM-PON Using 4x 25-Gb/s Optical Duobinary Channels with 16-GHz APD and Receiver-Side Post-Equalization. In ECOC, paper Mo.3.4.4., Valencia, Spain, 2015.)

11.4.6 REMARKS

The DBMs have been demonstration by the transmission model testbed. The system performance is error free, with receiver sensitivity −60 dBm without receiver electronic noise. The nondispersion compensation transmission can reach 9–10 km for 40 Gb/s, equivalent to 150 km for 10 Gb/s. The Simulink model is successfully developed. We have also demonstrated, both experimentally and in modeling, the CSRZ-DPSK transmissions. These transmission performances are compared with noncompensating 50 km SSMF transmission of DBM formats with pulse width of 33% and 50%. The latter format offers at least 2 dB improvement in term of the BER and receiver sensitivity over that of CSRZ-DPSK. The 50% and 33% FWHM-DBM schemes offer simpler driving circuitry for the optical modulators owing to its lower swing voltage levels applied to the electrodes of the dual-drive MZIM. This is very important when the bit rate is in the multi-GHz region. Furthermore, the 50% and 33% pulse width and Gaussian profile will further reduce the demand on the bandwidth of optical modulators, that is, a 30 GHz or lower MZIM can be used. A Simulink model has also been developed for simulation of the transmission performance of the DBM formats. Simple driving condition can be achieved, and its effectiveness demonstrated.

We have demonstrated both experimentally and in modeling the CSRZ-DPSK and DB transmission. These transmission performances are compared with noncompensating 50 km SSMF transmission of DBM with pulse width of 50% and 67% (NRZ). The later formats (50% and 67%) offer at least 5 dB and 1.2 improvements respectively in term of the BER and receiver sensitivity over that of CSRZ-DPSK. The 67%, 50%, and 33% FWHM DBM schemes offer simpler driving circuitry for the optical modulators due to its lower swing voltage levels applied to the electrodes of the dual-drive MZIM. Simpler and narrower bandwidth of the electronic amplifier for 67% DB case would also improve the receiver sensitivity. We thus expect 33% DB would worst than the case of CSRZ-DPSK and hence no simulation is conducted for this scenario. This is very important when the bit rate is in the multi-GHz region. Furthermore the 67%, 50% pulse widths and Gaussian profile will further reduce the demand on the bandwidth of optical modulators, that is a 30 GHz or lower MZIM can be used. Simulink models have also been developed for simulation of the transmission performance of the DBM formats. The balanced receivers are modeled exactly as a real practical sub-system. This allows a fair comparison with the transmission performance of the CSRZ-DPSK formats. Simple driving condition can be achieved, and its effectiveness demonstrated.

Although we have not conducted the transmission of DB format pulse sequences of variable pulse width over the multi-span optically amplified optical transmission lines under nonlinear region, that is, the fiber under the SPM effects, it is expected that the CSRZ 67%, RZ 33%, and RZ 50% DB formats would outperform their DPSK counterparts. Further, it is also expected that the RZ 33% DB would suffer much less nonlinear distortion effects than its 50% and 67%RZ counterparts. The DSP-based receiver DBM transmission is also briefly described, in which a simple FFE is employed to increase the error-free transmission.

11.5 DWDM VSB MODULATION FORMAT OPTICAL TRANSMISSION

This section presents the transmission of optical multiplexed channels of 40 Gb/s using vestigial single sideband modulation format over a long-reach optical fiber transmission system. The effects on the Q factor of fibers dispersion, the passband, and roll-off frequency of the OFs and the channel spacing are described. It has been demonstrated that BER of 10^{-12} or better can be achieved across all channels and minimum degradation of the channels can be obtained under this modulation format. OFs are designed with asymmetric roll-off bands. Simulations of the transmission system are also given and compared for channel spacing of 20, 30, and 40 GHz. It is shown that the passband of 28 GHz and 20 dB cut-off band performs best for 40 GHz channel spacing [17].

Although optical communications have been extensively developed for ultrahigh capacity transport networks, the demand for high-speed communication system over ultralong reach and ultralong haul offering greater capacity is expected to offer challenges for further technical development for bandwidth-efficient networking. The global Internet traffic has been growing rapidly, typically doubling the backbone traffic each year. This requires higher channel speeds per Internet port and indeed the Internet backbone are currently moving to 10, 40 and 100 Gb/s bit- and symbol levels. It is foreseen that the information economy has continued its growth unabated, thence efficient transmission techniques for the ever increasing higher bit or symbol rate will be required in the very near future.

The most common modulation formats for 40 Gb/s optical systems are the RZ, NRZ, CSRZ, which can also be integrated with differential PSK coding [17–9.23]. The VSB-RZ modulation format can also be considered as the appropriate choice for Tb/s long-haul optical transmission systems owing to its half-band property and hence is highly tolerant to dispersion and nonlinearity. On the other hand, VSB with NRZ could provide higher spectral efficiency than VSB-RZ format, since NRZ occupies only half of the RZ bandwidth and requires a lower peak transmit power in order to maintain the same energy per bit. This would offer the same BER as the RZ format.

This chapter and/or section gives a numerical simulation of the VSB-NRZ modulation format for long-haul optical fibers transmission system and the effects of the OFs on its transmission performance. The OFs used for eliminating the unwanted optical sideband are critical in the generation of the desired format. The design of such filters is given, so as to alter the pass- and roll-off bands of the filters to investigate their effects on the transmission system performance. Furthermore, the channel spacing of the multiplexed channels is also important and plays a major role in the specification of OFs. Simulation is presented for 8-channel DWDM of 40 GHz channel spacing optical fibers transmission system. The channels are transmitted at 40 Gb/s over a dispersion-managed 100 km span. The effects of these filters on back-to-back transmission and dispersion tolerance are the principal objectives of this work.

11.5.1 Transmission System

The schematic diagram of the optical transmission employing VSB modulation format is shown in Figure 11.43. An optical source operating in the continuous mode is launched into an external optical modulator, which modulates the lightwave carrier via a random bit pattern generator and associated microwave power amplifiers. The external modulator, an X-cut LiNbO$_3$ MZ intensity modulator, is used to offer chirp-free modulated output. For 40 Gb/s, the stabilization and linearity of the external modulator are critical. Normally, two modulators would be used, one for generating the required NRZ format and the other for either carrier suppression or phase modulation, if required, depending on the transmission format.

The NRZ data format is used in this simulation. The VSB modulation technique is used and generated by the use of an OF, which filters the unwanted sideband of the information channel. After the filtering, a number of optical VSB channel can be multiplexed and then transmitted through a number of optically amplified dispersion-managed fibers spans. Each optically amplified-dispersion-compensating inline unit would consist of an inline OA, followed by a dispersion compensating module, and then another optical amplification booster that would then enhance the total average optical power for transmission to the next span. Eye diagrams are obtained, and the BER can be deduced.

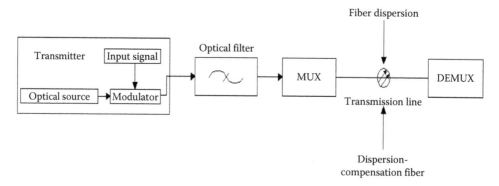

FIGURE 11.43 Schematic of the modeled VSB transmission system.

For example, the back-to-back eye diagram is observed when dispersion-compensation fibers are used with appropriate dispersion slope for equalization. Naturally, at the end of the transmission line, the multiplexed channels are separated via a demultiplexer. The principal objective of this chapter is to study the effects of the OF on the VSB system performance; thus, we do not include optical amplification noises in our modeling in the pre- and booster amplifiers located at each transmission span. Both RZ and the NRZ formats can be used. The main advantage of NRZ format is that it occupies half of the RZ bandwidth. The VSB modulation allows a small amount of the unwanted sideband existing at the output of modulator, depending on the roll of passband of the filter. Instead of eliminating the entire second sideband, such as in the case of SSB, the VSB modulation suppresses most of, but not completely, the second sideband. Using this technique, the difficulty in generating a very sharp cut-off can be overcome. The VSB modulation can be implemented with OFs to eliminate most of the second sideband. The spectrum of a VSB signal can be obtained as [20]:

$$x_c(t) = \frac{1}{2} AE \cos(\omega_c - \omega_1)t + \frac{1}{2} A(1 - E)\cos(\omega_c + \omega_1)t + \frac{1}{2} B \cos(\omega_c + \omega_2)t \qquad (11.10)$$

where:
E is magnitude of the optical field envelope
ω_c is the optical carrier radial frequency
ω_1 and ω_2 are the arbitrary radial frequencies of the signals

The signal can be demodulated by multiplying by $4\cos(\omega_c t)$ and applying the low-pass OF, leading to:

$$E(t) = AE \cos \omega_1 t + A(1 - E)\cos \omega_1 t + B \cos \omega_2 t$$
$$e(t) = A \cos \omega_1 t + B \cos \omega_2 t \qquad (11.11)$$

The basic characteristics of a VSB OF can be illustrated in Figure 11.44. We note that the roll-off band and passbands are asymmetric.

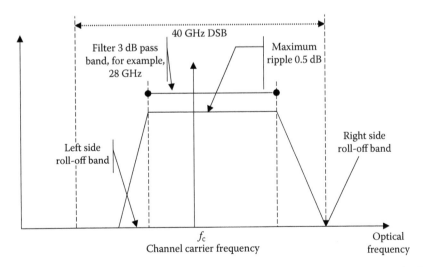

FIGURE 11.44 Spectral property of the VSB modulation format as compared with that of DSB (double sideband): details of optical VSB filter passband and roll-off bands.

11.5.2 VSB Filtering and DWDM Channels

To achieve the VSB-modulation-formatted signals, OF is implemented. In this chapter, a number of low-pass elliptic filters (LPEFs) are chosen, because this filter type offers the steepest transition region between the passband and stopband without suffering the unstable problems. The elliptic filter is a combination of the Chebyshev Type I and Chebyshev Type II filters and exhibits some amplitude response ripples in the passband and stopband. The main advantage of elliptic filter is that the width of the transition band is minimized for a finite ripple limit in the passband and a minimum attenuation in the stopband. Furthermore, these filters can be implemented using planar circuit technology [21]. The spectral response of the optical LPEF of order N is given by [21]:

$$|H_{LP}(j\omega)|^2 = \frac{1}{1+\varepsilon^2 E_N^2(\omega)} \tag{11.12}$$

where $E_N^2(\omega)$ is the Chebyshev rational function and can be determined from the specified ripple characteristics. Similarly, the s-domain transfer function of the LPEF of order N can be obtained as

$$H_{LP}(s) = \frac{H_0}{D(s)} \prod_{i=1}^{r} \frac{s^2 + A_{0i}}{s^2 + B_{1i}s + B_{0i}} \tag{11.13}$$

where $s = j\omega$, $r = N-1/2$ for odd N, and $r = N/2$ for even N. The definitions of all the bands of the filters can be referred to in Figure 11.44. Figure 11.45 illustrates a typical response of the LPEF. The carrier wavelength of 1550 nm is used as the center wavelength corresponding to the optical frequency of 193.41 THz. The characteristic of the elliptic filter is designed with the following properties: passband ripple = 0.5 dB or less; minimum stopband attenuation = 10 dB; passband region = 28 GHz; stopband region = 2 GHz; and 20 dB cut-off band = 10 GHz and 20 GHz for the left and right sides, respectively. The filter spectral characteristics are illustrated in Figures 11.45 and 11.46. The 8 multiplexed signals at the output of an LPEF filter are plotted in Figure 11.47. These type of OFs can be implemented with fiber Bragg gratings or multistage silica-on-silicon planar-integrated MZDI. We note that the MZDI can also be considered an all-pass (all-zero) filter

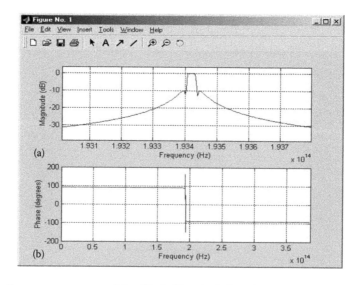

FIGURE 11.45 Frequency responses of the elliptic filter, including the amplitude (a) and phase (b).

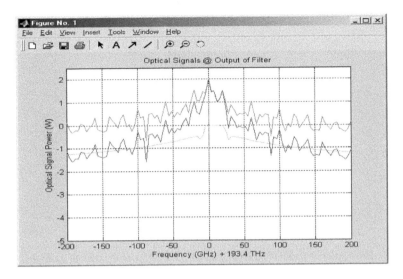

FIGURE 11.46 Filter characteristics (bottom) and signal spectra before filtering (most upper) and after filtering (middle curve). The filter frequency response is also included and indicated as the bottom graph.

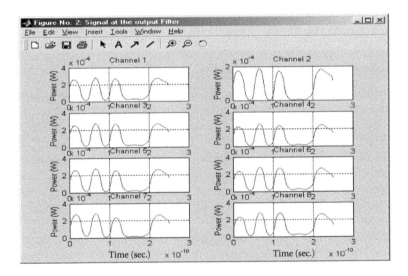

FIGURE 11.47 Signals of the DWDM channels at the output of the VSB filters.

that would offer no resonant peaks, and hence, its group delay is almost constant and thus no dispersion would be introduced by these OFs. In this work, we would take into account the dispersion characteristics deduced from the group delay of the design filter, and this would be compensated for by the dispersion compensating module.

The optical transmission system is simulated for 8 wavelength channels, in which the 1550 nm wavelength is taken as the center wavelength and the frequency spacing is taken on 20, 30, and 40 GHz spacing wavelength grid and so on. For simplicity, we use 1550 nm as the center wavelength rather than the exact international telecommunication union (ITU) spectral grid. After filtering out the unwanted sideband, the multichannels are multiplexed and then propagated through the single-mode low nonzero dispersion-shifted fibers (NZDSFs). The multiplexing is based on the WDM technique, in which multiple optical carriers at different wavelength are modulated using independent electrical bit streams and then will be transmitted over the same fibers. Our design employs

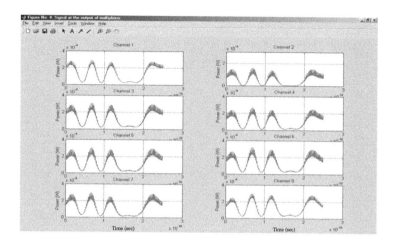

FIGURE 11.48   Time-domain signals of the channels at the outputs of the optical multiplexer.

variable channel spacing of 20–40 GHz. The signal spectra in the frequency and the time domain at the output of multiplexer are depicted in Figure 11.48. We observe that there are some ripples in the temporal signals. It is believed that these ripples appearing at the output of multiplexer are due to the crosstalks generated in multichannel transmission. At the output of the LPEF, the ripple is negligible, as the channels are filtered independently to each other.

11.5.3   TRANSMISSION DISPERSION AND COMPENSATION FIBERS

The SMFs, standard or NZ-DSF types, are naturally chosen as the transmission media in this simulation. This section briefly describes the design of the fibers so that their dispersion properties can be specified accurately instead of using the data provided by fiber manufacturers. Owing to chromatic dispersion, the GVD is frequency dependent. Consequently, different spectral components of the pulse travel at a different velocity and cause pulse dispersion that limits the performance of SMFs. Fiber dispersion consists of two components: material dispersion and waveguide dispersion. The transmission fibers and the matched DCFs, as described in Ref. [22], with matched dispersion factors and dispersion slopes for complete compensation of the channels across the waveband.

We have designed the transmission and dispersion compensating optical fibers with specific dispersion factor 17 ps/nm/km and dispersion slope 0.3 ps/(nm²km) for transmitting several channels. The dispersion and dispersion slope values of the DCF are designed with a factor of 5 times of those of the transmission fiber with a residual dispersion of about 1.5 ps/nm/km for the highest and lowest wavelength channels. Fibers parameters that are used in this simulation are fibers radius, $a = 1.6$ μm, and relative refractive index difference, $\Delta = 0.0339$. The dispersion factors of the centered channel (only channel 5 is illustrated) are indicated in Figure 11.49. The pulse broadening after transmitting through the fibers length L can be expressed as

$$\Delta\tau = D(\lambda)\sigma_\lambda L \tag{11.14}$$

where:
  $D(\lambda)$ is fibers spectral dispersion (ps/nm.km)
  σ_λ is the signal bandwidth when the laser linewidth is much smaller than that of the VSB signals
  L is the fibers transmission distance

Figure 11.49 shows the signals at the output of the demultiplexer for each channel (channel 1–channel 8) without dispersion compensation.

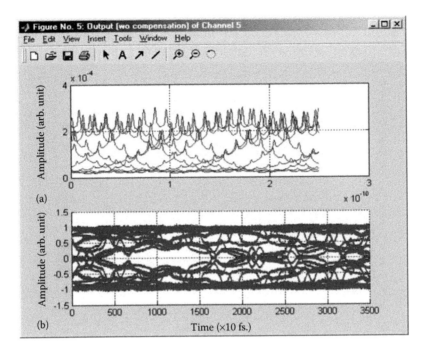

(a)

(b) Time (×10 fs.)

FIGURE 11.49 Signal outputs at the end of transmission fibers without dispersion compensation of the centered channel 5 at 1550 nm, with $D_T = 2.7455e{-}6$ ps/(nm.km): (a) time sequence and (b) eye diagram.

From Figure 11.49, the simulation results show that over 100 km span transmission line, the eye of the transmitted data closes completely, as expected, due to fibers dispersion. It is observed that the eye is still open at a BER of about 10^{-9} after 16 km designed fibers transmission. This significantly improves over the NRZ format.

We now turn to the DCFs. Since the GVD limits the system performance, it is essential to implement a DCF. The condition of dispersion compensation can be expressed as

$$D_1L_1 + D_2L_2 + \Gamma = 0 \tag{11.15}$$

where:
D_1 is total dispersion of the fibers transmission fibers section (ps/nm.km)
L_1 is the fibers transmission distance (km)
D_2 is the fibers dispersion compensation distance (ps/nm.km)
L_2 is the dispersion compensation distance (km)
Γ is the dispersion factor contributed by the VSB OF

Obviously, Equation 11.15 shows that the DCF must have its GVD negative at 1550 nm to compensate for the positive dispersion of the transmission fibers and the OF. The parameters of the DCF can be optimized for minimum nonlinear self-phase-modulation effect with: fibers radius, $a = 1.7$ µm, and relative refractive index difference, $\Delta = 0.0243$. Figure 11.50 shows the transmitted data at the end of the dispersion managed system with a dispersion compensation for channel 5 of the 8 channels. The effectiveness of dispersion compensation fibers can be seen from the improvement of BER versus receiver sensitivity due to its recovery of the data channels very close to their original sequences. The residual ripple of the compensated pulses clearly shows the effects of the filter passband on the time-domain pulses. These ripples contribute the penalty on the eye closure and hence the Q factor. An optical multiplexer then combines the modulated lightwaves to the transmitting fibers. At the receiver, the channels are separated into different channels by an optical

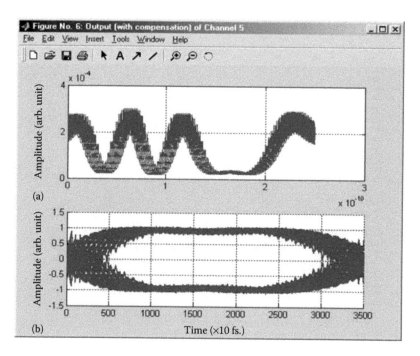

FIGURE 11.50 Signal outputs at the end of transmission line with dispersion compensation of the center channel—channel 5 at 1550 nm with DCF dispersion factor of −2.7918e−6 ps/(nm.km): (a) temporal sequence and (b) eye diagram.

demultiplexer. Figure 11.49 gives the simulation results of transmitted signals at the output of multiplexer without dispersion compensation, whereas Figure 11.50 shows the transmitted signals at the output of multiplexers, with DCF in cascade with the transmission fibers. We note that OAs at the input and booster amplifier at the output are not included in this simulation.

11.5.4 TRANSMISSION PERFORMANCE

The eye diagram is used to deduce the Q factor and hence the system BER. The random digital optical pulse sequence suffers distortion by noise, pulse broadening, and timing jitter errors introduced in the optically amplified fibers link. Assuming that the probability density functions of the "1" and "0" are equal and Gaussian, the eye diagram can be used to estimate the Q factor by

$$Q = \frac{\mu_1 - \mu_0}{\sigma_1 - \sigma_0}$$ (11.16)

where μ_1 and μ_0 are the means of the current at the output of the photodetector of the decision circuitry of the receiver at the sampling instant for symbol "1" and "0," respectively, whereas σ_1 and σ_0 are the standard deviations of the current at the decision circuit input at the sampling instant for symbol "1" and "0," respectively (Figure 11.51).

11.5.4.1 Effects of Channel Spacing on Q Factor

The capacity of the WDM system depends on how close channels can be packed into the wavelength domain. The minimum channel spacing is limited by interchannel crosstalk. The typical value of channel spacing should exceed four times that of the bit rate. The Q factor is obtained for each

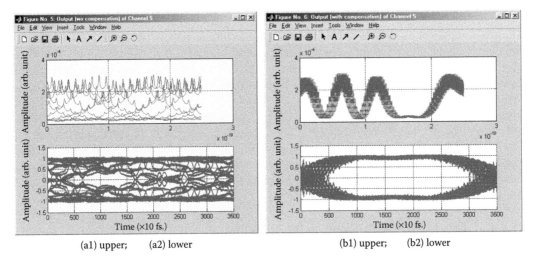

(a1) upper;   (a2) lower   (b1) upper;   (b2) lower

FIGURE 11.51 Signals received after transmission (a) without dispersion compensation and (b) with dispersion compensation. Upper curve (a1) and (b1) are temporal sequence and (a2) and (b2) are corresponding eye diagrams.

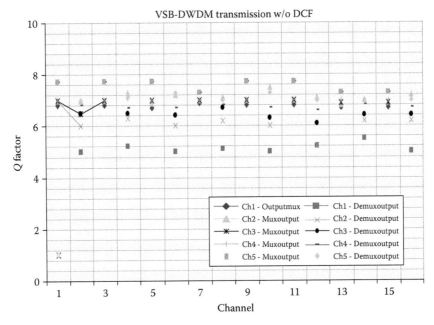

FIGURE 11.52 Channel spacing as a function of Q factor for the multiplexed channels 1–16 at the output of the optical mux (before transmission) and at the output of demux (after transmission). Horizontal axis—channel number.

channel derived at the outputs of the multiplexer as well as at the outputs of the demultiplexer which are located at the end of the 10-km designed-fiber transmission line without using the dispersion compensation as shown in Figure 11.52. The Q factor at the output of demultiplexer decreases significantly for channel spacing 20 and 30 GHz, whereas the Q factor for 40 GHz channel spacing decreases only slightly. This indicates the resilience of the VSB modulation format to the chromatic dispersion.

11.5.4.2   Effects of GVD on Q Factor

The effects of GVD on transmitted channels with and without dispersion-compensation fibers are evaluated with the Q factor as the reference performance parameter. Dispersion-compensation fibers play important role in optical transmission, since the GVD limits the transmission performance. By adding dispersion-compensation fibers, which have the opposite sign of the GVD, the effect of GVD can be considerably reduced. Figure 11.53 shows the dispersion tolerance of the transmission system, the DCF, has significantly improved the Q factor at the receiver to an equivalent BER close to an error-free 10^{-9} at a bit rate of 40 Gb/s. Shown also in this figure is the dispersion tolerance of RZ-DQPSK of 20 Gbauds/s. The performance of two modulation formats is very closely similar. However, the VSB format may offer an advantage of simplified transmitter with additional OF filter rather than another parallel transmitter arm for generation of the quadrature constellation in addition to the in-phase components.

11.5.4.3   Effects of Filter Passband on the Q Factor

The effect of VSB filtering on Q factor and its dispersion tolerance can be examined by varying the passband characteristic of the VSB filter and measuring the Q factor for each lightwave channel at the output of demultiplexer, as shown in Figure 11.54. It is clear that for 40 Gb/s NRZ data format, the system penalty is one unit of the Q factor that is equivalent to one decade of BER or about 10 dBQ when the passband of the OF is extended from 20 to 24 GHz. This penalty could be due to the cut-off of the signal band by the roll-off band of the OF, with more than half of the bandwidth eliminated. However, for 28 GHz passband, the Q factor is considerably increasing to 7.8 or BER = 10^{-15}, that is, error-free transmission when the sampling is at the center of the eye. This shows that the performance of VSB modulation format is better than double side band (DSB) modulation, since the VSB format eliminates most, but not all, of the redundant sideband. When DCF and pure-gain OAs are inserted at the front and after the DCF, the transmission distance can be extended to 15 spans, with a BER of 10^{-12} for all 16 wavelength channels.

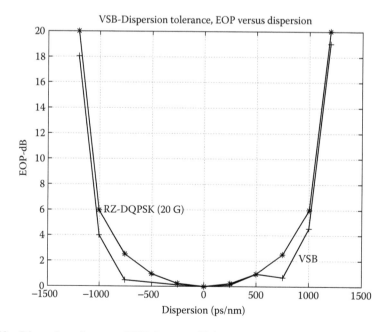

FIGURE 11.53   Dispersion tolerance of VSB format at 40 Gb/s (+) transmission as compared with 20 GBaud/s RZ-DQPSK (*). Eye-opening penalty version dispersion factor at BER of 1e−9 of channel 5.

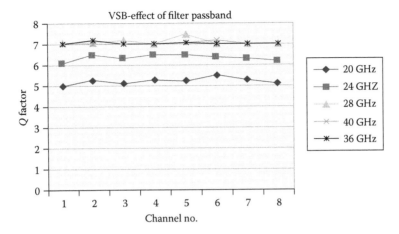

FIGURE 11.54 Variation of the Q factor as a function of the VSB filter passband.

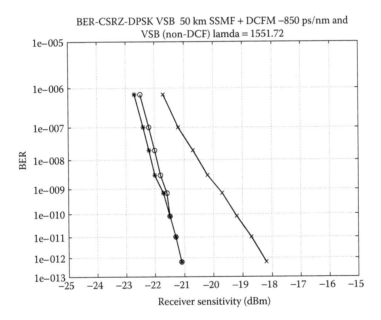

FIGURE 11.55 VSB experiment simulation and CSRZ-DPSK experimental values. Legend: (o VSB with Chebyshev type optical filter; (*) VSB raise cosine optical filter having the same cutoff passband as that of the Chebyshev filter and (+) CSRZ-DPSK (experiment).

Figure 11.55 shows the BER versus the receiver sensitivity of the VSB experiment simulation and measured experimental values of transmission of modulation format, the CSRZ-DPSK [18]. We observe that the OF with a raise-cosine shape and that of the Chebyshev type would offer nearly the same transmission performance of BER versus receiver sensitivity and about 2 dB at a BER of 10^{-9}, worse than CSRZ modulation format. This is reasonable because of the slow roll-off of the cut-off band of the filter. The advantage of the VSB modulation format of simpler transmitter structure can compensate for this penalty.

We could see that the roll-off band of the VSB optical filter is very critical. Unfortunately, the dispersion along the roll-off band is high and thus superimpose on the dispersive effects on the signal, resulting in higher error rate. It is commonly not recommended to employ VSB in the optical domain to increase the spectral efficiency owing to this critical effect.

11.6   SINGLE-SIBEBAND MODULATION

11.6.1   HILBERT TRANSFORM SSB MZIM SIMULATION

The Hilbert transform of signal $m(t)$ is defined to be the RF signal whose frequency components are all phase shifted by $\pi/2$ radians [19] (Figures 11.56 and 11.57).
    Therefore,

$$\hat{m}(t) = H\{m(t)\} \rightarrow m(t) = A\cos 2\pi f_o t \quad \text{then} \quad \hat{m}(t) = A\cos\left(2\pi f_o t - \frac{\pi}{2}\right) \tag{11.17}$$

Taking the Fourier transform we have:

$$M(f) = -j\,\text{sgn}(f)\frac{A}{2}[\delta(f+f_o)+\delta(f-f_o)] = \frac{A}{j2}[-\delta(f+f_o)+\delta(f-f_o)] \tag{11.18}$$

$$\text{Thus,}\ \hat{m}(t) = A\cos\left(2\pi f_o t - \frac{\pi}{2}\right) = A\sin 2\pi f_o t \tag{11.19}$$

11.6.2   SSB DEMODULATOR SIMULATION

The modulated signal is

$$u(t) = \frac{A_c}{2}\left\{\cos(\omega_c t + \gamma\pi + \alpha\pi\cos\omega_{rf}t) + \cos(\omega_c t + \alpha\pi\cos(\omega_{rf}t + \theta))\right\} \tag{11.20}$$

The demodulated signal is

$$s(t) = A_c\cos 2\pi f_c t * u(t) \tag{11.21}$$

The low-pass filter is used to filter high-frequency components in the signal. The only component left is the modulating signal (10 Gbps [binary * cosine signal]) after demodulation. It is required to multiply the demodulated signal by in-phase cosine signal, and also, there is high frequency signal, so that LPF is required to filter the high-frequency components in the signal (Figures 11.58 and 11.59).

11.7   CONCLUDING REMARKS

We have presented the simulation results of 40 Gb/s DWDM optical transmission systems with NRZ-VSB modulation format. The effects of filtering on the Q factor with respect to fibers dispersion and compensation, channel spectral spacing, and the effects of symmetry and asymmetry of the filter pass and cut-off bands are examined. We can draw the following conclusions: (1) The Q factor at the output of demultiplexer decreases significantly for channel spacing of 20 and 30 GHz because of noise and crosstalk interference. However, the Q factor decreases slightly for 40 GHz channel spacing. A BER of 10^{-12} can be achieved over all channels with these channel spacing. For dense and super-dense WDM optical transmissions, the demands on the roll-off, cut-off, and the passband of VSB OFs are high. (2) Fibers dispersion can reduce the system performance dramatically. Using dispersion-compensation fibers along the transmission line will improve the system performance significantly. Specific designed fibers with deterministic dispersion are used. (3) The 20 and 24 GHz passband OFs give low performance. The low performance could be caused by the unstable system, since about half of the bandwidth is eliminated. In contrast, for 28 GHz passband, the Q factor is considerably improved to 7.5. Thus, the performance of VSB modulation format is comparable and marginally enhanced as compared with that of the DSB NRZ-ASK modulation format, since the

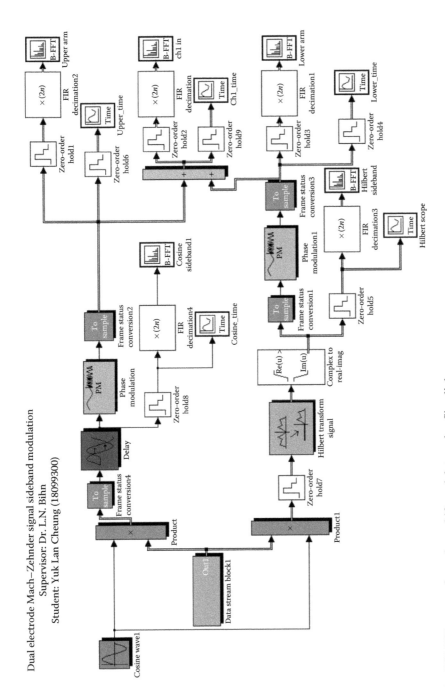

FIGURE 11.56   SSB Hilbert transform phase-shift modulator using Simulink.

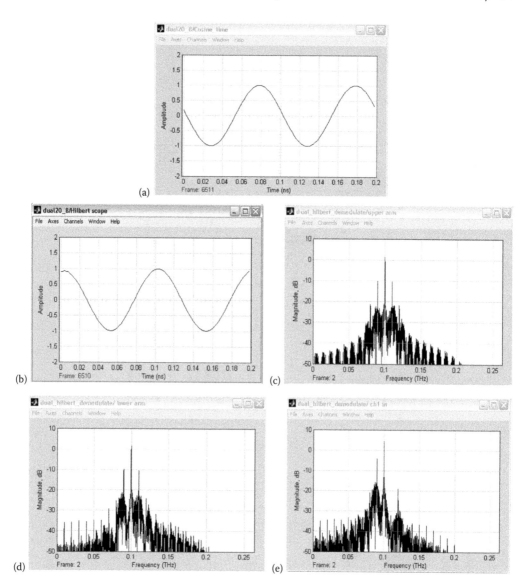

FIGURE 11.57 (a) Input RF signal (cosine signal) as monitored by a time scope. Hilbert transform monitored (b) of the input RF signal $H\{x(t)\} = A\cos(2\pi f_o t - \pi/2) = A\sin(2\pi f_o t)$. (c) At the upper arm of dual-electrode MZIM under a 10 ns delay in a Hilbert transform phase-shift system. (d) At the output of the lower arm of dual-electrode MZIM with Hilbert transform phase-shift $\theta = \pi/2$. (e) At the output of the dual-electrode MZIM SSB system, note the depletion of the lower SB.

VSB format eliminates most, but not all, of the redundant sideband. (4) Several types of OFs are examined, and the sharp roll-off band is expected to contribute to the improvement of the VSB transmission. However, the roll-off band must be higher than at least 20 dB the power level of the center of the spectrum of the signal and carrier frequency to avoid the pulse ripples. (5) The all-zero all-pass OFs can be designed and implemented in planar lightwave circuit technology to substitute for the two-poles elliptic filters to eliminate the filter contribution to the total dispersion of the transmission system.

Nonlinear effects such as the self-phase modulation, cross-phase modulation, Raman scattering, and patterning effects are not considered in this chapter.

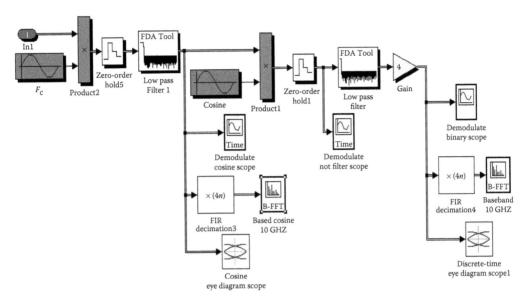

FIGURE 11.58   Single sideband demodulator platform using Simulink.

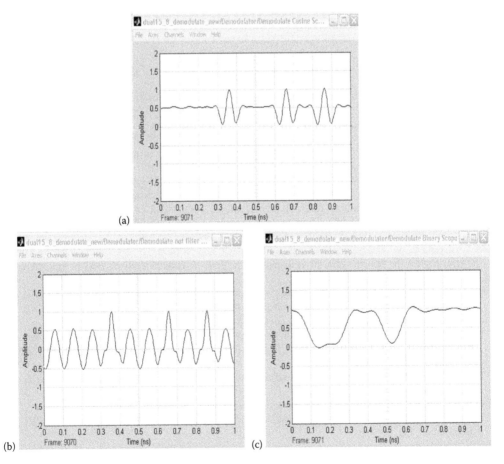

FIGURE 11.59   (a) Demodulate F_c signal time scope. (b) Demodulate cosine signal time scope before low-pass filter. (c) Demodulated signal time scope after low-pass filter.   (*Continued*)

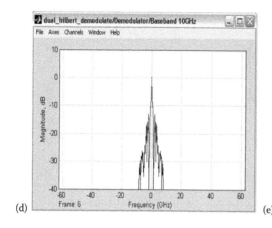

(d)

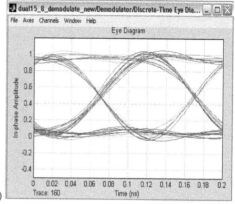

(e)

FIGURE 11.59 (Continued) (d) Demodulate signal spectrum. (e) Demodulated signal eye diagram NRZ-SSB with $Q = 8$ and BER = 10^{-15}.

REFERENCES

1. S.K. Kim, J. Lee, and J. Jeong, Transmission Performance of 10-Gb/s Optical DB Transmission Systems Considering Adjustable Chirp of Nonideal LiNbO3 MZIMs Due to Applied Voltage Ratio and Filter Bandwidth, *IEEE J. Lightwave Technol.*, 19(4), April 2004, 465–470.
2. H. Gnauk and P.J. Winzer, Optical Phase-Shift-Keyed Transmission, *IEEE J. Lightwave Technol.*, 23(1), January 2004, 115–118.
3. S. Zhang, *Advanced Optical Modulation Formats in High Speed Lightwave System*. University of Kansas, 2003, pp. 17–32.
4. K. Yonenaga, S. Kuwano, S. Norimatsu, and N. Shibata, Optical DB Transmission System with No Receiver Sensitivity Degradation, *Electron. Lett.*, 31(4), February 16, 1995, 736–737.
5. K. Yonenage, S. Kuwano, Dispersion-Tolerant Optical Transmission System Using DB Transmitter and Binary Receiver, *IEEE J. Lightwave Technol.*, 15(8), August 1997, 1530–1537.
6. T. Frank, P.B. Hansen, T.N. Nielsem, and L. Eskildson, DB Transmitter with Lower Intersymbol Interference, *IEEE Photon. Technol. Lett.*, 10(4), April 1998, 597–599.
7. F. Elrefaie, R.E Wagner, D.A. Atlas, and D.G. Daut, Chromatic Dispersion Limitations in Coherent Lightwave Transmission System, *IEEE J. Lightwave Technol.*, 6(5), May 1988, 704–710.
8. T. Ono, Y. Yuno, K. Fukuchi, T. Ito, H. Yamazaki, M. Yamaguchi, and K. Emura, Characteristics of Optical DB Signals in Terabit/s Capacity, High-Spectral Efficiency WDM Systems, *IEEE J. Lightwave Technol.*, 16(5), May 1998, 788–797.
9. Y. Miyamoto, M. Yoneyama, T. Otsuji, K. Yonenaga, and N. Shimizu, 40G-bits/s TDM Transmission Technologies Based on Ultra-High-Speed IC's, *IEEE J. Solid-St. Circ.*, 34(9), September 1999, 1246–1253.
10. O.V. Sinkin, J. Zweck, and C.R. Menyuk, A Comparative Study of Pulse Interactions in Optical Fiber Transmission systems with Different Modulation Formats, *The 14th Annual Meeting of the IEEE*, Lasers and Electro-Optics Society, San Diego, CA, 216–217, November 2001.
11. M. Yoneyama, K. Yonenaga, Y. Kisaka, and Y. Miyamoto, Differential Precoder IC Modules for 20- and 40-Gbits/s Optical DB Transmission System, *IEEE Trans. Microw. Theory Tech.*, 47(12), December 1999, 2263–2270.
12. K.P. Ho and J.M. Kahn, Spectrum of Externally Modulated Optical Signals, *IEEE J. Lightwave Technol.*, 22(2), February 2004, 658–663.
13. S. Haykin, *Communication Systems*, 4th edn. John Wiley & Sons, New York, 2001.
14. L.N. Binh, *Optical Fiber Communication Systems with MATLAB Simulink Model*, 2nd Ed., CRC Press, Boca Raton, FL, 2014.
15. Y. Zhu, Highly spectral efficient transmission with CSRZ-DQPSK. Presented at *IEEE Workshop on Advanced Modulation Formats*, San Francisco, CA, 2004.

16. Z. Ye, S. Li, N. Cheng, and X. Liu, Demonstration of High-Performance Cost-Effective 100-Gb/s TWDM-PON Using 4x 25-Gb/s Optical Duobinary Channels with 16-GHz APD and Receiver-Side Post-Equalization. In *ECOC*, paper Mo.3.4.4., Valencia, Spain, 2015.
17. L.N. Binh, Photonic Signal Processing – Part II.2: Tunable Photonic Filters using Cascaded All-Pole Micro-rings and All-Zero Interferometers, Monash University, Clayton, Australia, Technical Report No. MECSE-12-2005, 2005.
18. L.N. Binh, T.L. Huynh, K.-Y. Chin, and D. Sharma, Design of Dispersion Flattened and Compensating Fibers for Dispersion-Managed Optical Communications Systems, *Int. J. Wireless Opt. Comm.*, 1, 2004, 1–21.
19. L.W. Couch, Digital & Analog Communication Systems., International Edition, 8th edition, Pearson, Paris, 2013.
20. L.N. Binh, Advanced Digital Optical communications, CRC Press, Boca Raton, FL, 2015.

12 Superchannel Tera-Bits/s Transmission

This chapter presents the experimental platform, in prototype details, for Tbps optical transmission, whereby the generation of multicarrier lightwaves and modulation techniques for forms polarized-division multiplexed quadrature phase-shift keying (PDM-QPSK) channels closely spaced, so as to achieve the most effective spectral density for 1.12 Tbps or higher optical fiber transmission platform. Optical fiber transmission platforms of 1.12 Tbps and 2.24 Tbps are successfully developed, and transmission of superchannels over 3500 km of optically amplified SSMF spans with BER of 2×10^{-3} (hard FEC qualified) for 28 GBaud Nyquist QPSK and 2×10^{-2} (soft FEC qualified) at 32 GBaud. Furthermore, 20% FEC can be employed to achieve improvement in transmission distance up to 2500 km.

The Tbps superchannel platform consists of the following principal subsystems: (i) DSP-based DAC-PDM Nyquist QPSK optical transmitter, incorporating comb generator as the multicarrier light source; (ii) comb generators using recirculating frequency-shifting techniques and nonlinear driving of optical modulator to achieve Nx5 subcarriers and thence, by modulation, achieve the superchannel with a total capacity of 1 or 2 Tbps; and (iii) DSP algorithms are developed for DAC generation of Nyquist optical channel, as well as for offline processing and FEC coding and decoding, to further improve the sensitivity of the DSP-based optical coherent receiver.

A comparison of Tbps transmission system employing two different techniques of comb generation, thence Nyquist subchannels, is also given.

12.1 INTRODUCTION

12.1.1 OVERVIEW

The PDM-QPSK has been exploited as the 100 Gbps long-haul transmission commercial systems; the optimum technologies for 400 GE/1 TE transmission for next-generation optical networking have now attracted significant interests for deploying ultrahigh-capacity information over the global internet backbone networks. In our Tbps project, the development of the hardware platform of the 1 Tbps transmission system is critical for proving the design concept and conducting the field trial on Huawei's clients network owner such as Deutsch telecom or Vodafone. In a previous chapter, we have presented high-level design and a number of generic options for delivery of Tbps over optically amplified multispan link. Nyquist QPSK has been elected as the most effective format for delivery of high spectral efficiency and is effective in transmission and equalization at both the transmitter and the receiver.

Thus, in this chapter, we describe detailed design and experimental platform for delivery of Tbps using Nyquist QPSK at symbol rate of 28–32 GSa/s and 10 subcarriers. The generation of subcarriers has been demonstrated using either RFS or nonlinear driving of an IQ modulator to create five subcarriers per main carrier; thus, two main carriers are required. Techniques for evaluating the performance of the Tbps transmission system are described in details.

Nyquist pulse shaping is used for effectively packing multiplexed channels whose carriers are generated by comb-generation technique. Digital-to-analog converter with sampling rate varying from 56 G to 64 GS/s is used for generating Nyquist pulse shape includes the equalization of the transfer functions of the DAC and optical modulators.

12.1.2 Organization of the Chapter

The chapter is thus organized as follows: Section 12.2 gives an overview of the system requirements and comparisons of transmission systems, thence a generic architecture of the Tbps transmission systems. Section 12.3 gives further details of multichannel arrangement, so that Tbps superchannel can be achieved, especially the arrangement of comb generation of subcarriers and methods for modulation of these carriers and pulse shaping, so that spectral efficiency can be achieved. Section 12.4 then gives details of key hardware subsystems of the Tbps transmission systems, including the three principal blocks, the comb generators, and DAC-based modulation subsystem, as well as the coherent receiving and extraction of the probe channels for evaluation of the transmission performance. Technical details of the pattern generation from DAC are also given. Section 12.5 then summarizes the transmission performance of the Tbps transmission systems, especially the 1 Tbps and 2 Tbps using RCFS and nonlinear generation techniques. Finally, Section 12.6 gives some conclusions. Two appendices are included on the fiber technical parameters of G.652 and DAC operational parameters. Further technical details on optical modulators would be available on request.

12.2   SYSTEM REQUIREMENTS AND SPECIFICATIONS

12.2.1   Requirements and Specifications

Transmission distance: As next generation of backbone transport, the transmission distance
    should be comparable to the previous generation, namely 100 Gbps transmission system.
    As the most important requirement, we require that the 1 Tbps transmission for long haul
    be 1500–2000 km and for metro application be ~300 km.
CD tolerance: As SSMF fiber CD factor/coefficient 16.8 ps/nm is the largest among the current
    deployed fibers, CD tolerance should be up to 30,000 ps/nm at the central channel whose wave-
    length is approximated at 1550 nm. At the edge of C-band, this factor is expected to increase
    by about 0.092 ps/(nm².km) or about 32,760 ps/nm at 1560 nm and 26400 ps/nm at 1530 nm.[*]
PMD tolerance: The worse case of deployed fiber with 2000 km would have a DGD of 75 ps.
    Therefore, the PMD (mean all-order DGD) tolerance is 25 ps.
SOP rotation speed: According to the 100 Gbps experiments, SOP rotation can be up to
    10 KHz; we take the same specs as 100 G system.
Modulation format: PDM-QPSK for long-haul transmission; PDM-16-QAM for metro application.
Spectral efficiency: Compared with 100 G system with an increase of factor 2. Both Nyquist-
    WDM and CO-OFDM can fulfill this. However, it depends on technological and eco-
    nomical requirements that would determine the suitability of the technology for optical
    network deployment.

12.2.2   Generic System Architecture of a Superchannel 1 Tbps Transmission System

Single-channel 1 Tbps transmission is impossible to realize using only one carrier or subcarrier and modulation. If we keep the non-FEC baud rate close to 25 GBaud, we need PDM-4096QAM modulation scheme to achieve 1 Tbps per single lightwave carrier. Full specifications of the Tbps transmission system are given in Table 12.1. This is impossible under the constraint of maximum laser amplitude power available to date. Furthermore, as the OSNR requirement increases exponentially to the modulation format level, it is impossible to reach the transmission distance with limited laser power and the

---

[*] See Appendix A for technical specification of Corning fiber G.652 SSMF.

TABLE 12.1
1 Tbps Off-line System Specification

Parameter Technique	Superchannel RCFS Comb Gen	Superchannel Nonlinear Comb Gen	Some Specs 1.1.1	Remarks 1.1.2
Bit rate	1,2...N Tbps (whole C-band)	1 Tbps, 2,...N Tbps	~1.28 Tbps at 28–32 GB	20% OH for OTN, FEC
Number of ECLs	1	$N \times 2$		
Nyquist roll-off	0.1 or less	0.1 or less		DAC pre-equalization required
Baud rate (GBauds)	28–32	28–32	28, 30, or 31.5 GBaud	Pending on FEC coding allowance
Transmission distance	2500	2500	1200 (16 spans) ~2000 km (25 spans) 2500 km (30 spans) 500 km	20% FEC required for long-haul application Metro application
Modulation format	QPSK/16-QAM	QPSK/16-QAM	Multicarrier Nyquist WDM PDM-DQPSK/QAM	For long haul For long haul
			Multicarrier Nyquist WDM PDM-16-QAM	For metro
Channel spacing			4 × 50 GHz	For long haul
			2 × 50 GHz	For long haul
Launch power	<<0 dBm if 20 Tbps is used		~−3 to 1 dBm lower if $N > 2$	Depending on QPSK/16-QAM and Long haul/metro can be different
B2B ROSNR at 2e−2 (BOL) (dB)	14.5	14.5	15 dB for DQPSK 22 dB for 16-QAM	1 dB hardware penalty 1 dB narrow filtering penalty
Fiber type	SSMF G.652 (or 655)	SSMF G.652 (or 655)	G.652 SSMF	
Span loss	22	22	22 dB (80 km)	
Amplifier	EDFA (G > 22 dB); NF < 5 dB		EDFA (OAU or OBU)	
BER	2e−3	2e−3	Pre-FEC 2e−2 (20%) or 1e−3 classic FEC (7%)	
CD penalty (dB)			0 dB at +/−3000 ps/nm <0.3 dB at +/−30,000 ps/nm	16.8 ps/nm/km and 0.092 ps/(nm².km)
PMD penalty (DGD)			0.5 dB at 75 ps, 2.5 symbol periods	
SOP rotation speed	10 kHz	10 kHz	10 kHz	
Filters cascaded penalty	1.1.3	1.1.4	<1 dB at 12 pcs WSS	
Driver linearity	Required	Required	THD <3%	16 QAM even more strict

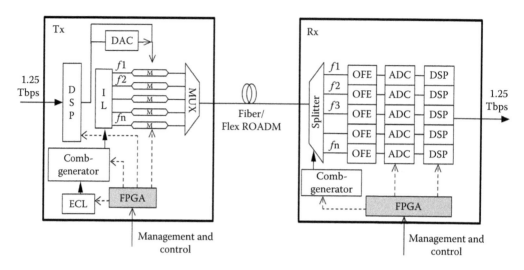

FIGURE 12.1 Architecture of a superchannel 1 Tbps transmission platform using DSP-based coherent reception systems.

threshold of the nonlinear level of the transmission fiber. Even PDM-16-QAM needs, theoretically, 7 dB more OSNR than PDM-QPSK to reach the same distance. To solve this problem, one can either increase the baud rate or employ more subcarriers, that is, a superchannel with multiple subcarriers. Increasing the baud rate leads to higher demand on the bandwidth for O/E components, thus challenging to technology advances and cost. Therefore, employing high-spectral-efficient superchannel with subchannel bandwidth of 50 GHz or below seems to be the most favorable choice.

Considering overhead of OTN framer and FEC of either 7% or 20% overhead, the total bit rate can thus be 28Gx4 × 10 → 1.12 Tbps or 32Gx4 × 10 → 1.28 Tbps (Figure 12.1).

For multiple subcarrier transmissions, we would need multicarrier modulation techniques and corresponding multicarrier demodulation and coherent detection (CoD). Currently, there are a few methods and apparatus to generate multicarriers. The commonly used one is a single ECL laser source and a recirculating shift loop and an MZM modulated by an RF signal. The frequency of the RF signal is the spacing of the subchannels. Tx-DSP and DAC are needed for shaping the signal pulse, for example, Nyquist pulse shaping, and pulse shaping for compensating nonideal O/E components transfer function, CD precompensation, and so on.

Interleaver can be used to separate the subcarriers for individual subchannel modulation. At the output, all the signals of subchannels are multiplexed together to form a superchannel 1.12 Tbps optical signal (Figure 12.2).

The principal question is how we can create superchannels under laboratory conditions to achieve the proof of concept of the Tbps transmission, that is, how to generate superchannels, transmission, and detection/demodulation at the receiving end? Thus, it is one of the main principles of this chapter to describe techniques employed to generate superchannels and probe channel to evaluate the performance of the transmission systems for Tbps superchannel.

12.3 MULTICARRIER NYQUIST TRANSMISSION SYSTEM

In Nyquist WDM, the goal is to place different optical subchannels as close as possible to their baud rate, the Nyquist rate, but minimizing channel crosstalk through the formation of approximately close-to-rectangular spectrum shape of each channel by either optical filtering or electrical filtering and equalization.

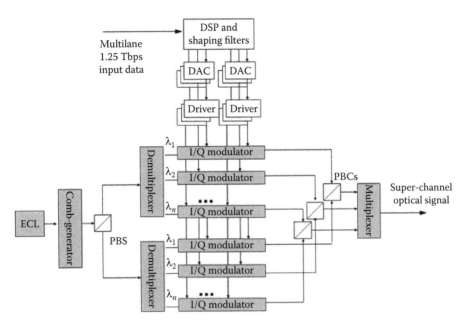

FIGURE 12.2 General architecture of a superchannel transmitter for coherent reception systems.

12.3.1 NYQUIST SIGNAL GENERATION USING DAC BY EQUALIZATION IN FREQUENCY DOMAIN

The rectangular spectrum has a sinc (i.e., $\sin x/x$) time-domain impulse response. At the sampling instants $t = kT$ ($k = 1, 2, \ldots N$ is nonzero integer), its amplitudes reach zero. This implies that at the ideal sampling instants, the ISI from neighboring symbols is negligible or free of ISI. Figure 12.3 depicts such Nyquist pulse and its spectrum for either a single channel or multiple channels. Note that the maximum of the next pulse raise is the minimum of the previous impulse of the consecutive Nyquist channel.

Considering one subchannel carrier 25 GBaud PDM-DQPSK signal, the resulting capacity is 100 Gbps for a subchannel; hence, to reach 1 Tbps, 10 subchannels would be required. To increase the spectral efficiency, the bandwidths of these 10 subchannels must be packed densely together. The most likely technique for packing the channel as close as possible in the frequency, with minimum ISI, is the Nyquist pulse shaping, which is described later in this section. Thus, the name Nyquist-WDM system is coined. However, in practice, such "brick-wall"-like spectrum, shown in Figure 12.3, is impossible to obtain, and hence, nonideal solution for non-ISI pulse shape should be found; however, the raise cosine pulse with some roll-off property condition should be met.

The raised-cosine filter is an implementation of a low-pass Nyquist filter, that is, one that has the property of vestigial symmetry. This means that its spectrum exhibits odd symmetry about $1/2T_s$, where T_s is the symbol period. Its frequency-domain representation is "brick-wall-like" function, given by

$$H(f) = \begin{cases} T_s & |f| \leq \dfrac{1-\beta}{2T_s} \\ \dfrac{T_s}{2}\left[1 + \cos\left(\dfrac{\pi T_s}{\beta}\left\{|f| - \dfrac{1-\beta}{2T_s}\right\}\right)\right] & \dfrac{1-\beta}{2T_s} < |f| \leq \dfrac{1+\beta}{2T_s} \\ 0 & \text{otherwise} \end{cases} \qquad (12.1)$$

with $0 \leq \beta \leq 1$

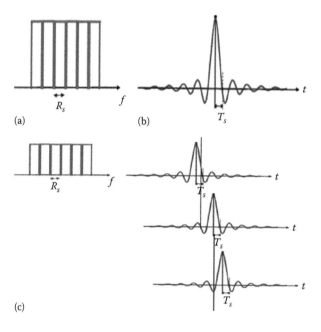

FIGURE 12.3 A superchannel Nyquist spectrum and its corresponding "impulse" response: (a) Spectrum; (b) impulse response in time domain of a single channel; and (c) sequence of pulse to obtain consecutive rectangular spectra. A superposition of these pulse sequences would form a rectangular "brick-wall-like" spectrum.

This frequency response is characterized by two values: β, the *roll-off factor*, and T_s, the reciprocal of the symbol rate in Sym/s, that is, $1/2T_s$ is the half bandwidth of the filter. The impulse response of such a filter can be obtained by analytically taking the inverse Fourier transformation of Equation 12.1, in terms of the normalized sinc function, as:

$$h(t) = \mathrm{sinc}\left(\frac{t}{T_s}\right)\frac{\cos\left(\dfrac{\pi\beta t}{T_s}\right)}{1-\left(2\dfrac{\pi\beta t}{T_s}\right)^2} \tag{12.2}$$

Where the roll-off factor, β, is a measure of the *excess bandwidth* of the filter, that is, the bandwidth occupied beyond the Nyquist bandwidth, as from the amplitude at $1/2T$. Figure 12.4 depicts the frequency spectra of raise cosine pulse with various roll-off factors. Their corresponding time-domain pulse shapes are given in Figure 12.4b.

When used to filter a symbol stream, a Nyquist filter has the property of eliminating ISI, as its impulse response is zero at all nT (where n is an integer), except when $n = 0$. Therefore, if the transmitted waveform is correctly sampled at the receiver, the original symbol values can be recovered completely. However, in many practical communications systems, a matched filter is used at the receiver, so as to minimize the effects of noises. For zero ISI, the net response of the product of the transmitting and receiving filters must equate to $H(f)$; thus, we can write:

$$H_R(f)H_T(f) = H(f) \tag{12.3}$$

Or, alternatively, we can rewrite that:

$$\left|H_R(f)\right| = \left|H_T(f)\right| = \sqrt{\left|H(f)\right|} \tag{12.4}$$

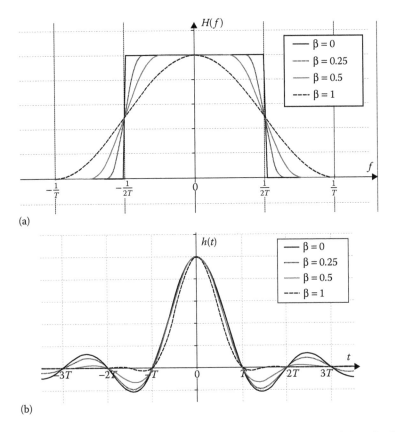

(a)

(b)

FIGURE 12.4 (a) Frequency response of raised-cosine filter with various values of the roll-off factor β and (b) Impulse response of raised-cosine filter with the roll-off factor β as a parameter.

The filters that can satisfy the conditions of Equation 12.4 are the root-raised-cosine filters. The main problem with root-raised-cosine filters is that they occupy larger-frequency bands of the Nyquist sinc pulse sequence. Thus, for the transmission system, we can split the overall raise-cosine filter with root-raise-cosine filter at both the transmitting and receiving ends, provided the system is linear. This linearity is to be specified accordingly. An optical fiber transmission system can be considered to be linear if the total power of all channels is under the nonlinear SPM threshold limit. When it is over, a weakly linear approximation of this threshold can be used.

The design of a Nyquist filter influences the performance of the overall transmission system. Oversampling factor, selection of roll-off factor for different modulation formats, and FIR Nyquist filter design are key parameters to be determined. If we take into account the transfer functions of the overall transmission channel, including fiber, WSS, and the cascade of the transfer functions of all O/E components, the total channel transfer function is more Gaussian-like. To compensate this effect in the Tx-DSP, one would thus need a special Nyquist filter to achieve the overall frequency response equivalent to that of the rectangular or raise cosine with roll-off factor, shown in Figures 12.5 and 12.6.

12.3.2 FUNCTION MODULES OF A NYQUIST-WDM SYSTEM

A generic schematic of the functional modules of Nyquist-WDM system is described in Figure 12.2. Two special features should be taken into account. At the Tx side, pulse shaping is so defined that the overall transfer function from driver to the analog-to-digital converter (ADC) on the Rx side

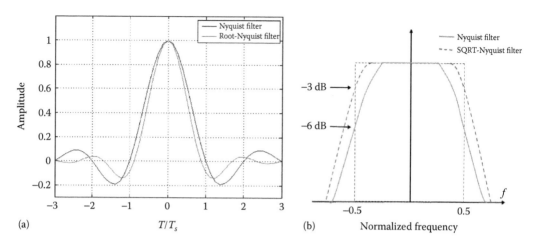

FIGURE 12.5 (a) Impulse and (b) corresponding frequency response of sinc Nyquist pulse shape or root-raise-cosine (RRC) Nyquist filters.

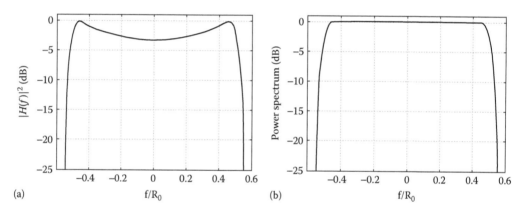

FIGURE 12.6 (a) Desired Nyquist filter for spectral equalization and (b) output spectrum of the Nyquist-filtered QPSK signal.

should ideally be rectangular or NRZ form. The implementation can be either using look-up table or Nyquist low-pass filter in frequency domain or time domain. Alternatively, the optical spectrum of the modulated lightwaves at the output of the optical modulator can be obtained via the port of the OSA. This is then equalized to achieve "brick-wall"-like spectrum. Thus, the required spectrum on the DAC to achieve this equalization is known and then used to modify the DAC driving output voltage levels at (H_I^+, H_I^-) and (H_Q^+, H_Q^-) and (V_I^+, V_I^-) and (V_Q^+, V_Q^-), where the equalized optical spectra can be obtained for the two polarized modes of the linearly polarized mode to be launched into the SSMF optically amplified transmission lines.

For each subchannel, 4x DACs are needed to convert the discrete digital signal to analog signal (X_I, X_Q, Y_I, Y_Q). In order to match the voltage amplitude (power) requirement of the I-Q modulator, 4x RF broadband drivers are required to provide appropriate RF signal amplitude for driving the MZM modulator, so as to obtain the phase difference between the in-phase and quadrature phase constellation points. At the output of the transmitter, all subchannels are multiplexed together to form a superchannel (Figure 12.7).

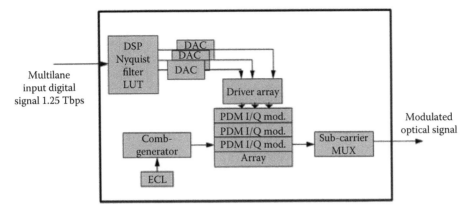

FIGURE 12.7   Transmitter architecture of a Nyquist-WDM system with electrical Nyquist filter.

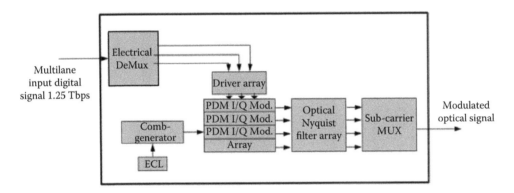

FIGURE 12.8   Schematic of the Nyquist optical transmitter architecture with optical Nyquist filter.

There is another way to generate the Nyquist signal, which is suitable when Tx-DSP is not available. The Nyquist spectrum is generated by an optical rectangular (or raise-cosine roll-off) filter, as shown in Figure 12.8. However, in practice, it is very difficult to achieve a sharp filter roll-off without much phase fluctuation at the edges of the filter.

At the receiver, multicarrier parallel demodulation is used. At first, the superchannel will be separated into individual subchannels. Each subchannel will be demodulated like a single channel reception: OFE downconverts the optical spectrum to baseband (homodyne coherent demodulation) and the four-lane analog signal will be converted into digital signal using 4xADC. Digital signal will be processed by ASIC-containing modules CD compensation, timing recovery, MIMO FIR filter, carrier recovery, and so on (Figure 12.9).

12.3.3   DSP Architecture

Single subchannel DSP can be either using the traditional FDEQ + TDEQ architecture or pure FDEQ architecture (Figure 12.10).

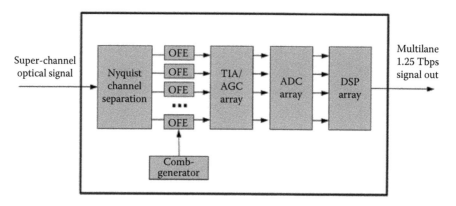

FIGURE 12.9   Receiver architecture of Nyquist WDM system.

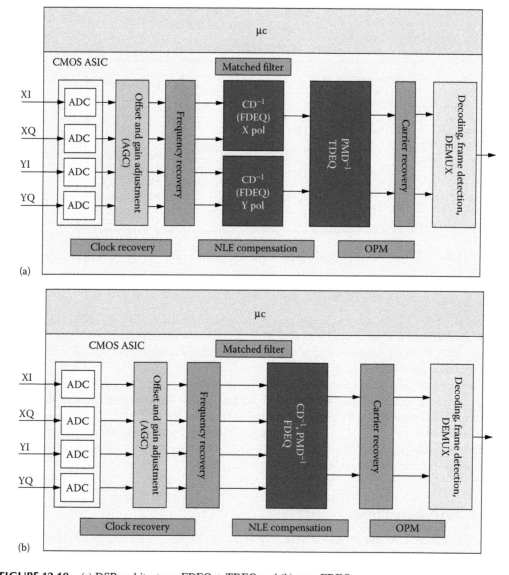

FIGURE 12.10   (a) DSP architecture, FDEQ + TDEQ and (b) pure FDEQ.

12.4   KEY HARDWARE SUBSYSTEMS FOR AN OFFLINE DEMO SYSTEM

12.4.1   COMB-GENERATION TECHNIQUES

12.4.1.1   Recirculating Frequency Shifting
The main modules of a multicarrier generator (comb generator) are illustrated as follows (Figure 12.11):

1. A CW ECL or multilaser bank of ECLs whose line width is sufficiently narrow, possibly in order of less than 100 KHz, can be employed as the original lightwave carrier/carriers
2. An MZM for pulse shaping (CSRZ), which will make the output spectrum of the super-channel flat
3. An MZM used as phase modulator, the phase-control signal coming from an RF signal generator
4. RF generator; its frequency is the spacing of the subchannels
5. Sinusoidal signal generator
6. 90° Phase shifter, which makes the spectrum to be only single-side band

12.4.1.2   Nonlinear Excitation Comb Generation and Multiplexed Laser Sources
Under the condition that the modulator, an IQ modulator, can be driven such that the amplitude swing to $2V_\pi$, the generation of first- and second-order frequency shifting components can be formed. We operate the modulator in the region such that no suppression of the primary carrier can be achieved, and thus, five subcarriers can be generated from one main carrier. Thus, using two main carriers, we can generate 10 subcarriers, and hence, the modulation of these subcarriers can form 10×100 Gb/s or 1 Tbps superchannels (Figure 12.12).

12.4.2   HARDWARE SET-UP OF COMB GENERATORS

12.4.2.1   RCFS Comb Generator
The general path flow of the experiment platform is shown in Figure 12.13. There are two separate paths that need to be noted (in order of importance). The main structure is the optical fiber or integrated optic loop, in which, besides the optical path interconnecting all optical components which

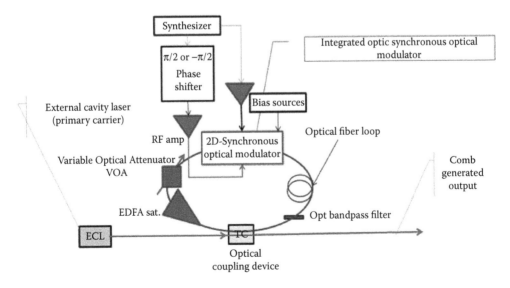

FIGURE 12.11   Block diagram of a recirculating frequency-shifting comb generator.

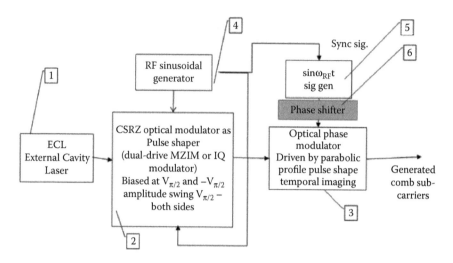

FIGURE 12.12   Comb generation using nonlinear driving condition on MZIM I/Q modulator.

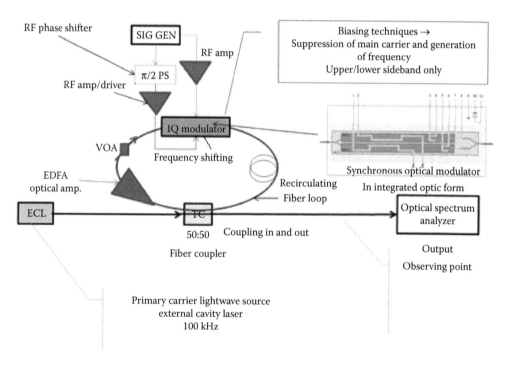

FIGURE 12.13   Principle of the experimental set-up of the comb generator using recirculating frequency shifting.

shift the frequency of the lightwaves and tapping part of its energy out to the output port. The optical coupler is used for both injecting the lightwaves from a sources operating at CW mode and tapping the frequency-shifted lightwaves out to the output port.

An optical modulator incorporated in the loop performs the frequency shifting. Single sideband operating is used by RF phase delay by $\pi/2$ with respect to each other, hence the Hilbert transformation and the suppression of the right or left sideband accordingly, depending on the relative phase shift between the in-phase and quadrature-phase electrodes of the IQ modulator. The insert of Figure 12.13 is the integrated optic plan view of the IQ_modulator waveguide and electrodes. There are two child interferometers with overlaid electrodes, which perform the suppression of the carrier or sideband by biasing and driving condition by applying RF waves (see Figure 12.13). Note that electrical phase shifters are inserted in both arms of the RF paths before applying to the electrodes to create $\pi/2$ phase shift, thence suppression of one sideband. See Figure 12.14b for the comb generator already packaged.

In addition, there is a parent interferometric optic structure in which both child interferometers are positioned. A pair of electrodes is also employed on one branch of this interferometer, so that phase shifting between the emerging lightwaves from the child interferometers can be altered with respect to each other. A variable optical attenuator is also used to ensure that the total loop gain is less than unity, so that lasing effects would not occur. This is also used to adjust the gain equalization in order to obtain different gain factor.

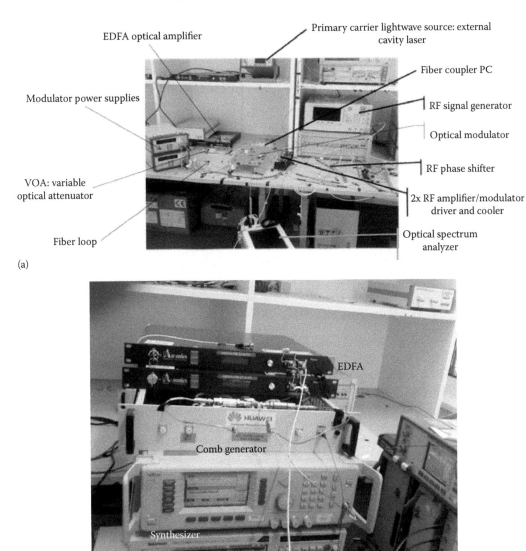

(a)

(b)

FIGURE 12.14 Laboratory set-up of a comb generator prototype employing recirculating frequency-shifting technique: (a) Comb generator under development; (b) packaged comb generator with external EDFA and signal synthesizer; and back plane of the comb generator.

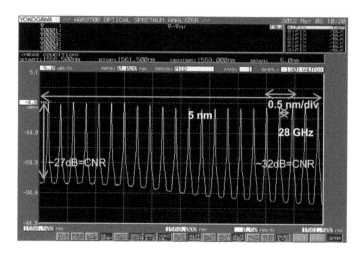

FIGURE 12.15   Spectrum of generated multicarriers.

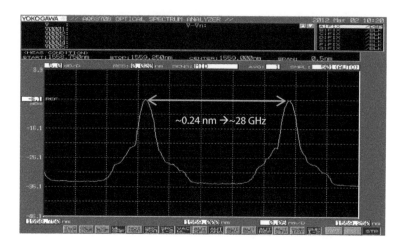

FIGURE 12.16   Measured subchannel spacing between two adjacent channels.

A sinusoidal RF wave generator is also required. However, in practice, such oscillator is simple, provided that the frequency is known. It is noted that the output optical port of the comb generator must be connected with an angled PM fiber connector, so that no feedback of optical comb waves back to the recirculating loop. If no back scattering occurs, the interference effects would be created and no noise disturbance on modulated signals would be observed, hence much higher BER.

By referring to Figures 12.15 and 12.16, we can observe: (i) The spectra of the generated subcarriers are equalized over a wide region of 5 nm; carrier noise ratio (CNR) reaches 27–32 dB. (ii) The spacing between two adjacent subcarriers can be varied by tuning the excitation RF frequency; shown is 28 GHz spacing.

12.4.2.2   Nonlinear Comb Generator

The schematic structure of the nonlinear comb generator is shown in Figure 12.14, and its packaged version is shown in Figure 12.17. As can be observed, the principal component of this comb generator is the IQ modulator driven into nonlinear region, mainly overdoubling the voltage level, such that it covers twice the range of the voltage required for shifting the phase of π. The packaged comb generator is shown with two RF phase shifters, so that the harmonic level

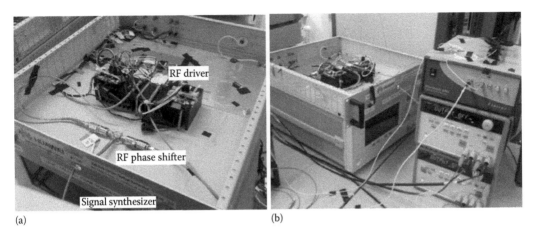

FIGURE 12.17   Nonlinear comb generator: (a) Structure, including optical modulator and RF phase shifters and two branches to RF port inputs and (b) overall packaged comb generator with power supply and signal generator.

can be adjusted to obtain equalized amplitudes with respect to the main carrier. The WSS can be used to split selected channels and to equalize the channel power level. In practice, it is required to have all channels of the same power level launched into the first spans and then all cascaded optically amplified fiber spans. A typical spectrum of a 1 × 5 subcarrier comb generator is shown in Figure 12.18, in which the fundamental/primary carrier is located at the middle of the spectrum and the first and second harmonics on the two sides of the bands can also be identified without much difficulty. The modulation of such five subchannels can be seen in Figure 12.18. However, if one channel can carry only Nyquist QPSK of 112 Gb/s, then there must be another primary carrier position at a distance from the first primary carrier, so that 10 subcarriers can be created, and thence, 1 Tbps channel can be generated via the modulation of the subcarriers. The $LiNbO_3$ modulator is transparent at these spectral regions and the electro-optic coefficients are quite constant as well, so the modulation of these sub carriers would result into more or less the same phase modulation. The spectrum of such 10 channels created from 10 subcarriers is depicted in Figure 12.19.

12.4.3   MULTICARRIER MODULATION

This section describes the generation and modulation of superchannels by (i) generation of multi-subcarriers from primary carriers by either nonlinear driving of optical modulators or RFS techniques. The formation of Tbps superchannels can be 10 × 100 G (1 Tbps) or 20 × 100 G (2 Tbps). It is noted that 5 × 200 G can also be used and the 200 G can be generated by using the high-level modulation format, for example, 16-QAM instead of 4-QAM such as QPSK.

12.4.3.1   Generation of Multisubcarriers for Tbps Superchannels

12.4.3.1.1   Recirculating Frequency-shifting Technique

Multicarrier modulation using comb generator is constructed and tested in the laboratory for advanced optical communication systems of Huawei ERC in Munich, Germany. The schematic of this comb generator is shown in Figure 12.20, in which a recirculating loop is the basic configuration, incorporating an optical modulator, an OA to compensate the insertion loss of various devices, and an optical attenuator whose coefficient can be adjusted so that the overall loop gain is slightly less than unity. This loop gain less than unity is required to ensure that no lasing would

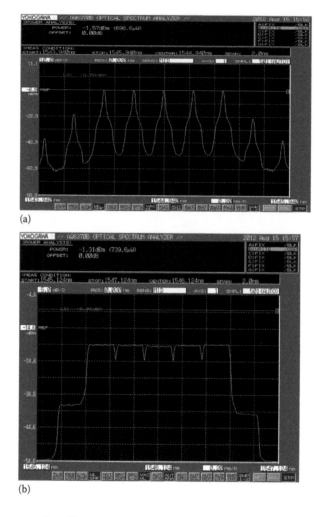

(a)

(b)

FIGURE 12.18 Spectrum of nonlinear comb generator 1 × 5 subcarriers: (a) Subcarriers at 30 GHz spacing and (b) modulated subchannels.

occur. An optical coupler with a coupling ratio of 50:50 or 3 dB split is employed to coupling into and out of the loop the lightwaves and combed subcarrier lines, respectively. A synthesizer, which is a synthesizer/signal generator, is used to generate a sinusoidal wave in RF domain, whose frequency determines the frequency shifting of the optical lines of the subcarriers of the comb generator.

The optical modulator incorporated in the loop is a synchronous modulator, which can be an IQ modulator biased such that synchronization of the interferometers in the two optical paths of the parent MZIM can be maximized. Two arms of the optical modulator are driven with a π/2 phase shift with respect to each other, so that an SSB shifting of the lightwave can be achieved at the output of the modulator with the loop open. The spectrum of the SSB lightwave at the output of the modulator is shown in Figure 12.21a. Note that this frequency shifting can be either shifting left or right, depending on the relative phase shift between the two arms of the synchronous modulator, as shown in Figure 12.21b.

A synthesizer is required to launch the RF waves to the two arms with RF amplifiers to boost the electrical signals to appropriate levels, so that maximum modulation depth can be achieved.

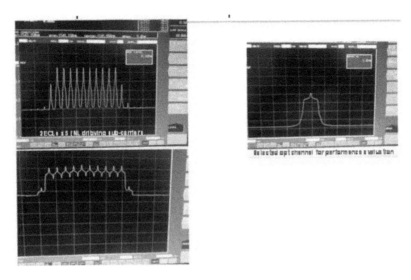

FIGURE 12.19   Spectrum of 10 subcarriers produced by nonlinear comb generation from two primary carriers and modulated channel.

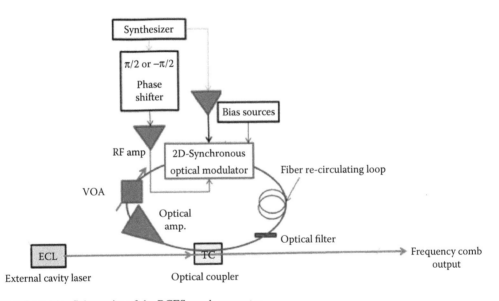

FIGURE 12.20   Schematics of the RCFS comb generator.

We note that although the modulator 3 dB bandwidth is specified at 23 GHz, it is still possible to operate the modulator at 33 GHz; however, some RF signal loss would be expected. This is because the driving signal is purely sinusoidal and very narrow band unlike data signals employed to drive the data modulator. We have successfully generated comb lines from 28 GHz to 33 GHz for channel spacing 28 GHz, 30 GHz, and 33 GHz (Figure 12.22).

The ECL is the primary lightwave, which is used to generate comb subchannels. This laser output power is set at about 13 dBm, which is the highest power level. Higher power level can be used if EDFA booster amplifier is used. Thus, we note that the power level of the SBS line in Figure 12.21 at −10 dBm and −4 dBm for the comb lines in Figure 12.23, which has also been passing through an EDFA at

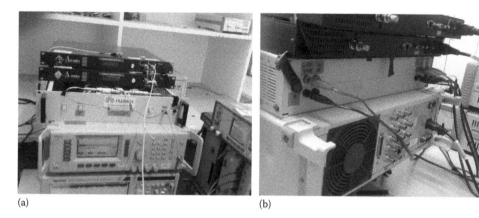

(a)   (b)

FIGURE 12.21 Image of the RCFS comb generator, including the optical loops and optical amplifiers connected externally, the RF synthesizer, and RF ports to be launched into the synchronous modulator. The top optical amplifier (EDFA) is used for amplifying the comb lines for data modulation and thence transmission. (a) Front panel and (b) back panel with DC power supply line inputs.

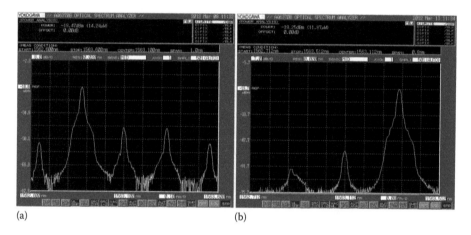

(a)   (b)

FIGURE 12.22 Spectrum of the SSB lightwave at output of the synchronous optical modulator: (a) shifting right in frequency scale 9 and (b) shifting right in frequency.

saturation power of 16 dBm, which is distributed to all subcarrier lines. The images of the front panel and back panel of the constructed RCFS comb generator are shown in Figure 12.21a and b, respectively.

The OA incorporated in the recirculating ring operates in the saturation mode only when the power of the lines is sufficient to boost it into the saturated region, otherwise it would operate in the linear region. For this reason, we observe that the frequency shifting lines would not be in the saturated level for the first few subcarrier lines, as seen in Figure 12.23b in the far most right side of the spectrum. The tuning of the DC bias supply voltage to the electrodes is quite sensitive and would be in the 10s of mV range.

One most important point we must point out is that the output fiber port must be angled connector type, so that no reflection of subcarrier combed lines can be feedback into the recirculating to avoid the noise oscillation of the generated comb lines. Note that the suppression of the other harmonic lines is more than 25 dB.

The superchannel spectra shown in Figure 12.24a and b display those of the main carrier under modulation, using 25 GSym/s QPSK and 28 GSym/s Nyquist QPSK, with spacing 50 GHz

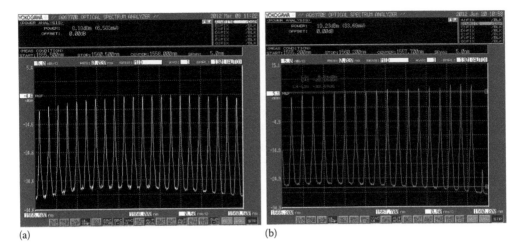

FIGURE 12.23 Spectrum of multisubcarriers generated by the RCFS loop. (a) Full-scale = 1555.5–1560.5 nm with 25 GHz shift between comb lines; and (b) full scale = 1555.2–1560.2 nm with 28 GHz shift between comb lines.

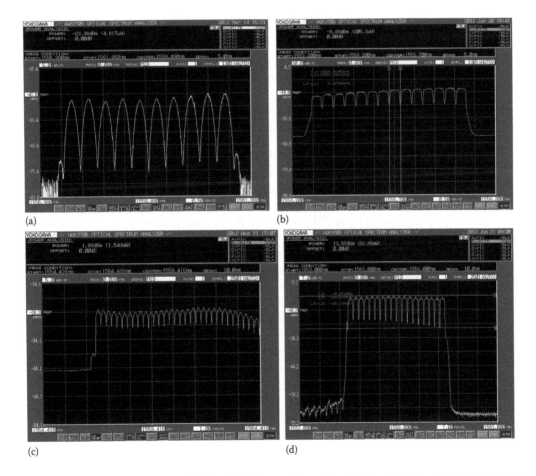

FIGURE 12.24 Spectrum of 10 × 25 GBaud QPSK superchannel modulation: (a) Under QPSK with 50 GHz spacing; (b) 14 channels under Nyquist QPSK after transmission over 2000 km optically amplified SSMF spans; (c) higher than 20 subcarriers under Nyquist QPSK modulation; and (d) 20 modulated lightwave superchannels after filtering.

and 30 GHz, respectively. The roll-off factor for Nyquist signals is 0.1 in optical domain after equalization embedded in electronic-domain DAC signals.

A typical spectrum of the generated multisubcarriers by the RCFS is shown in Figure 12.23a and b for all subcarriers and window of selected 20 subcarriers with amplitude equalization, respectively. While Figure 12.24a and b shows the spectra of modulated lightwaves, as observed by an OSA under the modulation formats QPSK and QPSK Nyquist at 28 GB/s, Figure 12.24c shows the spectrum of the modulated channels of the RCFS combed subcarriers.

A further note on the RCFS comb generator is that the RF drivers to both ports of the optical modulator should be a linear driver type, so that no harmonic distortion would be generated and interfere with the harmonics of the comb lines. It is noted that the magnitude of the third harmonic is suppressed to more than −30 dB with respect to that of the fundamental, the first and second, harmonics.

12.4.3.1.2   Operating the RCFS Comb Generator

The operation of the RCFS comb generator should follow the procedures: (i) Disconnect the fiber recirculating loop if it is already connected; (ii) connect the output optical port of the synchronous modulator and monitor it via an OSA; (iii) tune the DC power supplies to the two slave MZIMs of the synchronous modulator such that the suppression of the primary carrier can be observed to the minimum value. Once the minimum level is achieved by one electrode pair on an MZIM, then change to the other slave MZIM and repeat the above step so that minimum primary carrier is suppressed; (iv) Now, apply the RF signals into the input of the RF 3 dB splitter and monitor the optical spectrum. Adjust again the supplies to the electrodes of MZIM slaves and the electrode applied to the master MZIM, so that suppression of one sideband can be observed. Note that the RF level can be tuned, so that maximum generation of the sideband line can be achieved; (v) thence, tune the RF phase shifter to obtain the maximum level of the SSB line and maximum suppression of the other SSB lines. Note that adjustment of the DC supplies to the slave MZIM may be required to suppress the primary carrier.

12.4.3.1.3   Nonlinear Modulation for Multi-Subcarrier Generation

12.4.3.1.3.1   Nonlinear Driving Over $2V_\pi$
Alternating to the superchannel generation using RFS techniques described above, we also conducted the generation of higher-order subcarriers by driving the IQ modulator over its nonlinear regions, that is, the amplitudes of the RF waves cover over at least two $2V_\pi$ of the child MZM sections of the IQ optical modulator. The generation of multi-subcarriers can be implemented by using frequency double and triple by swinging the modulator with an amplitude of V_π and biasing at the minimum transmission point, resulting in frequency doubling and suppression of carrier. This scheme is shown in Figure 12.25a and b.

A group of 10 or 20 subcarriers can be generated by using two or four primary carriers tuned to appropriate spacing to accommodate the upper and lower higher-order subcarriers. For example, if there is 30 GHz spacing between subcarriers, then there should be 150 GHz spacing between two main primary carriers to accommodate for two lower sideband subcarriers of the higher-frequency source and two other upper sideband subcarriers of the lower-frequency light source. Figure 12.26a and b shows the spectra of 1×5 and 4×5 subcarriers from one and two primary carrier sources, respectively. Note that for the spectra of 20 subcarriers, one subcarrier is band stopped by passing through an optical notch (or band-stopped) filter, whose centered frequency is set such that it falls exactly at the frequency of the subcarrier. By doing this, one can arbitrarily extract a number of subcarriers as well as injecting subcarriers from different lightwave sources, so that independent channels can be generated to ensure the ability of decorrelating adjacent channels. Hence, the measurements of the quality of transmitted channels can reflect their true performance. It is also noted that the third-order subcarrier was suppressed to more than 30 dB. In case where the seven subcarriers need to be generated, the RF driving signal power level can be used; however, we doubt that the modulator would not stand the high voltage level developed across the electrical-optical transfer characteristics of the IQ modulator.

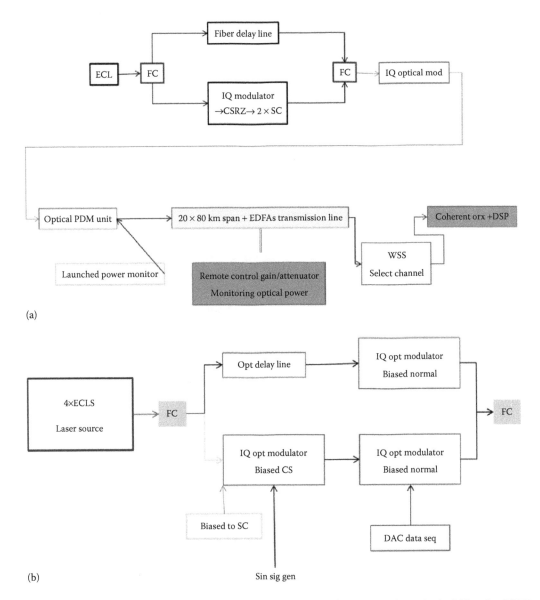

FIGURE 12.25 Schematic of optical modulation to generate 4 × 3 subchannels to obtain 1 Tbps by CSRZ frequency doubling by one path generating × 2 subcarriers by CSRZ driving the modulator and then combined with nonfrequency shifting optical path: (a) 1 × 3 generation and (b) 4 × 3 for 1 Tbps channel formation.

12.4.3.1.3.2 CSRZ and Combined Paths Another possibility to generate 1 Tbps, the combined four lasers are split into two optical paths. One path is fed into a CSRZ modulator to generate 2 × 4 subcarriers with frequency doubling or plus and minus first order. These eight subcarriers are then combined with nonfrequency shifted primary carriers; thus, the combined optical path would produce a set of 10 subcarriers, which are then modulated by an IQ modulator driven via the DAC output ports to obtain Nyquist QPSK channels. The combined optical path can also be fed into a WSS to separate two independent streams of carriers, which are then fed into two IQ modulators driven by independent PRBS from the DAC output ports to produce decorrelated modulated optical channels.

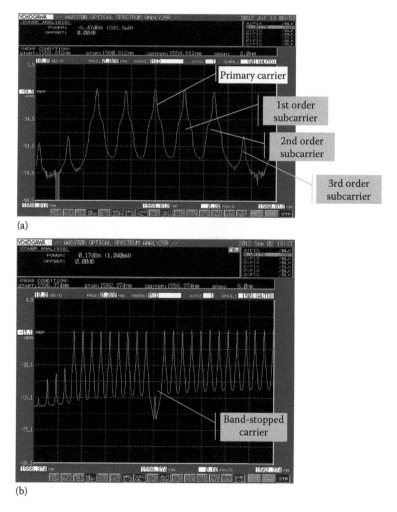

(a)

(b)

FIGURE 12.26 Spectrum of subcarriers generated from a primary laser source by nonlinear driving of IQ modulator: (a) 1 × 5 and (b) 4 × 5 subcarriers with one band-stopped.

12.4.3.1.3.3 Remarks Although both schemes described have been tested and implemented, the scheme using 4 × 5 to obtain 20 subcarriers are finally selected to feed into optical modulators to obtain 10 or 20 Tbps for the demonstration platform to field trials with Deutsche Telecom and Vodafone, respectively.

12.4.3.2 Supercomb Generator as Dummy Channels

Some subcarriers of the supercomb lines generated by the RCFS comb generators can be used as dummy channels. The generated comb lines can be filtered by band stopping of the WSS device, as shown in Figure 12.27. One subcarrier is filtered out and the rest of the supercomb lines are then fed to a PDM-IQ modulator, so that PDM QPSK channels can be generated, as shown in Figure 12.28.

12.4.3.3   2 Tbps and 1 Tbps Optical Transmitter at Different Symbol Rates

1 Tbps and 2 Tbps superchannels can be generated using either RCFS comb generator or the nonlinear comb generator, described in Section 12.4.3.1. The difference between these two schemes is that for the RCFS-comb-generated channels, the phase and frequencies of all subchannels are locked to

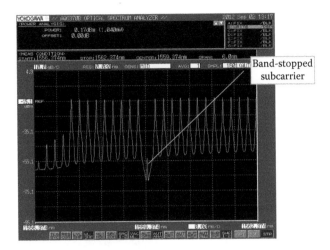

FIGURE 12.27 Band-stopped subcarrier of a multi-subcarrier line used for modulation and generation of dummy channels.

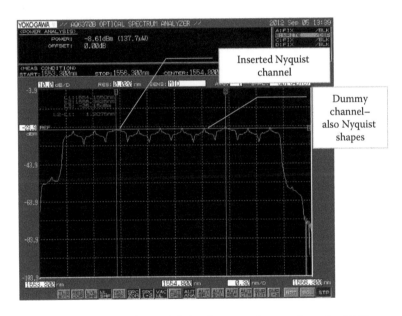

FIGURE 12.28 PDM QPSK channels by modulation for use as dummy channels of 1 Tbps superchannel.

the reference primary carrier. Thus, with the number of dummy channels employed for transmission, the synchronization of these channels would lead to high nonlinear SPM effects.

12.4.4 DIGITAL-TO-ANALOG CONVERTER

12.4.4.1 Structure

The principal subsystem of the superchannel transmission system is the digital signal processing based optical transmitter in which the DAC plays the central role in pulse shaping, equalization, and pattern generation (Figures 12.29 and 12.30). An external sinusoidal signal is required to be fed into the DAC, so that multiple clock sources can be generated for sampling at 56–64 GSa/s. Thus, the

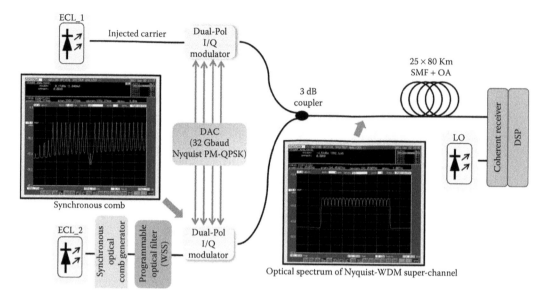

FIGURE 12.29 Arrangement of ECL1 and ECL2 for nonlinear comb generation of 10 subcarriers in association with Nyquist pulse shaping and modulation for 25 spans of optically amplifier fiber (non-DCF) transmission. The spacing between the lasers allows allocation of equally spaced—of 28 GHz to 32 GHz—four subchannels by nonlinear comb generation.

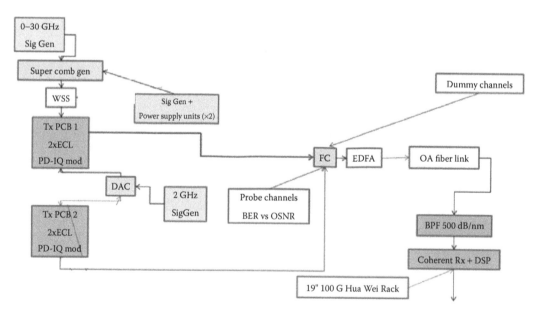

FIGURE 12.30 Schematic of structure of Tbps optical transmission systems using 4×5 subcarrier generation techniques.

noises and clock accuracy depend on the stability and noise of this synthesizer. An Agilent signal generator model N9310A is employed in our platform. However, a Rohde and Schwartz signal generator with low phase noise is preferred. Four DAC submodules are integrated in one IC with four pairs of eight outputs of $(V_I^+, V_Q^+)(H_I^+, H_Q^+)$ and $(V_I^-, V_Q^-)(H_I^-, H_Q^-)$ (Figures 12.30 and 12.31).

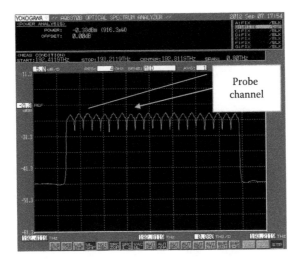

FIGURE 12.31 Spectrum of 2 Tbps optical channels using nonlinear comb generator with two probe channels and 18 dummy channels.

12.4.4.2 Generation of I- and Q-Components

Electrical outputs from the quad DACs are in pair of positive and negative and are complementary with each other. Thus, we would be able to form two sets of four output ports from the DAC development board. Each output can be independently generated with offline uploading of pattern scripts. The arrangement of the DAC and PDM-IQ optical modulator is depicted in Figure 12.32. Note that we require two PDM IQ modulators for generation of odd and even optical channels.

As Nyquist pulse-shaped sequences are required, a number of pressing steps are conducted: (i) Characterization of the DAC transfer functions; and (ii) pre-equalization in the RF domain to achieve equalized spectrum in the optical domain, that is, at the output of the PDM IQ modulator.

The characterization of the DAC is conducted by launching to the DAC sinusoidal wave at different frequencies and measure the waveforms at all eight output ports. As observable in the inserts of Figure 12.32, the electrical spectrum of the DAC is quite flat, provided that pre-equalization is done in the digital domain launching to the DAC. The spectrum of the DAC

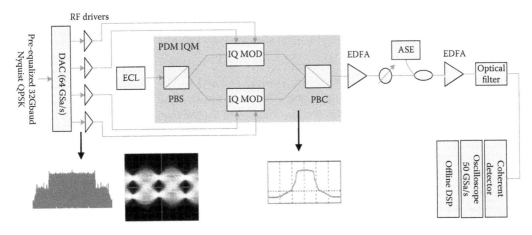

FIGURE 12.32 Experimental set-up of 128 Gb/s Nyquist PDM-QPSK transmitter and B2B performance evaluation.

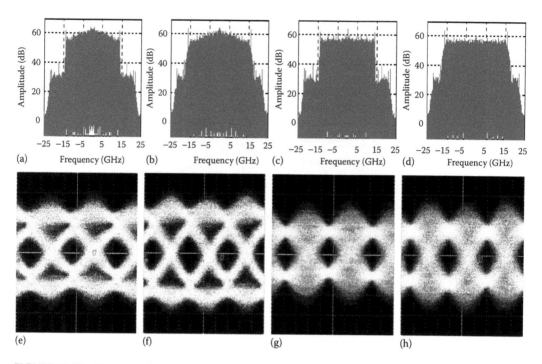

FIGURE 12.33 Spectrum (a) and eye diagram (e) of 28 Gbaud RF signals after DAC without pre-equalization and (b) and (f) of 32 Gbaud; spectrum (c) and eye diagram (g) of 28 Gbuad RF signals after DAC with pre-equalization and (d) and (h) of 32 Gbaud.

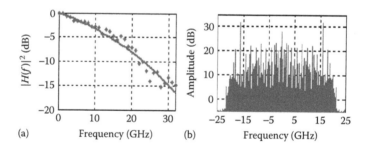

FIGURE 12.34 (a) Frequency transfer characteristics of the DAC. Note the near linear variation of the magnitude as a function of the frequency. (b) Noise spectral characteristics of the DAC.

output without equalization is shown in Figure 12.33a and b. The amplitude spectrum is not flat because of the transfer function of the DAC, as given in Figure 12.34, which is obtained by driving the DAC with sinusoidal waves of different frequencies. This shows that the DAC acts as a low-pass filter; the amplitude of its passband gradually decrease when the frequency is increased. This effect can come from the number of samples, which is reduced when the frequency is increased, as the sampling rate can only be set in the range of 56–64 GSa/s. The equalized RF spectra are depicted in Figure 12.33c and d. The time-domain waveforms corresponding to the RF spectra are shown in Figure 12.33e and f, and thence g and h, for the CoD after the conversion back to electrical domain from the optical modulator via the real-time sampling oscilloscope Tektronix DPO 73304A or DSA 720004B. Furthermore, the noise distribution of the DAC is shown in Figure 12.34b, which indicates that the sideband spectra of Figure 12.33 come from these noise sources.

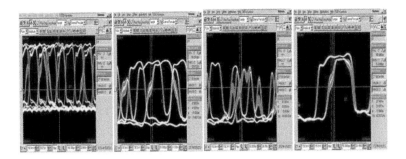

FIGURE 12.35 Various DAC time-domain waveforms at output ports of DAC, where nonoptimal driving conditions are applied.

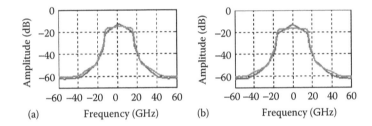

FIGURE 12.36 Optical spectrum after PDM IQM; gray line for without pre-equalization, light gray for with pre-equalization. (a) 28 GSym/s and (b) 32 GSym/s.

It noted that the driving condition for the DAC is very sensitive to the supply current and voltage levels, which are given in Appendix B, with resolution of even down to 10^{-3} V. With this sensitivity, care must be taken when new patterns are fed to DAC for driving the optical modulator. Optimal procedures must be conducted with the evaluation of the constellation diagram and BER derived from such constellation. We believe that the new version of the DAC supplied from Fujitsu Semiconductor Pty Ltd of England, UK has overcome somehow these sensitivity problems. However, we still recommend that care be taken and inspection of the constellation after the coherent receiver must be done to ensure that error free in the B2B connection. Various time-domain signal patterns obtained in the electrical time domain generated by DAC at the output ports are illustrated in Figure 12.35. Obviously, the variations of the in-phase and quadrature signals give rise to the noise and hence blurry constellations.

12.4.4.3 Optical Modulation

Modulation of the PDM IQ modulator can be implemented using RF signals of the output pairs of the DAC. Two sets of RF output ports are employed to provide the in-phase and quadrature signals to the modulators with traveling-wave electrode pair. The DC voltage power supplies are also applied to DC biasing connections. The integrated structure of the PDM IQ modulator is shown in Figure 12.37. There are a number of DC electrodes for biasing the IQ modulator, as shown in Figure 12.38. This set is duplicated for both polarized modulators integrated on the same LiNbO$_3$ chip. The S21 parameter frequency response shows that the 3 dB optical bandwidth of the modulator is 23 GHz (Figure 12.36). This is sufficiently wide for 28 and 32 GB digital sequences. The electrical to optical transfer characteristic is shown in Figure 12.39. We note that the DC bias electrodes are separate from the RF traveling-wave electrode; thus, the voltage level is quite high owing to the short length of the DC electrode pair. The electrode V_π may reach around 15–20 V when using these electrodes, so care should be taken to avoid the damage of these electrodes, that is, the air breakdown electric field of about 50 kV/m across the electrode spacing of only about 5 μm.

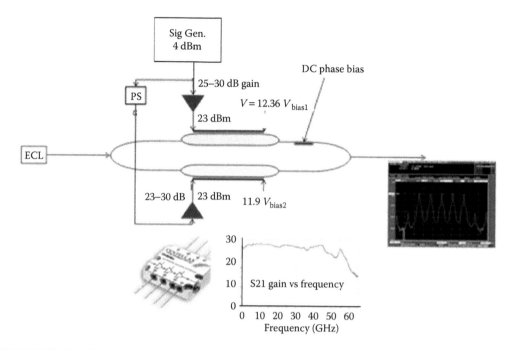

FIGURE 12.37   Plane view of the PDM IQ modulator and associated bias and RF driving circuitry.

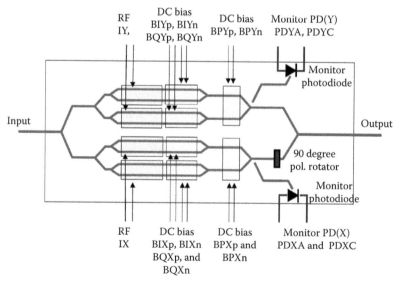

– RF terminals: Single-ended, GPPO-compliant interface, and internal termination.
– DC bias terminals: Separated from RF section and push–pull configuration
– Photodiodes: Located at each QPSK modulator sections, the phase of the photocurrent inverted
  to the optical output from QPSK modulator.

FIGURE 12.38   Schematic of optical waveguides and electrodes of the PDM IQ modulator. Two polarization-dependent IQ modulators are in parallel.

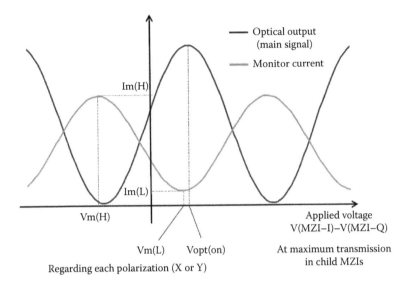

FIGURE 12.39 Electrical to optical transfer characteristics of the PDM IQ modulator.

12.4.4.4 Synchronization

Care must be taken to ensure that synchronization of the pulse sequences at the output of the DAC ports occurs, otherwise the constellation of the QPSK structure would be noisy and hence result in worst BER. This can be done by programmable delay time in the data sequence to be loaded into the DAC.

12.4.4.5 Modular Hardware Platform

The overall packed comb generator using nonlinear driving of the modulator is shown in Figure 12.40, with the comb packaged box on top and signal generator plus power supplies placed in the bottom and staked separately. The comb generator using RCFS structure is shown in Figure 12.41, in which two EDFAs are placed on the top, one for optical compensation inside the recirculating loop and the other for amplifying the signals power to be coupled to either WSS or the IQ modulator. As noted in the above section, the total optical power must be distributed to all comb lines; thus, typical power level for each comb line of the RCFS comb is about −5 to −7 dBm, depending on the number of

FIGURE 12.40 Nonlinear comb generator—view of with and without power supply connection.

FIGURE 12.41   RCFS optical comb generator.

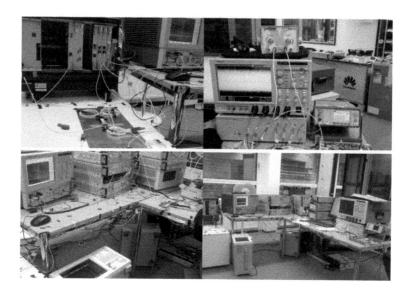

FIGURE 12.42   Overview of the Nyquist WDM QPSK transmission platform.

comb lines generated from the recirculating loop. Note that EDFAs are of PM type, including input and output fiber patch cords (Figure 12.42).

12.5   NON-DCF 1 Tbps AND 2 Tbps SUPERCHANNEL TRANSMISSION PERFORMANCE

12.5.1   Transmission Platform

The schematic structure of the transmission system is shown in Figure 12.43, which consists of an optically amplified multi-SSMF span fiber transmission, incorporating variable optical attenuators, optical transmitters, and coherent optical receiver with offline DSP. Optical bandpass filter is employed using either the D-40 demux or Yenista sharp roll-off (500 dB/nm) to extract the subchannel of the superchannel whose performance is to be measured, normally in term of BER versus OSNR.

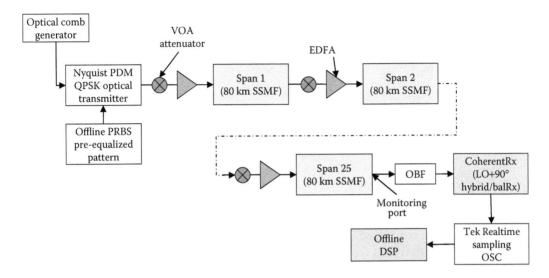

FIGURE 12.43 Schematic structure of the Nyquist QPSK PDM optically amplified multispan transmission line.

The multispan optically amplified fiber transmission line consists of two main parts: the fiber spools and banks of OAs and VOA, as shown in Figure 12.44a and b, respectively. The launched power to the first span is to be adjusted and thence at the VOA and amplifiers into each span, as the saturation level of EDFA varies with respect to the total launched power. Thus, in order to achieve the same launched power into each span, the remote control is employed to adjust these input launched power span by span, especially when the launched power is varied to obtain different OSNR. The B2B is measured by inserting noises at different levels to the signal power monitoring port. A D-40 demultiplexer is also available and incorporated in the 19″ rack, so that a subchannel of the superchannel can be extracted for measuring the transmission or B2B performance (Figure 12.45).

The transmitter employs Nyquist QPSK, as described in Sections 12.4.4 and 12.3.1, where the channel spacing and baud rate can be varied and the DAC can be optimized. Optical receiver is a coherent type with an external local oscillator whose wavelength is tuned to the right wavelength location of the subchannel whose performance is to be measured. In practice, this LO must be remotely or automatically locked to the wavelength of the subchannel.[*]

12.5.2 PERFORMANCE

12.5.2.1 Tbps Initial Transmission Using Three Subchannel Transmission Test
During the initial phase of the development of the Tbps transmission system, we considered a test of transmission of three channels for transmission only over 2000 km optically amplified multispan, as described in Section 12.5.1. This was done to ensure that the signal quality can be achieved using Nyquist QPSK with interference due to linear crosstalk caused by overlapping of subchannels, as they are packed so close to each other using Nyquist criteria. Therefore, we set up the generation of three subchannels by using an ECL source split 50:50 into two branches, one fed through a CSRZ optical modulator driven with a sinusoidal signal source of frequency equal to the subchannel spacing and the other branch modulated with Nyquist QPSK format combined with the two CSRZ carriers, which are also modulated by Nyquist QPSK but with different random patterns. Thus,

[*] L.N., Binh et al., "Optical PLL super combed carrier cloning: circuit and methodology for locking superchannel coherent receivers", patent under preparation and approval.

(a)

Fibers
in/out Tx/Rx

VOA + EDFA
Bank 1

VOA + EDFA
Bank 2

VOA + EDFA
Bank 3

Demux

(b)

FIGURE 12.44   Fiber spools (a) to be connected to optical amplifiers and variable optical attenuator and (b) arranged in bank in 19″ rack.

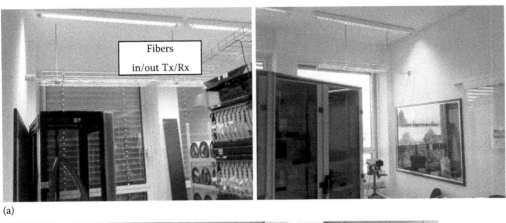

(a)

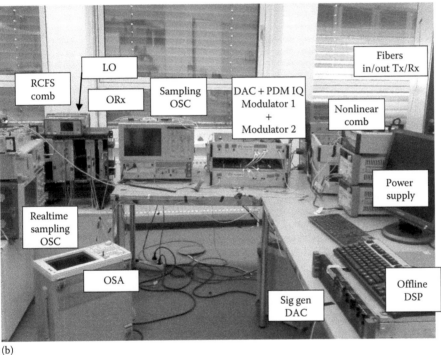

(b)

FIGURE 12.45 Fiber lines (a) running from optically amplified multifiber span transmission line to transmitter and (b) receiver platform.

we do have three subchannels in which are decorrelated, and thus, the BER versus OSNR can be obtained to ensure that the effects of overlapping can be justified. The schematic of this arrangement is shown in Figure 12.46. The transmission performance measured with BER against launched power and subchannel spacing of 28 and 30 GHz with 28 GBauds Nyquist QPSK transmitted over 1600 km line is shown in Figure 12.47.

The BER is optimum at launched power of −1 to 0 dBm for both single channel and probe channel, which are central of the three channels. This proves that the Nyquist pulse shaping can offer the performance close to that of a single-channel transmission over 1600 km optically amplified multispan and nondispersion compensating fiber line. These performance results gave much confidence

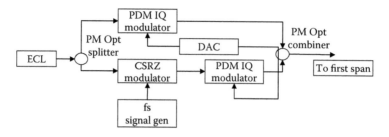

FIGURE 12.46   Schematic of generation of three subcarrier channel transmitters for test transmission in the initial phase for Tbps transmission system. PM = polarization maintaining.

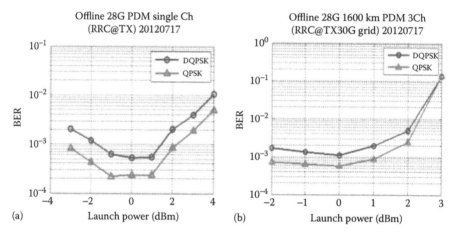

FIGURE 12.47   BER versus launched power (dBm) using three channels with central channel as probe channel over transmission distance of 1600 km non-DCF, 20 optically amplified 80 km SSMF spans, with VOA inserted in front of EDFA. 28 GBaud PDM-Nyquist QPSK: (a) single channel and (b) three channels.

in the pulse shaping and for proceeding to superchannel of 1 Tbps and 2 Tbps over a single dual-polarized mono-mode fiber (Figures 12.48 and 12.49).

Furthermore, we also use optical filter at the receiver, that is, before feeding into the hybrid coupler, the optical filter (the demux D-40 or a sharp roll-off Yenistar filter with 500 dB/nm) is employed to extract the probe channel. The transmission performance for a five-channel superchannel is shown in Figure 12.50. In general, the sharp roll-off filter would offer 1 dB improvement over the demux D-40, which commonly has a parabolic-like roll-off filtering characteristic.

12.5.2.2   1 Tbps, 2 Tbps, or N Tbps Transmission

We can now extend the transmission of Nyquist pulse-shaped subchannels with the total capacity of all channels of 1.12 Tbps and thence 2.24 Tbps. For 1.12 Tbps per second, we use nine subchannels plus one probe channel, which are all modulated and pulse-shaped, satisfying Nyquist criteria with raise cosine filter and pre-equalization. Thus, we have 10 channels generated using either RCFS or nonlinear comb generation. Then one channel is suppressed by using band stop filter of a WSS and then combined with another Nyquist-shaped QPSK channel independently driven by a decorrelated sequence generated from DAC ports. This inserted channel acts as the probe channel; hence, one selects to measure the performance of transmission for any subchannel of the superchannel, as required, instead of all independently driven subchannels, owing to the limitation

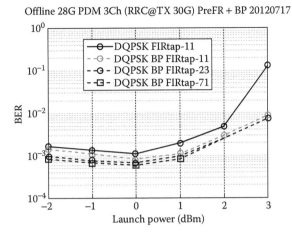

FIGURE 12.48 1600 km transmission Nyquist QPSK processed with FIR 11 to 23 taps: BER versus OSNR. Number of subchannels is 3, generated and modulated, as shown in Figure 12.47.

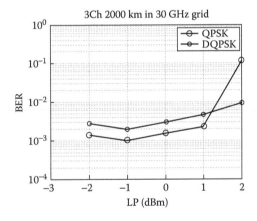

FIGURE 12.49 Three-channel test, with central channel as probe over 2000 km transmission by simulation 31.8 GHz channel spacing 28 GBaud QPSK with processing QPSK and DPQSK (to optimize for cycle slippage).

of independent sequence from one DAC subsystem. Figure 12.51 shows the generic scheme for generating such probe channel within the superchannel. We can select any channel spectral location to insert the probe channel by tuning the source wavelength and the WSS, depending on whether the RCFS or nonlinear comb generator is employed. Figure 12.52 shows the B2B performance of 10 subchannels (2×5 by nonlinear driving modulator, described above) under Nyquist pulse shaping and PDM QPSK modulation format. Individual in-phase and quadrature-phase components are also processed to see the effects of one component on other channels. Thus, total BER of QPSK is the total sum of all these individual components. The DQPSK processing offers 1 dB better in OSNR for the same BER and BER of 10^{-3} achieved for OSNR of 15.2 and 16 dB, respectively, for QPSK and DQPSK, respectively. The main reasons for evaluating all individual in-phase and quadrature received signals are to ensure that the DAC generates signals at the ports $(V_I^+, V_Q^+)(H_I^+, H_Q^+)$ and $(V_I^-, V_Q^-)(H_I^-, H_Q^-)$, enforcing no penalties on the coherent detected signals. With this scheme for generation of 1 and 2 Tbps, it is not difficult to extend to N_Tpbs, provided that the RCFS comb generator is used or one has to employ $N/5$ ECLs if nonlinear Comb generation technique is resorted.

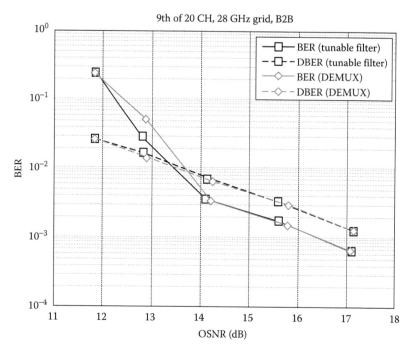

FIGURE 12.50 Performance of channel 9 of the superchannel transmission over 2000 km with D-40 demux or Yenistar filter.

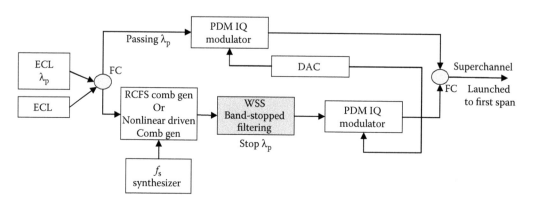

FIGURE 12.51 Generation of (N-1) subcarrier dummy channels and +1 probe channel for Tbps superchannel transmission systems.

The principal problems that we would face in this case would be the nonlinear distortion effects that would be described in the coming section.

Using nonlinear comb generator, 10 subchannels are generated, thence with one channel eliminated by passing through the WSS set by band-stopped filtering. These subcarriers are then modulated and combined with the probe channel and then amplified launched into the first span. The B2B performance is shown in Figure 12.52. The channel spacing is 30 GHz, with 28 Gbauds as the symbol rate and the BER obtained is 10^{-3} at 14.5 dB. The transmission performance of the 10 subchannel superchannel over 200 km of optically amplified non-DCF SSMF multispan transmission line is shown in Figures 12.52 and 12.53, in which B2B with four in-phase and quadrature components of the PDM-28G Nyquist QPSK is analyzed, with their BER plotted against OSNR at the initial phase

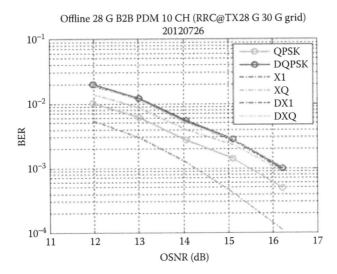

FIGURE 12.52 BER versus OSNR for the horizontal (I-)and vertical components (Q-) of QPSK Coherent reception systems of 10 subchannels of 112 Gb/s 28 GBauds Nyquist QPSK superchannel generation and transmission using nonlinear driven comb generator. 28 GBauds, 30 G grid subchannel spacing, 10 subchannels with one probe channel.

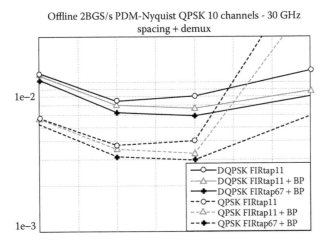

FIGURE 12.53 BER versus launched power (−2 dBm to 1.5 dB, with 0.5 dBm per division in horizontal scale) 28 GBauds Nyquist QPSK with 30 GHz spacing with and without back propagation additional processing and 11 tap FIR and 2000 km transmission non-DCF fiber multispan line probe channel selected using D40 demux.

before transmission and then over the complete link. The attenuation per span is 22 dB and the EDFA stages are optimized, so that the launched power into each span can be kept the same at each span. The optimum launched power is −1 to 0 dBm, with a BER of 2e−3, 11 tap FIR filter, and back propagation to moderately compensate for fiber nonlinearity.

The back propagation is conducted by propagating the received sampled signals at the receiver and then converting into the optical domain level and propagating through span by span with the nonlinear coefficient equal and in opposite sign with those of the SSMF. The back-propagation distance per span is about 22 km as the effective length of the fiber under nonlinear SPM effects (Figures 12.54 through 12.56).

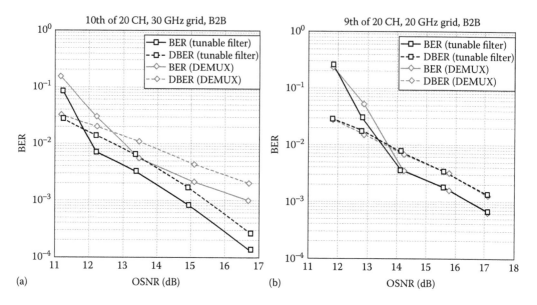

FIGURE 12.54   Transmission performance of (a) channel 9 and (b) channel 20 of the >2 Tbps superchannel with channel spacing of 30 GHz and 28 Gbauds Nyquist QPSK roll-off factor 0.1, using D-40 demux and Yenistar sharp roll-off filter (500 dB/nm).

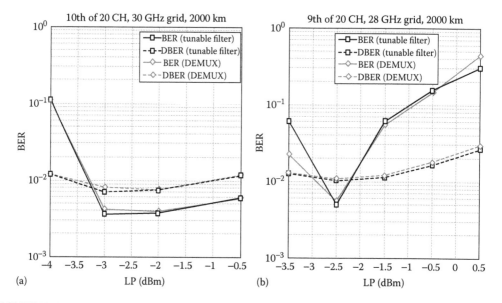

FIGURE 12.55   BER versus launched power of superchannel for probe channels. (a) channel 9 and (b) channel 20. Transmission distance is 2000 km consisting of 80-km-span multispan optically amplified non-DCF optical transmission line.

12.5.2.3   Tbps Transmission Incorporating FEC at Coherent DSP Receiver

Figure 12.57 shows the BER against OSNR for soft differential FEC processing and Nyquist pulse shaping QPSK modulation format. The symbol rate is 28 Gbauds with individual components of QPSK, that is, the real and imaginary parts (in-phase and quadrature-phase components) for system under B2B connection, without FEC and with FEC. Figure 12.58 shows the performance improvement under FEC by displaying the BER versus OPSNR for Nyquist pulse shaping QPSK with

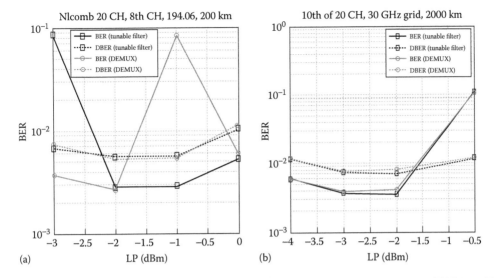

FIGURE 12.56 Five channels nonlinear comb generation subcarriers transmission over 2000 km: BER versus OSNR using 40 channel demultiplexer or sharp roll-off optical filter: (a) test channel is channel 8 and (b) channel 10.

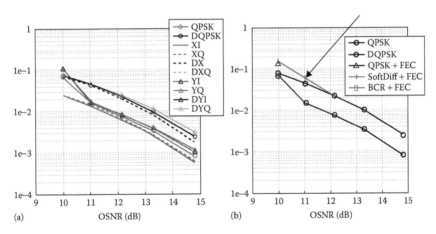

FIGURE 12.57 Soft differential FEC processing for Nyquist QPSK, BER versus OSNR. 28 Gbauds with individual components of QPSK, that is, real and imaginary parts (in-phase and quadrature-phase components), B2B set-up (a) without FEC and (b) with FEC.

BCJR and without FEC. Figure 12.59 displays BER versus OSNR with and without electronic precompensation in DAC, and back-to-back scenario for the system operating at 32 GBauds and under modulation format QPSK polarization multiplexing.

Figure 12.60 shows the performance of transmission systems under (a) optimization and (b) nonoptimized DAC generation of random sequence at 28 G and 32 GBauds, connection is under a back-to-back scenario, and the modulation format of Nyquist pulse shaping QPSK. Finally Figure 12.61 shows the performance of a 32 GB Nyquist QPSK pulse shaping with a roll-off factor of 0.1. The sub-channel spacing is used as a parameter and transmission distance is of 2000 km consisting of several 80km non-DCF SSMF optically amplified spans under scenarios of with and without FEC 20%, far left and far right curves of the figure, respectively.

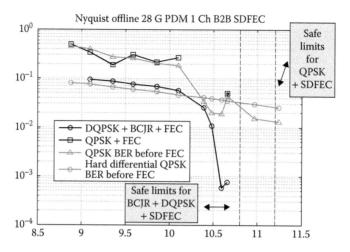

FIGURE 12.58 FEC improvement and BER versus OPSNR for Nyquist QPSK with BCJR and without FEC.

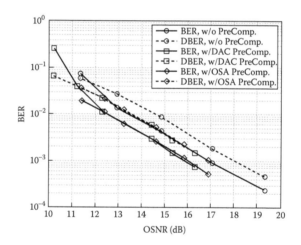

FIGURE 12.59 32 GBauds PDM Nyquist QPSK transmission, BER versus OSNR with and without electronic compensation in DAC, and back-to-back scenario.

12.5.2.4 Coding Gain of FEC and Transmission Simulation

Simulations are conducted to estimate the gain and BER against OSNR with error coding and noncoding, so as to assist with the experimental performance. The simulation results (Figure 12.62) proves that the coding gain can be achieved for transmission system operating under a QPSK modulation format incorporating Nyquist-pulse-shaping with a roll-off factor of 0.1. Under coding, we could see that the OSNR requires for 10^{-3} may be reduced down to 11 dB. Thus, an extra margin of 4 dB can be gained to allow the extension of the transmission reach of Nyquist QPSK to an extra 500 km. Figure 12.62a and b shows the effects of 30 and 28.5 GHz subchannel spacing on the BER of the Nyquist QPSK with LDPC and significant improvement of FEC on its performance. Similarly, for Figure 12.62c and d related to 33 and 32.5 subchannel spacing with 32 GBauds. Figure 12.62e and f then prove that improvement can be achieved when BCJR additional coding is superimposed to obtain further coding gain. Figure 12.62g displays all gain curves into one graph, and Figure 12.62h and i shows the spectra of subchannels (the odd channels only).

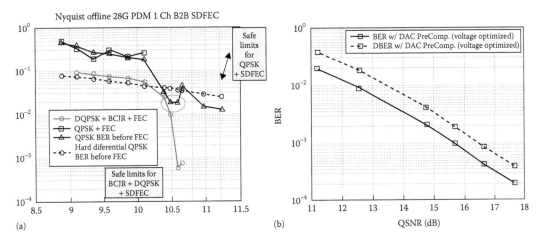

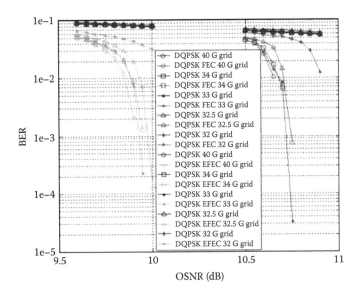

FIGURE 12.60 Performance of transmission systems under (a) optimization and (b) nonoptimized DAC generation of random sequence at 28 G and 32 GBauds, back-to-back scenario, and modulation format Nyquist QPSK.

FIGURE 12.61 32 GB Nyquist QPSK roll-off = 0.1 with subchannel spacing as parameter and transmission over 2000 km non-DCF optically amplified spans under scenarios of with and without FEC 20%, far left and far right of the graph, respectively.

12.5.2.5 MIMO Filtering Process to Extend Transmission Reach

The filter structure incorporated in the DSP processing is shown in Figure 12.63. Owing to whitening noise effects in the optical amplification stages in the link, the root-raised cosine filter is used at the receiver for match filtering; that is, the complete filtering process in the link forms a complete Nyquist filter and satisfies the Nyquist shaping criteria. Following the Rx imperfections compensation and CD compensation, a complex butterfly FIR structure is applied to compensate for PMD. Each of the four complex FIR filters is realized by a butterfly structure of corresponding real FIR filter. The recursive constant modulus algorithm (CMA) LMS algorithm continuously updates the filter taps, which guarantee the initial convergence and tracking of time-variant channel distortions. In the steady state, the complex butterfly structure is a digital, real representation of the inverse impulse response determined by the tap coefficients.

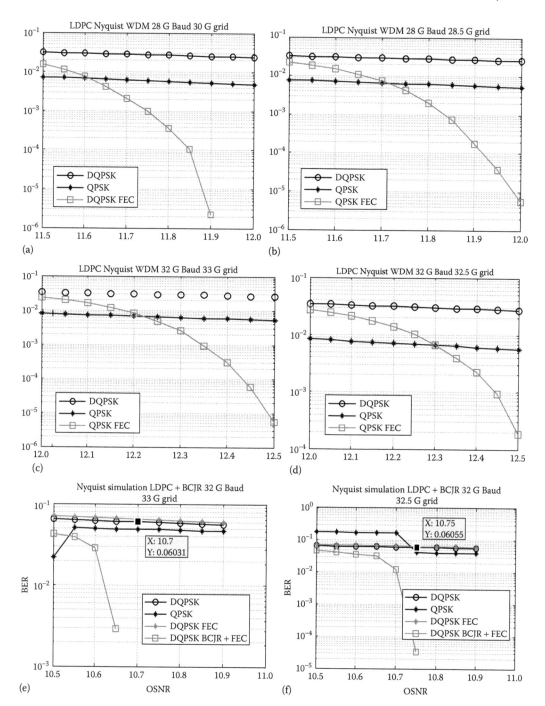

FIGURE 12.62 Simulation of Nyquist QPSK with and without error coding gain: (a and b) LDPC coding and gain at 28 GHz grid; (c and d) LDPC Nyquist QPSK with 33 GHz grid; (e and f) BCJR + LDPC and 3 Ghz grid.

(Continued)

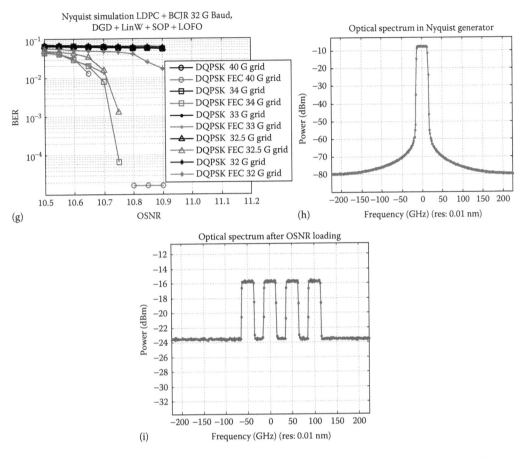

FIGURE 12.62 (Continued) Simulation of Nyquist QPSK with and without error coding gain: (g) Nyquist QPSK with and without coding gain with grid—frequency spacing as parameters (32–40 GHz) as summary of (a–f); (h); and (i) spectra of channels before and after ASE noise loading.

The output signals from the FIR equalization stages (x' and y') at time k are related to the input signal vectors (x and y) containing samples $k-L+1$ to k by

$$x'(k) = \mathbf{h}_{xx} \cdot \mathbf{x}(k) + \mathbf{h}_{yx} \cdot \mathbf{y}(k)$$

$$y'(k) = \mathbf{h}_{yy} \cdot \mathbf{y}(k) + \mathbf{h}_{xy} \cdot \mathbf{x}(k)$$

(12.5)

where $\mathbf{h}_{xx}, \mathbf{h}_{xy}, \mathbf{h}_{yy}, \mathbf{h}_{yx}$ are the $T/2$-spaced tap vectors (T is symbol period), for the FIR filter, and the dot "." denotes the vector dot product. The length L of the tap vector is equal to the impulse response of the distorted transmission medium to be compensated. Initial equalizer acquisition is performed on the first several thousand symbols, depending on the "learning/training" process. These symbols are subsequently discarded in the error counting. The equalizer tap vectors are then updated continuously throughout the processing of the data set in order to track channel changes.

The MIMO filter length in the commercial 100 G receivers without tailoring for Nyquist transmission is usually set between 7 and 11 as a trade-off between complexity and requirements, since the measured mean DGD in real long-haul optical links is around 25 ps. Further increasing the FIR complexity would enhance the gain in case of a longer pulse response to be employed, owing to the limited transfer function of the transmission system. This is verified in our experimental platform with the BER against OSNR, the launched power, and number of taps, as depicted in Figure 12.64a–c. It is noted that the tap number higher than 9 does not offer any gain improvement in performance,

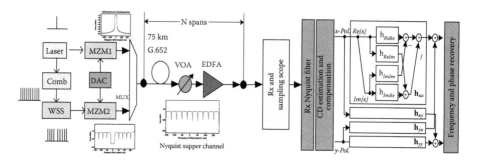

FIGURE 12.63 Nyquist superchannel experimental set-up and receiver structure; each of the four complex MIMO filters consists of four real FIR filters.

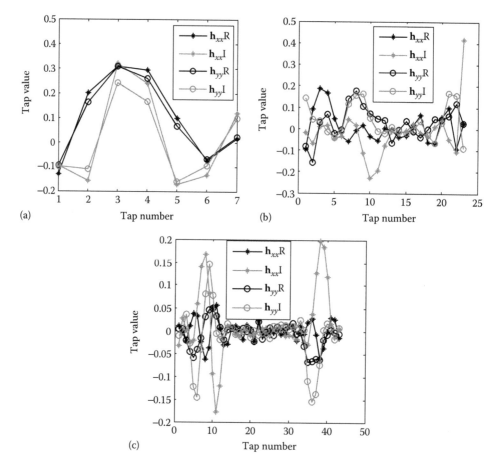

FIGURE 12.64 MIMO filter tap length extending effects in Nyquist WDM optical transmission experimental results: (a) Tap length 7, (b) tap length 23, and (c) tap length 43, $h_{xx}R$ for the real part of h_{xx}, and $h_{xx}I$ for the imaginary part of h_{xx}.

as depicted in Figure 12.64. The MIMO filter tap length extending effects in commercial WDM optical transmission experiments (a) back-to-back (B2B) performance (L stands for the filter tap length), (b) 1500 km transmission line (c) 2000 km transmission line. h_{xx} for the in-phase real part of h_{xx}, h_{xx} for the quadrature imaginary part of h_{xx}.

In our Nyquist pulse-shaping QPSK coherent transmission experimental platform, the pulse sequence is shaped by a Nyquist square root filter (RRC) of a roll-off factor of 0.1 which can be

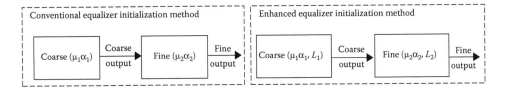

FIGURE 12.65 Conventional and enhanced MIMO filter tap convergence algorithm.

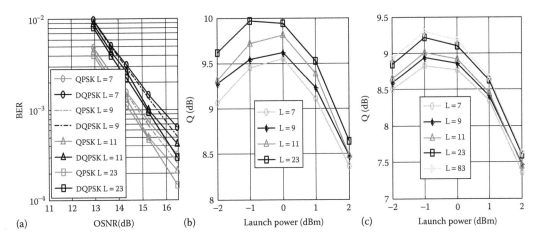

FIGURE 12.66 Enhanced MIMO filter tap length extending performance (BER versus OSNR) in Nyquist WDM optical transmission experiments: (a) B2B, (b) 1500 km transmission results, and (c) 2000 km transmission.

generated by a time-domain FIR filter with 65 taps. Simulation results do not show any penalty caused by our FIR tap settings. However, hardware imperfections likely require filter pulse response of more FIR taps to achieve the best performance. Furthermore, the convergence of the CMA algorithm would fail for FIR with longer tap length, as shown in Figure 12.65. With a linear FIR filter consisting of seven taps, we are able to acquire the channel performance for all channels. Therefore, the conventional CMA method is used, and the performance is shown in Figure 12.65a. The coarse step applies fast learning via larger values of a weighting factor μ_1 and a forgetting factor α_1. The fine step uses smaller values of these two parameters and improves the final performance.

To solve the convergence problem, we perform the acquisition procedure in two steps, with smaller and larger FIR filter lengths. First, a shorter FIR filter with a smaller number of taps L_1, for example, less than 9, and a larger μ_1 and α_1 can be used to find the main values of the starting taps. After the preconvergence phase, the filter is extended with the reiterative tap values obtained in the first procedure, while the extended taps are set to null in association with smaller values of μ_2 and α_2. This method ensures the filter convergence (Figure 12.66).

Using this new algorithm, the B2B performance can be improved by up to 0.7 dB at BER of 10^{-3}, with a tap extension from 7 to 23, as shown in Figure 12.65a. The performance of new algorithm is further verified over the transmission in 1500 km and 2000 km links. The tap extension from 7 to 23 enables a Q gain of almost 0.5 dB at two optimum launch powers of −1 and 0 dBm. To test the MIMO filter performance improvement by extending the filter length, we checked the filter of length 83 taps. The improvement could be only about 0.1 dB, indicating that further increasing the complexity brings only a negligible gain. Thus, a filter length of 23 is the FIR filter limit.

12.6 MULTICARRIER SCHEME COMPARISON

A preliminary comparison and analysis of Nyquist QPSK transmission scheme for superchannel Tbps transmission over long-haul and expected metro networks are shown in Table 12.2.

TABLE 12.2

Comparisons of Tbps Transmission Schemes

	Nyquist WDM	CO-OFDM	EOFDM
Hardware complexity	Tx: similar Rx: multi-OFE	Tx: similar Rx: need OFFT	Tx: similar Rx: multi-OFE
DSP complexity	More complex, possibly due to compensation—but no higher degree of complexity, possibly more processing time required for CPU	Normal	Normal
SE (theoretically)	Similar	Similar	Similar
Tx-DSP + DAC	Essential	Possibly not	essential
ADC sampling rate	1.2× Baud	2× Baud	2× Bandwidth
Bandwidth requirement on O/E components	Depending on subchannel spacing	Depending on subchannel spacing	Depending on subchannel spacing
Special requirements	DSP for sequence estimation (MAP, MLSE)	Orthogonal channel separation	Cyclic prefix and guard band for OFDM symbols
Flexibility	Medium	Medium	High

Hardware complexity: The transmitter side of all schemes requires a comb generator, which can be either using one ECL and then employing RCFS technique to generate N subcarriers of locked phase and frequency or a multiple factor of five subcarriers per ECL, using nonlinear driving method. Furthermore, a set of parallel PDM I/Q modulators and DAC with eight ports of positive and complementary signals, an optical subchannel DeMux, and a Mux would be necessary for generating independent sets of in-phase and quadrature signals for modulating the PDM-IQ modulators. At the receiver side, standard coherent receiving system can be used, with insertion of sharp optical filter at the front end of the optical hybrid coupler, which is required to separate the subchannels. This may require a set of optical filters, one for each subchannel. Thus, a technical solution should be provided to avoid this complex and expensive solution. This is the setback of the Nyquist WDM technique.

DSP complexity: For Nyquist WDM, because the Nyquist filtering results in ISI, at the receiver side, one needs sequence estimation algorithms such as MAP and MLSE. For Nyquist QPSK and different comb-generated subcarriers and subchannels, FIR with lower number of taps would be suitable, so that the complexity can be acceptable.

ECLs: For RCFS comb generation, there needs only one ECL source but additional modulator and optic components are required, whereas for nonlinear driving comb generation, the number of ECLs would be increased accordingly. To cover the whole C-band with superchannels of 30 GHz spacing, there would be around 12 ECLs. This number may be high, especially when they are to be packaged in the same line card.

Spectral efficiency: All Nyquist pulse-shaping schemes can enhance the spontaneous emission by a factor of about 2; the overlapping between subchannels would create minimum crosstalk if the roll-off factor is less than 0.1 and the third-order harmonics of the subchannels and comb generated subcarriers is more than 30 dB below the primary carrier.

Tx-DSP and DAC: Ideally, CO-OFDM may not need DSP, but for compensation of components and transmission impairments, it is preferred to use Tx-DSP. The other two schemes must use Tx-DSP and DAC to shape the pulse or generate the designed signal.

ADC sampling rate: This depends on channel spacing, ideally by a factor of 2× the single-sided signal bandwidth. As Nyquist WDM narrow filters the two-sided signal bandwidth to only 1.3 of Baud rate, for the same baud rate, it requires the lowest sampling rate of ADC.

ADC and DAC of 56 to 64 GS/s are available from Fujitsu, which allow the generation of random sequence for 28–32 GB/s.

Bandwidth requirements on O/E components: Similar to the sampling rate, Nyquist WDM needs the lowest bandwidth assuming the same baud rate. The other two schemes are similar.

Flexibility: All three schemes have flexibility in the number of subchannels, modulation formats, and bandwidth of subchannels. Although eOFDM can adjust some parameters in electrical domain by means of Tx-DSP, it has more flexibility than other two schemes.

12.7 CONCLUDING REMARKS

One to N Tbps transmission R&D activities are conducted in different departments of Huawei Technologies Co. Ltd. In order to cover all the possible schemes and have thorough investigations for the emerging technology, we have conducted and covered all possibilities of pulse shaping and spacing between subchannels with sharp roll-off as well as extending the spectral region over the whole C-band. The ERC team investigates Nyquist superchannel, ultradense WDM using comb generations of either RCFS or nonlinear driving techniques. The target is to have a best-in-class 1 Tbps Nyquist-WDM demonstration system until the end of 2012 and develop prototype model systems, which rare under field demonstration for Deutsche Telecom and Vodafone in Europe. The developed system is packaged in 19″ rack and can be deployed easily to field trails. We have completed the development of the solution of 10 × 28 GB/s (or 30 or 32 GB/s) PDM Nyquist QPSK with 28, 30, or 32 GHz spacing between adjacent subchannels. The transmission performance achieved so far has reached 2000 km, with BER of 10^{-3} and an OSNR of 14.7 dB. This performance will allow us to confidently conduct field trial with FEC coding to obtain further coding gain to combat additional ASE noises from an additional 500 km with six spans of 80 km length and 6 additional OAs to extend the transmission line to 2500 km reach.

In summary, the Tbps transmission systems work very well in the laboratory environment and would also perform in installed fiber multispan lines. Field trials are under current testings of the systems. Once the transmission performance from the Tbps transmission systems in laboratory environment and field trials is available and once we have the confirmation/comparison of the technical feasibility of Tbps optical coherent communication systems and, more importantly, the decision to provide the Tbps as a transmission product for Huawei, we will detail the technical specifications for design and productions.

APPENDIX A: TECHNICAL DATA OF STANDARD SINGLE-MODE OPTICAL FIBER

Corning® SMF-28™ Optical Fiber
Product Information

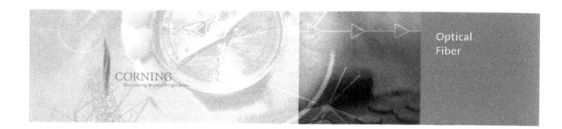

PI1036
Issued: April 2001
Supercedes: March 2001
ISO 9001 Registered

Corning® Single-Mode Optical Fiber

The Standard For Performance

Corning® SMF-28™ single-mode optical fiber has set the standard for value and performance for telephony, cable television, submarine, and utility network applications. Widely used in the transmission of voice, data, and/or video services, SMF-28 fiber is manufactured to the most demanding specifications in the industry. SMF-28 fiber meets or exceeds ITU-T Recommendation G.652, TIA/EIA-492CAAA, IEC Publication 60793-2 and GR-20-CORE requirements.

Taking advantage of today's high-capacity, low-cost transmission components developed for the 1310 nm window, SMF-28 fiber features low dispersion and is optimized for use in the 1310 nm wavelength region. SMF-28 fiber also can be used effectively with TDM and WDM systems operating in the 1550 nm wavelength region.

Features And Benefits

- Versatility in 1310 nm and 1550 nm applications.

- Outstanding geometrical properties for low splice loss and high splice yields.

- OVD manufacturing reliability and product consistency.

- Optimized for use in loose tube, ribbon, and other common cable designs.

The Sales Leader

Corning SMF-28 fiber is the world's best selling fiber. In 2000, SMF-28 fiber was deployed in over 45 countries around the world. All types of network providers count on this fiber to support network expansion into the 21st Century.

Protection And Versatility

SMF-28 fiber is protected for long-term performance and reliability by the CPC™ coating system. Corning's enhanced, dual acrylate CPC coatings provide excellent fiber protection and are easy to work with. CPC coatings are designed to be mechanically stripped and have an outside diameter of 245 μm. They are optimized for use in many single- and multi-fiber cable designs including loose tube, ribbon, slotted core, and tight buffer cables.

Patented Quality Process

SMF-28 fiber is manufactured using the Outside Vapor Deposition (OVD) process, which produces a totally synthetic ultra-pure fiber. As a result, Corning SMF-28 fiber has consistent geometric properties, high strength, and low attenuation. Corning SMF-28 fiber can be counted on to deliver excellent performance and high reliability, reel after reel. Measurement methods comply with ITU recommendations G.650, IEC 60793-1, and Bellcore GR-20-CORE.

Optical Specifications

Attenuation

Standard Attenuation Cells

Wavelength	Attenuation Cells (dB/km)	
(nm)	Premium *	Standard
1310	≤0.35	≤0.40
1550	≤0.25	≤0.30

* Lower attenuation available in limited quantities.

Point Discontinuity

No point discontinuity greater than 0.10 dB at either 1310 nm or 1550 nm.

Attenuation at the Water Peak

The attenuation at 1383 ± 3 nm shall not exceed 2.1 dB/km.

Attenuation vs. Wavelength

Range (nm)	Ref. λ (nm)	Max. α Difference (dB/km)
1285 - 1330	1310	0.05
1525 - 1575	1550	0.05

The attenuation in a given wavelength range does not exceed the attenuation of the reference wavelength (λ) by more than the value α.

Attenuation with Bending

Mandrel Diameter (mm)	Number of Turns	Wavelength (nm)	Induced Attenuation* (dB)
32	1	1550	≤0.50
50	100	1310	≤0.05
50	100	1550	≤0.10

*The induced attenuation due to fiber wrapped around a mandrel of a specified diameter.

Cable Cutoff Wavelength (λ_{ccf})

$\lambda_{ccf} \leq 1260$ nm

Mode-Field Diameter

9.2 ± 0.4 μm at 1310 nm

10.4 ± 0.8 μm at 1550 nm

Dispersion

Zero Dispersion Wavelength (λ_0):

1302 nm $\leq \lambda_0 \leq$ 1322 nm

Zero Dispersion Slope (S_0):

≤ 0.092 ps/(nm²·km)

$$\text{Dispersion} = D(\lambda) \colon \approx \frac{S_0}{4}\left[\lambda - \frac{\lambda_0^4}{\lambda^3}\right] \text{ ps/(nm·km)},$$
$$\text{for } 1200 \text{ nm} \leq \lambda \leq 1600 \text{ nm}$$

$\lambda =$ Operating Wavelength

Polarization Mode Dispersion

Fiber Polarization Mode Dispersion (PMD)

	Value (ps/√km)
PMD Link Value	≤ 0.1*
Maximum Individual Fiber	≤ 0.2

* Complies with IEC SC 86A/WG1, Method 1, September 1997.

The PMD link value is a term used to describe the PMD of concatenated lengths of fiber (also known as the link quadrature average). This value is used to determine a statistical upper limit for system PMD performance.

Individual PMD values may change when cabled. Corning's fiber specification supports network design requirements for a 0.5 ps/√km maximum PMD.

APPENDIX B: DAC OPERATING CONDITIONS

Power cable setup instruction sheet

1. The first task is to unplug the PSU board from the DK board.

2. Some DK boards have a soldered connection across all three pins of link LK4 (on lower side of PCB). This should be removed if present.

3. Others have a three pole jumper plug on LK4. This should be removed.

4. Place a standard two pole jumper across pins 2 and 3 of LK4 if one is not already present.

5. Next connect the cables supplied to the board, and set the power supplies for the following voltages;

Rail	Voltage	Current limiter setting
AVDDACNSY	1.248 v	3000 mA

Rail	Voltage	Current limiter setting
1V8A	1.95 v	2000 mA
1V8B	1.82 v	2000 mA
0V9A	0.998 v	3000 mA
0V9B	1.015 v	2500 mA
AVDNEGDAC	-0.936 v	800 mA
AVDNEGNSY	-0.973 v	800 mA
AVDEDAC	2.04 v	3000 mA
5V_SYN	5.0 v	500 mA

In the wiring harness supplied, red is +ve, black is common ground, blue is –ve wrt ground. All connectors are labelled with their rail names.

Note that the DAC RF supply has been taken from the 0V9B rail, so that is why it is higher than 0.9 volts. All other 0.9v loads have been placed on 0V9A via links on the DK board.

The following power-up sequence is suggested;

1. output_high(5V SYN);   //DK board fan and 5v rail to clock source module
   delay_ms(20);

2. output_high(1V8A);   //Feeds 1V8A rail, supplying VDDE rail to DAC via LK2
   delay_ms(20);

3. output_high(1V8B);   //Feeds 1V8B rail, supplying AVDEDACNSY rail to DAC via LK1. Also AVDESFIS51 rail to DAC via LK3.
   delay_ms(20);

4. output_high(0V9A);   //Feeds 0V9A rail, which supplies AVDDAC/AVDDAC_B rail to DAC via LK5. Also VDDI rail to DAC via LK4. Also AVDISFIS51

Index

Milton Keynes UK
Ingram Content Group UK Ltd.
UKHW052023071024
449327UK00027B/2401

9 780367 870294